THE DAWN OF EVERYTHING

ALSO BY DAVID GRAEBER

Toward an Anthropological Theory of Value: The False Coin of Our Own Dreams

Fragments of an Anarchist Anthropology

Lost People: Magic and the Legacy of Slavery in Madagascar

Direct Action: An Ethnography

Debt: The First 5,000 Years

The Democracy Project: A History, a Crisis, a Movement

The Utopia of Rules: On Technology, Stupidity, and the Secret Joys of Bureaucracy

Bullshit Jobs: A Theory

ALSO BY DAVID WENGROW

The Archaeology of Early Egypt: Social Transformations in North-East Africa, 10,000 to 2650 BC

What Makes Civilization? The Ancient Near East and the Future of the West

The Origins of Monsters: Image and Cognition in the First Age of Mechanical Reproduction

THE DAWN OF EVERYTHING

A New History of Humanity

DAVID GRAEBER *and* DAVID WENGROW

FARRAR, STRAUS AND GIROUX

NEW YORK

Farrar, Straus and Giroux
120 Broadway, New York 10271

Printed in the United States of America
Originally published in 2021 by Allen Lane, Great Britain
Published in the United States by Farrar, Straus and Giroux
First American edition, 2021

Library of Congress Cataloging-in-Publication Data
Names: Graeber, David, author. | Wengrow, D., author.
Title: The dawn of everything : a new history of humanity / David Graeber and David Wengrow.
Description: First American edition. | New York : Farrar, Straus and Giroux, 2021. | Includes bibliographical references and index. | Summary: "A trailblazing account of human history, challenging our most fundamental assumptions about social evolution—from the development of agriculture and cities to the emergence of 'the state,' political violence, and social inequality—and revealing new possibilities for human emancipation" —Provided by publisher.
Identifiers: LCCN 2021025790 | ISBN 9780374157357 (hardcover)
Subjects: LCSH: Civilization—Philosophy. | Social history. | World history.
Classification: LCC CB19 .G69 2021 | DDC 901—dc23
LC record available at https://lccn.loc.gov/2021025790

www.fsgbooks.com
www.twitter.com/fsgbooks • www.facebook.com/fsgbooks

5 7 9 10 8 6

Contents

List of Maps and Figures

Foreword and Dedication (by David Wengrow)

David Rolfe Graeber died aged fifty-nine on 2 September 2020, just over three weeks after we finished writing this book, which had absorbed us for more than ten years. It began as a diversion from our more 'serious' academic duties: an experiment, a game almost, in which an anthropologist and an archaeologist tried to reconstruct the sort of grand dialogue about human history that was once quite common in our fields, but this time with modern evidence. There were no rules or deadlines. We wrote as and when we felt like it, which increasingly became a daily occurrence. In the final years before its completion, as the project gained momentum, it was not uncommon for us to talk two or three times a day. We would often lose track of who came up with what idea or which new set of facts and examples; it all went into 'the archive', which quickly outgrew the scope of a single book. The result is not a patchwork but a true synthesis. We could sense our styles of writing and thought converging by increments into what eventually became a single stream. Realizing we didn't want to end the intellectual journey we'd embarked on, and that many of the concepts introduced in this book would benefit from further development and exemplification, we planned to write sequels: no less than three. But this first book had to finish somewhere, and at 9.18 p.m. on 6 August David Graeber announced, with characteristic Twitter-flair (and loosely citing Jim Morrison), that it was done: 'My brain feels bruised with numb surprise.' We got to the end just as we'd started, in dialogue, with drafts passing constantly back and forth between us as we read, shared and discussed the same sources, often into the small hours of the night. David was far more than an anthropologist. He was an activist and public intellectual of international repute who tried to live his ideas about social justice and liberation, giving hope to the oppressed and inspiring countless others to follow suit. The book is dedicated to the fond memory of David Graeber (1961–2020) and, as he wished, to the memory of his parents, Ruth Rubinstein Graeber (1917–2006) and Kenneth Graeber (1914–1996). May they rest together in peace.

Acknowledgements

Sad circumstances oblige me (David Wengrow) to write these acknowledgements in David Graeber's absence. He is survived by his wife Nika. David's passing was marked by an extraordinary outpouring of grief, which united people across continents, social classes and ideological boundaries. Ten years of writing and thinking together is a long time, and it is not for me to guess whom David would have wished to thank in this particular context. His co-travellers along the pathways that led to this book will already know who they are, and how much he treasured their support, care and advice. Of one thing I am certain: this book would not have happened – or at least not in anything remotely like its present form – without the inspiration and energy of Melissa Flashman, our wise counsel at all times in all things literary. In Eric Chinski of Farrar, Straus and Giroux and Thomas Penn of Penguin UK we found a superb editorial team and true intellectual partners. For their passionate engagements with and interventions in our thinking over many years, heartfelt thanks to Debbie Bookchin, Alpa Shah, Erhard Schüttpelz and Andrea Luka Zimmerman. For generous, expert guidance on different aspects of the book thanks to: Manuel Arroyo-Kalin, Elizabeth Baquedano, Nora Bateson, Stephen Berquist, Nurit Bird-David, Maurice Bloch, David Carballo, John Chapman, Luiz Costa, Philippe Descola, Aleksandr Diachenko, Kevan Edinborough, Dorian Fuller, Bisserka Gaydarska, Colin Grier, Thomas Grisaffi, Chris Hann, Wendy James, Megan Laws, Patricia McAnany, Barbara Alice Mann, Simon Martin, Jens Notroff, José R. Oliver, Mike Parker Pearson, Timothy Pauketat, Matthew Pope, Karen Radner, Natasha Reynolds, Marshall Sahlins, James C. Scott, Stephen Shennan and Michele Wollstonecroft.

A number of the arguments in this book were first presented as named lectures and in scholarly journals: an earlier version of Chapter Two appeared in French as 'La sagesse de Kandiaronk: la critique indigène, le mythe du progrès et la naissance de la Gauche' (*La Revue du MAUSS*); parts of Chapter Three were first presented as 'Farewell to the childhood of man: ritual, seasonality, and the origins of inequality' (The 2014 Henry Myers Lecture, *Journal of the Royal Anthropological Institute*); of Chapter Four as 'Many seasons ago: slavery and its rejection among foragers on the Pacific Coast of North America' (*American Anthropologist*); and of Chapter Eight as 'Cities before the state in early Eurasia' (The 2015 Jack Goody Lecture, *Max Planck Institute for Social Anthropology*).

Thanks to the various academic institutions and research groups that welcomed us to speak and debate on topics relating to this book, and especially to Enzo Rossi and Philippe Descola for memorable occasions at the University of Amsterdam and the Collège de France. James Thomson (formerly editor-in-chief at *Eurozine*) first helped us get our ideas out into the wider world with the essay 'How to change the course of human history (at least, the part that's already happened)', which he adopted with conviction when other publishing venues shied away; thanks also to the many translators who have extended its audience since; and to Kelly Burdick of *Lapham's Quarterly* for inviting our contribution to a special issue on the theme of democracy, where we aired some of the ideas to be found here in Chapter Nine.

From the very beginning, both David and I incorporated our work on this book into our teaching, respectively at the LSE Department of Anthropology and the UCL Institute of Archaeology, so on behalf of both of us I wish to thank our students of the last ten years for their many insights and reflections. Martin, Judy, Abigail and Jack Wengrow were by my side every step of the way. My last and deepest thanks to Ewa Domaradzka for providing both the sharpest criticism and the most devoted support a partner could wish for; you came into my life, much as David and this book did: 'Rain riding suddenly out of the air, Battering the bare walls of the sun . . . Rain, rain on dry ground!'

I

Farewell to Humanity's Childhood

Or, why this is not a book about the origins of inequality

'This mood makes itself felt everywhere, politically, socially, and philosophically. We are living in what the Greeks called the καιρός (Kairos) – the right time – for a "metamorphosis of the gods," i.e. of the fundamental principles and symbols.'

C. G. Jung, *The Undiscovered Self* (1958)

Most of human history is irreparably lost to us. Our species, *Homo sapiens*, has existed for at least 200,000 years, but for most of that time we have next to no idea what was happening. In northern Spain, for instance, at the cave of Altamira, paintings and engravings were created over a period of at least 10,000 years, between around 25,000 and 15,000 BC. Presumably, a lot of dramatic events occurred during this period. We have no way of knowing what most of them were.

This is of little consequence to most people, since most people rarely think about the broad sweep of human history anyway. They don't have much reason to. Insofar as the question comes up at all, it's usually when reflecting on why the world seems to be in such a mess and why human beings so often treat each other badly – the reasons for war, greed, exploitation, systematic indifference to others' suffering. Were we always like that, or did something, at some point, go terribly wrong?

It is basically a theological debate. Essentially the question is: are humans innately good or innately evil? But if you think about it, the question, framed in these terms, makes very little sense. 'Good' and 'evil' are purely human concepts. It would never occur to anyone to argue about whether a fish, or a tree, were good or evil, because 'good' and 'evil' are concepts humans made up in order to compare ourselves with one another. It follows that arguing about whether humans are fundamentally good or evil makes about as much sense as arguing about whether humans are fundamentally fat or thin.

Nonetheless, on those occasions when people do reflect on the lessons of prehistory, they almost invariably come back to questions of this kind. We are all familiar with the Christian answer: people once lived in a state of innocence, yet were tainted by original sin. We desired to be godlike and have been punished for it; now we live in a fallen state while hoping for future redemption. Today, the popular version of this story is typically some updated variation on Jean-Jacques Rousseau's *Discourse on the Origin and the Foundation of Inequality Among Mankind*, which he wrote in 1754. Once upon a time, the story goes, we were hunter-gatherers, living in a prolonged state of childlike innocence, in tiny bands. These bands were egalitarian; they could be for the very reason that they were so small. It was only after the 'Agricultural Revolution', and then still more the rise of cities, that this happy condition came to an end, ushering in 'civilization' and 'the state' – which also meant the appearance of written literature, science and philosophy, but at the same time, almost everything bad in human life: patriarchy, standing armies, mass executions and annoying bureaucrats demanding that we spend much of our lives filling in forms.

Of course, this is a very crude simplification, but it really does seem to be the foundational story that rises to the surface whenever anyone, from industrial psychologists to revolutionary theorists, says something like 'but of course human beings spent most of their evolutionary history living in groups of ten or twenty people,' or 'agriculture was perhaps humanity's worst mistake.' And as we'll see, many popular writers make the argument quite explicitly. The problem is that anyone seeking an alternative to this rather depressing view of history will quickly find that the only one on offer is actually even worse: if not Rousseau, then Thomas Hobbes.

Hobbes's *Leviathan*, published in 1651, is in many ways the founding text of modern political theory. It held that, humans being the selfish creatures they are, life in an original State of Nature was in no sense innocent; it must instead have been 'solitary, poor, nasty, brutish, and short' – basically, a state of war, with everybody fighting against everybody else. Insofar as there has been any progress from this benighted state of affairs, a Hobbesian would argue, it has been largely due to exactly those repressive mechanisms that Rousseau was complaining about: governments, courts, bureaucracies, police. This view of things has been around for a very long time as well. There's a reason why, in English, the words 'politics' 'polite' and 'police' all sound the same – they're all derived from the Greek word *polis*, or city, the Latin equivalent of which is *civitas*, which also gives us 'civility,' 'civic' and a certain modern understanding of 'civilization'.

Human society, in this view, is founded on the collective repression of our baser instincts, which becomes all the more necessary when humans are living in large numbers in the same place. The modern-day Hobbesian, then, would argue that, yes, we did live most of our evolutionary history in tiny bands, who could get along mainly because they shared a common interest in the survival of their offspring ('parental investment', as evolutionary biologists call it). But even these were in no sense founded on equality. There was always, in this version, some 'alpha-male' leader. Hierarchy and domination, and cynical self-interest, have always been the basis of human society. It's just that, collectively, we have learned it's to our advantage to prioritize our long-term interests over our short-term instincts; or, better, to create laws that force us to confine our worst impulses to socially useful areas like the economy, while forbidding them everywhere else.

As the reader can probably detect from our tone, we don't much like the choice between these two alternatives. Our objections can be classified into three broad categories. As accounts of the general course of human history, they:

1. simply aren't true;
2. have dire political implications;
3. make the past needlessly dull.

This book is an attempt to begin to tell another, more hopeful and more interesting story; one which, at the same time, takes better account of what the last few decades of research have taught us. Partly, this is a matter of bringing together evidence that has accumulated in archaeology, anthropology and kindred disciplines; evidence that points towards a completely new account of how human societies developed over roughly the last 30,000 years. Almost all of this research goes against the familiar narrative, but too often the most remarkable discoveries remain confined to the work of specialists, or have to be teased out by reading between the lines of scientific publications.

To give just a sense of how different the emerging picture is: it is clear now that human societies before the advent of farming were not confined to small, egalitarian bands. On the contrary, the world of hunter-gatherers as it existed before the coming of agriculture was one of bold social experiments, resembling a carnival parade of political forms, far more than it does the drab abstractions of evolutionary theory. Agriculture, in turn, did not mean the inception of private property, nor did it mark an irreversible step towards inequality. In fact, many of the first farming communities were relatively free of ranks and hierarchies. And far from setting class differences in stone, a surprising number of the world's earliest cities were organized on robustly egalitarian lines, with no need for authoritarian rulers, ambitious warrior-politicians, or even bossy administrators.

Information bearing on such issues has been pouring in from every quarter of the globe. As a result, researchers around the world have also been examining ethnographic and historical material in a new light. The pieces now exist to create an entirely different world history – but so far, they remain hidden to all but a few privileged experts (and even the experts tend to hesitate before abandoning their own tiny part of the puzzle, to compare notes with others outside their specific subfield). Our aim in this book is to start putting some of the pieces of the puzzle together, in full awareness that nobody yet has anything like a complete set. The task is immense, and the issues so important, that it will take years of research and debate even to begin to understand the real implications of the picture we're starting to see. But it's crucial that we set the process in motion. One thing that will quickly become clear is that the prevalent 'big picture' of history – shared by modern-day followers of Hobbes and Rousseau alike – has almost nothing to do with the facts. But to begin making sense of the new information that's now before our eyes, it is not enough to compile and sift vast quantities of data. A conceptual shift is also required.

To make that shift means retracing some of the initial steps that led to our modern notion of social evolution: the idea that human societies could be arranged according to stages of development, each with their own characteristic technologies and forms of organization (hunter-gatherers, farmers, urban-industrial society, and so on). As we will see, such notions have their roots in a conservative backlash against critiques of European civilization, which began to gain ground in the early decades of the eighteenth century. The origins of that critique, however, lie not with the philosophers of the Enlightenment (much though they initially admired and imitated it), but with indigenous commentators and observers of European society, such as the Native American (Huron-Wendat) statesman Kandiaronk, of whom we will learn much more in the next chapter.

Revisiting what we will call the 'indigenous critique' means taking seriously contributions to social thought that come from outside the European canon, and in particular from those indigenous peoples whom Western philosophers tend to cast either in the role of history's angels or its devils. Both positions preclude any real possibility of intellectual exchange, or even dialogue: it's just as hard to debate someone who is considered diabolical as someone considered divine, as almost anything they think or say is likely to be deemed either irrelevant or deeply profound. Most of the people we will be considering in this book are long since dead. It is no longer possible to have any sort of conversation with them. We are nonetheless determined to write prehistory as if it consisted of people one would have been able to talk to, when they were still alive – who don't just exist as paragons, specimens, sock-puppets or playthings of some inexorable law of history.

There are, certainly, tendencies in history. Some are powerful; currents so strong that they are very difficult to swim against (though there always seem to be some who manage to do it anyway). But the only 'laws' are those we make up ourselves. Which brings us on to our second objection.

WHY BOTH THE HOBBESIAN AND ROUSSEAUIAN VERSIONS OF HUMAN HISTORY HAVE DIRE POLITICAL IMPLICATIONS

The political implications of the Hobbesian model need little elaboration. It is a foundational assumption of our economic system that humans are at base somewhat nasty and selfish creatures, basing their decisions on cynical, egoistic calculation rather than altruism or co-operation; in which case, the best we can hope for are more sophisticated internal and external controls on our supposedly innate drive towards accumulation and self-aggrandizement. Rousseau's story about how humankind descended into inequality from an original state of egalitarian innocence seems more optimistic (at least there was somewhere better to fall from), but nowadays it's mostly deployed to convince us that while the system we live under might be unjust, the most we can realistically aim for is a bit of modest tinkering. The term 'inequality' is itself very telling in this regard.

Since the financial crash of 2008, and the upheavals that followed, the question of inequality – and with it, the long-term history of inequality – have become major topics for debate. Something of a consensus has emerged among intellectuals and even, to some degree, the political classes that levels of social inequality have got out of hand, and that most of the world's problems result, in one way or another, from an ever-widening gulf between the haves and the have-nots. Pointing this out is in itself a challenge to global power structures; at the same time, though, it frames the issue in a way that people who benefit from those structures can still find ultimately reassuring, since it implies no meaningful solution to the problem would ever be possible.

After all, imagine we framed the problem differently, the way it might have been fifty or 100 years ago: as the concentration of capital, or oligopoly, or class power. Compared to any of these, a word like 'inequality' sounds like it's practically designed to encourage half-measures and compromise. It's possible to imagine overthrowing capitalism or breaking the power of the state, but it's not clear what eliminating inequality would even mean. (Which kind of inequality? Wealth? Opportunity? Exactly how equal would people have to be in order for us to be able to say we've 'eliminated inequality'?) The term 'inequality' is a way of framing social problems appropriate to an age of technocratic reformers, who assume from the outset that no real vision of social transformation is even on the table.

Debating inequality allows one to tinker with the numbers, argue about Gini coefficients and thresholds of dysfunction, readjust tax regimes or social welfare mechanisms, even shock the public with figures showing just how bad things have become ('Can you imagine? The richest 1 per cent of the world's population own 44 per cent of the world's wealth!') – but it also allows one to do all this without addressing any of the factors that people actually object to about such 'unequal' social arrangements: for instance, that some manage to turn their wealth into power over others; or that other people end up being told their needs are not important, and their lives have no intrinsic worth. The last, we are supposed to believe, is just the inevitable effect of inequality; and inequality, the inevitable result of living in any large, complex, urban, technologically sophisticated society. Presumably it will always be with us. It's just a matter of degree.

Today, there is a veritable boom of thinking about inequality: since 2011, 'global inequality' has regularly featured as a top item for debate in the World Economic Forum at Davos. There are inequality indexes, institutes for the study of inequality, and a relentless stream of publications trying to project the current obsession with property distribution back into the Stone Age. There have even been attempts to calculate income levels and Gini coefficients for Palaeolithic mammoth hunters (they both turn out to be very low).[1] It's almost as if we feel some need to come up with mathematical formulae justifying the expression, already popular in the days of Rousseau, that in such societies 'everyone was equal, because they were all equally poor.'

The ultimate effect of all these stories about an original state of innocence and equality, like the use of the term 'inequality' itself, is to make wistful pessimism about the human condition seem like common sense: the natural result of viewing ourselves through history's broad lens. Yes, living in a truly egalitarian society might be possible if you're a Pygmy or a Kalahari Bushman. But if you want to create a society of true equality today, you're going to have to figure out a way to go back to becoming tiny bands of foragers again with no significant personal property. Since foragers require a pretty extensive territory to forage in, this would mean having to reduce the world's population by something like 99.9 per cent. Otherwise, the best we can hope for is to adjust the size of the boot that will forever be stomping on our faces; or, perhaps, to wangle a bit more wiggle room in which some of us can temporarily duck out of its way.

A first step towards a more accurate, and hopeful, picture of world history might be to abandon the Garden of Eden once and for all, and simply do away with the notion that for hundreds of thousands of years, everyone on earth shared the same idyllic form of social organization. Strangely enough, though, this is often seen as a reactionary move. 'So are you saying true equality has never been achieved? That it's therefore impossible?' It seems to us that such objections are both counterproductive and frankly unrealistic.

First of all, it's bizarre to imagine that, say, during the roughly 10,000 (some would say more like 20,000) years in which people painted on the walls of Altamira, no one – not only in Altamira, but anywhere on earth – experimented with alternative forms of social organization. What's the chance of that? Second of all, is not the capacity to experiment with different forms of social organization itself a quintessential part of what makes us human? That is, beings with the capacity for self-creation, even freedom? The ultimate question of human history, as we'll see, is not our equal access to material resources (land, calories, means of production), much though these things are obviously important, but our equal capacity to contribute to decisions about how to live together. Of course, to exercise that capacity implies that there should be something meaningful to decide in the first place.

If, as many are suggesting, our species' future now hinges on our capacity to create something different (say, a system in which wealth cannot be freely transformed into power, or where some people are not told their needs are unimportant, or that their lives have no intrinsic worth), then what ultimately matters is whether we can rediscover the freedoms that make us human in the first place. As long ago as 1936, the prehistorian V. Gordon Childe wrote a book called *Man*

Makes Himself. Apart from the sexist language, this is the spirit we wish to invoke. We are projects of collective self-creation. What if we approached human history that way? What if we treat people, from the beginning, as imaginative, intelligent, playful creatures who deserve to be understood as such? What if, instead of telling a story about how our species fell from some idyllic state of equality, we ask how we came to be trapped in such tight conceptual shackles that we can no longer even imagine the possibility of reinventing ourselves?

SOME BRIEF EXAMPLES OF WHY RECEIVED UNDERSTANDINGS OF THE BROAD SWEEP OF HUMAN HISTORY ARE MOSTLY WRONG (OR, THE ETERNAL RETURN OF JEAN-JACQUES ROUSSEAU)

When we first embarked on this book, our intention was to seek new answers to questions about the origins of social inequality. It didn't take long before we realized this simply wasn't a very good approach. Framing human history in this way – which necessarily means assuming humanity once existed in an idyllic state, and that a specific point can be identified at which everything started to go wrong – made it almost impossible to ask any of the questions we felt were genuinely interesting. It felt like almost everyone else seemed to be caught in the same trap. Specialists were refusing to generalize. Those few willing to stick their necks out almost invariably reproduced some variation on Rousseau.

Let's consider a fairly random example of one of these generalist accounts, Francis Fukuyama's *The Origins of Political Order: From Prehuman Times to the French Revolution* (2011). Here is Fukuyama on what he feels can be taken as received wisdom about early human societies: 'In its early stages human political organization is similar to the band-level society observed in higher primates like chimpanzees,' which Fukuyama suggests can be regarded as 'a default form of social organization'. He then goes on to assert that Rousseau was largely correct in pointing out that the origin of political inequality lay in the development of agriculture, since hunter-gatherer societies (according to Fukuyama) have no concept of private property, and so little incentive to mark out a piece of land and say, 'This is mine.' Band-level societies of this sort, he suggests, are 'highly egalitarian'.[2]

Jared Diamond, in *The World Until Yesterday: What Can We Learn from Traditional Societies?* (2012) suggests that such bands (in which he believes humans still lived 'as recently as 11,000 years ago') comprised 'just a few dozen individuals', most biologically related. These small groups led a fairly meagre existence, 'hunting and gathering whatever wild animal and plant species happen to live in an acre of forest'. And their social lives, according to Diamond, were enviably simple. Decisions were reached through 'face-to-face discussion'; there were 'few personal possessions' and 'no formal political leadership or strong economic specialization'.[3] Diamond concludes that, sadly, it is only within such primordial groupings that humans ever achieved a significant degree of social equality.

For Diamond and Fukuyama, as for Rousseau some centuries earlier, what put an end to that equality – everywhere and forever – was the invention of agriculture, and the higher population levels it sustained. Agriculture brought about a transition from 'bands' to 'tribes'. Accumulation of food surplus fed population growth, leading some 'tribes' to develop into ranked societies known as 'chiefdoms'. Fukuyama paints an almost explicitly biblical picture of this process, a departure from Eden: 'As little bands of human beings migrated and adapted to different environments, they began their exit out of the state of nature by developing new social institutions.'[4] They fought wars over resources. Gangly and pubescent, these societies were clearly heading for trouble.

It was time to grow up and appoint some proper leadership. Hierarchies began to emerge. There was no point in resisting, since hierarchy – according to Diamond and Fukuyama – is inevitable once humans adopt large, complex forms of organization. Even when the new leaders began acting badly – creaming off agricultural surplus to promote their flunkies and relatives, making status permanent and hereditary, collecting trophy skulls and harems of slave-girls, or tearing out rivals' hearts with obsidian knives – there could be no going back. Before long, chiefs had managed to convince others they should be referred to as 'kings', even 'emperors'. As Diamond patiently explains to us:

> Large populations can't function without leaders who make the decisions, executives who carry out the decisions, and bureaucrats who administer the decisions and laws. Alas for all of you readers who are anarchists and dream of living without any state government, those are the reasons why your dream is unrealistic: you'll have to find some tiny band or tribe willing to accept you, where no one is a stranger, and where kings, presidents, and bureaucrats are unnecessary.[5]

A dismal conclusion, not just for anarchists but for anybody who ever wondered if there might be a viable alternative to the current status quo. Still, the truly remarkable thing is that, despite the self-assured tone, such pronouncements are not actually based on any kind of scientific evidence. As we will soon be discovering, there is simply no reason to believe that small-scale groups are especially likely to be egalitarian – or, conversely, that large ones must necessarily have kings, presidents or even bureaucracies. Statements like these are just so many prejudices dressed up as facts, or even as laws of history.[6]

ON THE PURSUIT OF HAPPINESS

As we say, it's all just an endless repetition of a story first told by Rousseau in 1754. Many contemporary scholars will quite literally say that Rousseau's vision has been proved correct. If so, it is an extraordinary coincidence, since Rousseau himself never suggested that the innocent State of Nature really happened. On the contrary, he insisted he was engaging in a thought experiment: 'One must not take the kind of research which we enter into as the pursuit of truths of history, but solely as hypothetical and conditional reasonings, better fitted to clarify the nature of things than to expose their actual origin . . .'[7]

Rousseau's portrayal of the State of Nature and how it was overturned by the coming of agriculture was never intended to form the basis for a series of evolutionary stages, like the ones Scottish philosophers such as Smith, Ferguson or Millar (and later on, Lewis Henry

Morgan) were referring to when they spoke of 'Savagery' and 'Barbarism'. In no sense was Rousseau imagining these different states of being as levels of social and moral development, corresponding to historical changes in modes of production: foraging, pastoralism, farming, industry. Rather, what Rousseau presented was more of a parable, by way of an attempt to explore a fundamental paradox of human politics: how is it that our innate drive for freedom somehow leads us, time and again, on a 'spontaneous march to inequality'?[8]

Describing how the invention of farming first leads to private property, and property to the need for civil government to protect it, this is how Rousseau puts things: 'All ran towards their chains, believing that they were securing their liberty; for although they had reason enough to discern the advantages of a civil order, they did not have experience enough to foresee the dangers.'[9] His imaginary State of Nature was primarily invoked as a way of illustrating the point. True, he didn't invent the concept: as a rhetorical device, the State of Nature had already been used in European philosophy for a century. Widely deployed by natural law theorists, it effectively allowed every thinker interested in the origins of government (Locke, Grotius and so on) to play God, each coming up with his own variant on humanity's original condition, as a springboard for speculation.

Hobbes was doing much the same thing when he wrote in *Leviathan* that the primordial state of human society would necessarily have been a '*Bellum omnium contra omnes*', a war of all against all, which could only be overcome by the creation of an absolute sovereign power. He wasn't saying there had actually been a time when everyone lived in such a primordial state. Some suspect that Hobbes's state of war was really an allegory for his native England's descent into civil war in the mid seventeenth century, which drove the royalist author into exile in Paris. Whatever the case, the closest Hobbes himself came to suggesting this state really existed was when he noted how the only people who weren't under the ultimate authority of some king were the kings themselves, and they always seemed to be at war with one another.

Despite all this, many modern writers treat *Leviathan* in the same way others treat Rousseau's *Discourse* – as if it were laying the groundwork for an evolutionary study of history; and although the two have completely different starting points, the result is rather similar.[10]

'When it came to violence in pre-state peoples,' writes the psychologist Steven Pinker, 'Hobbes and Rousseau were talking through their hats: neither knew a thing about life before civilization.' On this point, Pinker is absolutely right. In the same breath, however, he also asks us to believe that Hobbes, writing in 1651 (apparently through his hat), somehow managed to guess right, and come up with an analysis of violence and its causes in human history that is 'as good as any today'.[11] This would be an astonishing – not to mention damning – verdict on centuries of empirical research, if it only happened to be true. As we'll see, it is not even close.[12]

We can take Pinker as our quintessential modern Hobbesian. In his magnum opus, *The Better Angels of Our Nature: Why Violence Has Declined* (2012), and subsequent books like *Enlightenment Now: The Case for Reason, Science, Humanism, and Progress* (2018) he argues that today we live in a world which is, overall, far less violent and cruel than anything our ancestors had ever experienced.[13]

Now, this may seem counter-intuitive to anyone who spends much time watching the news, let alone who knows much about the history of the twentieth century. Pinker, though, is confident that an objective statistical analysis, shorn of sentiment, will show us to be living in an age of unprecedented peace and security. And this, he suggests, is the logical outcome of living in sovereign states, each with a monopoly over the legitimate use of violence within its borders, as opposed to the 'anarchic societies' (as he calls them) of our deep evolutionary past, where life for most people was, indeed, typically 'nasty, brutish, and short'.

Since, like Hobbes, Pinker is concerned with the origins of the state, his key point of transition is not the rise of farming but the emergence of cities. 'Archaeologists', he writes, 'tell us that humans lived in a state of anarchy until the emergence of civilization some five thousand years ago, when sedentary farmers first coalesced into cities and states and developed the first governments.'[14] What follows is, to put it bluntly, a modern psychologist making it up as he goes along. You might hope that a passionate advocate of science would approach the topic scientifically, through a broad appraisal of the evidence – but this is precisely the approach to human prehistory that Pinker seems to find uninteresting. Instead he relies on anecdotes, images and individual sensational discoveries, like the headline-making find, in 1991, of 'Ötzi the Tyrolean Iceman'.

'What is it about the ancients,' Pinker asks at one point, 'that they couldn't leave us an interesting corpse without resorting to foul play?' There is an obvious response to this: doesn't it rather depend on which corpse you consider interesting in the first place? Yes, a little over 5,000 years ago someone walking through the Alps left the world of the living with an arrow in his side; but there's no particular reason to treat Ötzi as a poster child for humanity in its original condition, other than, perhaps, Ötzi suiting Pinker's argument. But if all we're doing is cherry-picking, we could just as easily have chosen the much earlier burial known to archaeologists as Romito 2 (after the Calabrian rock-shelter where it was found). Let's take a moment to consider what it would mean if we did this.

Romito 2 is the 10,000-year-old burial of a male with a rare genetic disorder (acromesomelic dysplasia): a severe type of dwarfism, which in life would have rendered him both anomalous in his community and unable to participate in the kind of high-altitude hunting that was necessary for their survival. Studies of his pathology show that, despite generally poor levels of health and nutrition, that same community of hunter-gatherers still took pains to support this individual through infancy and into early adulthood, granting him the same share of meat as everyone else, and ultimately according him a careful, sheltered burial.[15]

Neither is Romito 2 an isolated case. When archaeologists undertake balanced appraisals of hunter-gatherer burials from the Palaeolithic, they find high frequencies of health-related disabilities – but also surprisingly high levels of care until the time of death (and beyond, since some of these funerals were remarkably lavish).[16] If we did want to reach a general conclusion about what form human societies originally took, based on statistical frequencies of health indicators from ancient burials, we would have to reach the exact opposite conclusion to Hobbes (and Pinker): in origin, it might be claimed, our species is a nurturing and care-giving species, and there was simply no need for life to be nasty, brutish or short.

We're not suggesting we actually do this. As we'll see, there is reason to believe that during the Palaeolithic, only rather unusual individuals were buried at all. We just want to point out how easy it would be to play the same game in the other direction – easy, but frankly not too enlightening.[17] As we get to grips with the actual evidence, we always find that the realities of early human social life were far more complex, and a good deal more interesting, than any modern-day State of Nature theorist would ever be likely to guess.

When it comes to cherry-picking anthropological case studies, and putting them forward as representative of our 'contemporary ancestors' – that is, as models for what humans might have been like in a State of Nature – those working in the tradition of Rousseau tend to prefer African foragers like the Hadza, Pygmies or !Kung. Those who follow Hobbes prefer the Yanomami.

The Yanomami are an indigenous population who live largely by growing plantains and cassava in the Amazon rainforest, their traditional homeland, on the border of southern Venezuela and northern Brazil. Since the 1970s, the Yanomami have acquired a reputation as the quintessential violent savages: 'fierce people', as their most famous ethnographer, Napoleon Chagnon, called them. This seems decidedly unfair to the Yanomami since, in fact, statistics show they're not particularly violent – compared with other Amerindian groups, Yanomami homicide rates turn out average-to-low.[18] Again, though, actual statistics turn out to matter less than the availability of dramatic images and anecdotes. The real reason the Yanomami are so famous, and have such a colourful reputation, has everything to do with Chagnon himself: his 1968 book *Yanomamö: The Fierce People*, which sold millions of copies, and also a series of films, such as *The Ax Fight*, which offered viewers a vivid glimpse of tribal warfare. For a while all this made Chagnon the world's most famous anthropologist, in the process turning the Yanomami into a notorious case study of primitive violence and establishing their scientific importance in the emerging field of sociobiology.

We should be fair to Chagnon (not everyone is). He never claimed the Yanomami should be treated as living remnants of the Stone Age; indeed, he often noted that they obviously weren't. At the same time, and somewhat unusually for an anthropologist, he tended to define them primarily in terms of things they lacked (e.g. written language, a police force, a formal judiciary), as opposed to the positive features of their culture, which has rather the same effect of setting them up as quintessential primitives.[19] Chagnon's central argument was that adult Yanomami men achieve both cultural and reproductive advantages by killing other adult men; and that this feedback between violence and biological fitness – if generally representative of the early human condition – might have had evolutionary consequences for our species as a whole.[20]

This is not just a big 'if' – it's enormous. Other anthropologists started raining down questions, not always friendly.[21] Allegations of professional misconduct were levelled at Chagnon (mostly revolving around ethical standards in the field), and everyone took sides. Some of these accusations appear baseless, but the rhetoric of Chagnon's defenders grew so heated that (as another celebrated anthropologist, Clifford Geertz, put it) not only was he held up as the epitome of rigorous, scientific anthropology, but all who questioned him or his social Darwinism were excoriated as 'Marxists', 'liars', 'cultural anthropologists from the academic left', 'ayatollahs' and 'politically correct bleeding hearts'. To this day, there is no easier way to get anthropologists to begin denouncing each other as extremists than to mention the name of Napoleon Chagnon.[22]

The important point here is that, as a 'non-state' people, the Yanomami are supposed to exemplify what Pinker calls the 'Hobbesian trap', whereby individuals in tribal societies find themselves caught in repetitive cycles of raiding and warfare, living fraught and precarious lives, always just a few steps away from violent death on the tip of a sharp weapon or at the end of a vengeful club. That, Pinker tells us, is the kind of dismal fate ordained for us by evolution. We have only escaped it by virtue of our willingness to place ourselves under the common protection of nation states, courts of law and police forces; and also by embracing virtues of reasoned debate and self-control that Pinker sees as the exclusive heritage of a European 'civilizing process', which produced the Age of Enlightenment (in other words, were it not for Voltaire, and the police, the knife-fight over Chagnon's findings would have been physical, not just academic).

There are many problems with this argument. We'll start with the most obvious. The idea that our current ideals of freedom, equality and democracy are somehow products of the 'Western tradition' would in fact have come as an enormous surprise to someone like Voltaire. As we'll soon see, the Enlightenment thinkers who propounded such ideals almost invariably put them in the mouths of foreigners, even 'savages' like the Yanomami. This is hardly surprising, since it's almost impossible to find a single author in that Western tradition, from Plato to Marcus Aurelius to Erasmus, who did not make it clear that they would have been opposed to such ideas. The word 'democracy' might have been invented in Europe (barely, since Greece at the time was much closer culturally to North Africa and the Middle East than it was to, say, England), but it's almost impossible to find a single European author before the nineteenth century who suggested it would be anything other than a terrible form of government.[23]

For obvious reasons, Hobbes's position tends to be favoured by those on the right of the political spectrum, and Rousseau's by those leaning left. Pinker positions himself as a rational centrist, condemning what he considers to be the extremists on either side. But why then insist that all significant forms of human progress before the twentieth century can be attributed only to that one group of humans who used to refer to themselves as 'the white race' (and now, generally, call themselves by its more accepted synonym, 'Western civilization')? There is simply no reason to make this move. It would be just as easy (actually, rather easier) to identify things that can be interpreted as the first stirrings of rationalism, legality, deliberative democracy and so forth all over the world, and only then tell the story of how they coalesced into the current global system.[24]

Insisting, to the contrary, that all good things come only from Europe ensures one's work can be read as a retroactive apology for genocide, since (apparently, for Pinker) the enslavement, rape, mass murder and destruction of whole civilizations – visited on the rest of the world by European powers – is just another example of humans comporting themselves as they always had; it was in no sense unusual. What was really significant, so this argument goes, is that it made possible the dissemination of what he takes to be 'purely' European notions of freedom, equality before the law, and human rights to the survivors.

Whatever the unpleasantness of the past, Pinker assures us, there is every reason to be optimistic, indeed happy, about the overall path our species has taken. True, he does concede there is scope for some serious tinkering in areas like poverty reduction, income inequality or indeed peace and security; but on balance – and relative to the number of people living on earth today – what we have now is a spectacular improvement on anything our species accomplished in its history so far (unless you're Black, or live in Syria, for example). Modern life is, for Pinker, in almost every way superior to what came before; and here he does produce elaborate statistics which purport to show how every day in every way – health, security, education, comfort, and by almost any other conceivable parameter – everything is actually getting better and better.

It's hard to argue with the numbers, but as any statistician will tell you, statistics are only as good as the premises on which they are based. Has 'Western civilization' really made life better for everyone? This ultimately comes down to the question of how to measure human happiness, which is a notoriously difficult thing to do. About the only dependable way anyone has ever discovered to determine whether one way of living is really more satisfying, fulfilling, happy or otherwise preferable to any other is to allow people to fully experience both, give them a choice, then watch what they actually do. For instance, if Pinker is correct, then any sane person who had to choose between (a) the violent chaos and abject poverty of the 'tribal' stage in human development and (b) the relative security and prosperity of Western civilization would not hesitate to leap for safety.[25]

But empirical data *is* available here, and it suggests something is very wrong with Pinker's conclusions.

Over the last several centuries, there have been numerous occasions when individuals found themselves in a position to make precisely this choice – and they almost never go the way Pinker would have predicted. Some have left us clear, rational explanations for why they made the choices they did. Let us consider the case of Helena Valero, a Brazilian woman born into a family of Spanish descent, whom Pinker mentions as a 'white girl' abducted by Yanomami in 1932 while travelling with her parents along the remote Rio Dimití.

For two decades, Valero lived with a series of Yanomami families, marrying twice, and eventually achieving a position of some importance in her community. Pinker briefly cites the account Valero later gave of her own life, where she describes the brutality of a Yanomami raid.[26] What he neglects to mention is that in 1956 she abandoned the Yanomami to seek her natal family and live again in 'Western civilization,' only to find herself in a state of occasional hunger and constant dejection and loneliness. After a while, given the ability to make a fully informed decision, Helena Valero decided she preferred life among the Yanomami, and returned to live with them.[27]

Her story is by no means unusual. The colonial history of North and South America is full of accounts of settlers, captured or adopted by indigenous societies, being given the choice of where they wished to stay and almost invariably choosing to stay with the latter.[28] This even applied to abducted children. Confronted again with their biological parents, most would run back to their adoptive kin for protection.[29] By contrast, Amerindians incorporated into European society by adoption or marriage, including those who – unlike the unfortunate Helena Valero – enjoyed considerable wealth and schooling, almost invariably did just the opposite: either escaping at the earliest opportunity, or – having tried their best to adjust, and ultimately failed – returning to indigenous society to live out their last days.

Among the most eloquent commentaries on this whole phenomenon is to be found in a private letter written by Benjamin Franklin to a friend:

> When an Indian Child has been brought up among us, taught our language and habituated to our Customs, yet if he goes to see his relations and make one Indian Ramble with them there is no persuading him ever to return, and that this is not natural merely as Indians, but as men, is plain from this, that when white persons of either sex have been taken prisoner young by the Indians, and lived awhile among them, tho' ransomed by their Friends, and treated with all imaginable tenderness to prevail with them to stay among the English, yet in a Short time

> they become disgusted with our manner of life, and the care and pains that are necessary to support it, and take the first opportunity of escaping again into the Woods, from whence there is no reclaiming them. One instance I remember to have heard, where the person was to be brought home to possess a good Estate; but finding some care necessary to keep it together, he relinquished it to a younger brother, reserving to himself nothing but a gun and match-Coat, with which he took his way again to the Wilderness.[30]

Many who found themselves embroiled in such contests of civilization, if we may call them that, were able to offer clear reasons for their decisions to stay with their erstwhile captors. Some emphasized the virtues of freedom they found in Native American societies, including sexual freedom, but also freedom from the expectation of constant toil in pursuit of land and wealth.[31] Others noted the 'Indian's' reluctance ever to let anyone fall into a condition of poverty, hunger or destitution. It was not so much that they feared poverty themselves, but rather that they found life infinitely more pleasant in a society where no one else was in a position of abject misery (perhaps much as Oscar Wilde declared he was an advocate of socialism because he didn't like having to look at poor people or listen to their stories). For anyone who has grown up in a city full of rough sleepers and panhandlers – and that is, unfortunately, most of us – it is always a bit startling to discover there's nothing inevitable about any of this.

Still others noted the ease with which outsiders, taken in by 'Indian' families, might achieve acceptance and prominent positions in their adoptive communities, becoming members of chiefly households, or even chiefs themselves.[32] Western propagandists speak endlessly about equality of opportunity; these seem to have been societies where it actually existed. By far the most common reasons, however, had to do with the intensity of social bonds they experienced in Native American communities: qualities of mutual care, love and above all happiness, which they found impossible to replicate once back in European settings. 'Security' takes many forms. There is the security of knowing one has a statistically smaller chance of getting shot with an arrow. And then there's the security of knowing that there are people in the world who will care deeply if one is.

HOW THE CONVENTIONAL NARRATIVE OF HUMAN HISTORY IS NOT ONLY WRONG, BUT QUITE NEEDLESSLY DULL

One gets the sense that indigenous life was, to put it very crudely, just a lot more interesting than life in a 'Western' town or city, especially insofar as the latter involved long hours of monotonous, repetitive, conceptually empty activity. The fact that we find it hard to imagine how such an alternative life could be endlessly engaging and interesting is perhaps more a reflection on the limits of our imagination than on the life itself.

One of the most pernicious aspects of standard world-historical narratives is precisely that they dry everything up, reduce people to cardboard stereotypes, simplify the issues (are we inherently selfish and violent, or innately kind and co-operative?) in ways that themselves undermine, possibly even destroy, our sense of human possibility. 'Noble' savages are, ultimately, just as boring as savage ones; more to the point, neither actually exist. Helena Valero was herself adamant on this point. The Yanomami were not devils, she insisted, neither were they angels. They were human, like the rest of us.

Now, we should be clear here: social theory always, necessarily, involves a bit of simplification. For instance, almost any human action might be said to have a political aspect, an economic aspect, a psychosexual aspect and so forth. Social theory is largely a game of make-believe in which we pretend, just for the sake of argument, that there's just one thing going on: essentially, we reduce everything to a cartoon so as to be able to detect patterns that would be otherwise invisible. As a result, all real progress in social science has been rooted in the courage to say things that are, in the final analysis, slightly ridiculous: the work of Karl Marx, Sigmund Freud or Claude Lévi-Strauss being only particularly salient cases in point. One must simplify the world to discover something new about it. The problem comes when, long after the discovery has been made, people continue to simplify.

Hobbes and Rousseau told their contemporaries things that were startling, profound and opened new doors of the imagination. Now their ideas are just tired common sense. There's nothing in them that justifies the continued simplification of human affairs. If social scientists today continue to reduce past generations to simplistic, two-dimensional caricatures, it is not so much to show us anything original, but just because they feel that's what social scientists are expected to do so as to appear 'scientific'. The actual result is to impoverish history – and as a consequence, to impoverish our sense of possibility. Let us end this introduction with an illustration, before moving on to the heart of the matter.

Ever since Adam Smith, those trying to prove that contemporary forms of competitive market exchange are rooted in human nature have pointed to the existence of what they call 'primitive trade'. Already tens of thousands of years ago, one can find evidence of objects – very often precious stones, shells or other items of adornment – being moved around over enormous distances. Often these were just the sort of objects that anthropologists would later find being used as 'primitive currencies' all over the world. Surely this must prove capitalism in some form or another has always existed?

The logic is perfectly circular. If precious objects were moving long distances, this is evidence of 'trade' and, if trade occurred, it must have taken some sort of commercial form; therefore, the fact that, say, 3,000 years ago Baltic amber found its way to the Mediterranean, or shells from the Gulf of Mexico were transported to Ohio, is proof that we are in the presence of some embryonic form of market economy. Markets are universal. Therefore, there must have been a market. Therefore, markets are universal. And so on.

All such authors are really saying is that they themselves cannot personally imagine any other way that precious objects might move about. But lack of imagination is not itself an argument. It's almost as if these writers are afraid to suggest anything that seems original, or, if they do, feel obliged to use vaguely scientific-sounding language ('trans-regional interaction spheres', 'multi-scalar networks of exchange') to avoid having to speculate about what precisely those things might be. In fact, anthropology provides endless illustrations of how valuable objects might travel long distances in the absence of anything that remotely resembles a market economy.

The founding text of twentieth-century ethnography, Bronisław

Malinowski's 1922 *Argonauts of the Western Pacific*, describes how in the 'kula chain' of the Massim Islands off Papua New Guinea, men would undertake daring expeditions across dangerous seas in outrigger canoes, just in order to exchange precious heirloom arm-shells and necklaces for each other (each of the most important ones has its own name, and history of former owners) – only to hold it briefly, then pass it on again to a different expedition from another island. Heirloom treasures circle the island chain eternally, crossing hundreds of miles of ocean, arm-shells and necklaces in opposite directions. To an outsider, it seems senseless. To the men of the Massim it was the ultimate adventure, and nothing could be more important than to spread one's name, in this fashion, to places one had never seen.

Is this 'trade'? Perhaps, but it would bend to breaking point our ordinary understandings of what that word means. There is, in fact, a substantial ethnographic literature on how such long-distance exchange operates in societies without markets. Barter does occur: different groups may take on specialities – one is famous for its feather-work, another provides salt, in a third all women are potters – to acquire things they cannot produce themselves; sometimes one group will specialize in the very business of moving people and things around. But we often find such regional networks developing largely for the sake of creating friendly mutual relations, or having an excuse to visit one another from time to time;[33] and there are plenty of other possibilities that in no way resemble 'trade'.

Let's list just a few, all drawn from North American material, to give the reader a taste of what might really be going on when people speak of 'long-distance interaction spheres' in the human past:

1. **Dreams or vision quests:** among Iroquoian-speaking peoples in the sixteenth and seventeenth centuries it was considered extremely important literally to realize one's dreams. Many European observers marvelled at how Indians would be willing to travel for days to bring back some object, trophy, crystal or even an animal like a dog that they had dreamed of acquiring. Anyone who dreamed about a neighbour or relative's possession (a kettle, ornament, mask and so on) could normally demand it; as a result, such objects would

often gradually travel some way from town to town. On the Great Plains, decisions to travel long distances in search of rare or exotic items could form part of vision quests.[34]

2. **Travelling healers and entertainers**: in 1528, when a shipwrecked Spaniard named Álvar Núñez Cabeza de Vaca made his way from Florida across what is now Texas to Mexico, he found he could pass easily between villages (even villages at war with one another) by offering his services as a magician and curer. Curers in much of North America were also entertainers, and would often develop significant entourages; those who felt their lives had been saved by the performance would, typically, offer up all their material possessions to be divided among the troupe.[35] By such means, precious objects could easily travel very long distances.
3. **Women's gambling**: women in many indigenous North American societies were inveterate gamblers; the women of adjacent villages would often meet to play dice or a game played with a bowl and plum stone, and would typically bet their shell beads or other objects of personal adornment as the stakes. One archaeologist versed in the ethnographic literature, Warren DeBoer, estimates that many of the shells and other exotica discovered in sites halfway across the continent had got there by being endlessly wagered, and lost, in inter-village games of this sort, over very long periods of time.[36]

We could multiply examples, but assume that by now the reader gets the broader point we are making. When we simply guess as to what humans in other times and places might be up to, we almost invariably make guesses that are far less interesting, far less quirky – in a word, far less human than what was likely going on.

ON WHAT'S TO FOLLOW

In this book we will not only be presenting a new history of humankind, but inviting the reader into a new science of history, one that restores our ancestors to their full humanity. Rather than asking how we ended up unequal, we will start by asking how it was that 'inequality' became such an issue to begin with, then gradually build up an alternative narrative that corresponds more closely to our current state of knowledge. If humans did not spend 95 per cent of their evolutionary past in tiny bands of hunter-gatherers, what were they doing all that time? If agriculture, and cities, did not mean a plunge into hierarchy and domination, then what did they imply? What was really happening in those periods we usually see as marking the emergence of 'the state'? The answers are often unexpected, and suggest that the course of human history may be less set in stone, and more full of playful possibilities, than we tend to assume.

In one sense, then, this book is simply trying to lay down foundations for a new world history, rather as Gordon Childe did when, back in the 1930s, he invented phrases like 'the Neolithic Revolution' or 'the Urban Revolution'. As such it is necessarily uneven and incomplete. At the same time, this book is also something else: a quest to discover the right questions. If 'what is the origin of inequality?' is not the biggest question we should be asking about history, what then should it be? As the stories of one-time captives escaping back to the woods again make clear, Rousseau was not entirely mistaken. Something *has* been lost. He just had a rather idiosyncratic (and ultimately, false) notion of what it was. How do we characterize it, then? And how lost is it really? What does it imply about possibilities for social change today?

For about a decade now, we – that is, the two authors of this book – have been engaged in a prolonged conversation with each other about exactly these questions. This is the reason for the book's somewhat unusual structure, which begins by tracing the historical roots of the question ('what is the origin of social inequality?') back to a series of encounters between European colonists and Native American intellectuals in the seventeenth century. The impact of those encounters upon what we now term the Enlightenment, and indeed our basic conceptions of human history, is both more subtle and profound than we usually care to admit. Revisiting them, as we discovered, has startling implications for how we make sense of the human past today, including the origins of farming, property, cities, democracy, slavery and civilization itself. In the end, we decided to write a book that would echo, to some degree at least, that evolution in our own thought. In those conversations, the real breakthrough moment came when we decided to move away from European thinkers like Rousseau entirely, and instead consider perspectives that derive from those indigenous thinkers who ultimately inspired them.

So let us begin right there.

2

Wicked Liberty

The indigenous critique and the myth of progress

Jean-Jacques Rousseau left us a story about the origins of social inequality that continues to be told and retold, in endless variations, to this day. It is the story of humanity's original innocence, and unwitting departure from a state of pristine simplicity on a voyage of technological discovery that would ultimately guarantee both our 'complexity' and our enslavement. How did this ambivalent story of civilization come about?

Intellectual historians have never really abandoned the Great Man theory of history. They often write as if all important ideas in a given age can be traced back to one or other extraordinary individual – whether Plato, Confucius, Adam Smith or Karl Marx – rather than seeing such authors' writings as particularly brilliant interventions in debates that were already going on in taverns or dinner parties or public gardens (or, for that matter, lecture rooms), but which otherwise might never have been written down. It's a bit like pretending William Shakespeare had somehow invented the English language. In fact, many of Shakespeare's most brilliant turns of phrase turn out to have been common expressions of the day, which any Elizabethan Englishman or woman would be likely to have thrown into casual conversation, and whose authors remain as obscure as those of knock-knock jokes – even if, were it not for Shakespeare, they'd probably have passed out of use and been forgotten long ago.

All this applies to Rousseau. Intellectual historians sometimes write as if Rousseau had personally kicked off the debate about social inequality with his 1754 *Discourse on the Origin and the Foundation of Inequality Among Mankind*. In fact, he wrote it to submit to an essay contest on the subject.

IN WHICH WE SHOW HOW CRITIQUES OF EUROCENTRISM CAN BACKFIRE, AND END UP TURNING ABORIGINAL THINKERS INTO 'SOCK-PUPPETS'

In March 1754, the learned society known as the Académie des Sciences, Arts et Belles-Lettres de Dijon announced a national essay competition on the question: 'what is the origin of *inequality* among men, and is it authorized by natural law?' What we'd like to do in this chapter is ask: why is it that a group of scholars in *Ancien Régime* France, hosting a national essay contest, would have felt this was an appropriate question in the first place? The way the question is put, after all, assumes that social inequality did *have* an origin; that is, it takes for granted that there was a time when human beings were equals – and that something then happened to change this situation.

That is actually quite a startling thing for people living under an absolutist monarchy like that of Louis XV to think. After all, it's not as if anyone in France at that time had much personal experience of living in a society of equals. This was a culture in which almost every aspect of human interaction – whether eating, drinking, working or socializing – was marked by elaborate pecking orders and rituals of social deference. The authors who submitted their essays to this competition were men who spent their lives having all their needs attended to by servants. They lived off the patronage of dukes and archbishops, and rarely entered a building without knowing the precise order of importance of everyone inside. Rousseau was one such man: an ambitious young philosopher, he was at the time engaged in an elaborate project of trying to sleep his way into influence at court. The closest he'd likely ever come to experiencing social equality himself was someone doling out equal slices of cake at a dinner party. Yet everyone at the time also agreed that this situation was somehow unnatural; that it had not always been that way.

If we want to understand why *that* was, we need to look not only at France, but also at France's place in a much larger world.

Fascination with the question of social inequality was relatively new in the 1700s, and it had everything to do with the shock and confusion that followed Europe's sudden integration into a global economy, where it had long been a very minor player.

In the Middle Ages, most people in other parts of the world who actually knew anything about northern Europe at all considered it an obscure and uninviting backwater full of religious fanatics who, aside from occasional attacks on their neighbours ('the Crusades'), were largely irrelevant to global trade and world politics.[1] European intellectuals of that time were just rediscovering Aristotle and the ancient world, and had very little idea what people were thinking and arguing about anywhere else. All this changed, of course, in the late fifteenth century, when Portuguese fleets began rounding Africa and bursting into the Indian Ocean – and especially with the Spanish conquest of the Americas. Suddenly, a few of the more powerful European kingdoms found themselves in control of vast stretches of the globe, and European intellectuals found themselves exposed, not only to the civilizations of China and India but to a whole plethora of previously unimagined social, scientific and political ideas. The ultimate result of this flood of new ideas came to be known as the 'Enlightenment'.

Of course, this isn't usually the way historians of ideas tell this story. Not only are we taught to think of intellectual history as something largely produced by individuals writing great books or thinking great thoughts, but these 'great thinkers' are assumed to perform both these activities almost exclusively with reference to each other. As a result, even in cases where Enlightenment thinkers openly insisted they were getting their ideas from foreign sources (as the German philosopher Gottfried Wilhelm Leibniz did when he urged his compatriots to adopt Chinese models of statecraft), there's a tendency for contemporary historians to insist they weren't really serious; or else that when they said they were embracing Chinese, or Persian, or indigenous American ideas these weren't really Chinese, Persian or indigenous American ideas at all but ones they themselves had made up and merely attributed to exotic Others.[2]

These are remarkably arrogant assumptions – as if 'Western thought' (as it later came to be known) was such a powerful and monolithic body of ideas that no one else could possibly have any meaningful influence on it. It's also pretty obviously untrue. Just consider the case of Leibniz: over the course of the eighteenth and nineteenth centuries,

European governments gradually came to adopt the idea that every government should properly preside over a population of largely uniform language and culture, run by a bureaucratic officialdom trained in the liberal arts whose members had succeeded in passing competitive exams. It might seem surprising that they did so, since nothing remotely like that had existed in any previous period of European history. Yet it was almost exactly the system that had existed for centuries in China.

Are we really to insist that the advocacy of Chinese models of statecraft by Leibniz, his allies and followers really had *nothing* to do with the fact that Europeans did, in fact, adopt something that looks very much like Chinese models of statecraft? What is really unusual about this case is that Leibniz was so honest about his intellectual influences. When he lived, Church authorities still wielded a great deal of power in most of Europe: anyone making an argument that non-Christian ways were in any way superior might find themselves facing charges of atheism, which was potentially a capital offence.[3]

It is much the same with the question of inequality. If we ask, not 'what are the origins of social inequality?' but 'what are the origins of the *question* about the origins of social inequality?' (in other words, how did it come about that, in 1754, the Académie de Dijon would think this an appropriate question to ask?), then we are immediately confronted with a long history of Europeans arguing with one another about the nature of faraway societies: in this case, particularly in the Eastern Woodlands of North America. What's more, a lot of those conversations make reference to arguments that took place between Europeans and indigenous Americans about the nature of freedom, equality or for that matter rationality and revealed religion – indeed, most of the themes that would later become central to Enlightenment political thought.

Many influential Enlightenment thinkers did in fact claim that some of their ideas on the subject were directly taken from Native American sources – even though, predictably, intellectual historians today insist this cannot really be the case. Indigenous people are assumed to have lived in a completely different universe, inhabited a different reality, even; anything Europeans said about them was simply a shadow-play projection, fantasies of the 'noble savage' culled from the European tradition itself.[4]

Of course, such historians typically frame this position as a critique of Western arrogance ('how can you suggest that genocidal imperialists were actually listening to those whose societies they were in the process of stamping out?'), but it could equally well be seen as a form of Western arrogance in its own right. There is no contesting that European traders, missionaries and settlers did actually engage in prolonged conversations with people they encountered in what they called the New World, and often lived among them for extended periods of time – even as they also colluded in their destruction. We also know that many of those living in Europe who came to embrace principles of freedom and equality (principles barely existing in their countries a few generations before) claimed that accounts of these encounters had a profound influence on their thinking. To deny any possibility that they were right is, effectively, to insist that indigenous people could not possibly have any real impact on history. It is, in fact, a way of infantilizing non-Westerners: a practice denounced by these very same authors.

In recent years, a growing number of American scholars, most themselves of indigenous descent, have challenged these assumptions.[5] Here we follow in their footsteps. Basically, we are going to retell the story, starting from the assumption that all parties to the conversation between European colonists and their indigenous interlocutors were adults, and that, at least occasionally, they actually listened to each other. If we do this, even familiar histories suddenly begin to look very different. In fact, what we'll see is not only that indigenous Americans – confronted with strange foreigners – gradually developed their own, surprisingly consistent critique of European institutions, but that these critiques came to be taken very seriously in Europe itself.

Just how seriously can hardly be overstated. For European audiences, the indigenous critique would come as a shock to the system, revealing possibilities for human emancipation that, once disclosed, could hardly be ignored. Indeed, the ideas expressed in that critique came to be perceived as such a menace to the fabric of European society that an entire body of theory was called into being, specifically to refute them. As we will shortly see, the whole story we summarized in the last chapter – our standard historical meta-narrative about the ambivalent progress of human civilization, where freedoms are lost as societies grow bigger and more complex – was invented largely for the purpose of neutralizing the threat of indigenous critique.

The first thing to emphasize is that 'the origin of social inequality' is not a problem which would have made sense to anyone in the Middle Ages. Ranks and hierarchies were assumed to have existed from the very beginning. Even in the Garden of Eden, as the thirteenth-century philosopher Thomas Aquinas observed, Adam clearly outranked Eve. 'Social equality' – and therefore, its opposite, inequality – simply did not exist as a concept. A recent survey of medieval literature by two Italian scholars in fact finds no evidence that the Latin terms *aequalitas* or *inaequalitas* or their English, French, Spanish, German and Italian cognates were used to describe social relations at all before the time of Columbus. So one cannot even say that medieval thinkers rejected the notion of social equality: the idea that it might exist seems never to have occurred to them.[6]

In fact, the terms 'equality' and 'inequality' only began to enter common currency in the early seventeenth century, under the influence of natural law theory. And natural law theory, in turn, arose largely in the course of debates about the moral and legal implications of Europe's discoveries in the New World.

It's important to remember that Spanish adventurers like Cortés and Pizarro carried out their conquests largely without authorization from higher authorities; afterwards, there were intense debates back home over whether such unvarnished aggression against people who, after all, posed no threat to Europeans could really be justified.[7] The key problem was that – unlike non-Christians of the Old World, who could be assumed to have had the opportunity to learn the teachings of Jesus, and therefore to have actively rejected them – it was fairly obvious that the inhabitants of the New World simply never had any exposure to Christian ideas. So they couldn't be classed as infidels.

The conquistadors generally finessed this question by reading a declaration in Latin calling on all the Indians to convert before attacking them. Legal scholars in universities like Salamanca in Spain were not impressed by this expedient. At the same time, attempts to write off the inhabitants of the Americas as so utterly alien that they fell outside the bounds of humanity entirely, and could be treated literally like animals, also didn't find much purchase. Even cannibals, the jurists noted, had governments, societies and laws, and were able to construct arguments to defend the justice of their (cannibalistic) social arrangements; therefore they were clearly humans, vested by God with powers of reason.

The legal and philosophical question then became: what rights do human beings have simply by dint of being human – that is, what rights could they be said to have 'naturally', even if they existed in a State of Nature, innocent of the teachings of written philosophy and revealed religion, and without codified laws? The matter was hotly debated. We need not linger here on the exact formulae that natural law theorists came up with (suffice to say, they did allow that Americans had natural rights, but ended up justifying their conquest anyway, provided their subsequent treatment was not *too* violent or oppressive), but what is important, in this context, is that they opened a conceptual door. Writers like Thomas Hobbes, Hugo Grotius or John Locke could skip past the biblical narratives everyone used to start with, and begin instead with a question such as: what might humans have been like in a State of Nature, when all they had was their humanity?

Each of these authors populated the State of Nature with what they took to be the simplest societies known in the Western Hemisphere, and thus they concluded that the original state of humanity was one of freedom and equality, for better or worse (Hobbes, for example, definitely felt it was worse). It's important to stop here for a moment and consider why they came to this verdict – because it was by no means an obvious or inevitable conclusion.

First of all, while it may seem obvious to us, the fact that natural law theorists in the seventeenth century fixed on apparently simple societies as exemplars of primordial times – societies like the Algonkians of North America's Eastern Woodlands, or the Caribs and Amazonians, rather than urban civilizations like the Aztecs or Inca – would not have seemed obvious at the time.

Earlier authors, confronted by a population of forest dwellers with no king and employing only stone tools, were unlikely to have seen them as in any way primordial. Sixteenth-century scholars, such as the Spanish missionary José de Acosta, were more likely to conclude they were looking at the fallen vestiges of some ancient civilization, or refugees who had, in the course of their wanderings, forgotten the arts of metallurgy and civil governance. Such a conclusion would have made obvious common sense for people who assumed that all truly important knowledge had been revealed by God at the beginning of time, that cities had existed before the Flood, and that saw their own intellectual life largely as attempts to recover the lost wisdom of ancient Greeks and Romans.

History, in Renaissance Europe of the fifteenth to sixteenth centuries, was not a story of progress. It was largely a series of disasters. Introducing the concept of a State of Nature didn't exactly flip all this around, at least not immediately, but it did allow political philosophers after the seventeenth century to imagine people without the trappings of civilization as something other than degenerate savages; as a kind of humanity 'in the raw'. And this, in turn, allowed them to ask a host of new questions about what it meant to be human. What social forms would still exist, even among people who had no recognizable form of law or government? Would marriage exist? What forms might it take? Would Natural Man tend to be naturally gregarious, or would people tend to avoid one another? Was there such a thing as natural religion?

But the question still remains: why is it that by the eighteenth century, European intellectuals had come to fix on the idea of primordial freedom or, especially, equality, to such an extent that it seemed perfectly natural to ask a question like 'what is the origin of *inequality* among men?' This seems particularly odd considering how, prior to that time, most did not even consider social equality possible.

First of all, a qualification is in order. A certain folk egalitarianism already existed in the Middle Ages, coming to the fore during popular festivals like carnival, May Day or Christmas, when much of society revelled in the idea of a 'world turned upside down', where all powers and authorities were knocked to the ground or made a mockery of. Often the celebrations were framed as a return to some primordial 'age of equality' – the Age of Cronus, or Saturn, or the land of Cockaygne. Sometimes, too, these ideals were invoked in popular revolts.

True, it's never entirely clear how far such egalitarian ideals are merely a side effect of hierarchical social arrangements that obtained at ordinary times. Our notion that everyone is equal before the law, for instance, originally traces back to the idea that everyone is equal before the king, or emperor: since if one man is invested with absolute power, then obviously everyone else is equal in comparison. Early Christianity similarly insisted that all believers were (in some ultimate sense) equal in relation to God, whom they referred to as 'the Lord'. As this illustrates, the overarching power under which ordinary mortals are all de facto equals need not be a real flesh-and-blood human; one of the whole points of creating a 'carnival king' or 'May queen' is that they exist in order to be dethroned.[8]

Europeans educated in classical literature would also have been familiar with speculation about long-ago, happy, egalitarian orders that appear in Greco-Roman sources; and notions of equality, at least among Christian nations, were to be found in the concept of *res publica*, or commonwealth, which again looked to ancient precedents. All this is only to say that a state of equality was not utterly inconceivable to European intellectuals before the eighteenth century. None of it, however, explains *why* they came almost universally to assume that human beings, innocent of civilization, would ever exist in such a state. True, there were classical precedents for such ideas, but there were classical precedents for the opposite as well.[9] For answers, we must return to arguments deployed to establish that the inhabitants of the Americas were fellow humans to begin with: to assert that, however exotic or even perverse their customs might seem, Native Americans were capable of making logical arguments in their own defence.

What we're going to suggest is that American intellectuals – we are using the term 'American' as it was used at the time, to refer to indigenous inhabitants of the Western Hemisphere; and 'intellectual' to refer to anyone in the habit of arguing about abstract ideas – actually played a role in this conceptual revolution. It is very strange that this should be considered a particularly radical idea, but among mainstream intellectual historians today it is almost a heresy.

What makes this especially odd is that no one denies that many European explorers, missionaries, traders, settlers and others who sojourned on American shores spent years learning native languages and perfecting their skills in conversation with native speakers; just as indigenous Americans did the work of learning Spanish, English, Dutch or French. Neither, we think, would anyone who has ever learned a truly alien language deny that doing so takes a great deal of imaginative work, trying to grasp unfamiliar concepts. We also know that missionaries typically conducted long philosophical debates as part of their professional duties; many others, on both sides, argued with one another either out of simple curiosity, or because they had immediate practical reasons to understand the other's point of view. Finally, no one would deny that travel literature, and missionary relations – which often contained summaries of, or even extracts from, these exchanges – were popular literary genres, avidly followed by educated Europeans. Any middle-class household in eighteenth-century Amsterdam or Grenoble would have been likely to have on its shelves at the very least a copy of the *Jesuit Relations of New France* (as France's North American colonies were then known), and one or two accounts written by voyagers to faraway lands. Such books were appreciated largely because they contained surprising and unprecedented ideas.[10]

Historians are aware of all this. Yet the overwhelming majority still conclude that even when European authors explicitly say they are borrowing ideas, concepts and arguments from indigenous thinkers, one should not take them seriously. It's all just supposed to be some kind of misunderstanding, fabrication, or at best a naive projection of pre-existing European ideas. American intellectuals, when they appear in European accounts, are assumed to be mere representatives of some Western archetype of the 'noble savage' or sock-puppets, used as plausible alibis to an author who might otherwise get into trouble for presenting subversive ideas (deism, for example, or rational materialism, or unconventional views on marriage).[11]

Certainly, if one encounters an argument ascribed to a 'savage' in a European text that even remotely resembles anything to be found in Cicero or Erasmus, one is automatically supposed to assume that no 'savage' could possibly have really said it – or even that the conversation in question never really took place at all.[12] If nothing else, this habit of thought is very convenient for students of Western literature, themselves trained in Cicero and Erasmus, who might otherwise be forced to actually try to learn something about what indigenous people thought about the world, and above all what they made of Europeans.

We intend to proceed in the opposite direction.

We will examine early missionary and travel accounts from New France – especially the Great Lakes region – since these were the accounts Rousseau himself was most familiar with, to get a sense of what its indigenous inhabitants did actually think of French society, and how they came to think of their own societies differently as a result. We will argue that indigenous Americans did indeed develop a very strong critical view of their invaders' institutions: a view which focused first on these institutions' lack of freedom, and only later, as they became more familiar with European social arrangements, on equality.

One of the reasons that missionary and travel literature became so popular in Europe was precisely because it exposed its readers to this kind of criticism, along with providing a sense of social possibility: the knowledge that familiar ways were not the only ways, since – as these books showed – there were clearly societies in existence that did things very differently. We will suggest that there is a reason why so many key Enlightenment thinkers insisted that their ideals of individual liberty and political equality were inspired by Native American sources and examples. Because it was true.

IN WHICH WE CONSIDER WHAT THE INHABITANTS OF NEW FRANCE MADE OF THEIR EUROPEAN INVADERS, ESPECIALLY IN MATTERS OF GENEROSITY, SOCIABILITY, MATERIAL WEALTH, CRIME, PUNISHMENT AND LIBERTY

The 'Age of Reason' was an age of debate. The Enlightenment was rooted in conversation; it took place largely in cafés and salons. Many classic Enlightenment texts took the form of dialogues; most cultivated an easy, transparent, conversational style clearly inspired by the salon. (It was the Germans, back then, who tended to write in the obscure style for which French intellectuals have since become famous.) Appeal to 'reason' was above all a style of argument. The ideals of the French Revolution – liberty, equality and fraternity – took the form they did in the course of just such a long series of debates and conversations. All we're going to suggest here is that those conversations stretched back further than Enlightenment historians assume.

Let's begin by asking: what did the inhabitants of New France make of the Europeans who began to arrive on their shores in the sixteenth century?

At that time, the region that came to be known as New France was inhabited largely by speakers of Montagnais-Naskapi, Algonkian and Iroquoian languages. Those closer to the coast were fishers, foresters and hunters, though most also practised horticulture; the Wendat (Huron),[13] concentrated in major river valleys further inland, growing maize, squash and beans around fortified towns. Interestingly, early French observers attached little importance to such economic distinctions, especially since foraging or farming was, in either case, largely women's work. The men, they noted, were primarily occupied in hunting and, occasionally, war, which meant they could in a sense be considered natural aristocrats. The idea of the 'noble savage' can be traced back to such estimations. Originally, it didn't refer to nobility of character but simply to the fact that the Indian men concerned themselves with hunting and fighting, which back at home were largely the business of noblemen.

But if French assessments of the character of 'savages' tended to be decidedly mixed, the indigenous assessment of French character was distinctly less so. Father Pierre Biard, for example, was a former theology professor assigned in 1608 to evangelize the Algonkian-speaking Mi'kmaq in Nova Scotia, who had lived for some time next to a French fort. Biard did not think much of the Mi'kmaq, but reported that the feeling was mutual: 'They consider themselves better than the French: "For," they say, "you are always fighting and quarrelling among yourselves; we live peaceably. You are envious and are all the time slandering each other; you are thieves and deceivers; you are covetous, and are neither generous nor kind; as for us, if we have a morsel of bread we share it with our neighbour." They are saying these and like things continually.'[14] What seemed to irritate Biard the most was that the Mi'kmaq would constantly assert that they were, as a result, 'richer' than the French. The French had more material possessions, the Mi'kmaq conceded; but they had other, greater assets: ease, comfort and time.

Twenty years later Brother Gabriel Sagard, a Recollect Friar,[15] wrote similar things of the Wendat nation. Sagard was at first highly critical of Wendat life, which he described as inherently sinful (he was obsessed with the idea that Wendat women were all intent on seducing him), but by the end of his sojourn he had come to the conclusion their social arrangements were in many ways superior to those at home in France. In the following passages he was clearly echoing Wendat opinion: 'They have no lawsuits and take little pains to acquire the goods of this life, for which we Christians torment ourselves so much, and for our excessive and insatiable greed in acquiring them we are justly and with reason reproved by their quiet life and tranquil dispositions.'[16] Much like Biard's Mi'kmaq, the Wendat were particularly offended by the French lack of generosity to one another: 'They reciprocate hospitality and give such assistance to one another that the necessities of all are provided for without there being any indigent beggar in their towns and villages; and they considered it a very bad thing when they heard it said that there were in France a great many of these needy beggars, and thought that this was for lack of charity in us, and blamed us for it severely.'[17]

Wendat cast a similarly jaundiced eye at French habits of conversation. Sagard was surprised and impressed by his hosts' eloquence and powers of reasoned argument, skills honed by near-daily public discussions of communal affairs; his hosts, in contrast, when they did get to see a group of Frenchmen gathered together, often remarked on the way they seemed to be constantly scrambling over each other and cutting each other off in conversation, employing weak arguments, and overall (or so the subtext seemed to be) not showing themselves to be particularly bright. People who tried to grab the stage, denying others the means to present their arguments, were acting in much the same way as those who grabbed the material means of subsistence and refused to share it; it is hard to avoid the impression that Americans saw the French as existing in a kind of Hobbesian state of 'war of all against all'. (It's probably worthy of remark that especially in this early contact period, Americans were likely to have known Europeans largely through missionaries, trappers, merchants and soldiers – that is, groups almost entirely composed of men. There were at first very few French women in the colonies, and fewer children. This probably had the effect of making the competitiveness and lack of mutual care among them seem all the more extreme.)

Sagard's account of his stay among the Wendat became an influential bestseller in France and across Europe: both Locke and Voltaire cited *Le grand voyage du pays des Hurons* as a principal source for their descriptions of American societies. The multi-authored and much more extensive *Jesuit Relations*, which appeared between 1633 and 1673, were also widely read and debated in Europe, and include many a similar remonstrance aimed at the French by Wendat observers. One of the most striking things about these seventy-one volumes of missionary field reports is that neither the Americans, nor their French interlocutors, appear to have had very much to say about 'equality' per se – for example, the words *égal* or *égalité* barely appear, and on those very few occasions when they do it's almost always in reference to 'equality of the sexes' (something the Jesuits found particularly scandalous).

This appears to be the case, irrespective of whether the Jesuits in question were arguing with the Wendat – who might not seem egalitarian in anthropological terms, since they had formal political offices and a stratum of war captives whom the Jesuits, at least, referred to as 'slaves' – or the Mi'kmaq or Montagnais-Naskapi, who were organized into what later anthropologists would consider egalitarian bands of hunter-gatherers. Instead, we hear a multiplicity of American voices complaining about the competitiveness and selfishness of the French – and even more, perhaps, about their hostility to freedom.

That indigenous Americans lived in generally free societies, and that Europeans did not, was never really a matter of debate in these exchanges: both sides agreed this was the case. What they differed on was whether or not individual liberty was desirable.

This is one area in which early missionary or travellers' accounts of the Americas pose a genuine conceptual challenge to most readers today. Most of us simply take it for granted that 'Western' observers, even seventeenth-century ones, are simply an earlier version of ourselves; unlike indigenous Americans, who represent an essentially alien, perhaps even unknowable Other. But in fact, in many ways, the authors of these texts were nothing like us. When it came to questions of personal freedom, the equality of men and women, sexual mores or popular sovereignty – or even, for that matter, theories of depth psychology[18] – indigenous American attitudes are likely to be far closer to the reader's own than seventeenth-century European ones.

These differing views on individual liberty are especially striking. Nowadays, it's almost impossible for anyone living in a liberal democracy to say they are against freedom – at least in the abstract (in practice, of course, our ideas are usually much more nuanced). This is one of the lasting legacies of the Enlightenment and of the American and French Revolutions. Personal freedom, we tend to believe, is inherently good (even if some of us also feel that a society based on total individual liberty – one which took it so far as to eliminate police, prisons or any sort of apparatus of coercion – would instantly collapse into violent chaos). Seventeenth-century Jesuits most certainly did *not* share this assumption. They tended to view individual liberty as animalistic. In 1642, the Jesuit missionary Le Jeune wrote of the Montagnais-Naskapi:

> They imagine that they ought by right of birth, to enjoy the liberty of wild ass colts, rendering no homage to any one whomsoever, except when they like. They have reproached me a hundred times because we fear our Captains, while they laugh at and make sport of theirs. All the authority of their chief is in his tongue's end; for he is powerful in so far as he is eloquent; and, even if he kills himself talking and haranguing, he will not be obeyed unless he pleases the Savages.[19]

In the considered opinion of the Montagnais-Naskapi, however, the French were little better than slaves, living in constant terror of their superiors. Such criticism appears regularly in Jesuit accounts; what's more, it comes not just from those who lived in nomadic bands, but equally from townsfolk like the Wendat. The missionaries, moreover, were willing to concede that this wasn't all just rhetoric on the Americans' part. Even Wendat statesmen couldn't compel anyone to do anything they didn't wish to do. As Father Lallemant, whose correspondence provided an initial model for *The Jesuit Relations*, noted of the Wendat in 1644:

> I do not believe that there is any people on earth freer than they, and less able to allow the subjection of their wills to any power whatever – so much so that Fathers here have no control over their children, or

> Captains over their subjects, or the Laws of the country over any of them, except in so far as each is pleased to submit to them. There is no punishment which is inflicted on the guilty, and no criminal who is not sure that his life and property are in no danger . . .[20]

Lallemant's account gives a sense of just how politically challenging some of the material to be found in the *Jesuit Relations* must have been to European audiences of the time, and why so many found it fascinating. After expanding on how scandalous it was that even murderers should get off scot-free, the good father did admit that, when considered as a means of keeping the peace, the Wendat system of justice was not ineffective. Actually, it worked surprisingly well. Rather than punish culprits, the Wendat insisted the culprit's entire lineage or clan pay compensation. This made it everyone's responsibility to keep their kindred under control. 'It is not the guilty who suffer the penalty,' Lallemant explains, but rather 'the public that must make amends for the offences of individuals.' If a Huron had killed an Algonquin or another Huron, the whole country assembled to agree the number of gifts due to the grieving relatives, 'to stay the vengeance that they might take'.

Wendat 'captains', as Lallemant then goes on to describe, 'urge their subjects to provide what is needed; no one is compelled to it, but those who are willing bring publicly what they wish to contribute; it seems as if they vied with one another according to the amount of their wealth, and as the desire of glory and of appearing solicitous for the public welfare urges them to do on like occasions.' More remarkable still, he concedes: 'this form of justice restrains all these peoples, and seems more effectually to repress disorders than the personal punishment of criminals does in France,' despite being 'a very mild proceeding, which leaves individuals in such a spirit of liberty that they never submit to any Laws and obey no other impulse than that of their own will'.[21]

There are a number of things worth noting here. One is that it makes clear that some people were indeed considered wealthy. Wendat society was not 'economically egalitarian' in that sense. However, there was a difference between what we'd consider economic resources – like land, which was owned by families, worked by women, and whose products were largely disposed of by women's collectives – and the kind of 'wealth' being referred to here, such as *wampum* (a word applied to strings and belts of beads, manufactured from the shells of Long Island's quahog clam) or other treasures, which largely existed for political purposes.

Wealthy Wendat men hoarded such precious things largely to be able to give them away on dramatic occasions like these. Neither in the case of land and agricultural products, nor that of *wampum* and similar valuables, was there any way to transform access to material resources into power – at least, not the kind of power that might allow one to make others work for you, or compel them to do anything they did not wish to do. At best, the accumulation and adroit distribution of riches might make a man more likely to aspire to political office (to become a 'chief' or 'captain' – the French sources tend to use these terms in an indiscriminate fashion); but as the Jesuits all continually emphasized, merely holding political office did not give anyone the right to give anybody orders either. Or, to be completely accurate, an office holder could give all the orders he or she liked, but no one was under any particular obligation to follow them.

To the Jesuits, of course, all this was outrageous. In fact, their attitude towards indigenous ideals of liberty is the exact opposite of the attitude most French people or Canadians tend to hold today: that, in principle, freedom is an altogether admirable ideal. Father Lallemant, though, was willing to admit that in practice such a system worked quite well; it created 'much less disorder than there is in France' – but, as he noted, the Jesuits were opposed to freedom in principle:

> This, without doubt, is a disposition quite contrary to the spirit of the Faith, which requires us to submit not only our wills, but our minds, our judgments, and all the sentiments of man to a power unknown to our senses, to a Law that is not of earth, and that is entirely opposed to the laws and sentiments of corrupt nature. Add to this that the laws of the Country, which to them seem most just, attack the purity of the Christian life in a thousand ways, especially as regards their marriages . . . [22]

The *Jesuit Relations* are full of this sort of thing: scandalized missionaries frequently reported that American women were considered to have full control over their own bodies, and that therefore unmarried women had sexual liberty and married women could divorce at will. This, for the Jesuits, was an outrage. Such sinful conduct, they believed, was just the extension of a more general principle of freedom, rooted in natural dispositions, which they saw as inherently pernicious. The 'wicked liberty of the savages', one insisted, was the single greatest impediment to their 'submitting to the yoke of the law of God'.[23] Even finding terms to translate concepts like 'lord', 'commandment' or 'obedience' into indigenous languages was extremely difficult; explaining the underlying theological concepts, well-nigh impossible.

IN WHICH WE SHOW HOW EUROPEANS LEARNED FROM (NATIVE) AMERICANS ABOUT THE CONNECTION BETWEEN REASONED DEBATE, PERSONAL FREEDOMS AND THE REFUSAL OF ARBITRARY POWER

In political terms, then, French and Americans were not arguing about equality but about freedom. About the only specific reference to political equality that appears in the seventy-one volumes of *The Jesuit Relations* occurs almost as an aside, in an account of an event in 1648. It happened in a settlement of Christianized Wendat near the town of Quebec. After a disturbance caused by a shipload of illegal liquor finding its way into the community, the governor persuaded Wendat leaders to agree to a prohibition of alcoholic beverages, and published an edict to that effect – crucially, the governor notes, backed up by threat of punishment. Father Lallemant, again, records the story. For him, this was an epochal event:

> 'From the beginning of the world to the coming of the French, the Savages have never known what it was so solemnly to forbid anything to their people, under any penalty, however slight. They are free people, each of whom considers himself of as much consequence as the others; and they submit to their chiefs only in so far as it pleases them.'[24]

Equality here is a direct extension of freedom; indeed, is its expression. It also has almost nothing in common with the more familiar

(Eurasian) notion of 'equality before the law', which is ultimately equality before the sovereign – that is, once again, equality in common subjugation. Americans, by contrast, were equal insofar as they were equally free to obey or disobey orders as they saw fit. The democratic governance of the Wendat and Five Nations of the Haudenosaunee, which so impressed later European readers, was an expression of the same principle: if no compulsion was allowed, then obviously such social coherence as did exist had to be created through reasoned debate, persuasive arguments and the establishment of social consensus.

Here we return to the matter with which we began: the European Enlightenment as the apotheosis of the principle of open and rational debate. We've already mentioned Sagard's grudging respect for the Wendat facility in logical argumentation (a theme that also runs through most Jesuit accounts). At this point, it is important to bear in mind that the Jesuits were the intellectuals of the Catholic world. Trained in classical rhetoric and techniques of disputation, Jesuits had learned the Americans' languages primarily so as to be able to argue with them, to persuade them of the superiority of the Christian faith. Yet they regularly found themselves startled and impressed by the quality of the counterarguments they had to contend with.

How could such rhetorical facility have come to those with no awareness of the works of Varro and Quintilian? In considering the matter, the Jesuits almost always noted the openness with which public affairs were conducted. So, Father Le Jeune, Superior of the Jesuits in Canada in the 1630s: 'There are almost none of them incapable of conversing or reasoning very well, and in good terms, on matters within their knowledge. The councils, held almost every day in the Villages, and on almost all matters, improve their capacity for talking.' Or, in Lallemant's words: 'I can say in truth that, as regards intelligence, they are in no wise inferior to Europeans and to those who dwell in France. I would never have believed that, without instruction, nature could have supplied a most ready and vigorous eloquence, which I have admired in many Hurons; or more clear-sightedness in public affairs, or a more discreet management in things to which they are accustomed.'[25] Some Jesuits went further, remarking – not without a trace of frustration – that New World savages seemed rather cleverer overall than the people they were used to dealing with at home (e.g. 'they nearly all show more intelligence in their business, speeches, courtesies, intercourse, tricks, and subtleties, than do the shrewdest citizens and merchants in France').[26]

Jesuits, then, clearly recognized and acknowledged an intrinsic relation between refusal of arbitrary power, open and inclusive political debate and a taste for reasoned argument. It's true that Native American political leaders, who in most cases had no means to compel anyone to do anything they had not agreed to do, were famous for their rhetorical powers. Even hardened European generals pursuing genocidal campaigns against indigenous peoples often reported themselves reduced to tears by their powers of eloquence. Still, persuasiveness need not take the form of logical argumentation; it can just as easily involve appeal to sentiment, whipping up passions, deploying poetic metaphors, appealing to myth or proverbial wisdom, employing irony and indirection, humour, insult, or appeals to prophecy or revelation; and the degree to which one privileges any of these has everything to do with the rhetorical tradition to which the speaker belongs, and the presumed dispositions of their audience.

It was largely the speakers of Iroquoian languages such as the Wendat, or the five Haudenosaunee nations to their south, who appear to have placed such weight on reasoned debate – even finding it a form of pleasurable entertainment in own right. This fact alone had major historical repercussions. Because it appears to have been exactly this form of debate – rational, sceptical, empirical, conversational in tone – which before long came to be identified with the European Enlightenment as well. And, just like the Jesuits, Enlightenment thinkers and democratic revolutionaries saw it as intrinsically connected with the rejection of arbitrary authority, particularly that which had long been assumed by the clergy.

Let's gather together the strands of our argument so far.

By the mid seventeenth century, legal and political thinkers in Europe were beginning to toy with the idea of an egalitarian State of Nature; at least in the minimal sense of a default state that might be shared by societies which they saw as lacking government, writing, religion, private property or other significant means of distinguishing themselves from one another. Terms like 'equality' and 'inequality' were just beginning to come into common usage in intellectual circles – around the time, indeed, that the first French missionaries set out to evangelize the inhabitants of what are now Nova Scotia and Quebec.[27] Europe's reading public was growing increasingly curious about what such primordial societies might have been like. But they had no particular disposition to imagine men and women living in a State of Nature as especially 'noble', let alone as rational sceptics and champions of individual liberty.[28] This latter perspective was the product of a dialogic encounter.

As we've seen, at first neither side – not the colonists of New France, nor their indigenous interlocutors – had much to say about 'equality'. Rather, the argument was about liberty and mutual aid, or what might even be better called freedom and communism. We should be clear about what we mean by the latter term. Since the early nineteenth century, there have been lively debates about whether there was ever a thing that might legitimately be referred to as 'primitive communism'. At the centre of these debates, almost invariably, were the indigenous societies of the Northeast Woodlands – ever since Friedrich Engels used the Iroquois as a prime example of primitive communism in his *The Origin of the Family, Private Property and the State* (1884). Here, 'communism' always refers to communal ownership, particularly of productive resources. As we've already observed, many American societies could be considered somewhat ambiguous in this sense: women owned and worked the fields individually, even though they stored and disposed of the products collectively; men owned their own tools and weapons individually, even if they typically shared out the game and spoils.

However, there's another way to use the word 'communism': not as a property regime but in the original sense of 'from each according to their abilities, to each according to their needs'. There's also a certain minimal, 'baseline' communism which applies in all societies; a feeling that if another person's needs are great enough (say, they are drowning), and the cost of meeting them is modest enough (say, they are asking for you to throw them a rope), then of course any decent person would comply. Baseline communism of this sort could even be considered the very grounds of human sociability, since it is only one's bitter enemies who would not be treated this way. What varies is just how far it is felt such baseline communism should properly extend.

In many societies – and American societies of that time appear to have been among them – it would have been quite inconceivable to refuse a request for food. For seventeenth-century Frenchmen in North America, this was clearly not the case: their range of baseline communism appears to have been quite restricted, and did not extend to food and shelter – something which scandalized Americans. But just as we earlier witnessed a confrontation between two very different concepts of equality, here we are ultimately witnessing a clash between very different concepts of individualism. Europeans were constantly squabbling for advantage; societies of the Northeast Woodlands, by contrast, guaranteed one another the means to an autonomous life – or at least ensured no man or woman was subordinated to any other. Insofar as we can speak of communism, it existed not in opposition to but in support of individual freedom.

The same could be said of indigenous political systems that Europeans encountered across much of the Great Lakes region. Everything operated to ensure that no one's will would be subjugated to that of anyone else. It was only over time, as Americans learned more about Europe, and Europeans began to consider what it would mean to translate American ideals of individual liberty into their own societies, that the term 'equality' began to gain ground as a feature of the discourse between them.

IN WHICH WE INTRODUCE THE WENDAT PHILOSOPHER-STATESMAN KANDIARONK, AND EXPLAIN HOW HIS VIEWS ON HUMAN NATURE AND SOCIETY TOOK ON NEW LIFE IN THE SALONS OF ENLIGHTENMENT EUROPE (INCLUDING AN ASIDE ON THE CONCEPT OF 'SCHISMOGENESIS')

In order to understand how the indigenous critique – that consistent moral and intellectual assault on European society, widely voiced by Native American observers from the seventeenth century onwards – evolved, and its full impact on European thinking, we first need to understand something about the role of two men: an impoverished French aristocrat named Louis-Armand de Lom d'Arce, Baron de la Hontan, and an unusually brilliant Wendat statesman named Kandiaronk.

In 1683, Lahontan (as he came to be known), then seventeen years old, joined the French army and was posted to Canada. Over the course of the next decade he took part in a number of campaigns and exploratory expeditions, eventually attaining the rank of deputy to the Governor-General, the Comte de Frontenac. In the process he became fluent in both Algonkian and Wendat, and – by his own account at least – good friends with a number of indigenous political figures. Lahontan later claimed that, because he was something of a sceptic in religious matters and a political enemy of the Jesuits, these figures were willing to share with him their actual opinions about Christian teachings. One of them was Kandiaronk.

A key strategist of the Wendat Confederacy, a coalition of four Iroquoian-speaking peoples, Kandiaronk (his name literally meant 'the muskrat' and the French often referred to him simply as 'Le Rat') was at that time engaged in a complex geopolitical game, trying to play the English, French and Five Nations of the Haudenosaunee off against each other, with the initial aim of averting a disastrous Haudenosaunee assault on the Wendat, but with the long-term goal of creating a comprehensive indigenous alliance to hold off the settler advance.[29] Everyone who met him, friend or foe, admitted he was a truly remarkable individual: a courageous warrior, brilliant orator and unusually skilful politician. He was also, to the very end of his life, a staunch opponent of Christianity.[30]

Lahontan's own career came to a bad end. Despite having successfully defended Nova Scotia against an English fleet, he ran foul of its governor and was forced to flee French territory. Convicted in absentia of insubordination, he spent most of the next decade in exile, wandering about Europe trying, unsuccessfully, to negotiate a return to his native France. By 1702, Lahontan was living in Amsterdam and very much down on his luck, described by those who met him as penniless vagrant and freelance spy. All that was to change when he published a series of books about his adventures in Canada.

Two were memoirs of his American adventures. The third, entitled *Curious Dialogues with a Savage of Good Sense Who Has Travelled*

(1703), comprised a series of four conversations between Lahontan and Kandiaronk, in which the Wendat sage – voicing opinions based on his own ethnographic observations of Montreal, New York and Paris – casts an extremely critical eye on European mores and ideas about religion, politics, health and sexual life. These books won a wide audience, and before long Lahontan had become something of a minor celebrity. He settled at the court of Hanover, which was also the home base for Leibniz, who befriended and supported him before Lahontan fell ill and died, around 1715.

Most criticism of Lahontan's work simply assumes as a matter of course that the dialogues are made up, and that the arguments attributed to 'Adario' (the name given there to Kandiaronk) are the opinions of Lahontan himself.[31] In a way, this conclusion is unsurprising. Adario claims not only to have visited France, but expresses opinions on everything from monastic politics to legal affairs. In the debate on religion, he often sounds like an advocate of the deist position that spiritual truth should be sought in reason, not revelation, embracing just the sort of rational scepticism that was becoming popular in Europe's more daring intellectual circles at the time. It is also true that the style of Lahontan's dialogues seems partly inspired by the ancient Greek writings of the satirist Lucian; and also that, given the prevalence of Church censorship in France at the time, the easiest way for a freethinker to get away with publishing an open attack on Christianity probably would have been to compose a dialogue pretending to defend the faith from the attacks of an imaginary foreign sceptic – and then make sure one loses all the arguments.

In recent decades, however, indigenous scholars returned to the material in light of what we know about Kandiaronk himself – and came to very different conclusions.[32] The real-life Adario was famous not only for his eloquence, but was known for engaging in debates with Europeans of just the sort recorded in Lahontan's book. As Barbara Alice Mann remarks, despite the almost unanimous chorus of Western scholars insisting the dialogues are imaginary, 'there is excellent reason for accepting them as genuine.' First, there are the first-hand accounts of Kandiaronk's oratorical skills and dazzling wit. Father Pierre de Charlevoix described Kandiaronk as so 'naturally eloquent' that 'no one perhaps ever exceeded him in mental capacity.' An exceptional council speaker, 'he was not less brilliant in conversation in private, and [councilmen and negotiators] often took pleasure in provoking him to hear his repartees, always animated, full of wit, and generally unanswerable. He was the only man in Canada who was a match for the [governor] Count de Frontenac, who often invited him to his table to give his officers this pleasure.'[33]

During the 1690s, in other words, the Montreal-based governor and his officers (presumably including his sometime deputy, Lahontan) hosted a proto-Enlightenment salon, where they invited Kandiaronk to debate exactly the sort of matters that appeared in the *Dialogues*, and in which it was Kandiaronk who took the position of rational sceptic.

What's more, there is every reason to believe that Kandiaronk actually *had* been to France; that's to say, we know the Wendat Confederation did send an ambassador to visit the court of Louis XIV in 1691, and Kandiaronk's office at the time was Speaker of the Council, which would have made him the logical person to send. While the intimate knowledge of European affairs and understanding of European psychology attributed to Adario might seem implausible, Kandiaronk was a man who had been engaged in political negotiations with Europeans for years, and regularly ran circles around them by anticipating their logic, interests, blind spots and reactions. Finally, many of the critiques of Christianity, and European ways more generally, attributed to Adario correspond almost exactly to criticisms that are documented from other speakers of Iroquoian languages around the same time.[34]

Lahontan himself claimed to have based the *Dialogues* on notes jotted down during or after a variety of conversations he'd had with Kandiaronk at Michilimackinac, on the strait between Lakes Huron and Michigan; notes that he later reorganized with the governor's help and which were supplemented, no doubt, by reminiscences both had of similar debates held over Frontenac's own dinner table. In the process the text was no doubt augmented and embellished, and probably tweaked again when Lahontan produced his final edition in Amsterdam. There is, however, every reason to believe the basic arguments were Kandiaronk's own.

Lahontan anticipates some of these arguments in his *Memoirs*, when he notes that Americans who had actually been to Europe – here, he was very likely thinking primarily of Kandiaronk himself, as well as a number of former captives who had been put to work as galley slaves – came back contemptuous of European claims to cultural superiority. Those Native Americans who had been in France, he wrote,

> . . . were continually teasing us with the faults and disorders they observed in our towns, as being occasioned by money. There's no point in trying to remonstrate with them about how useful the distinction of property is for the support of society: they make a joke of anything you say on that account. In short, they neither quarrel nor fight, nor slander one another; they scoff at arts and sciences, and laugh at the difference of ranks which is observed with us. They brand us for slaves, and call us miserable souls, whose life is not worth having, alleging that we degrade ourselves in subjecting ourselves to one man [the king] who possesses all the power, and is bound by no law but his own will.

In other words, we find here all the familiar criticisms of European society that the earliest missionaries had to contend with – the squabbling, the lack of mutual aid, the blind submission to authority – but with a new element added in: the organization of private property. Lahontan continues: 'They think it unaccountable that one man should have more than another, and that the rich should have more respect than the poor. In short, they say, the name of savages, which we bestow upon them, would fit ourselves better, since there is nothing in our actions that bears an appearance of wisdom.'

Native Americans who had the opportunity to observe French society from up close had come to realize one key difference from their own, one which may not otherwise have been apparent. Whereas in their own societies there was no obvious way to convert wealth into power over others (with the consequence that differences of wealth had little effect on individual freedom), in France the situation could not have been more different. Power over possessions could be directly translated into power over other human beings.

But here let us give the floor to Kandiaronk himself. The first of the *Dialogues* is about religious matters, in which Lahontan allows his foil calmly to pick apart the logical contradictions and incoherence of the Christian doctrines of original sin and redemption, paying particular attention to the concept of hell. As well as casting doubt on the historicity of scripture, Kandiaronk continually emphasizes the fact that Christians are divided into endless sects, each convinced they are entirely right and that all the others are hell-bound. To give a sense of its flavour:

> **Kandiaronk**: Come on, my brother. Don't get up in arms ... It's only natural for Christians to have faith in the holy scriptures, since, from their infancy, they've heard so much of them. Still, it is nothing if not reasonable for those born without such prejudice, such as the Wendats, to examine matters more closely.
>
> However, having thought long and hard over the course of a decade about what the Jesuits have told us of the life and death of the son of the Great Spirit, any Wendat could give you twenty reasons against the notion. For myself, I've always held that, if it were possible that God had lowered his standards sufficiently to come down to earth, he would have done it in full view of everyone, descending in triumph, with pomp and majesty, and most publicly ... He would have gone from nation to nation performing mighty miracles, thus giving everyone the same laws. Then we would all have had exactly the same religion, uniformly spread and equally known throughout the four corners of the world, proving to our descendants, from then till ten thousand years into the future, the truth of this religion. Instead, there are five or six hundred religions, each distinct from the other, of which according to you, the religion of the French, alone, is any good, sainted, or true.[35]

The last passage reflects perhaps Kandiaronk's most telling point: the extraordinary self-importance of the Jesuit conviction that an all-knowing and all-powerful being would freely choose to entrap himself in flesh and undergo terrible suffering, all for the sake of a single species, designed to be imperfect, only some of which were going to be rescued from damnation anyway.[36]

There follows a chapter on the subject of law, where Kandiaronk takes the position that European-style punitive law, like the religious doctrine of eternal damnation, is not necessitated by any inherent corruption of human nature, but rather by a form of social organization that encourages selfish and acquisitive behaviour. Lahontan objects:

true, reason is the same for all humans, but the very existence of judges and punishment shows that not everyone is capable of following its dictates:

> **Lahontan:** This is why the wicked need to be punished, and the good need to be rewarded. Otherwise, murder, robbery and defamation would spread everywhere, and, in a word, we would become the most miserable people upon the face of the earth.
>
> **Kandiaronk:** For my own part, I find it hard to see how you could be much more miserable than you already are. What kind of human, what species of creature, must Europeans be, that they have to be forced to do good, and only refrain from evil because of fear of punishment? . . .
>
> You have observed that we lack judges. What is the reason for that? Well, we never bring lawsuits against one another. And why do we never bring lawsuits? Well, because we made a decision neither to accept or make use of money. And why do we refuse to allow money into our communities? The reason is this: we are determined not to have laws – because, since the world was a world, our ancestors have been able to live contentedly without them.

Given that the Wendat most certainly did have a legal code, this might seem disingenuous on Kandiaronk's part. By laws, however, he is clearly referring to laws of a coercive or punitive nature. He goes on to dissect the failings of the French legal system, dwelling particularly on judicial persecution, false testimony, torture, witchcraft accusations and differential justice for rich and poor. In conclusion, he swings back to his original observation: the whole apparatus of trying to force people to behave well would be unnecessary if France did not also maintain a contrary apparatus that encourages people to behave badly. That apparatus consisted of money, property rights and the resultant pursuit of material self-interest:

> **Kandiaronk:** I have spent six years reflecting on the state of European society and I still can't think of a single way they act that's not inhuman, and I genuinely think this can only be the case, as long as you stick to your distinctions of 'mine' and 'thine'. I affirm that what you call money is the devil of devils; the tyrant of the French, the source of

> all evils; the bane of souls and slaughterhouse of the living. To imagine one can live in the country of money and preserve one's soul is like imagining one could preserve one's life at the bottom of a lake. Money is the father of luxury, lasciviousness, intrigues, trickery, lies, betrayal, insincerity, – of all the world's worst behaviour. Fathers sell their children, husbands their wives, wives betray their husbands, brothers kill each other, friends are false, and all because of money. In the light of all this, tell me that we Wendat are not right in refusing to touch, or so much as to look at silver?

For Europeans in 1703, this was heady stuff.

Much of the subsequent exchange consists of the Frenchman trying to convince Kandiaronk of the advantages of adopting European civilization, and Kandiaronk countering that the French would do much better to adopt the Wendat way of life. Do you seriously imagine, he says, that I would be happy to live like one of the inhabitants of Paris, to take two hours every morning just to put on my shirt and make-up, to bow and scrape before every obnoxious galoot I meet on the street who happened to have been born with an inheritance? Do you really imagine I could carry a purse full of coins and not immediately hand them over to people who are hungry; that I would carry a sword but not immediately draw it on the first band of thugs I see rounding up the destitute to press them into naval service? If, on the other hand, Lahontan were to adopt an American way of life, Kandiaronk tells him, it might take a while to adjust – but in the end he'd be far happier. (Kandiaronk had a point, as we've seen in the last chapter; settlers adopted into indigenous societies almost never wanted to go back.)

Kandiaronk is even willing to propose that Europe would be better off if its whole social system was dismantled:

> **Lahontan:** Try for once in your life to actually listen. Can't you see, my dear friend, that the nations of Europe could not survive without gold and silver – or some similar precious symbol. Without it, nobles, priests, merchants and any number of others who lack the strength to work the soil would simply die of hunger. Our kings would not be kings; what soldiers would we have? Who would work for kings, or anybody

else? . . . It would plunge Europe into chaos and create the most dismal confusion imaginable.

Kandiaronk: You honestly think you're going to sway me by appealing to the needs of nobles, merchants and priests? If you abandoned conceptions of mine and thine, yes, such distinctions between men would dissolve; a levelling equality would then take its place among you as it now does among the Wendat. And yes, for the first thirty years after the banishing of self-interest, no doubt you would indeed see a certain desolation as those who are only qualified to eat, drink, sleep and take pleasure would languish and die. But their progeny would be fit for our way of living. Over and over I have set forth the qualities that we Wendat believe ought to define humanity – wisdom, reason, equity, etc. – and demonstrated that the existence of separate material interests knocks all these on the head. A man motivated by interest cannot be a man of reason.

Here, finally, 'equality' is invoked as a self-conscious ideal – but only as the result of a prolonged confrontation between American and European institutions and values, and as a calculated provocation, turning European civilizing discourse backwards on itself.

One reason why modern commentators have found it so easy to dismiss Kandiaronk as the ultimate 'noble savage' (and, therefore, as a mere projection of European fantasies) is because many of his assertions are so obviously exaggerated. It's not really true that the Wendat, or other American societies, had no laws, never quarrelled and knew no inequalities of wealth. At the same time, as we've seen, Kandiaronk's basic line of argument is perfectly consistent with what French missionaries and settlers in North America had been hearing from other indigenous Americans. To argue that because the *Dialogues* romanticize, they can't really reflect what he said, is to assume that people are incapable of romanticizing themselves – despite the fact that this is what any skilful debater is likely to do under such circumstances, and all sources concur that Kandiaronk was perhaps the most skilful they'd ever met.

Back in the 1930s, the anthropologist Gregory Bateson coined the term 'schismogenesis' to describe people's tendency to define themselves against one another.[37] Imagine two people getting into an argument about some minor political disagreement but, after an hour, ending up taking positions so intransigent that they find themselves on completely opposite sides of some ideological divide – even taking extreme positions they would never embrace under ordinary circumstances, just to show how much they completely reject the other's points. They start out as moderate social democrats of slightly different flavours; before a few heated hours are over, one has somehow become a Leninist, the other an advocate of the ideas of Milton Friedman. We know this kind of thing can happen in arguments. Bateson suggested such processes can become institutionalized on a cultural level as well. How, he asked, do boys and girls in Papua New Guinea come to behave so differently, despite the fact that no one ever explicitly instructs them about how boys and girls are supposed to behave? It's not just by imitating their elders; it's also because boys and girls each learn to find the behaviour of the opposite sex distasteful and try to be as little like them as possible. What start as minor learned differences become exaggerated until women come to think of themselves as, and then increasingly actually become, everything that men are not. And, of course, men do the same thing towards women.

Bateson was interested in psychological processes within societies, but there's every reason to believe something similar happens *between* societies as well. People come to define themselves against their neighbours. Urbanites thus become more urbane, as barbarians become more barbarous. If 'national character' can really be said to exist, it can only be as a result of such schismogenetic processes: English people trying to become as little as possible like French, French people as little like Germans, and so on. If nothing else, they will all definitely exaggerate their differences in arguing with one another.

In a historical confrontation of civilizations like that taking place along the east coast of North America in the seventeenth century, we can expect to see two contradictory processes. On the one hand, it is only to be expected that people on both sides of the divide will learn from one another and adopt each other's ideas, habits and technologies (Americans began using European muskets; European settlers began to adopt more indulgent American approaches to disciplining children). At the same time, they will also almost invariably do the opposite, picking out certain points of contrast and exaggerating or idealizing them – eventually even trying to act, in some respects, as little like their new neighbours as possible.

Kandiaronk's focus on money is typical of such situations. To this day, indigenous societies incorporated into the global economy, from Bolivia to Taiwan, almost invariably frame their own traditions, as Marshall Sahlins puts it, by opposition to the white man's 'living in the way of money'.[38]

All these would be rather trivial concerns had Lahontan's books not been so successful; but they were to have an enormous impact on European sensibilities. Kandiaronk's opinions were translated into German, English, Dutch and Italian, and continued in print, in multiple editions, for over a century. Any self-respecting intellectual of the eighteenth century would have been almost certain to have read them. They also inspired a flood of imitations. By 1721, Parisian theatregoers were flocking to Delisle de la Drevetière's comedy *L'Arlequin sauvage*: the story of a Wendat brought to France by a young sea captain, featuring a long series of indignant monologues in which the hero 'attributes the ills of [French] society to private property, to money, and in particular to the monstrous inequality which makes the poor the slaves of the rich'.[39] The play was revived almost yearly for the next two decades.[40]

Even more strikingly, just about every major French Enlightenment figure tried their hand at a Lahontan-style critique of their own society, from the perspective of some imagined outsider. Montesquieu chose a Persian; the Marquis d'Argens a Chinese; Diderot a Tahitian; Chateaubriand a Natchez; Voltaire's *L'Ingénu* was half Wendat and half French.[41] All took up and developed themes and arguments borrowed directly from Kandiaronk, supplemented by lines from other 'savage critics' in travellers' accounts.[42] Indeed, a strong case can be made for the real origins of the 'Western gaze' – that rational, supposedly objective way of looking at strange and exotic cultures which came to characterize later European anthropology – lying not in travellers' accounts, but rather in European accounts of precisely these imaginary sceptical natives: gazing inwards, brows furrowed, at the exotic curiosities of Europe itself.

Perhaps the single most popular work of this genre, published in

1747, was *Letters of a Peruvian Woman* by the prominent *saloniste* Madame de Graffigny, which viewed French society through the eyes of an imaginary kidnapped Inca princess. The book is considered a feminist landmark, in that it may well be the first European novel about a woman which does not end with the protagonist either marrying or dying. Graffigny's Inca heroine, Zilia, is as critical of the vanities and absurdities of European society as she is of patriarchy. By the nineteenth century, the novel was remembered in some quarters as the first work to introduce the notion of state socialism to the general public, Zilia wondering why the French king, despite levying all sorts of heavy taxes, cannot simply redistribute the wealth in the same manner as the Sapa Inca.[43]

In 1751, preparing a second edition of her book, Madame de Graffigny sent letters to a variety of friends asking for suggested changes. One of these correspondents was a twenty-three-year-old seminary student and budding economist, A. R. J. Turgot, and we happen to have a copy of his reply – which was long and highly (if constructively) critical. Turgot's text could hardly be more important, since it marks a key moment in his own intellectual development: the point where he began to turn his most lasting contribution to human thought – the idea of material economic progress – into a general theory of history.

IN WHICH WE EXPLAIN THE DEMIURGIC POWERS OF A. R. J. TURGOT, AND HOW HE TURNED THE INDIGENOUS CRITIQUE OF EUROPEAN CIVILIZATION ON ITS HEAD, LAYING THE BASIS FOR MOST MODERN VIEWS OF SOCIAL EVOLUTION (OR: HOW AN ARGUMENT ABOUT 'FREEDOM' BECAME ONE ABOUT 'EQUALITY')

The Inca Empire could hardly be described as 'egalitarian' – indeed, it was an empire – but Madame de Graffigny represented it as a benevolent despotism; one in which all are ultimately equal before the king.

Zilia's critique of France, like that of all imaginary outsiders writing in the tradition of Kandiaronk, focuses on the lack of individual freedom in French society and its violent inequalities.[44] But Turgot found such thinking disturbing, even dangerous.

Yes, Turgot acknowledged, 'we all love the idea of freedom and equality' – in principle. But we must consider a larger context. In reality, he ventured, the freedom and equality of savages is not a sign of their superiority; it's a sign of inferiority, since it is only possible in a society where each household is largely self-sufficient and, therefore, where everyone is equally poor. As societies evolve, Turgot reasoned, technology advances. Natural differences in talents and capacities between individuals (which have always existed) become more significant, and eventually they form the basis for an ever more complex division of labour. We progress from simple societies like those of the Wendat to our own complex 'commercial civilization', in which the poverty and dispossession of some – however lamentable it may be – is nonetheless the necessary condition for the prosperity of society as a whole.

There is no avoiding such inequality, concluded Turgot in his reply to Madame de Graffigny. The only alternative, according to him, would be massive, Inca-style state intervention to create a uniformity of social conditions: an enforced equality which could only have the effect of crushing all initiative and, therefore, result in economic and social catastrophe. In light of all this, Turgot suggested Madame de Graffigny rewrite her novel in such a way as to have Zilia realize these terrible implications at the end of the book.

Unsurprisingly, Graffigny ignored his advice.

A few years later, Turgot would elaborate these same ideas in a series of lectures on world history. He had already been arguing – for some years – for the primacy of technological progress as a driver for overall social improvement. In these lectures, he developed this argument into an explicit theory of stages of economic development: social evolution, he reasoned, always begins with hunters, then moves on to a stage of pastoralism, then farming, and only then finally passes to the contemporary stage of urban commercial civilization.[45] Those who still remain hunters, shepherds or simple farmers are best understood as vestiges of our own previous stages of social development.

In this way, theories of social evolution – now so familiar that we rarely dwell on their origins – first came to be articulated in Europe: as a direct response to the power of indigenous critique. Within a few years, Turgot's breakdown of all societies into four stages was appearing in the lectures of his friend and intellectual ally Adam Smith in Glasgow, and was worked into a general theory of human history by Smith's colleagues: men like Lord Kames, Adam Ferguson and John Millar. The new paradigm soon began to have a profound effect on how indigenous people were imagined by European thinkers, and by the European public more generally.

Observers who had previously considered the modes of subsistence and division of labour in North American societies to be trivial matters, or of at best secondary importance, now began assuming that they were the only thing that really mattered. Everyone was to be sorted along the same grand evolutionary ladder, depending on their primary mode of acquiring food. 'Egalitarian' societies were banished to the bottom of this ladder, where at best they could provide some insight on how our distant ancestors might have lived; but certainly could no longer be imagined as equal parties to a dialogue about how the inhabitants of wealthy and powerful societies should conduct themselves in the present.

Let's pause for a moment to take stock. In the years between 1703 and 1751, as we've seen, the indigenous American critique of European society had an enormous impact on European thought. What began as widespread expressions of outrage and distaste by Americans (when first exposed to European mores) eventually evolved, through a thousand conversations, conducted in dozens of languages from Portuguese to Russian, into an argument about the nature of authority, decency, social responsibility and, above all, freedom. As it became clear to French observers that most indigenous Americans saw individual autonomy and freedom of action as consummate values – organizing their own lives in such a way as to minimize any possibility of one human being becoming subordinated to the will of another, and hence viewing French society as essentially one of fractious slaves – they reacted in a variety of different ways.

Some, like the Jesuits, condemned the principle of freedom outright.

Others – settlers, intellectuals and members of the reading public back home – came to see it as a provocative and appealing social proposition. (Their conclusions on this matter, incidentally, bore no particular relation to their feelings about indigenous populations themselves, whom they were often happy to see exterminated – though, in fairness, there were public figures on both sides of the intellectual divide who strongly opposed aggression against foreign peoples.) In fact, the indigenous critique of European institutions was seen as so powerful that anyone objecting to existing intellectual and social arrangements would tend to deploy it as a weapon of choice: a game, as we've seen, played by pretty much every one of the great Enlightenment philosophers.

In the process – and we've seen how this was already happening with Lahontan and Kandiaronk – an argument about freedom also became, increasingly, an argument about equality. Above all, though, all these appeals to the wisdom of 'savages' were still ways of challenging the arrogance of received authority: that medieval certainty which maintained that the judgments of the Church and the establishment it upheld, having embraced the correct version of Christianity, were necessarily superior to those of anyone else on earth.

Turgot's case reveals just how much those particular notions of civilization, evolution and progress – which we've come to think of as the very core of Enlightenment thought – are, in fact, relative latecomers to that critical tradition. Most importantly, it shows how the development of these notions came in direct response to the power of the indigenous critique. Indeed, it was to take an enormous effort to salvage that very sense of European superiority which Enlightenment thinkers had aimed to upend, unsettle and de-centre. Certainly, over the next century and more, such ideas became a remarkably successful strategy for doing so. But they also created a welter of contradictions: for instance, the peculiar fact that European colonial empires, unlike almost any other in history, were forced to espouse their own ephemerality, claiming to be mere temporary vehicles to speed up their subjects' march to civilization – at least those subjects who, unlike the Wendat, they hadn't largely wiped off the map.

At this point we find ourselves back full circle with Rousseau.

HOW JEAN-JACQUES ROUSSEAU, HAVING WON ONE PRESTIGIOUS ESSAY COMPETITION, THEN LOST ANOTHER (COMING IN OVER THE PERMITTED WORD LENGTH), BUT FINALLY WENT ON TO CONQUER THE WHOLE OF HUMAN HISTORY

The exchange between Madame de Graffigny and Turgot gives us a sense of intellectual debate in France in the early 1750s; at least, in the *saloniste* circles with which Rousseau was familiar. Were freedom and equality universal values, or were they – at least in their pure form – inconsistent with a regime based on private property? Did the progress of arts and sciences lead to improved understanding of the world, and therefore to moral progress as well? Or was the indigenous critique correct, and the wealth and power of France simply a perverse side effect of unnatural, even pathological, social arrangements? These were the questions on every debater's lips at the time.

If we know anything about those debates today, it's largely because of their influence on Rousseau's essay. The *Discourse on the Origins of Social Inequality* has been taught, debated and picked apart in a thousand classrooms – which is odd, because in many ways it is very much an eccentric outlier, even by the standards of its time.

In the early part of his life, Rousseau was known mainly as an aspiring composer. His rise to prominence as a social thinker began in 1750, when he took part in a contest sponsored by the same learned society, the Académie de Dijon, on the question, 'Has the restoration of the sciences and arts contributed to moral improvement?'[46] Rousseau won first prize, and national fame, with an essay in which he argued with great passion that they had not. Our elementary moral intuitions, he asserted, are fundamentally decent and sound; civilization merely corrupts by encouraging us to value form over content. Almost all the examples in this *Discourse on the Arts and Sciences* are taken from classical Greek and Roman sources – but in his footnotes, Rousseau hints at other sources of inspiration:

> I don't dare speak of those happy nations who do not know even the names of the vices which we have such trouble controlling, of those American savages whose simple and natural ways of keeping public order Montaigne does not hesitate to prefer, not merely to the laws of Plato, but even to anything more perfect which philosophy will ever be able to dream up for governing a people. He cites a number of striking examples of these for those who understand how to admire them. What's more, he says, they don't wear breeches![47]

Rousseau's victory sparked something of a scandal. It was considered controversial, to say the least, for an academy dedicated to the advancement of the arts and sciences to award top honours to an argument stating that the arts and sciences were entirely counterproductive. As for Rousseau, he spent much of the next several years writing well-publicized responses to criticisms of the piece (as well as using his new fame to produce a comic opera, *The Village Soothsayer*, which became popular at the French court). When in 1754 the Académie de Dijon announced a new contest on the origins of social inequality, they clearly felt they had to put the upstart in his place.

Rousseau took the bait. He submitted an even more elaborate treatise, clearly designed to shock and confound. Not only did it fail to win the prize (which was bestowed on a very conventional essay by a representative of the religious establishment named the Abbé Talbert, who attributed our current unequal condition largely to original sin), but the judges announced that, since Rousseau's submission went far over the word limit, they had not even read it all the way through.

Rousseau's essay is undoubtedly odd. It's also not exactly what it's often claimed to be. Rousseau does not, in fact, argue that human society begins in a state of idyllic innocence. He argues, rather confusingly, that the first humans were essentially good, but nonetheless systematically avoided one another for fear of violence. As a result, human beings in a State of Nature were solitary creatures, which allows him to make a case that 'society' itself – that is, any form of ongoing association between individuals – was necessarily a restraint on human freedom. Even language marked a compromise. But the real innovation Rousseau introduces comes at the key moment of humanity's 'fall from grace', a moment triggered, he argues, by the emergence of property relations.

Rousseau's model of human society – which, he repeatedly emphasizes, is not meant to be taken literally, but is simply a thought experiment – involves three stages: a purely imaginary State of Nature, when individuals lived in isolation from one another; a stage of Stone Age savagery, which followed the invention of language (in which he includes most of the modern inhabitants of North America and other actually observable 'savages'); then finally, civilization, which followed the invention of agriculture and metallurgy. Each marks a moral decline. But, as Rousseau is careful to emphasize, the entire parable is a way to understand what made it possible for human beings to accept the notion of private property in the first place:

> The first man who, having enclosed a piece of land, thought of saying, 'This is mine', and found people simple enough to believe him, was the real founder of civil society. How many crimes, wars and murders, how much misery and horror the human race would have been spared if someone had pulled up the stakes and filled in the ditch and cried out to his fellow men: 'Beware of listening to this impostor. You are lost if you forget that the fruits of the earth belong to everyone, and that the earth itself belongs to no one!' But it is highly probable that by this time things had reached a point beyond which they could not go on as they were; for the idea of property, depending on many prior ideas which could only have arisen in successive stages, was not formed all at once in the human mind.[48]

Here, Rousseau asks exactly the same question that puzzled so many indigenous Americans. How is it that Europeans are able to turn wealth into power; turn a mere unequal distribution of material goods – which exists, at least to some degree, in any society – into the ability to tell others what to do, to employ them as servants, workmen or grenadiers, or simply to feel that it was no concern of theirs if they were left dying in a feverish bundle on the street?

While Rousseau does not cite Lahontan or the *Jesuit Relations* directly, he was clearly familiar with them,[49] as any intellectual of the time would have been, and his work is informed by the same critical questions: why are Europeans so competitive? Why do they not share food? Why do they submit themselves to other people's orders? Rousseau's long excursus on *pitié* – the natural sympathy that, he argues, savages have for one another and the quality that holds off the worst depredations of civilization in its second phase – only makes sense in light of the constant indigenous exclamations of dismay to be found in those books: that Europeans just don't seem to care about each other; that they are 'neither generous nor kind'.[50]

The reason for the essay's astonishing success, then, is that for all its sensationalist style, it's really a kind of clever compromise between two or perhaps even three contradictory positions on the most urgent social and moral concerns of eighteenth-century Europe. It manages to incorporate elements of the indigenous critique, echoes of the biblical narrative of the Fall, and something that at least looks a great deal like the evolutionary stages of material development that were only just being propounded, around that time, by Turgot and Scottish Enlightenment thinkers. Rousseau agrees, in essence, with Kandiaronk's view that civilized Europeans were, by and large, atrocious creatures, for all the reasons that the Wendat had outlined; and he agrees that property is the root of the problem. The one – major – difference between them is that Rousseau, unlike Kandiaronk, cannot really envisage society being based on anything else.

In translating the indigenous critique into terms that French philosophers could understand, this sense of possibility is precisely what was lost. To Americans like Kandiaronk, there was no contradiction between individual liberty and communism – that's to say, communism in the sense we've been using it here, as a certain presumption of sharing, that people who aren't actual enemies can be expected to respond to one another's needs. In the American view, the freedom of the individual was assumed to be premised on a certain level of 'baseline communism', since, after all, people who are starving or lack adequate clothes or shelter in a snowstorm are not really free to do much of anything, other than whatever it takes to stay alive.

The European conception of individual freedom was, by contrast, tied ineluctably to notions of private property. Legally, this association traces back above all to the power of the male household head in ancient Rome, who could do whatever he liked with his chattels and possessions, including his children and slaves.[51] In this view, freedom was always defined – at least potentially – as something exercised to the cost of others. What's more, there was a strong emphasis in ancient

Roman (and modern European) law on the self-sufficiency of households; hence, true freedom meant autonomy in the radical sense, not just autonomy of the will, but being in no way dependent on other human beings (except those under one's direct control). Rousseau, who always insisted he wished to live without being dependent on others' help (even as he had all his needs attended to by mistresses and servants), played out this very same logic in the conduct of his own life.[52]

When our ancestors, Rousseau wrote, made the fatal decision to divide the earth into individually owned plots, creating legal structures to protect their property, then governments to enforce those laws, they imagined they were creating the means to preserve their liberty. In fact, they 'ran headlong to their chains'. This is a powerful image, but it is unclear what Rousseau felt this lost liberty would actually have looked like; especially if, as he insisted, any ongoing human relationship, even one of mutual aid, is itself a restraint on liberty. It's hardly surprising that he ends up inventing a purely imaginary age in which each individual wandered alone among the trees; more surprising, perhaps, that his imaginary world has come so often to define the arc of our own horizons. How did this happen?

IN WHICH WE CONSIDER RELATIONSHIPS BETWEEN THE INDIGENOUS CRITIQUE, THE MYTH OF PROGRESS AND THE BIRTH OF THE LEFT

As we've mentioned before, in the wake of the French Revolution conservative critics blamed Rousseau for almost everything. Many held him personally responsible for the guillotine. The dream of restoring the ancient state of liberty and equality, they argued, led to exactly the effects Turgot had predicted: an Inca-style totalitarianism that could only be enforced through revolutionary terror.

It is true that political radicals at the time of the American and French Revolutions embraced Rousseau's ideas. Here, for example, is an extract purportedly from a manifesto written in 1776 which almost perfectly reproduces Rousseau's fusion of evolutionism and critique of private property as leading directly to the origins of the state:

> As families multiplied, the means of subsistence began to fail; the *nomad* (or roaming) life ceased, and PROPERTY started into existence; men chose habitations; agriculture made them intermix. Language became universal; living together, one man began to measure his strength with another, and the weaker were distinguished from the stronger. This undoubtedly created the idea of mutual defence, of one individual governing diverse families reunited, and of thus defending their persons and their fields against the invasion of an enemy; but hence LIBERTY was ruined in its foundation, and EQUALITY disappeared.[53]

These words are drawn from the purported manifesto of the Secret Order of the Illuminati, a network of revolutionary cadres organized within the Freemasons by a Bavarian law professor named Adam Weishaupt. The organization did exist in the late eighteenth century; its purpose was apparently to educate an enlightened international, or even anti-national, elite to work for the restoration of freedom and equality.

Conservatives almost immediately denounced the Order, leading to it being banned in 1785, less than ten years after its foundation, but right-wing conspiracists insisted it continued to exist, and that the Illuminati were the hidden hands pulling the strings behind the French Revolution (or later even the Russian). This is silly, but one reason the fantasy was possible is that the Illuminati were perhaps the first to propose that a revolutionary vanguard, trained in the correct interpretation of doctrine, would be able to understand the overall direction of human history – and, therefore, be capable of intervening to speed up its progress.[54]

It may seem ironic that Rousseau, who began his career by taking what we would now consider an arch-conservative position – that seeming progress leads only to moral decay – would end up becoming the supreme bête noire of so many conservatives.[55] But a special vitriol is always reserved for traitors.

Many conservative thinkers see Rousseau as having gone full circle from a promising start to creating what we now think of as the political left. Nor are they entirely wrong in this. Rousseau was indeed a crucial figure in the formation of left-wing thought. One reason intellectual debates of the mid eighteenth century seem so strange to us nowadays is precisely that what we understand as left/right divisions had not yet crystallized. At the time of the American Revolution, the terms 'left' and 'right' themselves did not yet exist. A product of the decade immediately following, they originally referred to the respective seating positions of aristocratic and popular factions in the French National Assembly of 1789.

Let us emphasize (we really shouldn't have to) that Rousseau's effusions on the fundamental decency of human nature and lost ages of freedom and equality were in no sense themselves responsible for the French Revolution. It's not as if he somehow caused the *sans culottes* to rise up by putting such ideas into their heads (as we've noted, for most of European history intellectuals seem to have been the only class of people who *weren't* capable of imagining that other worlds might be brought into being). But we can argue that, in folding together the indigenous critique and the doctrine of progress originally developed to counter it, Rousseau did in fact write the founding document of the left as an intellectual project.

For the same reason, right-wing thought has from the beginning been suspicious not just about ideas of progress, but also the entire tradition that emerged from the indigenous critique. Today, we assume that it is largely those on the political left who speak about the 'myth of the noble savage', and that any early European account that idealizes faraway people, or even attributes to them cogent opinions, is really just a romantic projection of European fantasies on to people the authors could never genuinely understand. The racist denigration of the savage, and naive celebration of savage innocence, are always treated as two sides of the same imperialist coin.[56] Yet originally this was an explicitly right-wing position, as explained by Ter Ellingson, the contemporary anthropologist who has reviewed the subject most comprehensively. Ellingson concluded there never was a 'noble savage' myth; at least not in the sense of a stereotype of simple societies living in an age of happy primordial innocence. Rather, travellers' accounts tend to supply a much more ambivalent picture, describing alien societies as a complicated, sometimes (to them) incoherent, mix of virtues and vices. What needs to be investigated, instead, might better be called the 'myth of the myth of the noble savage': why is it that certain

Europeans began attributing such a naive position to others? The answer isn't pretty. The phrase 'noble savage' was in fact popularized a century or so after Rousseau, as a term of ridicule and abuse. It was deployed by a clique of outright racists, who in 1859 – as the British Empire reached its height of power – took over the British Ethnological Society and called for the extermination of inferior peoples.

The original exponents of the idea blamed Rousseau, but before long students of literary history were scouring the archives looking for traces of the 'noble savage' everywhere. Almost all the texts discussed in this chapter came under scrutiny; all were dismissed as dangerous, romantic fantasies. At first, however, these dismissals came from the political right. Ellingson makes a particular example of Gilbert Chinard, whose 1913 volume *L'Amérique et le rêve exotique dans la littérature française au XVIIe et au XVIIIe siècle* (*America and the Exotic Dream in French Literature of the Seventeenth and Eighteenth Centuries*) was primarily responsible for establishing the notion of the 'noble savage' as a Western literary trope in American universities, since he was perhaps the least shy about his political agenda.

Citing Lahontan as the key figure in the formation of this notion, Chinard argued that Rousseau borrowed specific arguments either from Lahontan's *Memoirs* or his *Dialogues* with Kandiaronk. In a broader sense, he detects an affinity of temperament:

> It is Jean-Jacques [Rousseau], more than any other author, that the author of the *Dialogues with a Savage* resembles. With all his faults, his fundamentally ignoble motives, he has put into his style a passion, an enthusiasm which has no equivalent except in the *Discourse on Inequality*. Like Rousseau, he is an anarchist; like him, he is bereft of moral sensibility, and to a considerably greater degree; like him, he imagines himself to be the prey of persecutions of the human race leagued against himself; like him, he is indignant about the sufferings of the miserable and, even more than him, he throws out the call to arms; and like him, above all, he attributes to property all the evils that we suffer. In this, he permits us to establish a direct connection between the Jesuit missionaries and Jean-Jacques.[57]

According to Chinard, even the Jesuits (Lahontan's ostensible enemies) were ultimately playing the same game of introducing deeply subversive notions through the back door. Their motives in quoting the exasperated observations of their interlocutors were not innocent. Commenting directly on the above passage, Ellingson quite reasonably asks what on earth Chinard is actually talking about here: some kind of anarchist movement perpetrated by Lahontan, the Jesuits and Rousseau? A conspiracy theory to explain the French Revolution? Yes, concludes Ellingson, it almost is. The Jesuits, according to Chinard, have promoted 'dangerous ideas' in giving us the impression of the good qualities of 'savages', and 'this impression seems to have been contrary to the interests of the monarchical state and religion.' In fact, Chinard's fundamental characterization of Rousseau is as 'un continuateur des missionaires Jésuites', and he holds the missionaries responsible for giving rise to 'the revolutionary spirits [who] would transform our society and, inflamed by reading their relations, bring us back to the state of the American savages'.[58]

For Chinard, whether or not European observers were reporting the views of their indigenous interlocutors accurately is irrelevant. For indigenous Americans were, as Chinard puts it, 'a race different from our own' with whom no meaningful relation was possible: one might as well, he implies, record the political opinions of a leprechaun.[59] What really matters, he emphasizes, are the motives of the white people involved – and these people were clearly malcontents and troublemakers. He accuses one early observer on the customs of the Greenland Inuit of inserting a mix of socialism and 'illuminism' into his descriptions – that is, viewing savage customs through a lens that might as well have been borrowed from the Secret Order of the Illuminati.[60]

BEYOND THE 'MYTH OF THE STUPID SAVAGE' (WHY ALL THESE THINGS MATTER SO MUCH FOR OUR PROJECT IN THIS BOOK)

This is not the place to document how a right-wing critique morphed into a left-wing critique. To some degree, one can probably just put it down to the laziness of scholars schooled in the history of French or English literature, faced with the prospect of having to seriously engage with what a seventeenth-century Mi'kmaq might have actually been thinking. To say Mi'kmaq thought is unimportant would be racist; to say it's unknowable because the sources were racist, however, does rather let one off the hook.

To some degree, too, such reluctance to engage with indigenous sources is based on completely legitimate protests on the part of those who have, historically, been romanticized. Many have remarked that, to those on the receiving end, being told you are an inferior breed and that therefore anything you say can be ignored, and being told you are an innocent child of nature or the embodiment of ancient wisdom, and that therefore everything you say must be treated as ineffably profound are almost equally annoying. Both attitudes appear designed to prevent any meaningful conversation.

As we noted in our first chapter, when we set out to write this book we imagined ourselves making a contribution to the burgeoning literature on the origins of social inequality – except this time, one based on the actual evidence. As our research proceeded, we came to realize just how strange a question 'what are the origins of social inequality?' really was. Quite apart from the implications of primordial innocence, this way of framing the problem suggests a certain diagnosis of what is wrong with society, and what can and can't be done about it; and as we've seen, it often has very little to do with what people living in those societies we've come to call 'egalitarian' actually feel makes them different from others.

Rousseau sidestepped the question by reducing his savages to mere thought experiments. He was just about the only major figure of the French Enlightenment who *didn't* write a dialogue or other imaginative work attempting to look at European society from a foreign point of view. In fact, he strips his 'savages' of any imaginative powers of their own; their happiness is entirely derived from their inability to imagine things otherwise, or to project themselves into the future in any way at all.[61] They are thus also utterly lacking in philosophy. This is presumably why no one could foresee the disasters that would ensue when they first staked out property and began to form governments to protect it; by the time human beings were even capable of thinking that far ahead, the worst damage had already been done.

Back in the 1960s, the French anthropologist Pierre Clastres suggested that precisely the opposite was the case. What if the sort of people we like to imagine as simple and innocent are free of rulers, governments, bureaucracies, ruling classes and the like, not because they are lacking in imagination, but because they're actually *more* imaginative than we are? We find it difficult to picture what a truly free society would be like; perhaps they have no similar trouble picturing what arbitrary power and domination would be like. Perhaps they can not only imagine it, but consciously arrange their society in such a way as to avoid it. As we'll see in the next chapter, Clastres's argument has its limits. But by insisting that the people studied by anthropologists are just as self-conscious, just as imaginative, as the anthropologists themselves, he did more to reverse the damage than anyone before or since.

Rousseau has been accused of many crimes. He is innocent of most of them. If there is really a toxic element in his legacy, it is this: not his promulgation of the image of the 'noble savage', which he didn't really do, but his promulgation of what we might call the 'myth of the stupid savage' – even if one he considered blissful in its state of stupidity. Nineteenth-century imperialists adopted the stereotype enthusiastically, merely adding on a variety of ostensibly scientific justifications – from Darwinian evolutionism to 'scientific' racism – to elaborate on that notion of innocent simplicity, and thus provide a pretext for pushing the remaining free peoples of the world (or increasingly, as European imperial expansion continued, the formerly free peoples) into a conceptual space where their judgements no longer seemed threatening. This is the work we are trying to undo.

'Liberty, Equality, Fraternity' was the rallying cry of the French Revolution.[62] Today there are whole disciplines – sub-branches of philosophy and political science and legal studies – which take 'equality' as their principal subject matter. Everyone agrees that equality is a value; no one seems to agree on what the term actually refers to. Equality of opportunity? Equality of condition? Formal equality before the law?

Similarly, societies like the seventeenth-century Mi'kmaq, Algonkians or Wendat are regularly referred to as 'egalitarian societies'; or, if not, then as 'band' or 'tribal' societies, which is usually presumed to mean the same thing. It's never entirely clear exactly what the term is supposed to refer to. Are we talking about an ideology, the belief that everyone in society *should* be the same – obviously not in all ways, but in certain respects that are considered particularly important? Or should it be one in which people actually *are* the same? What might either of these actually mean in practice? That all members of society have equal access to land, or treat each other with equal dignity, or are equally free to make their opinions known in public assemblies; or are we talking about some scale of measurement that can be imposed by the observer: cash income, political power, calorie intake, house size, number and quality of personal possessions?

Would equality mean the effacement of the individual, or the celebration of the individual? (After all, to an outside observer, a society where everyone was exactly the same, and one where they were all so completely different as to preclude any sort of comparison, would seem equally 'egalitarian'.) Can one speak of equality in a society where elders are treated like gods and make all important decisions, if everyone in that society who survives past, say, fifty will eventually become an elder? What about gender relations? Many societies referred to as 'egalitarian' are only really egalitarian between adult men. Sometimes relations between men and women in such societies are anything but equal. At other times things are more ambiguous.

It may be, for instance, that men and women in a given society are not only expected to perform different sorts of work, but hold different opinions about why work (or what sorts of work) is important in the first place, and therefore feel they have a higher status; or perhaps that their respective roles are so different, it makes no sense to compare them. Many of the societies encountered by the French in North America fit this description. They could be seen as matriarchal from one perspective, patriarchal from another.[63] In such cases, can we speak of gender equality? Or would we only be able to do so if men and women were also equal according to some minimal external criterion: being equally free from the threat of domestic violence, for example, or having equal access to resources, or equal say in communal affairs?

Since there is no clear and generally accepted answer to any of these questions, use of the term 'egalitarian' has led to endless arguments. In fact, it remains entirely unclear what 'egalitarian' even means. Ultimately the idea is employed not because it has any real analytical substance, but rather for the same reason seventeenth-century natural law theorists speculated about equality in the State of Nature: 'equality' is a default term, referring to that kind of protoplasmic mass of humanity one imagines as being left over when all the trappings of civilization are stripped away. 'Egalitarian' people are those without princes, judges, overseers or hereditary priests, and usually without cities or writing, or preferably even farming. They are societies of equals only in the sense that all the most obvious tokens of inequality are missing.

It follows that any historical work which purports to be about the origins of social inequality is really an inquiry into the origins of civilization; one which in turn implies a vision of history like that of Turgot, which conceives 'civilization' as a system of social complexity, guaranteeing greater overall prosperity, but at the same time ensuring that certain compromises will necessarily have to be made in the areas of freedom and equality. We will be trying to write a different kind of history, which will also require a different understanding of 'civilization'.

To be clear, it's not that we consider the fact that princes, judges, overseers or hereditary priests – or for that matter, writing, cities and farming – only emerge at a certain point in human history to be uninteresting or insignificant. Quite on the contrary: in order to understand our current predicament as a species, it is absolutely crucial to understand how these things first came about. However, we would also insist that, in order to do so, we should reject the impulse to treat our distant ancestors as some sort of primordial human soup. Evidence accumulating from archaeology, anthropology and related fields suggests that – just like seventeenth-century Amerindians and Frenchmen – the people of prehistoric times had very specific ideas about what was important in their societies; that these varied considerably; and that describing such societies as uniformly 'egalitarian' tells us almost nothing about them.

No doubt there was usually a degree of equality by default; an assumption that humans are all equally powerless in the face of the gods; or a strong feeling that no one's will should be permanently subordinated to another's. Presumably there must have been, if only to ensure that permanent princes, judges, overseers or hereditary priests did not emerge for such long periods of time. But self-conscious ideas of 'equality', putting equality forward as an explicit value (as opposed to an ideology of freedom, or dignity, or participation that applies equally to all) appear to have been relative latecomers to human history. And even when they do appear, they rarely apply to everyone.

Ancient Athenian democracy, to take just one example, was based on political equality among its citizens – even if these were only somewhere between 10 and 20 per cent of the overall population – in the sense that each had the same rights to participate in public decision-making. We are taught to see this notion of equal civic participation as a milestone in political development, revived and expanded some 2,000 years later (as it happens, the political systems labelled 'democracies' in nineteenth-century Europe had almost nothing to do with ancient Athens, but this is not really the point). What's more to the point is that Athenian intellectuals at the time, who were mostly of aristocratic background, tended to consider the whole arrangement a tawdry business, and most of them much preferred the government of Sparta, ruled by an even smaller percentage of the total population, who lived collectively off the labours of serfs.

Spartan citizens, in turn, referred to themselves as the *Homoioi*, which could be translated either as 'the Equals' or 'Those Who Are All the Same' – they all underwent the same rigorous military training, adopted the same haughty disdain for both effeminate luxuries and individual idiosyncrasies, ate in communal mess halls and spent most of their lives practising for war.

This is not, then, a book about the origins of inequality. But it aims to answer many of the same questions in a different way. There is no doubt that something has gone terribly wrong with the world. A very small percentage of its population do control the fates of almost everyone else, and they are doing it in an increasingly disastrous fashion. To understand how this situation came about, we should trace the problem back to what first made possible the emergence of kings, priests, overseers and judges. But we no longer have the luxury of assuming we already know in advance what the precise answers will turn out to be. Taking guidance from indigenous critics like Kandiaronk, we need to approach the evidence of the human past with fresh eyes.

3
Unfreezing the Ice Age

In and out of chains: the protean possibilities of human politics

Most societies imagine a mythic age of creation. Once upon a time, the story goes, the world was different: fish and birds could talk, animals could turn into humans and humans into animals. It was possible, in such a time, for things to come into being that were entirely new, in a way that cannot really happen any more: fire, or cooking, or the institution of marriage, or the keeping of pets. In these lesser days, we are reduced to endlessly repeating the great gestures of that time: lighting our own particular fires, arranging our own particular marriages, feeding our particular pets – without ever being able to change the world in quite the same way.

In some ways, accounts of 'human origins' play a similar role for us today as myth did for ancient Greeks or Polynesians, or the Dreamtime for indigenous Australians. This is not to cast aspersions on the scientific rigour or value of these accounts. It is simply to observe that the two fulfil somewhat similar functions. If we think on a scale of, say, the last 3 million years, there actually was an age in which the lines between (what we today think of as) human and animal were still indistinct; and when someone, after all, did have to light a fire, cook a meal or perform a marriage ceremony for the first time. We know these things happened. Still, we really don't know how. It is very difficult to resist the temptation to make up stories about what might have happened: stories which necessarily reflect our own fears, desires, obsessions and concerns. As a result, such distant times can become a vast canvas for the working out of our collective fantasies.

This canvas of human prehistory is distinctively modern. The renowned theorist of culture W. J. T. Mitchell once remarked that dinosaurs are the quintessential modernist animal, since in Shakespeare's time no one knew such creatures had ever existed. In a similar way, until quite recently most Christians assumed anything worth knowing about early humans could be found in the Book of Genesis. Up until the early years of the nineteenth century, 'men of letters' – scientists included – still largely assumed that the universe did not even exist prior to late October, 4004 BC, and that all humans spoke the same language (Hebrew) until the dispersal of humanity, after the fall of the Tower of Babel sixteen centuries later.[1]

At that time there was as yet no 'prehistory'. There was only history, even if some of that history was wildly wrong. The term 'prehistory' only came into common use after the discoveries at Brixham Cave in Devon in 1858, when stone axes, which could only have been fashioned by humans, were found alongside remains of cave bear, woolly rhinoceros and other extinct species, all together under a sealed casing of rock. This, and subsequent archaeological findings, sparked a complete rethinking of existing evidence. Suddenly, 'the bottom dropped out of human history.'[2]

The problem is that prehistory turns out to be an extremely long period of time: more than 3 million years, during which we know our ancestors were, at least sometimes, using stone tools. For most of this period, evidence is extremely limited. There are phases of literally thousands of years for which the only evidence of hominin activity we possess is a single tooth, and perhaps a handful of pieces of shaped flint. While the technology we are capable of bringing to bear on such remote periods improves dramatically each decade, there's only so much you can do with sparse material. As a result, it's difficult to resist the temptation to fill in the gaps, to claim we know more than we really do. When scientists do this the results often bear a suspicious resemblance to those very biblical narratives modern science is supposed to have cast aside.

Let's take just one example. Back in the 1980s, there was a great deal of buzz about a 'mitochondrial Eve', the putative common ancestor of our entire species. Granted, no one was claiming to have actually found the physical remains of such an ancestor; but sequencing the DNA in mitochondria – the tiny cell-motors we inherit from our mothers – demonstrated that such an Eve must have existed, perhaps as recently as 120,000 years ago. And while no one imagined we'd ever find Eve herself, the discovery of a variety of other fossil skulls rescued from the East African Rift Valley (a natural 'preservation trap' for Palaeolithic remains, long since swept to oblivion in more exposed settings) seemed to provide a suggestion as to what Eve might have looked like and where she might have lived. While scientists continued debating the ins and outs, popular magazines were soon carrying stories about a modern counterpart to the Garden of Eden, the original incubator of humanity, the savannah-womb that gave life to us all.

Many of us probably still have something resembling this picture of human origins in our mind. More recent research, though, has shown it couldn't possibly be accurate. In fact, biological anthropologists and geneticists are now converging on an entirely different picture. Rather than everyone starting out the same, then dispersing from East Africa in some Tower-of-Babel moment to become the diverse nations and peoples of the earth, early human populations in Africa appear to have been far more physically diverse than anything we are familiar with today.

We modern-day humans tend to exaggerate our differences. The results of such exaggeration are often catastrophic. Between war, slavery, imperialism and sheer day-to-day racist oppression, the last several centuries have seen so much human suffering justified by minor differences in human appearance that we can easily forget just how minor these differences really are. By any biologically meaningful standard, living humans are barely distinguishable. Whether you go to Bosnia, Japan, Rwanda or the Baffin Islands, you can expect to see people with the same small and gracile faces, chin, globular skull and roughly the same distribution of body hair. Not only do we look the same, in many ways we act the same as well (for instance, everywhere from the Australian outback to Amazonia, rolling one's eyes is a way of saying 'what an idiot!'). The same applies to cognition. We might think different groups of humans realize their cognitive capacities in very different ways – and to some extent, of course, we do – but again, much of this perceived difference results from our having no real basis for comparison: there's no human language, for instance, that doesn't have nouns, verbs and adjectives; and while humans may enjoy very different forms of music and dance, there's no known human population that does not enjoy music and dancing at all.

Rewind a few hundred millennia and all this was most definitely *not* the case.

For most of our evolutionary history, we did indeed live in Africa – but not just the eastern savannahs, as previously thought: our biological ancestors were distributed everywhere from Morocco to the Cape.[3] Some of those populations remained isolated from each another for tens or even hundreds of thousands of years, cut off from their nearest relatives by deserts and rainforests. Strong regional traits developed.[4] The result probably would have struck a modern observer as something more akin to a world inhabited by hobbits, giants and elves than anything we have direct experience of today, or in the more recent past. Those elements that make up modern humans – the relatively uniform 'us' referred to above – seem only to have come together quite late in the process. In other words, if we think humans are different from each other now, it's largely illusory; and even such differences as do exist are utterly trivial and cosmetic, compared with what must have been happening in Africa during most of prehistory.

Ancestral humans were not only quite different from each other; they also coexisted with smaller-brained, more ape-like species such as *Homo naledi*. What were these ancestral societies like? At this point, at least, we should be honest and admit that, for the most part, we don't have the slightest idea. There's only so much you can reconstruct from cranial remains and the occasional piece of knapped flint – which is basically all we have. Most of the time we don't even really know what was going on below the neck, let alone with pigmentation, diet or anything else. What we do know is that we are composite products of this original mosaic of human populations, which interacted with one another, interbred, drifted apart and came together mostly in ways we can still only guess at.[5] It seems reasonable to assume that behaviours like mating and child-rearing practices, the presence or absence of dominance hierarchies or forms of language and proto-language must have varied at least as much as physical types, and probably far more.

Perhaps the only thing we can say with real certainty is that, in terms of ancestry, we are all Africans.

Modern humans first appeared in Africa. When they began expanding out of Africa into Eurasia, they encountered other populations such as Neanderthals and Denisovans – less different, but still different – and these various groups interbred.[6] Only after those other populations became extinct can we really begin talking about a single, human 'us' inhabiting the planet. What all this brings home is just how radically different the social and even physical world of our remote ancestors would have seemed to us – and this would have been true at least down to around 40,000 BC. The range of flora and fauna surrounding them was quite unlike anything that exists today. All of which makes it extremely difficult to draw analogies. There's simply nothing in the historical or ethnographic record that resembles a situation in which different subspecies of human interbred, interacted, co-operated, but sometimes also killed each other – and even if there were, the archaeological evidence is too thin and sporadic to test whether remote prehistory was really anything like that or not.[7]

The only thing we can reasonably infer about social organization among our earliest ancestors is that it's likely to have been extraordinarily diverse. Early humans inhabited a wide range of natural environments, from coastlands and tropical forest to mountains and savannah. They were far, far more physically diverse than humans are today; and presumably their social differences were even greater than their physical ones. In other words, there is no 'original' form of human society. Searching for one can only be a matter of myth-making, whether the resultant myths take the form of 'killer ape' fantasies that emerged in the 1960s, seared into collective consciousness by movies like Stanley Kubrick's *2001: A Space Odyssey*; or the 'aquatic ape'; or even the highly amusing but fanciful 'stoned ape' (the theory that consciousness emerged from the accidental ingestion of psychedelic mushrooms). Myths like these entertain YouTube watchers to this day.

We should be clear: there's nothing wrong with myths. Likely as not, the tendency to make up stories about the distant past as a way of reflecting on the nature of our species is itself, like art and poetry, one of those distinctly human traits that began to crystallize in deep prehistory. And no doubt some of these stories – for instance, feminist theories that see distinctly human sociability as originating in collective child-rearing practices – can indeed tell us something important about the paths that converged in modern humanity.[8] But such insights can only ever be partial because there was no Garden of Eden, and a single Eve never existed.

WHY THE 'SAPIENT PARADOX' IS A RED HERRING; AS SOON AS WE WERE HUMAN, WE STARTED DOING HUMAN THINGS

Human beings, today, are a fairly uniform species. This uniformity is not, in evolutionary terms, particularly old. Its genetic basis was established around half a million years ago, but it is almost certainly misguided to think we could ever specify a single, more recent point in time when *Homo sapiens* 'emerged' – that is, when all the various elements of the modern human condition converged, definitively, in some stupendous moment of creation.

Consider the first direct evidence of what we'd now call complex symbolic human behaviour, or simply 'culture'. Currently, it dates back no more than 100,000 years. Where exactly on the African continent this evidence for culture crops up is determined largely by conditions of preservation, and by the countries that have so far been most accessible for archaeological investigation. Rock shelters around the coastlands of South Africa are a key source, trapping prehistoric sediments that yield evidence of hafted tools and the expressive use of shell and ochre around 80,000 BC.[9] Comparably ancient finds are also known from other parts of Africa, but it's not until later, around 45,000 years ago – by which time our species was busily colonizing Eurasia – that similar evidence starts appearing much more widely, and in greater quantities.

In the 1980s and 1990s it was widely assumed that something profound happened, some kind of sudden creative efflorescence, around 45,000 years ago, variously referred to in the literature as the 'Upper Palaeolithic Revolution' or even the 'Human Revolution'.[10] But in the last two decades it has become increasingly clear to researchers that this is most likely an illusion, created by biases in our evidence.

Here's why. Much of the evidence for this 'revolution' is restricted to a single part of the world: Europe, where it is associated with replacement of Neanderthals by *Homo sapiens* around 40,000 BC. It includes more advanced toolkits for hunting and handicrafts, the first clear evidence for the making of images in bone, ivory and clay – including the famous sculpted 'female figurines',[11] dense clusters of carved and painted animal figures in caves, often observed with breathtaking accuracy; more elaborate ways of clothing and decorating the human body; the first attested use of musical instruments like bone flutes; regular exchange of raw materials over great distances, and also what are usually taken as the earliest proofs of social inequality, in the form of grand burials.

All this is impressive, and gives the impression of a lack of synchrony between the ticking of our genetic and cultural clocks. It seems to ask the question: why do so many tens of thousands of years stand between the biological origins of humanity and the widespread appearance of typically human forms of behaviour; between when we became capable of creating culture and when we finally got round to doing it? What were we actually doing in the interim? Many researchers have puzzled over this and have even coined a phrase for it: 'the sapient paradox'.[12] A few go so far as to postulate some late mutation in the human brain to explain the apparently superior cultural capacities of Upper Palaeolithic Europeans, but such views can no longer be taken seriously.

In fact, it's becoming increasingly clear that the whole problem is a mirage. The reason archaeological evidence from Europe is so rich is that European governments tend to be rich; and that European professional institutions, learned societies and university departments have been pursuing prehistory far longer on their own doorstep than in other parts of the world. With each year that passes, new evidence accumulates for early behavioural complexity elsewhere: not just Africa, but also the Arabian Peninsula, Southeast Asia and the Indian subcontinent.[13] Even as we write, a cave site on the coast of Kenya called Panga ya Saidi is yielding evidence of shell beads and worked pigments stretching back 60,000 years;[14] and research on the islands of Borneo and Sulawesi is opening vistas on to an unsuspected world of cave art, many thousands of years older than the famous images of Lascaux and Altamira, on the other side of Eurasia.[15] No doubt still earlier examples of complex pictorial art will one day be found somewhere on the continent of Africa.

If anything, then, Europe was late to the party. Even after its initial colonization by modern humans – starting around 45,000 BC – the continent was still thinly populated, and the new arrivals coexisted there, albeit fairly briefly, with more established Neanderthal populations (themselves engaged in complex cultural activities of various sorts).[16] Why there appears to be such a sudden cultural efflorescence, shortly after their arrival, may have something to do with climate and demography. To put it bluntly: with the movement of the ice sheets, human populations in Europe were living in harsher and more confined spaces than our species had encountered before. Game-rich valleys and steppe were bounded by tundra to the north and dense coastal forests to the south. We have to picture our ancestors moving between relatively enclosed environments, dispersing and gathering, tracking the seasonal movements of mammoth, bison and deer herds. While the absolute number of people may still have been startlingly small,[17] the density of human interactions seems to have radically increased, especially at certain times of year. And with this came remarkable bursts of cultural expression.[18]

WHY EVEN VERY SOPHISTICATED RESEARCHERS STILL FIND WAYS TO CLING TO THE IDEA THAT SOCIAL INEQUALITY HAS AN 'ORIGIN'

As we will see in a moment, the societies that resulted in what archaeologists call the Upper Palaeolithic period (roughly 50,000–15,000 BC) – with their 'princely' burials and grand communal buildings – seem to completely defy our image of a world made up of tiny egalitarian forager bands. The disconnect is so profound that some archaeologists have begun taking the opposite tack, describing Ice Age Europe as populated by 'hierarchical' or even 'stratified' societies. In this, they make common cause with evolutionary psychologists who insist that dominance behaviour is hardwired in our genes, so much so that the moment society goes beyond tiny bands, it must necessarily take the form of some ruling over others.

Almost everyone who isn't a Pleistocene archaeologist – that is, who is not forced to confront the evidence – simply ignores it and carries on exactly as they had before, writing as if hunter-gatherers can be assumed to have lived in a state of primordial innocence. As

Christopher Boehm puts it, we seem doomed to play out an endless recycling of the war between 'Hobbesian hawks and Rousseauian doves': those who view humans as either innately hierarchical or innately egalitarian.

Boehm's own work is revealing in this regard. An evolutionary anthropologist and a specialist in primate studies, he argues that while humans do have an instinctual tendency to engage in dominance-submissive behaviour, no doubt inherited from our simian ancestors, what makes societies distinctively human is our ability to make the conscious decision *not* to act that way. Carefully working through ethnographic accounts of existing egalitarian foraging bands in Africa, South America and Southeast Asia, Boehm identifies a whole panoply of tactics collectively employed to bring would-be braggarts and bullies down to earth – ridicule, shame, shunning (and in the case of inveterate sociopaths, sometimes even outright assassination)[19] – none of which have any parallel among other primates.

For instance, while gorillas do not mock each other for beating their chests, humans do so regularly. Even more strikingly, while the bullying behaviour might well be instinctual, counter-bullying is not: it's a well-thought-out strategy, and forager societies who engage in it display what Boehm calls 'actuarial intelligence'. That's to say, they understand what their society might look like if they did things differently: if, for instance, skilled hunters were *not* systematically belittled, or if elephant meat was *not* portioned out to the group by someone chosen at random (as opposed to the person who actually killed the beast). This, he concludes, is the essence of politics: the ability to reflect consciously on different directions one's society could take, and to make explicit arguments why it should take one path rather than another. In this sense, one could say Aristotle was right when he described human beings as 'political animals' – since this is precisely what other primates never do, at least not to our knowledge.

This is a brilliant and important argument – but, like so many authors, Boehm seems strangely reluctant to consider its full implications. Let's do so now.

If the very essence of our humanity consists of the fact that we are self-conscious political actors, and therefore capable of embracing a wide range of social arrangements, would that not mean human beings should actually have explored a wide range of social arrangements over the greater part of our history? In the end, confusingly, Boehm assumes that all human beings until very recently chose instead to follow exactly the same arrangements – we were strictly 'egalitarian for thousands of generations before hierarchical societies began to appear' – thereby casually tossing early humans back into the Garden of Eden once again. Only with the beginnings of agriculture, he suggests, did we all collectively flip back to hierarchy. Before 12,000 years ago, Boehm insists, humans were basically egalitarian, living in what he calls 'societies of equals, and outside the family there were no dominators'.[20]

So, according to Boehm, for about 200,000 years political animals all chose to live just one way; then, of course, they began to rush headlong into their chains, and ape-like dominance patterns re-emerged. The solution to the battle between 'Hobbesian hawks and Rousseauian doves' turns out to be: our genetic nature is Hobbesian, but our political history is pretty much exactly as described by Rousseau. The result? An odd insistence that for many tens of thousands of years, nothing happened. This is an unsettling conclusion, especially when we consider some of the actual archaeological evidence for the existence of 'Palaeolithic politics'.

IN WHICH WE OBSERVE HOW GRAND MONUMENTS, PRINCELY BURIALS AND OTHER UNEXPECTED FEATURES OF ICE AGE SOCIETIES HAVE UPENDED OUR ASSUMPTIONS OF WHAT HUNTER-GATHERERS ARE LIKE, AND CONSIDER WHAT IT MIGHT MEAN TO SAY THERE WAS 'SOCIAL STRATIFICATION' SOME 30,000 YEARS AGO

Let's start with rich hunter-gatherer burials. Examples can be found across much of western Eurasia, from the Dordogne to the Don. They include discoveries in rock shelters and open-air settlements. Some of the earliest come from sites like Sunghir in northern Russia and Dolní

Věstonice in the Moravian basin, south of Brno, and date from between 34,000 and 26,000 years ago. What we find here are not cemeteries but isolated burials of individuals or small groups, their bodies often placed in striking postures and decorated – in some cases, almost saturated – with ornaments. In the case of Sunghir that meant many thousands of beads, laboriously worked from mammoth ivory and fox teeth. Originally, such beads would have decorated clothing made of fur and animal skins. Some of the most lavish costumes are from the conjoined burials of two boys, flanked by great lances made of straightened mammoth tusks.[21]

At Dolní Věstonice, one triple burial contains two young men with elaborate headdresses, posed either side of an older man, all lying on a bed of soil stained red with ochre.[22] Of similar antiquity is a group of cave burials unearthed on the coast of Liguria, near the modern border between Italy and France. Complete bodies of young or adult men, including one especially lavish interment known to archaeologists as *Il Principe* ('the Prince'), were laid out in striking poses and suffused with jewellery, including beads made of marine shell and deer canines, as well as blades of exotic flint. *Il Principe* bears that name because he's also buried with what looks to the modern eye like royal regalia: a flint sceptre, elk antler batons and an ornate headdress lovingly fashioned from perforated shells and deer teeth. Moving further west, to the Dordogne, we encounter a 16,000-year-old burial of a young woman, the so-called 'Lady of Saint-Germain-de-la-Rivière', which contains a rich assemblage of stomach and pelvic ornaments made of shell and stag teeth. The teeth are taken from deer hunted in the Spanish Basque country 190 miles away.[23]

Such findings have completely altered the specialist view of human societies in prehistory. The pendulum has swung so far away from the old notion of egalitarian bands that some archaeologists now argue that, thousands of years before the origins of farming, human societies were already divided along lines of status, class and inherited power. As we'll see, this is highly unlikely, but the evidence these archaeologists point to is real enough: for instance, the extraordinary outlays of labour involved in making grave goods (10,000 work hours for the Sunghir beads alone, by some estimates); the highly advanced and standardized methods of production, possibly suggesting specialized craftspeople; or the way in which exotic, prestigious materials were transported from very distant locations; and, most suggestive of all, a few cases where such wealth was buried with children, maybe implying some kind of inherited status.[24]

Another unexpected result of recent archaeological research, causing many to revise their view of prehistoric hunter-gatherers, is the appearance of monumental architecture. In Eurasia, the most famous examples are the stone temples of the Germuş Mountains, overlooking the Harran Plain in southeast Turkey. In the 1990s, German archaeologists, working on the plain's northern frontier, began uncovering extremely ancient remains at a place known locally as Göbekli Tepe.[25] What they found has since come to be regarded as an evolutionary conundrum. The main source of puzzlement is a group of twenty megalithic enclosures, initially raised there around 9000 BC, and then repeatedly modified over many centuries. These enclosures were established at a time when the surrounding plain was a mixture of woodland and steppe, teeming with wild plant and animal species that colonized the Middle East as the last Ice Age was drawing to a close.

The enclosures at Göbekli Tepe are massive. They comprise great T-shaped pillars, some over sixteen feet high and weighing up to a ton, which were hewn from the site's limestone bedrock or nearby quarries. The pillars, at least 200 in total, were raised into sockets and linked by walls of rough stone. Each is a unique work of sculpture, carved with images from the world of dangerous carnivores and poisonous reptiles, as well as game species, waterfowl and small scavengers. Animal forms project from the rock in varying depths of relief: some hover coyly on the surface, others emerge boldly into three dimensions. These often nightmarish creatures follow divergent orientations, some marching to the horizon, others working their way down into the earth. In places, the pillar itself becomes a sort of standing body, with human-like limbs and clothing.

The creation of these remarkable buildings implies strictly coordinated activity on a really large scale, even more so if multiple enclosures were constructed simultaneously, according to an overall plan (a current point of debate).[26] But the larger question remains: who made them? While groups of humans not too far away had already begun cultivating crops at the time, to the best of our knowledge those who built Göbekli Tepe had not. Yes, they harvested and processed wild cereals and other plants in season, but there is no compelling reason to see them as 'proto-farmers', or to suggest they had any interest in orienting their livelihoods around the domestication of crops. Indeed, there was no particular reason why they should, given the availability of fruits, berries, nuts and edible wild fauna in their vicinity. (In fact, there are good reasons to think the builders of Göbekli Tepe were different, in some quite startling ways, from nearby groups who were beginning to take up farming, but this will have to wait for a later chapter; for the moment, we're just interested in the monuments.)

To some, the raised location and orientation of the buildings at Göbekli Tepe suggest an astronomical or chronometric function, each chain of pillars aligned with a particular cycle of celestial movements. Archaeologists remain sceptical, pointing out that the structures may once have been roofed, and that their layout was subject to many alterations over time. But what has mostly intrigued scholars of different disciplines so far is something else: the apparent proof they offer that 'hunter-gatherer societies had evolved institutions to support major public works, projects, and monumental constructions, and thus had a complex social hierarchy prior to their adoption of farming.'[27] Again, matters are not so simple, because these two phenomena – hierarchy and the measure of time – were closely interwoven.

While Göbekli Tepe is often presented as an anomaly, there is in fact a great deal of evidence for monumental construction of different sorts among hunter-gatherers in earlier periods, extending back into the Ice Age.

In Europe, between 25,000 and 12,000 years ago public works were already a feature of human habitation across an area reaching from Kraków to Kiev. Along this transect of the glacial fringe, remains of impressive circular structures have been found that are clearly distinguishable from ordinary camp-dwellings in their scale (the largest were over thirty-nine feet in diameter), permanence, aesthetic qualities and prominent locations in the Pleistocene landscape. Each was erected on a framework made of mammoth tusks and bones, taken from many tens of these great animals, which were arranged in alternating sequences and patterns that go beyond the merely functional to produce structures that would have looked quite striking to our eyes, and magnificent indeed to people at the time. Great wooden enclosures of up to 130 feet in length also existed, of which only the post-holes and sunken floors remain.[28] Göbekli Tepe too is likely to have had its wooden counterparts.

Monumentality is always to some degree a relative concept; that's to say, a building or structure is 'monumental' only in comparison to other buildings and structures a viewer has actually experienced. Obviously, the Ice Age produced nothing on the scale of the Pyramids of Giza or the Roman Colosseum – but, by the standards of their day, the kind of structures we've been describing can only have been considered public works, involving sophisticated design and the co-ordination of labour on an impressive scale. Research at the Russian site of Yudinovo suggests that 'mammoth houses', as they are often called, were not in fact dwellings at all, but monuments in the strict sense: carefully planned and constructed to commemorate the completion of a great mammoth hunt (and the solidarity of the extended hunting group), using whatever durable parts remained once carcasses had been processed for their meat and hides; and later covered with sediment to create a durable marker in the landscape.[29] We are talking here about really staggering quantities of meat: for each structure (there were five at Yudinovo), there was enough mammoth to feed hundreds of people for around three months.[30] Open-air settlements like Yudinovo, Mezhirich and Kostenki, where such mammoth monuments were erected, often became central places whose inhabitants exchanged amber, marine shells and animal pelts over impressive distances.

So what are we to make of all this evidence for stone temples, princely burials, mammoth monuments and bustling centres of trade and craft production, stretching back far into the Ice Age? What are they doing there, in a Palaeolithic world where – at least on some accounts – nothing much is ever supposed to have happened, and human societies can best be understood by analogy with troops of chimps or bonobos? Unsurprisingly, perhaps, some have responded by completely abandoning the idea of an egalitarian Golden Age, concluding instead that this must have been a society dominated by powerful leaders, even dynasties – and, therefore, that self-aggrandizement and coercive power have always been the enduring forces behind human social evolution. But this doesn't really work either.

Evidence of institutional inequality in Ice Age societies, whether grand burials or monumental buildings, is sporadic. Richly costumed burials appear centuries, and often hundreds of miles, apart. Even if we put this down to the patchiness of the evidence, we still have to ask why the evidence is so patchy in the first place: after all, if any of these Ice Age 'princes' had behaved like, say, Bronze Age (let alone Renaissance Italian) princes, we'd also be finding all the usual trappings of centralized power: fortifications, storehouses, palaces. Instead, over tens of thousands of years, we see monuments and magnificent burials, but little else to indicate the growth of ranked societies, let alone anything remotely resembling 'states'. To understand why the early record of human social life is patterned in this strange, staccato fashion we first have to do away with some lingering preconceptions about 'primitive' mentalities.

IN WHICH WE DISPOSE OF LINGERING ASSUMPTIONS THAT 'PRIMITIVE' FOLK WERE SOMEHOW INCAPABLE OF CONSCIOUS REFLECTION, AND DRAW ATTENTION TO THE HISTORICAL IMPORTANCE OF ECCENTRICITY

In the last chapter, we suggested that the really insidious element of Rousseau's legacy is not so much the idea of the 'noble savage' as that of the 'stupid savage'. We may have got over the overt racism of most nineteenth-century Europeans, or at least we think we have, but it's not unusual to find even very sophisticated contemporary thinkers who feel it's more appropriate to compare 'bands' of hunter-gatherers with chimps or baboons than with anyone they'd ever be likely to meet. Consider the following passage from the historian Yuval Noah

Harari's *Sapiens: A Brief History of Humankind* (2014). Harari starts off with a perfectly reasonable observation: that our knowledge of early human history is extremely limited, and social arrangements probably varied a great deal from place to place. True, he overstates his case (he suggests we can really know nothing, even about the Ice Age), but the basic point is well taken. Then we get this:

> The sociopolitical world of the foragers is another area about which we know next to nothing ... scholars cannot even agree on the basics, such as the existence of private property, nuclear families and monogamous relationships. It's likely that different bands had different structures. Some may have been as hierarchical, tense and violent as the nastiest chimpanzee group, while others were as laid-back, peaceful and lascivious as a bunch of bonobos.

So not only was everyone living in bands until farming came along, but these bands were basically ape-like in character. If this seems unfair to the author, remember that Harari could just as easily have written 'as tense and violent as the nastiest biker gang', and 'as laid-back, peaceful and lascivious as a hippie commune'. One might have imagined the obvious thing to compare one group of human beings with would be ... another group of human beings. Why, then, did Harari choose chimps instead of bikers? It's hard to escape the impression that the main point of difference is that bikers *choose* to live the way they do. Such choices imply political consciousness: the ability to argue and reflect about the proper way to live – which is precisely, as Boehm reminds us, what apes don't do. Yet Harari, like so many others, chooses to compare early humans with apes anyway.

In this way, the 'sapient paradox' returns. Not as something real, but as a side effect of the weird way we read the evidence: insisting either that for countless millennia we had modern brains, but for some reason decided to live like monkeys anyway; or that we had the ability to overcome our simian instincts and organize ourselves in an endless variety of ways, but for some equally obscure reason only ever chose one way to organize ourselves.

Perhaps the real question here is what it means to be a 'self-conscious political actor'. Philosophers tend to define human consciousness in terms of self-awareness; neuroscientists, on the other hand, tell us we spend the overwhelming majority of our time effectively on autopilot, working out habitual forms of behaviour without any sort of conscious reflection. When we are capable of self-awareness, it's usually for very brief periods of time: the 'window of consciousness', during which we can hold a thought or work out a problem, tends to be open on average for roughly seven seconds. What neuroscientists (and it must be said, most contemporary philosophers) almost never notice, however, is that the great exception to this is when we're talking to someone else. In conversation, we can hold thoughts and reflect on problems sometimes for hours on end. This is of course why so often, even if we're trying to figure something out by ourselves, we imagine arguing with or explaining it to someone else. Human thought is inherently dialogic. Ancient philosophers tended to be keenly aware of all this: that's why, whether they were in China, India or Greece, they tended to write their books in the form of dialogues. Humans were only fully self-conscious when arguing with one another, trying to sway each other's views, or working out a common problem. True individual self-consciousness, meanwhile, was imagined as something that a few wise sages could perhaps achieve through long study, exercise, discipline and meditation.

What we'd now call political consciousness was always assumed to come first. In this sense, the Western philosophical tradition has taken a rather unusual direction over the last few centuries. Around the same time as it abandoned dialogue as its typical mode of writing, it also began imagining the isolated, rational, self-conscious individual not as a rare achievement, something typically accomplished – if at all – after literally years of living isolated in a cave or monastic cell, or on top of a pillar in a desert somewhere, but as the normal default state of human beings anywhere.

Even stranger, over the course of the eighteenth and nineteenth centuries it was *political* self-consciousness that European philosophers came to see as some kind of amazing historical achievement: as a phenomenon which only really became possible with the Enlightenment itself, and the subsequent American and French Revolutions. Before that, it was assumed, people blindly followed traditions, or what they assumed to be the will of God. Even when peasants or popular rebels rose up to try to overthrow oppressive regimes they couldn't admit they were doing so, but convinced themselves they were restoring 'ancient customs' or acting on some kind of divine inspiration. To Victorian intellectuals, the notion of people self-consciously imagining a social order more to their liking and then trying to bring it into being was simply not applicable before the modern age – and most were deeply divided as to whether it would even be a good idea in their own time.

All this would have come as a great surprise to Kandiaronk, the seventeenth-century Wendat philosopher-statesman whose impact on European political thought we discussed in the previous chapter. Like many North American peoples of his time, Kandiaronk's Wendat nation saw their society as a confederation created by conscious agreement; agreements open to continual renegotiation. But by the late nineteenth and early twentieth centuries, many in Europe and America had reached the point of arguing that someone like Kandiaronk could never have really existed in the first place. 'Primitive' folk, they argued, were not only incapable of political self-consciousness, they were not even capable of fully conscious thought on the individual level – or at least conscious thought worthy of the name. That is, just as they pretended a 'rational Western individual' (say, a British train guard or French colonial official) could be assumed to be fully self-aware all the time (a clearly absurd assumption), they argued that anyone classified as a 'primitive' or 'savage' operated with a 'pre-logical mentality', or lived in a mythological dreamworld. At best, they were mindless conformists, bound in the shackles of tradition; at worst, they were incapable of fully conscious, critical thought of any kind.

Such theories might be considered the high-water mark of the reaction against the indigenous critique of European society. The arguments attributed to figures like Kandiaronk could be written off as simple projections of Western 'noble savage' fantasies, because real savages were assumed to live in an entirely different mental universe. Nowadays no reputable scholar would make such claims: everyone at least pays lip service to the psychic unity of mankind. But in practice, as we've seen, little has changed. Scholars still write as if those living in earlier stages of economic development, and especially those who are classified as 'egalitarian', can be treated as if they were literally all the same, living in some collective group-think: if human differences show up in any form – different 'bands' being different from each other – it is only in the same way that bands of great apes might differ. Political self-consciousness, or certainly anything we'd now call visionary politics, would have been impossible.

And if certain hunter-gatherers turn out not to have been living perpetually in 'bands' at all, but instead congregating to create grand landscape monuments, storing large quantities of preserved food and treating particular individuals like royalty, contemporary scholars are at best likely to place them in a new stage of development: they have moved up the scale from 'simple' to 'complex' hunter-gatherers, a step closer to agriculture and urban civilization. But they are still caught in the same Turgot-like evolutionary straitjacket, their place in history defined by their mode of subsistence, and their role blindly to enact some abstract law of development which we understand but they do not; certainly, it rarely occurs to anyone to ask what sort of worlds they *thought* they were trying to create.[31]

Now, admittedly, there have always been exceptions to this rule. Anthropologists who spend years talking to indigenous people in their own languages, and watching them argue with one another, tend to be well aware that even those who make their living hunting elephants or gathering lotus buds are just as sceptical, imaginative, thoughtful and capable of critical analysis as those who make their living by operating tractors, managing restaurants or chairing university departments. A few, such as the early-twentieth-century scholar Paul Radin in his 1927 book *Primitive Man as Philosopher*, ended up concluding that at least those he knew best – Winnebago and other Native North Americans – were actually, on average, rather more thoughtful.

Radin himself was considered something of an oddball by his contemporaries (he always avoided getting a proper academic job; the legend in Chicago was that when once given a teaching fellowship there, he was so intimidated before his first lecture that he immediately marched out to a nearby highway and contrived to get his leg broken by a car, then spent the rest of the term happily reading in the hospital). Perhaps not coincidentally, what really struck him about the 'primitive' societies he was most familiar with was their tolerance of eccentricity. This, he concluded, was simply the logical extension of that same rejection of coercion that so impressed the Jesuits in Quebec. If, he noted, a Winnebago decided that gods or spirits did not really exist and refused to perform rituals meant to appease them, or even if he declared the collective wisdom of the elders wrong and invented his own personal cosmology (and both these things did, quite regularly, happen), such a sceptic would definitely be made fun of, while his closest friends and family might worry lest the gods punish him in some way. However, it would never occur to *them* to punish him, or that anyone should try to force him into conformity – for instance, by blaming him for a bad hunt and therefore refusing to share food with him until he agreed to perform the usual rituals.

There is every reason to believe that sceptics and non-conformists exist in every human society; what varies is how others react to them.[32] Radin was interested in the intellectual consequences, the kind of speculative systems of thought such out-of-sync characters might create. Others have noted the political implications. It's often people who are just slightly odd who become leaders; the truly odd can become spiritual figures, but, even more, they can and often do serve as a kind of reserve of potential talent and insight that can be called on in the event of a crisis or unprecedented turn of affairs. Thomas Beidelman, for instance, observes that among the early-twentieth-century Nuer – a cattle-keeping people of South Sudan, famous for their rejection of anything that resembled government – there were politicians and village 'bulls' ('operator types' we'd now call them) who played fast and loose with the rules, but also 'earth priests' who mediated local disputes, and finally prophets. The politicians were often unconventional: for instance, it was not uncommon for the local 'bull' actually to be a woman whose parents had declared her a man for social purposes; the priests were always outsiders to the region; but the prophet was an altogether more extreme kind of figure. He might dribble, drool, maintain a vacant stare, act like an epileptic; or engage in long but pointless tasks such as spending hours arranging shells into designs on the ground in the bush; or long periods in the wilderness; or he may even eat excrement or ashes. Prophets, as Beidelman notes, 'may speak in tongues, go into trances, fast, balance on their head, wear feathers in their hair, be active by night rather than by day, and may perch on rooftops. Some sit with tethering pegs up their anuses.'[33] Many, too, were physically deformed. Some were cross-dressers, or given to unconventional sexual practices.

In other words, these were seriously unorthodox people. The impression one gets from the literature is that any Nuer settlement of pre-colonial times was likely to be complemented by a minor penumbra of what might be termed extreme individuals; ones who in our own society would likely be classified as anything from highly eccentric or defiantly queer to neurodivergent or mentally ill. Normally, prophets were treated with bemused respect. They were ill; but the illness was a direct consequence of being touched by God. As a result, when great calamities or unprecedented events occurred – a plague, a foreign invasion – it was among this penumbra that everyone looked for a charismatic leader appropriate to the occasion. As a result, a person who might otherwise have spent his life as something analogous to the village idiot would suddenly be found to have remarkable powers of foresight and persuasion; even to be capable of inspiring new social movements among the youth or co-ordinating elders across Nuerland to put aside their differences and mobilize around some common goal; even, sometimes, to propose entirely different visions of what Nuer society might be like.

WHAT CLAUDE LÉVI-STRAUSS LEARNED FROM THE NAMBIKWARA ABOUT THE ROLE OF CHIEFS, AND SEASONAL VARIATIONS OF SOCIAL LIFE

Claude Lévi-Strauss is one of the few mid-twentieth-century anthropologists to take seriously the idea that early humans were our intellectual equals; hence his famous argument in *The Savage Mind* that mythological thought, rather than representing some sort of prelogical haze, is better conceived as a kind of 'neolithic science' as sophisticated as our own, just built on different principles. Less well known – but more relevant to the problems we are grappling with here – are some of his early writings on politics.

In 1944, Lévi-Strauss published an essay about politics among the

Nambikwara, a small population of part-time farmers, part-time foragers inhabiting a notoriously inhospitable stretch of savannah in northwest Mato Grosso, Brazil. The Nambikwara then had a reputation as extremely simple folk, given their very rudimentary material culture. For this reason, many treated them almost as a direct window on to the Palaeolithic. This, Lévi-Strauss pointed out, was a mistake. People like the Nambikwara live in the shadow of the modern state, trading with farmers and city people and sometimes hiring themselves out as labourers. Some might even be descendants of runaways from cities or plantations. Still, he noted, their ways of organizing their lives could be seen as a source of insights into more general features of the human condition, especially as these pertain to politics.

For Lévi-Strauss, what was especially instructive about the Nambikwara was that, for all that they were averse to competition (they had little wealth to compete over anyway), they did appoint chiefs to lead them. The very simplicity of the resulting arrangement, he felt, might expose 'some basic functions' of political life that 'remain hidden in more complex and elaborate systems of government'. Not only was the role of the chief socially and psychologically quite similar to that of a national politician or statesman in European society, he noted, it also attracted similar personality types: people who 'unlike most of their companions, enjoy prestige for its own sake, feel a strong appeal to responsibility, and to whom the burden of public affairs brings its own reward'.[34]

Modern politicians play the role of wheelers and dealers, brokering alliances or negotiating compromises between different constituencies or interest groups. In Nambikwara society this didn't happen much because there weren't really many differences in wealth or status. However, chiefs did play an analogous role, brokering between two entirely different social and ethical systems, which obtained at different times of year. Allow us to explain. In the 1940s, the Nambikwara lived in what were effectively two very different societies. During the rainy season, they occupied hilltop villages of several hundred people and practised horticulture; during the rest of the year they dispersed into small foraging bands. Chiefs made or lost their reputations by acting as heroic leaders during the 'nomadic adventures' of the dry season, during which times they typically gave orders, resolved crises and behaved in what would at any other time be considered an unacceptably authoritarian manner; in the wet season, a time of much greater ease and abundance, they relied on those reputations to attract followers to settle around them in villages, where they employed only gentle persuasion and led by example to guide their followers in the construction of houses and tending of gardens. In doing so they cared for the sick and needy, mediated disputes and never imposed anything on anyone.

How should we think about these chiefs? They were not patriarchs, Lévi-Strauss concluded; neither were they petty tyrants (even though for certain limited periods they were allowed to act as such); and there was no sense in which they were invested with mystical powers. More than anything, they resembled modern politicians operating tiny embryonic welfare states, pooling resources and doling them out to those in need. What impressed Lévi-Strauss above all was their political maturity. It was the chiefs' skill in directing small bands of dry-season foragers, of making snap decisions in crises (crossing a river, directing a hunt) that later qualified them to play the role of mediators and diplomats in the village plaza. But in doing so they were effectively moving back and forth, each year, between what evolutionary anthropologists (in the tradition of Turgot) insist on thinking of as totally different stages of social development: from hunters and foragers to farmers and back again.

It was precisely this quality that made the Nambikwara chief such a peculiarly familiar political figure: the calm sophistication with which he shifted between what were in effect two different social systems, all the while balancing a sense of personal ambition with the common good. These chiefs were in every sense self-conscious political actors. And it was their flexibility and adaptability that enabled them to take such a distanced perspective on whichever system obtained at any given time.

Although Lévi-Strauss went on to become the world's most renowned anthropologist and perhaps the most famous intellectual in France, his early essay on Nambikwara leadership fell into almost instant obscurity. To this day, very few outside the field of Amazonian studies have heard of it. One reason is that in the post-war decades, Lévi-Strauss was moving in exactly the opposite direction to the rest of his discipline. Where he emphasized similarities between the lives of hunters, horticulturalists and modern industrial democracies, almost everyone else – and particularly everyone interested in foraging societies – was embracing new variations on Turgot, though with updated language and backed up by a flood of hard scientific data. Throwing away old-fashioned distinctions between 'savagery', 'barbarism' and 'civilization', which were beginning to sound a little too condescending, they settled on a new sequence, which ran from 'bands' to 'tribes' to 'chiefdoms' to 'states'. The culmination of this trend was the landmark *Man the Hunter* symposium, held at the University of Chicago in 1966. This framed hunter-gatherer studies in terms of a new discipline which its attendees proposed to call 'behavioural ecology', starting with rigorously quantified studies of African savannah and rainforest groups – the Kalahari San, Eastern Hadza and Mbuti Pygmies – including calorie counts, time allocation studies and all sorts of data that simply hadn't been available to earlier researchers.

The new studies overlapped with a sudden upswing of popular interest in just these same African societies: for instance, the famous short films about the Kalahari Bushmen by the Marshalls (an American family of anthropologists and film-makers), which became fixtures of introductory anthropology courses and educational television across the world, along with best-selling books like Colin Turnbull's *The Forest People*. Before long, it was simply assumed by almost everyone that foragers represented a separate stage of social development, that they 'live in small groups', 'move around a lot', reject any social distinctions other than those of age and gender, and resolve conflicts by 'fission' rather than arbitration or violence.[35] The fact that these African societies were, in some cases at least, refugee populations living in places no one else wanted, or that many foraging societies documented in the ethnographic record (who had by this time been largely wiped out by European settler colonialism and were thus no longer available for quantitative analysis) were nothing like this, was occasionally acknowledged. But it was rarely treated as particularly relevant. The image of tiny egalitarian bands corresponded perfectly to what those weaned on the legacy of Rousseau felt hunter-gatherers *ought* to have been like. Now there seemed to be hard, quantifiable scientific data (and also movies!) to back it up.

In this new reality, Lévi-Strauss's Nambikwara were simply irrelevant. After all, in evolutionary terms they weren't even really foragers, since they only roamed about in foraging bands for seven or eight months a year. So the apparent paradox that their larger village settlements were egalitarian while their foraging bands were anything but could be ignored, lest it tarnish this crisp new picture. The kind of political self-consciousness which seemed so self-evident in Nambikwara chiefs, let alone the wild improvisation expected of Nuer prophets, had no place in the revised framework of human social evolution.

IN WHICH WE RETURN TO PREHISTORY, AND CONSIDER EVIDENCE FOR BOTH 'EXTREME INDIVIDUALS' AND SEASONAL VARIATIONS OF SOCIAL LIFE IN THE ICE AGE AND BEYOND

The twentieth-century Nambikwara, Winnebago or Nuer cannot provide us with direct windows on the past. What they can do is suggest angles of investigation we might not otherwise have thought to look for. After considering their social systems, it seems self-evident to ask if, in early human societies, there is evidence for seasonal variations of social structure; or if highly anomalous individuals were not only treated with respect, but played important political roles in the Palaeolithic period. As it turns out, the answer in both cases is 'yes'. In fact, the evidence is overwhelming.

Let's return to those rich Upper Palaeolithic burials, so often interpreted as evidence for the emergence of 'inequality', or even hereditary nobility of some sort. For some odd reason, those who make such arguments never seem to notice – or, if they do, to attach much significance to the fact – that a quite remarkable number of these skeletons (indeed, a majority) bear evidence of striking physical anomalies that could only have marked them out, clearly and dramatically, from their social surroundings.[36] The adolescent boys in both Sunghir and Dolní Věstonice, for instance, had pronounced congenital deformities; the bodies in the Romito Cave in Calabria were unusually short, with at least one case of dwarfism; while those in Grimaldi Cave were extremely tall even by our standards, and must have seemed veritable giants to their contemporaries.

All this seems very unlikely to be a coincidence. In fact, it makes one wonder whether even those bodies, which appear from their skeletal remains to be anatomically typical, might have been equally striking in some other way; after all, an albino, for example, or an epileptic prophet given to dividing his time between hanging upside down and arranging and rearranging snail shells would not be identifiable as such from the archaeological record. We can't know much about the day-to-day lives of Palaeolithic individuals buried with rich grave goods, other than that they seem to have been as well fed and cared for as anybody else; but we can at least suggest they were seen as the ultimate individuals, about as different from their peers as it was possible to be.

What does all this really tell us about social inequality in the last Ice Age? Well, first of all it suggests we might have to shelve any premature talk of the emergence of hereditary elites. It seems extremely unlikely that Palaeolithic Europe produced a stratified elite that just happened to consist largely of hunchbacks, giants and dwarfs. Second, we don't know how much the treatment of such individuals after death had to do with their treatment in life. Another important point here is that we are not dealing with a case of some people being buried with rich grave goods and others being buried with none. Rather it is a case of some people being buried with rich grave goods, and most others not being buried at all.[37] The very practice of burying bodies intact, and clothed, appears to have been exceptional in the Upper Palaeolithic. Most corpses were treated in completely different ways: de-fleshed, broken up, curated, or even processed into jewellery and artefacts. (In general, Palaeolithic people were clearly much more at home with human body parts than we are.)

The corpse in its complete and articulated form – and the clothed corpse even more so – was clearly something unusual and, one would presume, inherently strange. Some important circumstantial evidence reinforces this. In many such cases, an effort was made to contain the bodies of the Upper Palaeolithic dead by covering them with heavy objects: mammoth scapulae, wooden planks, stones or tight bindings.

Perhaps saturating them with clothing, weapons and ornaments was an extension of these concerns, celebrating but also containing something potentially dangerous. This too makes sense. The ethnographic record abounds with examples of anomalous beings – human or otherwise – treated as both exalted and dangerous; or one way in life, another in death.

Much here is speculation. There are any number of other interpretations that could be placed on the evidence – though the idea that these tombs mark the emergence of some sort of hereditary aristocracy seems the least likely of all. Those interred were extraordinary, 'extreme' individuals. The way they were treated – and here we are speaking not only about the ostentatious display of riches, but that their corpses were decorated, displayed and buried to begin with – marked them out as equally extraordinary in death. Anomalous in almost every respect, such burials can hardly be interpreted as proxies for social structure among the living. On the other hand, they clearly have something to do with all the contemporary evidence for music, sculpture, painting and complex architecture. What is one to make of them?

This is where seasonality comes into the picture.

Almost all the Ice Age sites with extraordinary burials and monumental architecture were created by societies that lived a little like Lévi-Strauss's Nambikwara, dispersing into foraging bands at one time of year, gathering together in concentrated settlements at another. True, they didn't gather to plant crops. Rather, the large Upper Palaeolithic sites are linked to migrations and seasonal hunting of game herds – woolly mammoth, steppe bison or reindeer – as well as cyclical fish-runs and nut harvests. This seems to be the explanation for those hubs of activity found in eastern Europe at places like Dolní Věstonice, where people took advantage of an abundance of wild resources to feast, engage in complex rituals and ambitious artistic projects, and trade minerals, marine shells and furs. In western Europe, equivalents would be the great rock shelters of the French Périgord and the Cantabrian coast, with their deep records of human activity, which similarly formed part of an annual round of seasonal congregation and dispersal.[38]

Archaeology also shows that patterns of seasonal variation lie behind the monuments of Göbekli Tepe. Activities around the stone temples correspond with periods of annual superabundance, between midsummer and autumn, when large herds of gazelle descended on to the Harran Plain. At such times, people also gathered at the site to process massive quantities of nuts and wild cereal grasses, making these into festive foods, which presumably fuelled the work of construction.[39] There is some evidence to suggest that each of these great structures had a relatively short lifespan, culminating in an enormous feast, after which its walls were rapidly filled in with leftovers and other refuse: hierarchies raised to the sky, only to be swiftly torn down again. Ongoing research is likely to complicate this picture, but the overall pattern of seasonal congregation for festive labour seems well established.

Such oscillating patterns of life endured long after the invention of agriculture. To take just one example, they may be key to understanding the famous Neolithic monuments of Salisbury Plain in England, and not just because the arrangements of standing stones themselves seem to function (among other things) as giant calendars. Stonehenge, framing the midsummer sunrise and the midwinter sunset, is the most famous of these. It turns out to have been the last in a long sequence of ceremonial structures, erected over the course of centuries in timber as well as stone, as people converged on the plain from remote corners of the British Isles at significant times of year. Careful excavation shows that many of these structures – now plausibly interpreted as monuments to the ancestors of a Neolithic aristocracy – were dismantled just a few generations after their construction.[40]

Still more striking, the people who built Stonehenge were not farmers, or not in the usual sense. They had once been; but the practice of erecting and dismantling grand monuments coincides with a period when the peoples of Britain, having adopted the Neolithic farming economy from continental Europe, appear to have turned their backs on at least one crucial aspect of it: abandoning the cultivation of cereals and returning, from around 3300 BC, to the collection of hazelnuts as their staple source of plant food. On the other hand, they kept hold of their domestic pigs and herds of cattle, feasting on them seasonally at nearby Durrington Walls, a prosperous town of some thousands of people – with its own Woodhenge – in winter, but largely empty and abandoned in summer. The builders of Stonehenge seem to have been neither foragers nor herders, but something in between.[41]

All this is crucial because it's hard to imagine how giving up agriculture could have been anything but a self-conscious decision. There is no evidence that one population displaced another, or that farmers were somehow overwhelmed by powerful foragers who forced them to abandon their crops. The Neolithic inhabitants of England appear to have taken the measure of cereal-farming and collectively decided that they preferred to live another way. How could such a decision have been made? We'll never know, but Stonehenge itself provides something of a hint since it is built of extremely large stones, some of which (the 'bluestones') were transported from as far away as Wales, while many of the cattle and pigs consumed at Durrington Walls were laboriously herded there from other distant locations.[42]

In other words, and remarkable as it may seem, even in the third millennium BC co-ordination of some sort was clearly possible across large parts of the British Isles. If Stonehenge was a shrine to exalted founders of a ruling clan – as some archaeologists now argue – it seems likely that members of their lineage claimed significant, even cosmic roles by virtue of their involvement in such events. On the other hand, patterns of seasonal aggregation and dispersal raise another question: if there were kings and queens at Stonehenge, exactly what sort could they have been? After all, these would have been kings whose courts and kingdoms existed for only a few months of the year, and otherwise dispersed into small communities of nut gatherers and stock herders. If they possessed the means to marshal labour, pile up food resources and provender armies of year-round retainers, what sort of royalty would consciously elect *not* to do so?

CONCERNING 'BUFFALO POLICE' (IN WHICH WE REDISCOVER THE ROLE OF SEASONALITY IN HUMAN SOCIAL AND POLITICAL LIFE)

Recall that for Lévi-Strauss, there was a clear link between seasonal variations of social structure and a certain kind of political freedom. The fact that one structure applied in the rainy season and another in the dry allowed Nambikwara chiefs to view their own social arrangements at one remove: to see them as not simply 'given', in the natural order of things, but as something at least partially open to human intervention. The case of the British Neolithic – with its alternating phases of dispersal and monumental construction – indicates just how far such intervention could sometimes go.

Writing in the midst of the Second World War, Lévi-Strauss probably didn't think he was saying anything all that extraordinary. For anthropologists in the first half of the twentieth century, it was common knowledge that societies doing a great deal of hunting, herding or foraging were often arranged in such a 'double morphology' (as Lévi-Strauss's great predecessor Marcel Mauss put it).[43] Lévi-Strauss was simply highlighting some of the political implications. But these implications are important. What the existence of similar seasonal patterns in the Palaeolithic suggests is that from the very beginning, or at least as far back as we can trace such things, human beings were self-consciously experimenting with different social possibilities. It might be useful here to look back at this forgotten anthropological literature, with which Lévi-Strauss would have been intimately familiar, to get a sense of just how dramatic these seasonal differences might be.

The key text here is Marcel Mauss and Henri Beuchat's (1903) 'Seasonal Variations of the Eskimo'. The authors begin by observing that the circumpolar Inuit 'and likewise many other societies . . . have two social structures, one in summer and one in winter, and that in parallel they have two systems of law and religion'. In the summer, Inuit dispersed into bands of roughly twenty or thirty people to pursue freshwater fish, caribou and reindeer, all under the authority of a single male elder. During this period, property was possessively marked and patriarchs exercised coercive, sometimes even tyrannical power over their kin – much more so than the Nambikwara chiefs in the dry season. But in the long winter months, when seals and walrus flocked to the Arctic shore, there was a dramatic reversal. Then, Inuit gathered together to build great meeting houses of wood, whale rib and stone; within these houses, virtues of equality, altruism and collective life prevailed. Wealth was shared, and husbands and wives exchanged partners under the aegis of Sedna, the Goddess of the Sea.[44]

Mauss thought the Inuit were an ideal case study because, living in the Arctic, they were facing some of the most extreme environmental constraints it was possible to endure. Yet even in sub-Arctic conditions, Mauss calculated, physical considerations – availability of game, building materials and the like – explained at best 40 per cent of the picture. (Other circumpolar peoples, he noted, including close neighbours of the Inuit facing near-identical physical conditions, organized themselves quite differently.) To a large extent, he concluded, Inuit lived the way they did because they felt that's how humans ought to live.

Around the same time that Marcel Mauss was combing French libraries for everything that had ever been written about the Inuit, the German ethnologist Franz Boas was carrying out research on the Kwakiutl, indigenous hunter-gatherers of Canada's Northwest Coast. Here, Boas discovered, it was winter – not summer – that was the time when society crystallized into its most hierarchical forms, and spectacularly so. Plank-built palaces sprang to life along the coastline of British Columbia, with hereditary nobles holding court over compatriots classified as commoners and slaves, and hosting the great banquets known as *potlatch*. Yet these aristocratic courts broke apart for the summer work of the fishing season, reverting to smaller clan formations – still ranked, but with entirely different and much less formal structures. In this case, people actually adopted different names in summer and winter – literally becoming someone else, depending on the time of year.[45]

Emigrating to the US, Boas went on to become a professor at New York's Columbia University, where he ended up training virtually everyone who was to make a name for themselves in American anthropology for the next half-century. One of his students, a Viennese-born ethnographer named Robert Lowie (who was also a close friend of Paul Radin, author of *Primitive Man as Philosopher*) did fieldwork among the Mandan-Hidatsa and Crow people of what are now Montana and Wyoming, and spent much of his career thinking through the political implications of seasonal variation among nineteenth-century tribal confederacies on the Great Plains.

Plains nations were one-time farmers who had largely abandoned cereal agriculture, after re-domesticating escaped Spanish horses and adopting a largely nomadic mode of life. In late summer and early autumn, small and highly mobile bands of Cheyenne and Lakota would congregate in large settlements to make logistical preparations for the buffalo hunt. At this most sensitive time of year they appointed a police force that exercised full coercive powers, including the right to imprison, whip or fine any offender who endangered the proceedings. Yet, as Lowie observed, this 'unequivocal authoritarianism' operated on a strictly seasonal and temporary basis. Once the hunting season – and the collective Sun Dance rituals that followed – were complete, such authoritarianism gave way to what he called 'anarchic' forms of organization, society splitting once again into small, mobile bands. Lowie's observations are startling:

> In order to ensure a maximum kill, a police force – either coinciding with a military club, or appointed *ad hoc*, or serving by virtue of clan affiliation – issued orders and restrained the disobedient. In most of the tribes they not only confiscated game clandestinely procured, but whipped the offender, destroyed his property, and, in case of resistance, killed him. The very same organisation which in a murder case would merely use moral suasion turned into an inexorable State agency during a buffalo drive. However . . . coercive measures extended considerably beyond the hunt: the soldiers also forcibly restrained braves intent on starting war parties that were deemed inopportune by the chief; directed mass migrations; supervised the crowds at a major festival; and might otherwise maintain law and order.[46]

'During a large part of the year,' Lowie continued, 'the tribe simply did not exist as such; and the families or minor unions of familiars that jointly sought a living required no special disciplinary organization. The soldiers were thus a concomitant of numerically strong aggregations, hence functioned intermittently rather than continuously.' But the soldiers' sovereignty, he stressed, was no less real for its temporary nature. As a result, Lowie insisted that Plains Indians did in fact know something of state power, even though they never actually developed a state.

It's easy to see why the neo-evolutionists of the 1950s and 1960s might not have known quite what to do with this legacy of fieldwork observations. They were arguing for the existence of discrete stages of political organization – successively: bands, tribes, chiefdoms, states – and held that the stages of political development mapped, at least very roughly, on to similar stages of economic development: hunter-gatherers, gardeners, farmers, industrial civilization. It was confusing enough that people like the Nambikwara seemed to jump back and forth, over the course of the year, between economic categories. The Cheyenne, Crow, Assiniboine or Lakota would appear to jump regularly from one end of the political spectrum to the other. They were a kind of band/state amalgam. In other words, they threw everything askew.

Still, Lowie is absolutely unequivocal on this point, and he was by no means the only anthropologist to observe it.[47] Most interestingly for our own perspective, he too stressed that the Plains Indians were conscious political actors, keenly aware of the possibilities and dangers of authoritarian power. Not only did they dismantle all means of exercising coercive authority the moment the ritual season was over, they were also careful to rotate which clan or warrior clubs got to wield it: anyone holding sovereignty one year would be subject to the authority of others in the next.[48]

Scholarship does not always advance. Sometimes it slips backwards. A hundred years ago, most social scientists understood that those who live mainly from wild resources were not normally restricted to tiny 'bands'. As we've seen, the assumption that they were only gained ground in the 1960s. In this regard, our earlier invocation of biker gangs and hippie communes wasn't entirely whimsical. These were the images being bounced around in the popular imagination at that time, and invoked in debates about human nature. It's surely no coincidence that the most popular ethnographic films of the post-war era either focused on the Kalahari Bushmen and Mbuti Pygmies ('band' societies, which could be imagined as roughly resembling hippie communes); or on the Yanomami or 'fierce people' (Amazonian horticulturalists who, in Napoleon Chagnon's version of reality – but not, let's recall, in Helena Valero's – do bear a rather disturbing resemblance to Hell's Angels).

Since in this new, evolutionist narrative 'states' were defined above all by their monopoly on the 'legitimate use of coercive force', the nineteenth-century Cheyenne or Lakota would have been seen as evolving from the 'band' level to the 'state' level roughly every November, and then devolving back again come spring. Obviously, this is silly. No one would seriously suggest such a thing. Still, it's worth pointing out because it exposes the much deeper silliness of the initial assumption: that societies must necessarily progress through a series of evolutionary stages to begin with. You can't speak of an evolution from band to tribe to chiefdom to state if your starting points are groups that move fluidly between them as a matter of habit.

Seasonal dualism also throws into chaos more recent efforts at classifying hunter-gatherers into either 'simple' or 'complex' types, since what have been identified as the diagnostic features of 'complexity' – territoriality, social ranks, material wealth or competitive display – appear during certain seasons of the year, only to be brushed aside in others by the exact same population. Admittedly, most professional anthropologists nowadays have come to recognize that these categories are hopelessly inadequate, but the main effect of this acknowledgement has just been to cause them to change the subject, or suggest that perhaps we shouldn't really be thinking about the broad sweep of human history at all any more. Nobody has yet proposed an alternative.

Meanwhile, as we've seen, archaeological evidence is piling up to suggest that in the highly seasonal environments of the last Ice Age, our remote ancestors were behaving much like the Inuit, Nambikwara or Crow. They shifted back and forth between alternative social arrangements, building monuments and then closing them down again, allowing the rise of authoritarian structures during certain times of year then dismantling them – all, it would seem, on the understanding that no particular social order was ever fixed or immutable. The same individual could experience life in what looks to us sometimes like a band, sometimes a tribe, and sometimes like something with at least some of the characteristics we now identify with states.

With such institutional flexibility comes the capacity to step outside the boundaries of any given structure and reflect; to both make and unmake the political worlds we live in. If nothing else, this explains the 'princes' and 'princesses' of the last Ice Age, who appear to show up, in such magnificent isolation, like characters in some kind of fairy tale or costume drama. Maybe they were almost literally so. If they reigned at all, then perhaps it was, like the ruling clans of Stonehenge, just for a season.[49]

WHY THE REAL QUESTION IS NOT 'WHAT ARE THE ORIGINS OF SOCIAL INEQUALITY?' BUT 'HOW DID WE GET STUCK?'

If we are right, and if human beings really have spent most of the last 40,000 or so years moving back and forth between different forms of social organization, building up hierarchies then dismantling them again, the implications are profound. For one thing, it suggests that Pierre Clastres was quite right when he proposed that, rather than being less politically self-conscious than people nowadays, people in stateless societies might actually have been considerably more so.

Clastres was another product of the 1960s. A student of Lévi-Strauss, he took to heart his master's view of Amazonian chiefs as mature political actors. But Clastres was also an anarchist (he was ultimately kicked out of Lévi-Strauss's research group on a flimsy pretext, involving unauthorized use of official stationery), and he took the argument much further. It wasn't just that Amazonian chiefs were calculating politicians. They were calculating politicians forced to manoeuvre in a social environment apparently designed to ensure they could never exercise real political power. In the winter, the groups they led were tiny and inconsequential. In the summer, they didn't 'lead' at all. Yes, their houses might have resembled social service dispensaries in modern welfare states; but as a result, in terms of material wealth, they were actually the poorest men in the village, since chiefs were expected constantly to give everything away. They were also expected to set an example by working much harder than everybody else. Even where they did have special privileges, like the Tupi or Nambikwara chiefs, who were the only men in their villages allowed to have multiple wives, the privilege was distinctly double-edged. The wives were held to be necessary to prepare feasts for the village. If any of those wives looked to other lovers, which it appears they ordinarily did, there was nothing much the chief could do about it, since he had to keep himself in everyone's good graces to remain chief.

Chiefs found themselves in this situation, Clastres argued, because they weren't the only ones who were mature and insightful political actors; almost everyone was. Rather than being trapped in some sort of Rousseauian innocence, unable to imagine more complex forms of organization, people were generally more capable than we are of imagining alternative social orders, and therefore had created 'societies against the state'. They had self-consciously organized in such a way that the forms of arbitrary power and domination we associate with 'advanced political systems' could never possibly emerge.

Clastres's argument was, as one might imagine, highly controversial. Some of the criticism directed at him was entirely justified (he had, for example, an enormous blind spot when it came to gender). Still, most of it was based on firm Rousseauian ground, insisting Clastres was ascribing too much imagination to 'primitive' or 'archaic' people who, almost by definition, shouldn't have any. How could one possibly claim, so such criticism went, that stateless societies were self-consciously organizing themselves to prevent the emergence of something they'd never actually experienced?

There are a lot of possible ways in which to respond to this objection. Were Amazonians of centuries past, for instance, entirely unaware of the great Andean empires to their west? People used to get around. It's unlikely they simply had no idea of developments in neighbouring parts of the continent. As we'll see in Chapter Seven, there is also now a good deal of evidence for the existence of large polities in Amazonia itself, in much earlier times. Perhaps these were the children of rebels who fled or even overthrew such ancient kingdoms. But the most obvious objection is that, if the Amazonians in question were anything like the Nambikwara, they actually *did* experience relations of arbitrary command during their yearly 'adventures' as foraging bands. Yet, oddly, Clastres himself never pointed this out. In fact, he never really talks about seasonality at all.

This is a curious omission. It's also an important one because, by leaving it out, Clastres really put the final nail in the coffin of that earlier tradition running from Marcel Mauss through to Robert Lowie; a tradition which treated 'primitive' societies as inherently flexible, and typically characterized by multiple forms of organization. Now, both the neo-evolutionists who saw 'primitive' folk as

Rousseauian *naïfs* and the radicals who insisted they were self-conscious egalitarians equally took it for granted they were stuck in a single, very simple mode of social existence.

In Clastres's case it's all the more surprising, because in his original statement on the powerlessness of Indian chiefs, published in 1962, he is quite candid in admitting he pinched almost his entire argument from Lowie. Fourteen years earlier, Lowie had argued that most indigenous American societies, from Montreal to Tierra del Fuego, were effectively anarchists.[50] His argument that the 'typical Indian chief is not a lawgiver, executive, or judge, but a pacifier, a benefactor of the poor, and a prolix Polonius' (that is, the actual functions of chiefly office are to (1) mediate quarrels, (2) provide for the needy, and (3) to entertain with beautiful speeches) is precisely echoed, point by point, in Clastres's account. So is Lowie's conclusion that, since the chiefly office is effectively designed so it can never be turned into a means of compulsion, the only way state-like authority could possibly have emerged was from religious visionaries of one sort or another.

Recall, though, that Lowie's original piece included one additional section, on the 'evolutionary germs' of top-down authority, which describes the seasonal 'police' and 'soldiers' of the Plains societies in detail. Clastres simply left it out. Why?

The answer is probably a simple one: seasonality was confusing. In fact, it's kind of a wild card. The societies of the Great Plains created structures of coercive authority that lasted throughout the entire season of hunting and the rituals that followed, dissolving when they dispersed into smaller groups. But those of central Brazil dispersed into foraging bands as a way of asserting a political authority that was ineffectual in village settings. Among the Inuit, fathers ruled in the summertime; but in winter gatherings patriarchal authority and even norms of sexual propriety were challenged, subverted or simply melted away. The Kwakiutl were hierarchical at both times of year, but nonetheless maintained different forms of hierarchy, giving effective police powers to performers in the Midwinter Ceremonial (the 'bear dancers' and 'fool dancers') that could be exercised only during the actual performance of the ritual. At other times, aristocrats commanded great wealth but couldn't give their followers direct orders. Many Central African forager societies are egalitarian all year round, but appear to alternate monthly between a ritual order dominated by men and another dominated by women.[51]

In other words, there is no single pattern. The only consistent phenomenon is the very fact of alteration, and the consequent awareness of different social possibilities. What all this confirms is that searching for 'the origins of social inequality' really is asking the wrong question.

If human beings, through most of our history, have moved back and forth fluidly between different social arrangements, assembling and dismantling hierarchies on a regular basis, maybe the real question should be 'how did we get stuck?' How did we end up in one single mode? How did we lose that political self-consciousness, once so typical of our species? How did we come to treat eminence and subservience not as temporary expedients, or even the pomp and circumstance of some kind of grand seasonal theatre, but as inescapable elements of the human condition? If we started out just playing games, at what point did we forget that we were playing?

We'll be tackling such questions in the chapters to come. For the moment, the main thing to stress is that this flexibility, and potential for political self-consciousness, was never entirely lost. Mauss pointed out much the same thing. Seasonality is still with us – even if it is a pale, contracted shadow of its former self. In the Christian world, for instance, there is still the midwinter 'holiday season' in which values and forms of organization do, to a limited degree, reverse themselves: the same media and advertisers who for most of the year peddle rabid consumerist individualism suddenly start announcing that social relations are what's really important, and that to give is better than to receive. (And in enlightened countries like Mauss's France, there's also the summer *grandes vacances* in which everybody downs tools for a month and flees the cities.)

There is a direct historical connection here. We've already seen how, among societies like the Inuit or Kwakiutl, times of seasonal congregation were also ritual seasons, almost entirely given over to dances, rites and dramas. Sometimes these could involve creating temporary kings or even ritual police with real coercive powers (though often, peculiarly, these ritual police doubled as clowns).[52] In other cases, they involved dissolving norms of hierarchy and propriety, as in the Inuit midwinter orgies. This dichotomy can still be observed in festive life almost everywhere. In the European Middle Ages, to take a familiar example, saints' days alternated between solemn pageants where all the elaborate ranks and hierarchies of feudal life were made manifest (much as they still are in, say, a college graduation ceremony, when we temporarily revert to medieval garb), and crazy carnivals in which everyone played at 'turning the world upside down'. In carnival, women might rule over men, children be put in charge of government, servants could demand work from their masters, ancestors could return from the dead, 'carnival kings' could be crowned and then dethroned, giant monuments like wicker dragons built and set on fire, or all formal ranks might even disintegrate into one or other form of Bacchanalian chaos.[53]

Just as with seasonality, there's no consistent pattern. Ritual occasions can either be much more stiff and formal, or much more wild and playful, than ordinary life. Alternatively, like funerals and wakes, they can slip back and forth between the two. The same seems to be true of festive life almost everywhere, whether it's Peru, Benin or China. This is why anthropologists often have such trouble defining what a 'ritual' even is. If you start from the solemn ones, ritual is a matter of etiquette, propriety: High Church ritual, for example, is really just a very elaborate version of table manners. Some have gone so far as to argue that what we call 'social structure' only really exists during rituals: think here of families that only exist as a physical group during marriages and funerals, during which times questions of rank and priority have to be worked out by who sits at which table, who speaks first, who gets the topmost cut of the hump of a sacrificed water buffalo, or the first slice of wedding cake.

But sometimes festivals are moments where entirely different social structures take over, such as the 'youth abbeys' that seem to have existed across medieval Europe, with their Boy Bishops, May Queens, Lords of Misrule, Abbots of Unreason and Princes of Sots, who during the Christmas, Mayday or carnival season temporarily took over many of the functions of government and enacted a bawdy parody of government's everyday forms. So there's another school of thought which says that rituals are really exactly the opposite. The really powerful ritual moments are those of collective chaos, effervescence, liminality or creative play, out of which new social forms can come into the world.[54]

There is also a centuries-long, and frankly not very enlightening, debate over whether the most apparently subversive popular festivals were really as subversive as they seem; or if they are really conservative, allowing common folk a chance to blow off a little steam and give vent to their baser instincts before returning to everyday habits of obedience.[55] It strikes us that all this rather misses the point.

What's really important about such festivals is that they kept the old spark of political self-consciousness alive. They allowed people to imagine that other arrangements are feasible, even for society as a whole, since it was always possible to fantasize about carnival bursting its seams and becoming the new reality. In the popular Babylonian story of Semiramis, the eponymous servant girl convinces the Assyrian king to let her be 'Queen for a Day' during some annual festival, promptly has him arrested, declares herself empress and leads her new armies to conquer the world. May Day came to be chosen as the date for the international workers' holiday largely because so many British peasant revolts had historically begun on that riotous festival. Villagers who played at 'turning the world upside' would periodically decide they actually preferred the world upside down, and took measures to keep it that way.

Medieval peasants often found it much easier than medieval intellectuals to imagine a society of equals. Now, perhaps, we begin to understand why. Seasonal festivals may be a pale echo of older patterns of seasonal variation – but, for the last few thousand years of human history at least, they appear to have played much the same role in fostering political self-consciousness, and as laboratories of social possibility. The first kings may well have been play kings. Then they became real kings. Now most (but not all) existing kings have been reduced once again to play kings – as least insofar as they mainly perform ceremonial functions and no longer wield real power. But even if all monarchies, including ceremonial monarchies, were to disappear, some people would still play at being kings.

Even in the European Middle Ages, in places where monarchy was unquestioned as a mode of government, 'Abbots of Unreason', Yuletide Kings and the like tended to be chosen either by election or by sortition (lottery), the very forms of collective decision-making that resurfaced, apparently out of nowhere, in the Enlightenment. (What's more, such figures tended to exercise power much in the manner of indigenous American chiefs: either limited to very circumscribed contexts, like the war chiefs who could give orders only during military expeditions; or like village chiefs who were arrayed with formal honours but couldn't tell anybody what to do.) For a great many societies, the festive year could be read as a veritable encyclopaedia of possible political forms.

WHAT BEING SAPIENS REALLY MEANS

Let us end this chapter where we began it. For far too long we have been generating myths. As a result, we've been mostly asking the wrong questions: are festive rituals expressions of authority, or vehicles for social creativity? Are they reactionary or progressive? Were our earliest ancestors simple and egalitarian, or complex and stratified? Is human nature innocent or corrupt? Are we, as a species, inherently co-operative or competitive, kind or selfish, good or evil?

Perhaps all these questions blind us to what really makes us human in the first place, which is our capacity – as moral and social beings – to negotiate between such alternatives. As we've already observed, it makes no sense to ask any such questions of a fish or a hedgehog. Animals already exist in a state 'beyond good and evil', the very one that Nietzsche dreamed humans might also aspire to. Perhaps we are doomed always to be arguing about such things. But certainly, it is more interesting to start asking other questions as well. If nothing else, surely the time has come to stop the swinging pendulum that has fixated generations of philosophers, historians and social scientists, leading their gaze from Hobbes to Rousseau, from Rousseau to Hobbes and back again. We do not have to choose any more between an egalitarian or hierarchical start to the human story. Let us bid farewell to the 'childhood of Man' and acknowledge (as Lévi-Strauss insisted) that our early ancestors were not just our cognitive equals, but our intellectual peers too. Likely as not, they grappled with the paradoxes of social order and creativity just as much as we do; and understood them – at least the most reflexive among them – just as much, which also means just as little. They were perhaps more aware of some things and less aware of others. They were neither ignorant savages nor wise sons and daughters of nature. They were, as Helena Valero said of the Yanomami, just people, like us; equally perceptive, equally confused.

Be this as it may, it's becoming increasing clear that the earliest known evidence of human social life resembles a carnival parade of political forms, far more than it does the drab abstractions of evolutionary theory. If there is a riddle here it's this: why, after millennia of constructing and disassembling forms of hierarchy, did *Homo sapiens* – supposedly the wisest of apes – allow permanent and intractable systems of inequality to take root? Was this really a consequence of adopting agriculture? Of settling down in permanent villages and, later, towns? Should we be looking for a moment in time like the one Rousseau envisaged, when somebody first enclosed a tract of land, declaring: 'This is mine and always will be!' Or is that another fool's errand?

These are the questions to which we now turn.

4
Free People, the Origin of Cultures, and the Advent of Private Property

(Not necessarily in that order)

Changing your social identity with the changing seasons might sound like a wonderful idea, but it's not something anyone reading this book is ever likely to experience first-hand. Yet until very recently, the European continent was still littered with folk practices that echoed these ancient rhythmic oscillations of social structure. Folklorists have long puzzled over all the little brigades of people disguised as plants and animals, the Straw Bears and Green Men, who marched dutifully out each spring and autumn into village squares, everywhere from rural England to the Rhodope Mountains of southern Bulgaria: were they genuine traces of ancient practices, or recent revivals and reinventions? Or revivals of traces? Or traces of revivals? It's often impossible to tell.

Most of these rituals have been gradually brushed aside as pagan superstition or repackaged as tourist attractions (or both). For the most part, all we're left with as an alternative to our mundane lives are our 'national holidays': frantic periods of over-consumption, crammed in the gaps between work, in which we entertain solemn injunctions that consumption isn't really what matters about life. As we've seen, our remote forager ancestors were much bolder experimenters in social form, breaking apart and reassembling their societies at different scales, often in radically different forms, with different value systems, from one time of year to the next. The festive calendars of the great agrarian civilizations of Eurasia, Africa and the Americas turn out to be mere distant echoes of that world and the political freedoms it entailed.

Still, we could never have figured that out by material evidence alone. If all we had to go on were Palaeolithic 'mammoth buildings' on the Russian steppe, or the princely burials of the Ligurian Ice Age and their associated physical remains, scholars would no doubt be left scratching their heads until the sun explodes. Human beings may be (indeed, we've argued they are) fundamentally imaginative creatures, but no one is *that* imaginative. You would have to be either extremely naive or extremely arrogant to think anybody could simply logic such matters out. (And even if someone did manage to come up with anything like Nuer prophets, Kwakiutl clown-police or Inuit seasonal wife-swapping orgies, simply through logical extrapolation they'd probably be instantly written off as kooks.)

This is precisely why the ethnographic record is so important. The Nuer and Inuit should never have been seen as 'windows on to our ancestral past'. They are creations of the modern age just the same as we are – but they do show us possibilities we never would have thought of and prove that people are actually capable of enacting such possibilities, even building whole social systems and value systems around them. In short, they remind us that human beings are far more interesting than (other) human beings are sometimes inclined to imagine.

In this chapter, we'll do two things. First, we'll continue our story forwards in time from the Palaeolithic, looking at some of the extraordinary cultural arrangements that emerged across the world before our ancestors turned their hands to farming. Second, we'll start answering the question we posed in the last chapter: how did we get stuck? How did some human societies begin to move away from the flexible, shifting arrangements that appear to have characterized our earliest ancestors, in such a way that certain individuals or groups were able to claim permanent power over others: men over women; elders over youth; and eventually, priestly castes, warrior aristocracies and rulers who actually ruled?

IN WHICH WE DESCRIBE HOW THE OVERALL COURSE OF HUMAN HISTORY HAS MEANT THAT MOST PEOPLE LIVE THEIR LIVES ON AN EVER-SMALLER SCALE AS POPULATIONS GET LARGER

In order for these things to become possible, a number of other factors first had to fall into place. One is the very existence of what we would intuitively recognize as discrete 'societies' to begin with. It may not even make sense to describe the mammoth hunters of Upper Palaeolithic Europe as being organized into separate, bounded societies, in the way we talk about the nations of Europe, or for that matter First Nations of Canada like the Mohawk, Wendat or Montagnais-Naskapi.

Of course, we know almost nothing about the languages people were speaking in the Upper Palaeolithic, their myths, initiation rituals, or conceptions of the soul; but we do know that, from the Swiss Alps to Outer Mongolia, they were often using remarkably similar tools,[1] playing remarkably similar musical instruments, carving similar female figurines, wearing similar ornaments and conducting similar funeral rites. What's more, there is reason to believe that at certain points in their lives, individual men and women often travelled very long distances.[2] Surprisingly, current studies of hunter-gatherers suggest that this is almost exactly what one should expect.

Research among groups such as the East African Hadza or Australian Martu shows that while forager societies today may be numerically small, their composition is remarkably cosmopolitan. When forager bands gather into larger residential groups these are not, in any sense, made up of a tight-knit unit of closely related kin; in fact, primarily biological relations constitute on average a mere 10 per cent of total membership. Most members are drawn from a much wider pool of individuals, many from quite far away, who may not even speak the same first languages.[3] This is true even for contemporary groups that are effectively encapsulated in restricted territories, surrounded by farmers and pastoralists.

In earlier centuries, forms of regional organization might extend thousands of miles. Aboriginal Australians, for instance, could travel halfway across the continent, moving among people who spoke entirely different languages, and still find camps divided into the same kinds of totemic moieties that existed at home. What this means is that half the residents owed them hospitality, but had to be treated as 'brothers' and 'sisters' (so sexual relations were strictly prohibited); while another half were both potential enemies and marriage partners. Similarly, a North American 500 years ago could travel from the shores of the Great Lakes to the Louisiana bayous and still find settlements – speaking languages entirely unrelated to their own – with members of their own Bear, Elk or Beaver clans who were obliged to host and feed them.[4]

It's difficult enough to reconstruct how these forms of long-distance organization operated just a few centuries ago, before they were destroyed by the coming of European settlers. So we can really only guess how analogous systems might have worked some 40,000 years ago. But the striking material uniformities observed by archaeologists across very long distances attest to the existence of such systems. 'Society', insofar as we can comprehend it at that time, spanned continents.

Much of this seems counter-intuitive. We are used to assuming that advances in technology are continually making the world a smaller place. In a purely physical sense, of course, this is true: the domestication of the horse, and gradual improvements in seafaring, to take just two examples, certainly made it much easier for people to move around. But at the same time, increases in the sheer number of human beings seem to have pulled in the opposite direction, ensuring that, for much of human history, ever-diminishing proportions of people actually travelled – at least, over long distances or very far from home. If we survey what happens over time, the scale on which social relations operate doesn't get bigger and bigger; it actually gets smaller and smaller.

A cosmopolitan Upper Palaeolithic is followed by a complicated period of several thousand years, beginning around 12,000 BC, in which it first becomes possible to trace the outlines of separate 'cultures' based on more than just stone tools. Some foragers, after this time, continued following large mammal herds; others settled on the coast and became fisherfolk, or gathered acorns in forests. Prehistorians use the term 'Mesolithic' for these postglacial populations. Across large parts of Africa and East Asia, their technological innovations – including pottery, 'micro-lithic' tool kits and stone grinding tools – signal new ways of preparing and eating wild grains, roots and other vegetables: chopping, slicing, grating, grinding, soaking, draining, boiling, and also ways of storing, smoking and otherwise preserving meats, plant foods and fish.[5]

Before long these had spread everywhere, and paved the way for the creation of what we'd now call cuisine: the kind of soups, porridges, stews, broths and fermented beverages we're familiar with today. But cuisines are also, almost everywhere, markers of difference. People who wake up to fish stews every morning tend to see themselves as a different sort of people from those who breakfast on a porridge of berries and wild oats. Such distinctions were no doubt echoed by parallel developments that are much more difficult to reconstruct: different tastes in clothing, dancing, drugs, hairstyles, courtship rituals; different forms of kinship organization and styles of formal rhetoric. The 'culture areas' of these Mesolithic foragers were still extremely large. True, the Neolithic versions that soon developed alongside them – associated with the first farming populations – were typically smaller; but for the most part they still spread out over territories considerably larger than most modern nation states.

Only much later do we begin to encounter the kind of situation familiar to anthropologists of Amazonia or Papua New Guinea, where a single river valley might contain speakers of half a dozen different languages, with entirely distinct economic systems or cosmological beliefs. Sometimes, of course, this tendency towards micro-differentiation was reversed – as with the spread of imperial languages like English or Han Chinese. But the overall direction of history – at least until very recently – would seem to be the very opposite of globalization. It is one of increasingly local allegiances: extraordinary cultural inventiveness, but much of it aimed at finding new ways for people to set themselves off against each other. True, the larger regional networks of hospitality endured in some places.[6] Overall, though, what we observe is not so much the world as a whole getting smaller, but most peoples' social worlds growing more parochial, their lives and passions more likely to be circumscribed by boundaries of culture, class and language.

We might ask why all this has happened. What are the mechanisms that cause human beings to spend so much effort trying to demonstrate that they are different from their neighbours? This is an important question. We shall be considering it in much more detail in the following chapter.

For the moment, we simply note that the proliferation of separate social and cultural universes – confined in space and relatively bounded – must have contributed in various ways to the emergence of more durable and intransigent forms of domination. The mixed composition of so many foraging societies clearly indicates that individuals were routinely on the move for a plethora of reasons, including taking the first available exit route if one's personal freedoms were threatened at home. Cultural porosity is also necessary for the kind of seasonal demographic pulses that made it possible for societies to alternate periodically between different political arrangements, forming massive congregations at one time of year, then dispersing into a multitude of smaller units for the remainder.

That is one reason why the majestic theatre of Palaeolithic 'princely' burials – or even of Stonehenge – never seems to have gone too far beyond theatrics. Simply put, it's difficult to exercise arbitrary power in, say, January over someone you will be facing on equal terms again come July. The hardening and multiplication of cultural boundaries can only have reduced such possibilities.

IN WHICH WE ASK WHAT, PRECISELY, IS EQUALIZED IN 'EGALITARIAN' SOCIETIES?

The emergence of local cultural worlds during the Mesolithic made it more likely that a relatively self-contained society might abandon seasonal dispersal and settle into some kind of full-time, top-down, hierarchical arrangement. In our terms, to get stuck. But of course, this in itself hardly explains why any particular society did, in fact, get stuck in such arrangements. We are back to something not entirely different from the 'origins of social inequality' problem – but by now, we can at least focus a little more sharply on what the problem really is.

As we have repeatedly observed, 'inequality' is a slippery term, so slippery, in fact, that it's not entirely clear what the term 'egalitarian society' should even mean. Usually, it's defined negatively: as the absence of hierarchies (the belief that certain people or types of people are superior to others), or as the absence of relations of domination or exploitation. This is already quite complex, and the moment we try to define egalitarianism in positive terms everything becomes much more so.

On the one hand, 'egalitarianism' (as opposed to 'equality', let alone 'uniformity' or 'homogeneity') seems to refer to the presence of some kind of ideal. It's not just that an outside observer would tend to see all members of, say, a Semang hunting party as pretty much interchangeable, like the cannon-fodder minions of some alien overlord in a science fiction movie (this would, in fact, be rather offensive); but rather, that Semang themselves feel they ought to be the same – not in every way, since that would be ridiculous, but in the ways that really matter. It also implies that this ideal is, largely, realized. So, as a first approximation, we can speak of an egalitarian society if (1) most people in a given society feel they really ought to be the same in some specific way, or ways, that are agreed to be particularly important; and (2) that ideal can be said to be largely achieved in practice.

Another way to put this might be as follows. If all societies are organized around certain key values (wealth, piety, beauty, freedom, knowledge, warrior prowess), then 'egalitarian societies' are those where everyone (or almost everyone) agrees that the paramount values should be, and generally speaking are, distributed equally. If wealth is what's considered the most important thing in life, then everyone is more or less equally wealthy. If learning is most valued, then everyone has equal access to knowledge. If what's most important is one's relationship with the gods, then a society is egalitarian if there are no priests and everyone has equal access to places of worship.

You may have noticed an obvious problem here. Different societies sometimes have radically different systems of value, and what might be most important in one – or at least, what everyone insists is most important in one – might have very little to do with what's important in another. Imagine a society in which everyone is equal before the gods, but 50 per cent of the population are sharecroppers with no property and therefore no legal or political rights. Does it really make sense to call this an 'egalitarian society' – even if everyone, including the sharecroppers, insists that it's really only one's relation to the gods that is ultimately important?

There's only one way out of this dilemma: to create some sort of universal, objective standards by which to measure equality. Since the time of Jean-Jacques Rousseau and Adam Smith, this has almost invariably meant focusing on property arrangements. As we've seen, it was only at this point, in the mid to late eighteenth century, that European philosophers first came up with the idea of ranking human societies according to their means of subsistence, and therefore that hunter-gatherers should be treated as a distinct variety of human being. As we've also seen, this idea is very much still with us. But so is Rousseau's argument that it was only the invention of agriculture that introduced genuine inequality, since it allowed for the emergence of landed property. This is one of the main reasons people today continue to write as if foragers can be assumed to live in egalitarian bands to begin with – because it's also assumed that without the productive assets (land, livestock) and stockpiled surpluses (grain, wool, dairy products, etc.) made possible by farming, there was no real material basis for anyone to lord it over anyone else.

Conventional wisdom also tells us that the moment a material surplus does become possible, there will also be full-time craft specialists, warriors and priests laying claim to it, and living off some portions of that surplus (or, in the case of warriors, spending the bulk of their time trying to figure out new ways to steal it from each other); and before long, merchants, lawyers and politicians will inevitably follow. These new elites will, as Rousseau emphasized, band together to protect their assets, so the advent of private property will be followed, inexorably, by the rise of 'the state'.

We will scrutinize this conventional wisdom in more detail later. For now, suffice to say that while there is a broad truth here, it is so broad as to have very little explanatory power. For sure, only cereal-farming and grain storage made possible bureaucratic regimes like those of Pharaonic Egypt, the Maurya Empire or Han China. But to say that cereal-farming was responsible for the rise of such states is a little like saying that the development of calculus in medieval Persia is responsible for the invention of the atom bomb. It is true that without calculus atomic weaponry would never have been possible. One might even make a case that the invention of calculus set off a chain of events that made it likely someone, somewhere, would eventually create nuclear weapons. But to assert that Al-Tusi's work on polynomials in the 1100s *caused* Hiroshima and Nagasaki is clearly absurd. Similarly, with agriculture. Roughly 6,000 years stand between the appearance of the first farmers in the Middle East and the rise of what we are used to calling the first states; and in many parts of the world, farming never led to the emergence of anything remotely like those states.[7]

At this juncture, we need to focus on the very notion of a surplus, and the much broader – almost existential – questions it raises. As philosophers realized long ago, this is a concept that poses fundamental questions about what it means to be human. One of the things that sets us apart from non-human animals is that animals produce only and exactly what they need; humans invariably produce more. We are creatures of excess, and this is what makes us simultaneously the most creative, and most destructive, of all species. Ruling classes are simply those who have organized society in such a way that they can extract the lion's share of that surplus for themselves, whether through tribute, slavery, feudal dues or manipulating ostensibly free-market arrangements.

In the nineteenth century, Marx and many of his fellow radicals did imagine that it was possible to administer such a surplus collectively, in an equitable fashion (this is what he envisioned as being the norm under 'primitive communism', and what he thought could once again be possible in the revolutionary future), but contemporary thinkers tend to be more sceptical. In fact, the dominant view among anthropologists nowadays is that the only way to maintain a truly egalitarian society is to eliminate the possibility of accumulating any sort of surplus at all.

The greatest modern authority on hunter-gatherer egalitarianism is, by general consent, the British anthropologist James Woodburn. In the post-war decades Woodburn conducted research among the Hadza, a forager society of Tanzania. He also drew parallels between them and the San Bushmen and Mbuti Pygmies, as well as a number of other small-scale nomadic forager societies outside Africa, such as the Pandaram of south India or Batek of Malaysia.[8] Such societies are, Woodburn suggests, the only genuinely egalitarian societies we know of, since they are the only ones that extend equality to gender relations and, as much as is practicable, to relations between old and young.

Focusing on such societies allowed Woodburn to sidestep the

question of what is being equalized and what isn't, because populations like the Hadza appear to apply principles of equality to just about everything it is possible to apply them to: not just material possessions, which are constantly being shared out or passed around, but herbal or sacred knowledge, prestige (talented hunters are systematically mocked and belittled), and so on. All such behaviour, Woodburn insisted, is based on a self-conscious ethos, that no one should ever be in a relation of ongoing dependency to anybody else. This echoes what we heard in the last chapter from Christopher Boehm about the 'actuarial intelligence' of egalitarian hunter-gatherers, but Woodburn adds a twist: the real defining feature of such societies is, precisely, the lack of any material surplus.

Truly egalitarian societies, for Woodburn, are those with 'immediate return' economies: food brought home is eaten the same day or the next; anything extra is shared out, but never preserved or stored. All this is in stark contrast to most foragers, and all pastoralists or farmers, who can be characterized as having 'delayed return' economies, regularly investing their energies in projects that only bear fruit at some point in the future. Such investments, he argues, inevitably lead to ongoing ties that can become the basis for some individuals to exercise power over others; what's more, Woodburn assumes a certain 'actuarial intelligence' – Hadza and other egalitarian foragers understand all this perfectly well, and as a result they self-consciously avoid stockpiling resources or engaging in any long-term projects.

Far from rushing blindly for their chains like Rousseau's savages, Woodburn's 'immediate return hunter-gatherers' understand precisely where the chains of captivity loom, and organize much of their lives to keep away from them. This might sound like the basis of something hopeful or optimistic. Actually, it's anything but. What it suggests is, again, that any equality worth the name is essentially impossible for all but the very simplest foragers. What kind of future might we then have in store? At best, we could perhaps imagine (with the invention of *Star Trek* replicators or other immediate-gratification devices) that it might be possible, at some point in the distant future, to create something like a society of equals once more. But in the meantime, we are definitively stuck. In other words, this is the Garden of Eden narrative all over again – just, this time, with the bar for paradise set even higher.

What's really striking about Woodburn's vision is that the foragers he focuses on appear to have reached such profoundly different conclusions from Kandiaronk, and several generations of First Nation critics before him, all of whom had trouble even imagining that differences of wealth could be translated into systematic inequalities of power. Recall that the American indigenous critique, as we described it in Chapter Two, was initially about something very different: the perceived failure of European societies to promote mutual aid and protect personal liberties. Only later, once indigenous intellectuals had more exposure to the workings of French and English society, did it come to focus on inequalities of property. Perhaps we should follow their initial train of thought.

Few anthropologists are particularly happy with the term 'egalitarian societies', for reasons that should now be obvious; but it lingers on because no one has suggested a compelling alternative. The closest we're aware of is the feminist anthropologist Eleanor Leacock's suggestion that most members of what are called egalitarian societies seem less interested in equality per se than what she calls 'autonomy'. What matters to Montagnais-Naskapi women, for instance, is not so much whether men and women are seen to be of equal status but whether women are, individually or collectively, able to live their lives and make their own decisions without male interference.[9]

In other words, if there is a value these women feel should be distributed equally, it is precisely what we would refer to as 'freedom'. Perhaps the best thing, then, would be to call these 'free societies'; or even, following the Jesuit Father Lallemant's verdict on the Montagnais-Naskapi's Wendat neighbours, 'free people', each of whom 'considers himself of as much consequence as the others; and they submit to their chiefs only in so far as it pleases them.'[10] At first glance, Wendat society, with its elaborate constitutional structure of chiefs, speakers and other office holders, might not seem an obvious choice for inclusion on a list of 'egalitarian' societies. But 'chiefs' are not really chiefs if they have no means to enforce orders. Equality, in societies such as those of the Wendat, was a direct consequence of individual liberty. Of course, the same can be said in reverse: liberties are not really liberties if one cannot act on them. Most people today also believe they live in free societies (indeed, they often insist that, politically at least, this is what is most important about their societies), but the freedoms which form the moral basis of a nation like the United States are, largely, *formal* freedoms.

American citizens have the right to travel wherever they like – provided, of course, they have the money for transport and accommodation. They are free from ever having to obey the arbitrary orders of superiors – unless, of course, they have to get a job. In this sense, it is almost possible to say the Wendat had play chiefs[11] and real freedoms, while most of us today have to make do with real chiefs and play freedoms. Or to put the matter more technically: what the Hadza, Wendat or 'egalitarian' people such as the Nuer seem to have been concerned with were not so much *formal* freedoms as *substantive* ones.[12] They were less interested in the right to travel than in the possibility of actually doing so (hence, the matter was typically framed as an obligation to provide hospitality to strangers). Mutual aid – what contemporary European observers often referred to as 'communism' – was seen as the necessary condition for individual autonomy.

This might help explain at least some of the apparent confusion around the term egalitarianism: it is possible for explicit hierarchies to emerge, but to nonetheless remain largely theatrical, or to confine themselves to very limited aspects of social life. Let us return for a moment to the Sudanese Nuer. Ever since the Oxford social anthropologist E. E. Evans-Pritchard published his classic ethnography of them in the 1940s, the Nuer were held out as the very paradigm for 'egalitarian' societies in Africa. They had nothing even remotely resembling institutions of government and were notorious for the high value they placed on personal independence. But by the 1960s, feminist anthropologists like Kathleen Gough were showing that, again, you couldn't really speak of equality of status here: males in Nuer communities were divided between 'aristocrats' (with ancestral connections to the territories where they live), 'strangers' and lowly war captives taken by force in raids on other communities. Neither were these purely formal distinctions. While Evans-Pritchard had written off such differences as inconsequential, in reality, as Gough noted, difference in rank implied differential access to women. Only the aristocrats could easily assemble enough cattle to arrange what Nuer considered a 'proper' marriage – that is, one in which they could claim paternity over the children and thus be remembered as ancestors after their death.[13]

So was Evans-Pritchard simply wrong? Not exactly. In fact, while rank and differential access to cattle became relevant when people were arranging marriages, they had almost no bearing in any other circumstances. It would have been impossible, even at a formal event like a dance or sacrifice, to determine who was 'above' anyone else. Most importantly, differences in wealth (cattle) never translated into the ability to give orders, or to demand formal obeisance. In an often-cited passage Evans-Pritchard wrote:

> That every Nuer considers himself as good as his neighbour is evident in their every movement. They strut about like lords of the earth, which, indeed, they consider themselves to be. There is no master and no servant in their society, but only equals who regard themselves as God's noblest creation . . . even the suspicion of an order riles a man and he either does not carry it out or he carries it out in a casual and dilatory manner that is more insulting than a refusal.[14]

Evans-Pritchard is referring here to men. What about women?

While in everyday affairs, Gough found, women operated with much the same independence as men, the marriage system did efface women's freedom to a degree. If a man paid the forty cattle typically required for bridewealth, this meant above all that he not only had the right to claim paternity over a woman's children but also acquired exclusive sexual access, which in turn usually meant the right to interfere with his wife's affairs in other respects as well. However, most Nuer women were not 'properly' married. In fact, the complexities of the system were such that a large proportion found themselves officially married to ghosts, or to other women (who could be declared male for genealogical purposes) – in which case, how they went about becoming pregnant and raising their children was nobody's business but their own. Even in sexual life, then, for women as for men, individual freedom was assumed unless there was some specific reason to curtail it.

The freedom to abandon one's community, knowing one will be welcomed in faraway lands; the freedom to shift back and forth between social structures, depending on the time of year; the freedom to disobey authorities without consequence – all appear to have been simply assumed among our distant ancestors, even if most people find them barely conceivable today. Humans may not have begun their history in a state of primordial innocence, but they do appear to have begun it with a self-conscious aversion to being told what to do.[15] If this is so, we can at least refine our initial question: the real puzzle is not when chiefs, or even kings and queens, first appeared, but rather when it was no longer possible simply to laugh them out of court.

Now it is undoubtedly true that, over the broad sweep of history, we find ever larger and more settled populations, ever more powerful forces of production, ever larger material surpluses, and people spending ever more of their time under someone else's command. It seems reasonable to conclude there is some sort of connection between these trends. But the nature of that connection, and the actual mechanisms, are entirely unclear. In contemporary societies we consider ourselves free people largely because we lack political overlords. For us, it's simply assumed that what we call 'the economy' is organized entirely differently, on the basis not of freedom but 'efficiency', and therefore that offices and shop floors are typically arranged in strict chains of command. Unsurprising, then, that so much current speculation on the origins of inequality focuses on economic changes, and particularly the world of work.

Here too, we think, much of the available evidence has been widely misconstrued.

A focus on work is not precisely the same as a focus on property, though if one is trying to understand how control of property first came to be translated into power of command, the world of work would be the obvious place to look. By framing the stages of human development largely around the ways people went about acquiring food, men like Adam Smith and Turgot inevitably put work – previously considered a somewhat plebeian concern – centre stage. There was a simple reason for this. It allowed them to claim that their own societies were self-evidently superior, a claim that – at the time – would have been much harder to defend had they used any criterion other than productive labour.[16]

Turgot and Smith began writing this way in the 1750s. They referred to the apex of development as 'commercial society', in which a complex division of labour demanded the sacrifice of primitive liberties but guaranteed dazzling increases in overall wealth and prosperity. Over the next several decades, the invention of the spinning jenny, Arkwright loom and, eventually, steam and coal power – and finally the emergence of a permanent (and increasingly self-conscious) industrial working class – completely shifted the terms of debate. Suddenly, there existed forces of production previously undreamed of. But there was also a staggering increase in the number of hours that people were expected to work. In the new mills, twelve- to fifteen-hour days and six-day weeks were considered standard; holidays were minimal. (John Stuart Mill protested that 'All the labour-saving machinery that has hitherto been invented has not lessened the toil of a single human being.')

As a result, and over the course of the nineteenth century, almost everyone arguing about the overall direction of human civilization took it for granted that technological progress was the prime mover of history, and that if progress was the story of human liberation, this could only mean liberation from 'unnecessary toil': at some future time, science would eventually free us from at least the most degrading, onerous and soul-destroying forms of work. In fact, by the Victorian era many began arguing that this was already happening. Industrialized farming and new labour-saving devices, they claimed, were already leading us towards a world where everyone would enjoy an existence of leisure and affluence – and where we wouldn't have to spend most of our waking lives running about at someone else's orders.

Granted, this must have seemed a bizarre claim to radical trade unionists in Chicago who, as late as the 1880s, had to engage in pitched battles with police and company detectives in order to win an eight-hour day – that is, obtain the right to a daily work regime that the average medieval baron would have considered unreasonable to expect of his serfs.[17] Yet, perhaps as a riposte to such campaigns, Victorian intellectuals began arguing that exactly the opposite was true: 'primitive man', they posited, had been engaged in a constant struggle for his very existence; life in early human societies was a perpetual chore. European or Chinese or Egyptian peasants toiled from dawn till dusk to eke out a living. And so, it followed, even the awful work regimes of the Dickensian age were actually an improvement on what had come before. All we are arguing about, they insisted, is the pace of improvement. By the dawn of the twentieth century, such reasoning had become universally accepted as common sense.

That is what made Marshall Sahlins's 1968 essay 'The Original Affluent Society' such an epochal event, and is why we must now consider both some of its implications and its limitations. Probably the most influential anthropological essay ever written, it turned that old Victorian wisdom – still prevalent in the 1960s – on its head, creating instant discussion and debate, inspiring everyone from socialists to hippies. Whole schools of thought (Primitivism, Degrowth) would likely have never come about without it. But Sahlins was also writing at a time when archaeologists still knew relatively little about pre-agricultural peoples, at least compared to what we know now. It might be best, then, first to take a look at his argument before turning to the evidence we have today and seeing how the piece measures up against it.

IN WHICH WE DISCUSS MARSHALL SAHLINS'S 'ORIGINAL AFFLUENT SOCIETY' AND REFLECT ON WHAT CAN HAPPEN WHEN EVEN VERY INSIGHTFUL PEOPLE WRITE ABOUT PREHISTORY IN THE ABSENCE OF ACTUAL EVIDENCE

Marshall Sahlins started his career in the late 1950s as a neo-evolutionist. When 'The Original Affluent Society' was published, he was still most famous for his work with Elman Service which proposed four stages of human political development: from bands to tribes, chiefdoms and states. All these terms are still widely used today. In 1968, Sahlins accepted an invitation to spend a year in Claude Lévi-Strauss's *laboratoire* in Paris, where, he later reported, he used to eat lunch in the cafeteria each day with Pierre Clastres (who would go on to write *Society Against the State*), arguing about ethnographic data and whether or not society was ripe for revolution.

These were heady days in French universities, full of student mobilizations and street fighting that ultimately led up to the student/worker insurrection of May 1968 (during which Lévi-Strauss maintained a haughty neutrality, but Sahlins and Clastres became enthusiastic participants). In the midst of all this political ferment, the nature of work, the need for work, the refusal of work, the possibility of gradually eliminating work were all heated matters of debate in both political and intellectual circles.

Sahlins's essay, perhaps the last truly great example of that genre of 'speculative prehistory' invented by Rousseau, first appeared in Jean-Paul Sartre's journal *Les Temps modernes*.[18] It made the argument that, at least when it comes to working hours, the Victorian narrative of continual improvement is simply backwards. Technological evolution has not liberated people from material necessity. People are not working less. All the evidence, he argued, suggests that over the course of human history the overall number of hours most people spend working has tended instead to increase. Even more provocatively, Sahlins insisted that people in earlier ages were not, necessarily, poorer than modern-day consumers. In fact, he contended, for much of our early history humans might just as easily be said to have lived lives of great material abundance.

True, a forager might seem extremely poor by our standards – but to apply our standards was obviously ridiculous. 'Abundance' is not an absolute measure. It refers to a situation where one has easy access to everything one feels one needs to live a happy and comfortable life. By those standards, Sahlins argued, most known foragers are rich. The fact that many hunter-gatherers, and even horticulturalists, only seem to have spent somewhere between two and four hours a day doing anything that could be construed as 'work' was itself proof of how easy their needs were to satisfy.

Before continuing, it's worth saying that the broad picture Sahlins presented appears to be correct. As we pointed out above, the average oppressed medieval serf still worked less than a modern nine-to-five office or factory worker, and the hazelnut gatherers and cattle herders who dragged great slabs to build Stonehenge almost certainly worked, on average, less than that. It's only very recently that even the richest countries have begun to turn such things around (obviously, most of us are not working as many hours as Victorian stevedores, though the overall decline in working hours is probably not as dramatic as we think). And for much of the world's population, things are still getting worse instead of better.

What stands the test of time less well is the image that most readers take away from Sahlins's essay: of happy-go-lucky hunter-gatherers, spending most of their time lounging in the shade, flirting, forming drum circles or telling stories. And this has everything to do with the ethnographic examples he was drawing on, largely the San, Mbuti and Hadza.

In the last chapter, we suggested a number of reasons why !Kung San (Bushmen) on the margins of the Kalahari and Hadza of the Serengeti Plateau became so popular in the 1960s as exemplars of what early human society might have been like (despite being quite unusual, as foragers go). One reason was simply the availability of data: by the 1960s, they were among the only foraging populations left who still maintained something like their traditional mode of life. It was also in this decade that anthropologists started carrying out time-allocation studies, recording systematically what members of different societies do over the course of a typical day and how much time they spend doing it.[19] Such research with African foragers also seemed to resonate with the famous discoveries of fossil hominins then being made by Louis and Mary Leakey in other parts of the continent, such as Olduvai Gorge in Tanzania. Since some of these modern hunter-gatherers were living in savannah-like environments, not unlike the ones in which our species now appeared to have evolved, it was tempting to imagine that here – in these living populations – one might catch a glimpse of human society in something like its original state.

Moreover, the results of those early time-allocation studies came as an enormous surprise. It's worth bearing in mind that, in the post-war decades, most anthropologists and archaeologists still very much took for granted the old nineteenth-century narrative of humanity's primordial 'struggle for existence'. To our ears, much of the rhetoric commonplace at the time, even among the most sophisticated scholars, sounds startlingly condescending: 'A man who spends his whole life following animals just to kill them to eat,' wrote the prehistorian Robert Braidwood in 1957, 'or moving from one berry patch to another, is really living just like an animal himself.'[20] Yet these first quantitative studies comprehensively disproved such pronouncements. They showed that, even in quite inhospitable environments like the deserts of Namibia or Botswana, foragers could easily feed everyone in their group and still have three to five days per week left for engaging in such extremely human activities as gossiping, arguing, playing games, dancing or travelling for pleasure.

Researchers in the 1960s were also beginning to realize that, far from agriculture being some sort of remarkable scientific advance, foragers (who after all tended to be intimately familiar with all aspects of the growing cycles of food plants) were perfectly aware of how one might go about planting and harvesting grains and vegetables. They just didn't see any reason why they should. 'Why should we plant,' one !Kung informant put it – in a phrase cited ever since in a thousand treatises on the origins of farming – 'when there are so many mongongo nuts in the world?' Indeed, concluded Sahlins, what some prehistorians had assumed to be technical ignorance was really a self-conscious social decision: such foragers had 'rejected the Neolithic Revolution in order to keep their leisure'.[21] Anthropologists were still struggling to come to terms with all this when Sahlins stepped in to draw the larger conclusions.

The ancient forager ethos of leisure (the 'Zen road to affluence') only broke down, or so Sahlins surmised, when people finally – for whatever reasons – began to settle in one place and accept the toils of agriculture. They did so at a terrible cost. It wasn't just ever-increasing hours of toil that followed but, for most, poverty, disease, war and slavery – all fuelled by endless competition and the mindless pursuit of new pleasures, new powers and new forms of wealth. With one deft move, Sahlins's 'Original Affluent Society' used the results of time-allocation studies to pull the rug from under the traditional story of human civilization. Like Woodburn, Sahlins brushes aside Rousseau's version of the Fall – the idea that, too foolish to reflect on the likely consequences of our actions in assembling, stockpiling and guarding property, we 'ran blindly for our chains'[22] – and takes us straight back to the Garden of Eden. If rejecting farming was a conscious choice, then so was that act of embracing it. We chose to eat of the fruit of the tree of knowledge, and for this we were punished. As St Augustine put it, we rebelled against God, and God's judgment was to cause our own desires to rebel against our rational good sense; our punishment for original sin is the infinity of our new desires.[23]

If there is a fundamental difference here from the biblical story, it's that the Fall (according to Sahlins) didn't happen just once. We didn't collapse and then begin slowly to pull ourselves back up. When it comes to labour and affluence, every new technological breakthrough seems to cause us to fall yet further.

Sahlins's piece is a brilliant morality tale. There is, however, one obvious flaw. The whole argument for an 'original affluent society' rested on a single fragile premise: that most prehistoric humans really did live in the specific manner of African foragers. As Sahlins was perfectly willing to admit, this was just a guess. In closing his essay, he asked whether 'marginal hunters such as the Bushmen of the Kalahari' really were any more representative of the Palaeolithic condition than the foragers of California (who placed great value on hard work) or the Northwest Coast (with their ranked societies and stockpiles of wealth)? Perhaps not, Sahlins conceded.[24] This often overlooked observation is crucial. It's not that Sahlins is suggesting that his own phrase 'original affluent society' is incorrect. Rather, he acknowledges that, just as there might have been many ways for free peoples to be free, there might have been more than just one way for (original) affluent societies to be affluent.

Not all modern hunter-gatherers value leisure over hard work, just as not all share the easy-going attitudes towards personal possessions of the !Kung or Hadza. Foragers in northwestern California, for instance, were notorious for their cupidity, organizing much of their lives around the accumulation of shell money and sacred treasures and adhering to a stringent work ethic in order to do so. The fisher-foragers of the Canadian Northwest Coast, on the other hand, lived in highly stratified societies where commoners and slaves were famously industrious. According to one of their ethnographers, the

Kwakiutl of Vancouver Island were not only well housed and fed, but lavishly supplied: 'Each household made and possessed many mats, boxes, cedar-bark and fur blankets, wooden dishes, horn spoons, and canoes. It was as though in manufacturing as well as in food production there was no point at which further expenditure of effort in the production of more of the same items was felt to be superfluous.'[25] Not only did the Kwakiutl surround themselves with endless piles of possessions, but they also put endless creativity into designing and crafting them, with results so striking and intricately beautiful as to make them the pride of ethnographic museums the world over. (Lévi-Strauss remarked that turn-of-the-century Kwakiutl were like a society where a dozen different Picassos were operative all at the same time.) This, surely, is a kind of affluence. But one entirely different from that of the !Kung or Mbuti.

Which, then, more resembled the original state of human affairs: the easy-going Hadza, or the industrious foragers of northwestern California? By now it will be clear to the reader that this is just the kind of question we *shouldn't* be asking. There was no truly 'original' state of affairs. Anyone who insists that one exists is by definition trading in myths (Sahlins, at least, was fairly honest about this). Human beings had many tens of thousands of years to experiment with different ways of life, long before any of them turned their hands to agriculture. Instead we might do better to look at the overall direction of change, so as to understand how it bears on our question: how humans came largely to lose the flexibility and freedom that seems once to have characterized our social arrangements, and ended up stuck in permanent relations of dominance and subordination.

To do this means continuing the story begun in Chapter Three, following our foraging ancestors out of the Ice Age (or Pleistocene era) into a phase of warmer global climate known as the Holocene. This will also take us far outside Europe, to places like Japan and the Caribbean coast of North America, where entirely new and unsuspected pasts are beginning to emerge; ones which – despite the stubborn efforts of scholars to shoehorn them into neat evolutionary boxes – look about as far from small, nomadic, egalitarian 'bands' as one can possibly imagine.

IN WHICH WE SHOW HOW NEW DISCOVERIES CONCERNING ANCIENT HUNTER-GATHERERS IN NORTH AMERICA AND JAPAN ARE TURNING SOCIAL EVOLUTION ON ITS HEAD

In modern-day Louisiana there is a place with the dispiriting name of Poverty Point. Here you can still see the remains of massive earthworks erected by Native Americans around 1600 BC. With its plush green lawns and well-trained coppices, today the site looks like something halfway between a wildlife management area and a golf club.[26] Grass-covered mounds and ridges rise neatly from carefully tended meadows, forming concentric rings which suddenly vanish where the Bayou Macon has eroded them away (*bayou* being derived, via Louisiana French, from the Choctaw word *bayuk*: marshy rivulets spreading out from the main channel of the Mississippi). Despite nature's best efforts to obliterate these earthworks, and early European settlers' best efforts to deny their obvious significance (perhaps these were the dwellings of an ancient race of giants, they conjectured, or one of the lost tribes of Israel?), they endure: evidence for an ancient civilization of the Lower Mississippi and testimony to the scale of its accomplishments.

Archaeologists believe these structures at Poverty Point formed a monumental precinct that once extended over 200 hectares, flanked by two enormous earthen mounds (the so-called Motley and Lower Jackson Mounds) which lie respectively north and south. To clarify what this means, it's worth noting that the first Eurasian cities – early centres of civic life like Uruk in southern Iraq, or Harappa in the Punjab – began as settlements of roughly 200 hectares in total. Which is to say that their entire layout could fit quite comfortably within the ceremonial precinct of Poverty Point. Like those early Eurasian cities, Poverty Point sprang from a great river, since transport by water, particularly of bulk goods, was in early times infinitely easier than transport by land. Like them, it formed the core of a much larger sphere of cultural interaction. People and resources came to Poverty Point from hundreds of miles away, as far north as the Great Lakes and from the Gulf of Mexico to the south.

Seen from the air – a 'god's-eye' view – Poverty Point's standing remains look like some sunken, gargantuan amphitheatre; a place of crowds and power, worthy of any great agrarian civilization. Something approaching a million cubic metres of soil was moved to create its ceremonial infrastructure, which was most likely oriented to the skies, since some of its mounds form enormous figures of birds, inviting the heavens to bear witness to their presence. But the people of Poverty Point weren't farmers. Nor did they use writing. They were hunters, fishers and foragers, exploiting a superabundance of wild resources (fish, deer, nuts, waterfowl) in the lower reaches of the Mississippi. And they were not the first hunter-gatherers in this region to establish traditions of public architecture. These traditions can be traced back far beyond Poverty Point itself, to around 3500 BC – which is also roughly the time that cities first emerged in Eurasia.

As archaeologists often point out, Poverty Point is 'a Stone Age site in an area where there is no stone', so the staggering quantities of lithic tools, weapons, vessels and lapidary ornaments found there must all have been originally carried from somewhere else.[27] The scale of its earthworks implies thousands of people gathering at the site at particular times of year, in numbers outstripping any historically known hunter-gatherer population. Much less clear is what attracted them there with their native copper, flint, quartz crystal, soapstone and other minerals; or how often they came, and how long they stayed. We simply don't know.

What we do know is that Poverty Point arrows and spearheads come in rich hues of red, black, yellow and even blue stone, and these are only the colours we discern. Ancient classifications were no doubt more refined. If stones were being selected with such care, we can only begin to imagine what was going on with cords, fibres, medicines and any living thing in the landscape treated as potential food or poison. Another thing we can be quite sure of is that 'trade' is not a useful way to describe whatever was going on here. For one thing, trade goes two ways, and Poverty Point presents no clear evidence for exports, or indeed commodities of any sort. The absence is strikingly obvious to anyone who's studied the remains of early Eurasian cities like Uruk and Harappa, which do seem to have been engaged in lively trade relations: these sites are awash with industrial quantities of ceramic packaging, and the products of their urban crafts are found far and wide.

Despite its great cultural reach, there is nothing at all of this commodity culture at Poverty Point. In fact, it's not clear if anything much was going out from the site, at least in material terms, other than certain enigmatic clay items known as 'cooking balls', which can hardly be considered trade goods. Textiles and fabrics may have been important, but we also have to allow for the possibility that Poverty Point's greatest assets were intangible. Most experts today view its monuments as expressions of sacred geometry, linked to calendar counts and the movement of celestial bodies. If anything was being stockpiled at Poverty Point, it may well have been knowledge: the intellectual property of rituals, vision quests, songs, dances and images.[28]

We can't possibly know the details. But it's more than just speculation to say that ancient foragers were exchanging complex information across this entire region, and in a highly controlled fashion. Material proof comes from close examination of the earthen monuments themselves. Through the great valley of the Mississippi, and some considerable way beyond, there exist other smaller sites of the same period. The various configurations of their mounds and ridges adhere to strikingly uniform geometrical principles, based on standard units of measurement and proportion apparently shared by early peoples throughout a significant portion of the Americas. The underlying system of calculus appears to have been based on the transformational properties of equilateral triangles, figured out with the aid of cords and strings, and then extended to the laying-out of massive earthworks.

Published in 2004, this remarkable discovery by John E. Clark, an archaeologist and authority on the pre-Columbian societies of Mesoamerica,[29] has been greeted by the scholarly community with responses ranging from lukewarm acceptance to plain disbelief, although nobody appears to have actually refuted it. Many prefer simply to ignore it. Clark himself seems surprised by his results. We will return to some wider implications in Chapter Eleven, but for now we can simply note an assessment of Clark's findings by two specialists in the field, who accept the evidence he presents 'not only for a standard unit of measurement but also for geometrical layouts and spacing intervals among first-mound complexes from Louisiana to

Mexico and Peru, which incorporate multiples of that standard'. At most, finding the same system of measurement across such distances may prove to be 'one of contemporary archaeology's most provocative revelations', and at the very least, they conclude, 'those who built the works were not simple, ordinary foragers.'[30]

Putting aside the (by now irrelevant) notion that there ever was such a thing as 'simple, ordinary foragers', it has to be said that, even if Clark's theory were true only for the Lower Mississippi and surrounding parts of the Eastern Woodlands,[31] it would still be quite remarkable. For, unless we are dealing with some kind of amazing cosmic coincidence, it means that someone had to convey knowledge of geometric and mathematical techniques for making accurate spatial measurements, and related forms of labour organization, over very long distances. If this were the case, it seems likely that they also shared other forms of knowledge as well: cosmology, geology, philosophy, medicine, ethics, fauna, flora, ideas about property, social structure, and aesthetics.

In the case of Poverty Point, should this be conceived as a form of exchange of knowledge for material goods? Possibly. But the movement of objects and ideas might have been organized any number of other ways as well. All we know for sure is that the lack of an agricultural base does not seem to have stopped those who gathered on Poverty Point from creating something that to us would appear very much like little cities which, at least during certain times of year, hosted a rich and influential intellectual life.

Today, Poverty Point is a National Park and Monument and UNESCO World Heritage Site. Despite these designations of international importance, its implications for world history have hardly begun to be explored. A hunter-gatherer metropolis the size of a Mesopotamian city-state, Poverty Point makes the Anatolian complex of Göbekli Tepe look like little more than a 'potbelly hill' (which is, in fact, what 'Göbekli Tepe' means in Turkish). Yet outside a small community of academic specialists, and of course local residents and visitors, very few people have heard of it.

The obvious question at this juncture must surely be: why isn't Poverty Point better known to audiences the world over? Why doesn't it feature more prominently (or at all) in discussions on the origins of urban life, centralization and their consequences for human history?

One reason, no doubt, is that Poverty Point and its predecessors (like the much older mound complex at Watson's Brake, in the nearby Ouachita basin) have been placed in a phase of American prehistory known as the 'Archaic'. The Archaic period covers an immense span of time, between the flooding of the Beringia land bridge (which once linked Eurasia to the Americas) around 8000 BC, and the initial adoption and spread of maize-farming in certain parts of North America, down to around 1000 BC. One word, for seven millennia of indigenous history. Archaeologists who first gave the period its name – which is really more of a chronological slap in the face – were basically declaring, 'this is the period before anything particularly important was happening.' So when undeniable evidence began to appear that all sorts of important things were indeed happening, and not just in the Mississippi basin, it was almost something of an archaeological embarrassment.

On the shores of the Atlantic and around the Gulf of Mexico lie enigmatic structures: just as remarkable as Poverty Point, but even less well known. Formed out of shell in great accumulations, they range from small rings to massive U-shaped 'amphitheatres' like those of St Johns River valley in Northeast Florida. These were no natural features. They too were built spaces where hunter-gatherer publics once assembled in their thousands. Far to the north and west, on the other side of the continent, more surprises loom up from the windswept shores of British Columbia: settlements and fortifications of striking magnitude, dating back as far as 2000 BC, facing a Pacific already familiar with the spectacle of war and long-range commerce.[32]

On the matter of hunter-gatherer history, North America isn't the only part of the world where evolutionary expectations are heading for a titanic collision with the archaeological record. In Japan and neighbouring islands, another monolithic cultural designation – 'Jōmon' – holds sway over more than 10,000 years of forager history, from around 14,000 BC to 300 BC. Japanese archaeologists spend much time subdividing the Jōmon period in ways just as intricate as the more pioneering North American scholars now do with their 'Archaic'. Everyone else, however, whether museumgoers or readers of high-school textbooks, is still confronted with the stark singularity of the term 'Jōmon', which, covering the long ages before rice-farming came to Japan, leaves us with an impression of drab conservatism, a time when nothing really happened. New archaeological discoveries are now revealing just how wrong this is.

The creation of a new Japanese national past is a somewhat paradoxical side effect of modernization. Since Japan's economic take-off in the 1960s, many thousands of archaeological sites have been discovered, excavated and meticulously recorded, either as a result of construction projects for roads, railways, housing or nuclear plants or as part of immense rescue efforts undertaken in the wake of environmental catastrophes such as the 2011 Tōhoku earthquake. The result is an immense archive of archaeological information. What begins to emerge from this data-labyrinth is an entirely different picture of what society was like before irrigated rice cultivation came to Japan from the Korean Peninsula.

Across the Japanese archipelago, between 14,000 and 300 BC, centennial cycles of settlement nucleation and dispersal came and went; monuments shot up in wood and stone, and then were pulled down again or abandoned; elaborate ritual traditions, including opulent burials, flourished and declined; specialized crafts waxed and waned, including remarkable accomplishments in the arts of pottery, wood and lacquer. In traditions of wild food procurement, strong regional contrasts are evident, ranging from maritime adaptations to acorn-based economies, both using large storage facilities for gathered resources. Cannabis came into use, for fibres and recreational drug use. There were enormous villages with grand storehouses and what seem to be ritual precincts, such as those found at Sannai Maruyama.[33]

An entire, forgotten social history of pre-agricultural Japan is resurfacing, for now largely as a mass of data points and state heritage archives. In future, as the bits get pieced back together, who knows what will come into view?

Europe, too, bears witness to the vibrant and complex history of non-agricultural peoples after the Ice Age. Take the monuments called in Finnish *Jätinkirkko*, the 'Giants' Churches' of the Bothnian Sea between Sweden and Finland: great stone ramparts, some up to 195 feet long, raised up in their tens by coastal foragers between 3000 and 2000 BC. Or the 'Big Idol', a seventeen-foot-tall totem pole with elaborate carvings rescued from a peat bog on the shores of Lake Shigirskoe, on the eastern slopes of the Central Urals. Dating to around 8000 BC, the Idol is the lone survivor of a long-lost tradition of large-scale wooden forager art which once produced monuments that presided over northern skies. Then come the amber-soaked burials of Karelia and southern Scandinavia, with their elaborate grave goods and corpses staged in expressive poses, echoing some forgotten etiquette of Mesolithic vintage.[34] And, as we've seen, even the major building phases of Stonehenge, long associated with early farmers, are now dated to a time when cereal cultivation was virtually abandoned and hazelnut-gathering once again took over in the British Isles, alongside livestock-herding.

Back in North America, some researchers are beginning to talk, a little awkwardly, of the 'New Archaic', a hitherto unsuspected era of 'monuments without kings'.[35] But the truth is that we still know precious little of the political systems lying behind a now almost globally attested phenomenon of forager monumentality, or indeed whether some of those monumental projects might have involved kings or other kinds of leaders. What we do know is that this changes forever the nature of the conversation about social evolution in the Americas, Japan, Europe, and no doubt most other places too. Clearly, foragers didn't shuffle backstage at the close of the last Ice Age, waiting in the wings for some group of Neolithic farmers to reopen the theatre of history. Why, then, is this new knowledge so rarely integrated into our accounts of the human past? Why does almost everyone (everyone, at least, who is not a specialist on Archaic North America or Jōmon Japan) still write as if such things were impossible before the coming of agriculture?

Of course, those of us with no access to archaeological reports can be excused. What information exists more widely tends to be restricted to scattered, and sometimes sensationalized, news summaries that are very hard to put together into a single picture. Scholars and professional researchers, on the other hand, have to actually make a considerable effort to remain so ignorant. Let us consider for a moment some of the peculiar forms of intellectual acrobatics required.

HOW THE MYTH THAT FORAGERS LIVE IN A STATE OF INFANTILE SIMPLICITY IS KEPT ALIVE TODAY (OR, INFORMAL FALLACIES)

Let's first ask why even some experts apparently find it so difficult to shake off the idea of the carefree, idle forager band; and the twin assumption that 'civilization' properly so called – towns, specialized craftspeople, specialists in esoteric knowledge – would be impossible without agriculture. Why would anyone continue to write history as if places like Poverty Point could never have existed? It can't just be the whimsical result of airy academic terminologies ('Archaic', 'Jōmon' and so on). The real answer, we suggest, has more to do with the legacy of European colonial expansion; and in particular its impact on both indigenous and European systems of thought, especially with regard to the expression of rights of property in land.

Recall how – long before Sahlins's notion of the 'original affluent society' – indigenous critics of European civilization were already arguing that hunter-gatherers were really better off than other people because they could obtain the things they wanted and needed so easily. Such views can be found as early as the sixteenth century – remember, for instance, the Mi'kmaq interlocutors who annoyed Père Biard so much by insisting they were richer than the French, for exactly that reason. Kandiaronk made similar arguments, insisting 'the Savages of Canada, notwithstanding their Poverty, are richer than you, among whom all sorts of crimes are committed upon the score of Mine and Thine.'[36]

As we've seen, indigenous critics like Kandiaronk, caught in the rhetorical moment, would frequently overstate their case, even playing along with the idea that they were blissful, innocent children of nature. They did this in order to expose what they considered the bizarre perversions of the European lifestyle. The irony is that, in doing so, they often played into the hands of those who argued that – being blissful and innocent children of nature – they also had no natural rights to their land.

Here it's important to understand a little of the legal basis for dispossessing people who had the misfortune already to be living in territories coveted by European settlers. This was, almost invariably, what nineteenth-century jurists came to call the 'Agricultural Argument', a principle which has played a major role in the displacement of untold thousands of indigenous peoples from ancestral lands in Australia, New Zealand, sub-Saharan Africa and the Americas: processes typically accompanied by the rape, torture and mass murder of human beings, and often the destruction of entire civilizations.

Colonial appropriation of indigenous lands often began with some blanket assertion that foraging peoples really were living in a State of Nature – which meant that they were deemed to be part of the land but had no legal claims to own it. The entire basis for dispossession, in turn, was premised on the idea that the current inhabitants of those lands weren't *really* working. The argument goes back to John Locke's *Second Treatise of Government* (1690), in which he argued that property rights are necessarily derived from labour. In working the land, one 'mixes one's labour' with it; in this way it becomes, in a sense, an extension of oneself. Lazy natives, according to Locke's disciples, didn't do that. They were not, Lockeans claimed, 'improving landlords' but simply made use of the land to satisfy their basic needs with the minimum of effort. James Tully, an authority on indigenous rights, spells out the historical implications: land used for hunting and gathering was considered vacant, and 'if the Aboriginal peoples attempt to subject the Europeans to their laws and customs or to defend the territories that they have mistakenly believed to be their property for thousands of years, then it is they who violate natural law and may be punished or "destroyed" like savage beasts.'[37] In a similar way, the stereotype of the carefree, lazy native, coasting through a life free from material ambition, was deployed by thousands of European conquerors, plantation overseers and colonial officials in Asia, Africa, Latin America and Oceania as a pretext for the use of bureaucratic terror to force local people into work: everything from outright enslavement to punitive tax regimes, corvée labour and debt peonage.

As indigenous legal scholars have been pointing out for years, the 'Agricultural Argument' makes no sense, even on its own terms. There are many ways, other than European-style farming, in which to care for and improve the productivity of land. What to a settler's eye seemed savage, untouched wilderness usually turns out to be landscapes actively managed by indigenous populations for thousands of years through controlled burning, weeding, coppicing, fertilizing and pruning, terracing estuarine plots to extend the habitat of particular wild flora, building clam gardens in intertidal zones to enhance the reproduction of shellfish, creating weirs to catch salmon, bass and sturgeon, and so on. Such procedures were often labour-intensive, and regulated by indigenous laws governing who could access groves, swamps, root beds, grasslands and fishing grounds, and who was entitled to exploit what species at any given time of year. In parts of Australia, these indigenous techniques of land management were such that, according to one recent study, we should stop speaking of 'foraging' altogether, and refer instead to a different sort of farming.[38]

Such societies might not have recognized private property rights in the same sense as Roman Law or English Common Law, but it's absurd to argue they had no property rights at all. They simply had different conceptions of property. This is true, incidentally, even of people like the Hadza or !Kung; and, as we will see, many other foraging peoples actually had extraordinarily complex and sophisticated conceptions of ownership. Sometimes these indigenous property systems formed the basis for differential access to resources, with the result that something like social classes emerged.[39] Usually, though, this did not happen, because people made sure that it didn't, much as they made sure chiefs did not develop coercive power.

We should nonetheless recognize that the economic base of at least some foraging societies was capable of supporting anything from priestly castes to royal courts with standing armies. Let us take just one dramatic example to illustrate the point.

One of the first North American societies described by European explorers in the sixteenth century were the Calusa, a non-agricultural people who inhabited the west coast of Florida, from Tampa Bay to the Keys. There they had established a small kingdom, ruled from a capital town called Calos, which today is marked by a thirty-hectare complex of high shell mounds known as Mound Key. Fish, shellfish and larger marine animals comprised a major part of the Calusa diet, supplemented by deer, raccoon and a variety of birds. Calusa also maintained a fleet of war canoes with which they would launch military raids on nearby populations, extracting processed foods, skins, weapons, amber, metals and slaves as tribute. When Juan Ponce de León entered Charlotte Harbor on 4 June 1513 he was met by a well-organized flotilla of such canoes, manned by heavily armed hunter-gatherers.

Some historians resist calling the Calusa leader a 'king', preferring terms like 'paramount chief', but first-hand accounts leave no doubt about his exalted status. The man known as 'Carlos', the ruler of Calos at the time of initial European contact, even looked like a European king: he wore a gold diadem and beaded leg bands and sat on a wooden throne – and, crucially, he was the only Calusa allowed to do so. His powers seemed absolute. 'His will was law, and insubordination was punishable by death.'[40] He was also responsible for performing secret rituals that ensured the renewal of nature. His subjects always greeted him by kneeling and raising their hands in a gesture of obeisance, and he was typically accompanied by representatives of the ruling class of warrior nobles and priests who, like him, devoted themselves largely to the business of government. And he had at his disposal the services of specialized craftsmen, including court metallurgists who worked silver, gold and copper.

Spanish observers reported a traditional practice: that on the death of a Calusa ruler, or of his principal wife, a certain quota of their subjects' sons and daughters had to be put to death. By most definitions, all this would make Carlos not just a king, but a sacred king, perhaps divine.[41] We know less about the economic basis for these arrangements, but court life appears to have been made possible not only by complex systems of access to coastal fishing grounds, which were exceedingly rich, but also by canals and artificial ponds dug out of the coastal everglades. The latter, in turn, allowed for permanent – that is, non-seasonal – settlements (though most Calusa did still scatter to fishing and gathering sites at certain times of year, when the big towns grew decidedly smaller).[42]

By all accounts, then, the Calusa had indeed 'got stuck' in a single economic and political mode that allowed extreme forms of inequality to emerge. But they did so without ever planting a single seed or tethering a single animal. Confronted with such cases, adherents of the view that agriculture was a necessary foundation for durable inequalities have two options: ignore them, or claim they represent some kind of insignificant anomaly. Surely, they will say, foragers who do these kinds of things – raiding their neighbours, stockpiling wealth, creating elaborate court ceremonial, defending their territories and so on – aren't really foragers at all, or at least not *true* foragers. Surely they must be farmers by other means, effectively practising agriculture (just with wild crops), or perhaps somehow caught in a moment of transition, 'on the way' to becoming farmers, just not yet having quite arrived?

All these are excellent examples of what Antony Flew called the 'No True Scotsman' style of argument (also known to logicians as the 'ad hoc rescue' procedure). For those unfamiliar with it, it works like this:

> Imagine Hamish McDonald, a Scotsman, sitting down with his *Glasgow Morning Herald* and seeing an article about how the 'Brighton Sex Maniac Strikes Again'. Hamish is shocked and declares that 'No Scotsman would do such a thing.' The next day he sits down to read his *Glasgow Morning Herald* again; and, this time, finds an article about an Aberdeen man whose brutal actions make the Brighton sex maniac seem almost gentlemanly. This fact shows that Hamish was wrong in his opinion, but is he going to admit this? Not likely. This time he says: 'No *true* Scotsman would do such a thing.'[43]

Philosophers frown on this style of argumentation as a classic 'informal fallacy', or variety of circular argument. You simply assert a proposition (e.g. 'hunter-gatherers do not have aristocracies'), then protect it from any possible counter-examples by continually changing the definition. We prefer a consistent approach.

Foragers are populations which don't rely on biologically domesticated plants and animals as their primary sources of food. Therefore, if it becomes apparent that a good number of them have in fact possessed complex systems of land tenure, or worshipped kings, or practised slavery, this altered picture of their activities doesn't somehow magically turn them into 'proto-farmers'. Nor does it justify the invention of endless sub-categories like 'complex' or 'affluent' or 'delayed-return' hunter-gatherers, which is simply another way of ensuring such peoples are kept in what the Haitian anthropologist Michel-Rolph Trouillot called the 'savage slot', their histories defined and circumscribed by their mode of subsistence – as if they were people who really ought to be lazing around all day, but for some reason got ahead of themselves.[44] Instead, it means that the initial assertion was, like that of the apocryphal Hamish McDonald, simply wrong.

IN WHICH WE DISPOSE OF ONE PARTICULARLY SILLY ARGUMENT THAT FORAGERS WHO SETTLE IN TERRITORIES THAT LEND THEMSELVES WELL TO FORAGING ARE SOMEHOW UNUSUAL

In academic thought, there's another popular way of propping up the myth of the 'Agricultural Revolution', and thereby writing off people like the Calusa as evolutionary quirks or anomalies. This is to claim that they only behaved the way they did because they were living in 'atypical' environments. Usually, what's meant by 'atypical' are wetlands of various sorts – coasts and river valleys – as opposed to the remoter corners of tropical forests or desert margins, which is assumed to be where hunter-gatherers really ought to be living, since that is where most of them live today. It is a particularly weird argument, but a lot of very serious people make it, so we'll briefly have to take it on.

Anyone who was still living mainly by hunting animals and gathering wild foodstuffs in the early to mid twentieth century was almost certainly living on land no one else particularly wanted. That's why so many of the best descriptions of foragers come from places like the Kalahari Desert or Arctic Circle. Ten thousand years ago, this was obviously not the case. Everyone was a forager; overall population densities were low. Foragers were therefore free to live in pretty much any sort of territory they fancied. All things being equal, those living off wild resources would tend to cleave to places where they were abundant. You would think this is self-evident, but apparently it isn't.

Those who today describe people like the Calusa as 'atypical' because they had such a prosperous resource base want us to believe, instead, that ancient foragers chose to avoid locations of this kind, shunning the rivers and coasts (which also offered natural arteries for movement and communication), because they were so keen to oblige later researchers by resembling twentieth-century hunter-gatherers (the sort for which detailed scientific data is available today). We are asked to believe that it was only after they ran out of deserts and mountains and rainforests that they reluctantly started to colonize richer and more comfortable environments. We might call this the 'all the bad spots are taken!' argument.

In fact, there was nothing atypical about the Calusa. They were just one of many fisher-forager populations living around the Straits of Florida – including the Tequesta, Pojoy, Jeaga, Jobe and Ais (some apparently ruled by dynasties of their own) – with whom Calusa conducted regular trade, fought wars and arranged dynastic marriages. They were also among the first Native American societies to be destroyed since, for obvious reasons, coasts and estuaries were the first spots where Spanish colonizers landed, bringing epidemic diseases, priests, tribute and, eventually, settlers. This was a pattern repeated on every continent, from America to Oceania, where invariably the most attractive ports, harbours, fisheries and surrounding lands were first snapped up by British, French, Portuguese, Spanish, Dutch or Russian settlers, who also drained tidal salt marshes and coastal lagoons to farm cereals and cash crops.[45]

Such was the fate of the Calusa and their ancient fishing and hunting grounds. When Florida was ceded to the British in the mid eighteenth century, the last handful of surviving subjects from the kingdom of Calos were shipped off to the Caribbean by their Spanish masters.

*

For most of human history, fishers, hunters and foragers did not have to contend with expansive empires; therefore, they themselves tended to be the most active human colonizers of aquatic environments. Archaeological evidence increasingly bears this out. It was long thought, for instance, that the Americas were first settled by humans travelling mainly over land (the so-called 'Clovis people'). Around 13,000 years ago they were supposed to have followed an arduous crossing from Beringia, the land bridge between Russia and Alaska, passing south between terrestrial glaciers, over frozen mountains – all because, for some reason, it never occurred to any of them to build a boat and follow the coast.

More recent evidence suggests a very different picture (or, as one Navajo informant put it when faced with an archaeological map of the terrestrial route via Beringia: 'maybe some other guys came over like that, but us Navajos came a different way').[46]

In fact, Eurasian populations made a much earlier entry to what was then a genuinely 'New World', some 17,000 years ago. What was more, they did indeed think to build boats, following a coastal route that passed around the Pacific Rim, hopping between offshore islands and linear patches of kelp forest and ending somewhere on the southern coast of Chile. Early eastward crossings also took place.[47] Of course, it's possible that these first Americans, on arriving in such rich coastal habitats, quickly abandoned them, preferring for some obscure reason to spend the rest of their lives climbing mountains, hacking their way through forests and trekking across endless monotonous prairies. But it seems more plausible to assume that the bulk of them stayed exactly where they were, often forming dense and stable settlements in such locations.

The problem is, until recently this has always been an argument from silence, since rising sea levels long ago submerged the earliest records of shoreline habitation in most parts of the world. Archaeologists have tended to resist the conclusion that such habitations must have existed despite the lack of physical remains; but, with advances in the investigation of underwater environments, the case is growing stronger. A distinctly soggier (but also frankly more commonsensical) account of early human dispersal and settlement is finally becoming possible.[48]

IN WHICH WE FINALLY RETURN TO THE QUESTION OF PROPERTY, AND INQUIRE AS TO ITS RELATION TO THE SACRED

All this means that, of the many distinct cultural universes beginning to take shape across the world in the Early Holocene, most were likely centred on environments of abundance rather than scarcity: more like the Calusa's than the !Kung's. Does this also mean they were likely to have similar political arrangements to the Calusa? Here some caution is in order.

That the Calusa managed to maintain a sufficient economic surplus to support what looks to us like a miniature kingdom does not mean such an outcome is inevitable as soon as a society is capable of stockpiling a sufficient quantity of fish. After all, the Calusa were seafaring people; they would have undoubtedly been familiar with kingdoms ruled by divine monarchs like the Great Sun of the Natchez in nearby Louisiana and, likely as not, the empires of Central America. It's possible they were simply imitating more powerful neighbours. Or maybe they were just odd. Finally, we don't really know how much power even a divine king like Carlos really had. Here it's useful to consider the Natchez themselves: an agricultural group, much better documented than the Calusa, and with a spectacular and purportedly absolute monarch of their own.

The Natchez Sun, as the monarch was known, inhabited a village in which he appeared to wield unlimited power. His every movement was greeted by elaborate rituals of deference, bowing and scraping; he could order arbitrary executions, help himself to any of his subjects' possessions, do pretty much anything he liked. Still, this power was strictly limited by his own physical presence, which in turn was largely confined to the royal village itself. Most Natchez did not live in the royal village (indeed, most tended to avoid the place, for obvious reasons); outside it, royal representatives were treated no more seriously than Montagnais-Naskapi chiefs. If subjects weren't inclined to obey these representatives' orders, they simply laughed at them. In other words, while the court of the Natchez Sun was not pure empty theatre – those executed by the Great Sun were most definitely dead – neither was it the court of Suleiman the Magnificent or Aurangzeb. It seems to have been something almost precisely in between.

Was Calusa kingship a similar arrangement? Spanish observers clearly didn't think so (they regarded it as a more or less absolute monarchy), but since typically half the point of such deadly theatrics is to impress outsiders, that tells us very little in itself.[49]

What have we learned so far?

Most obviously, that we can now put a final nail in the coffin of the prevailing view that human beings lived more or less like Kalahari Bushmen, until the invention of agriculture sent everything askew. Even were it possible to write off Pleistocene mammoth hunters as some kind of strange anomaly, the same clearly cannot be said for the period that immediately followed the glaciers' retreat, when dozens of new societies began to form along resource-rich coasts, estuaries and river valleys, gathering in large and often permanent settlements, creating entirely new industries, building monuments according to mathematical principles, developing regional cuisines, and so on.

We have also learned that at least some of these societies developed a material infrastructure capable of supporting royal courts and standing armies – even though we have, as yet, no clear evidence that they actually did so. To construct the earthworks at Poverty Point, for instance, must have taken enormous amounts of human labour and a strict regime of carefully planned-out work, but we still have little idea how that labour was organized. Japanese archaeologists, surveying thousands of years' worth of Jōmon sites, have discovered all sorts of treasures, but they are yet to find indisputable evidence that those treasures were monopolized by any sort of aristocracy or ruling elite.

We cannot possibly know exactly which forms of ownership existed in these societies. What we can suggest, and there's plenty of evidence to support it, is that all the places in question – Poverty Point, Sannai Maruyama, the Kastelli Giant's Church in Finland, or indeed the earlier resting places of Upper Palaeolithic grandees – were in some sense sacred places. This might not seem like saying very much, but it's important: it tells us a lot more about the 'origins' of private property than is generally assumed. In rounding off this discussion, we will try to explain why.

Let's turn again to the anthropologist James Woodburn, and a less well-known insight from his work on 'immediate return' hunter-gatherers. Even among those forager groups, famous for their assertive egalitarianism, he notes, there was one striking exception to the rule that no adult should ever presume to give direct orders to another, and that individuals should not lay private claim to property. This exception came in the sphere of ritual, of the sacred. In Hadza religion and the religions of many Pygmy groups, initiation into male (and sometimes female) cults forms the basis of exclusive claims to ownership, usually of ritual privileges, that stand in absolute contrast to the minimization of exclusive property rights in everyday, secular life. These various forms of ritual and intellectual property, Woodburn observed, are generally protected by secrecy, by deception and often by the threat of violence.[50]

Here, Woodburn cites the sacred trumpets that initiated males of certain Pygmy groups keep hidden in secret places in the forest. Not only are women and children not supposed to know about such sacred treasures; should any follow the men to spy on them, they would be attacked or even raped.[51] Strikingly similar practices involving sacred trumpets, sacred flutes or other fairly obvious phallic symbols are commonplace in certain contemporary societies of Papua New Guinea and Amazonia. Very often there is a complex game of secrets, whereby the instruments are periodically taken out of their hiding places and men pretend they are the voices of spirits, or use them as part of costumed masquerades in which they impersonate spirits to terrify women and children.[52]

Now, these sacred items are, in many cases, the only important and exclusive forms of property that exist in societies where personal autonomy is taken to be a paramount value, or what we may simply call 'free societies'. It's not just relations of command that are strictly confined to sacred contexts, or even occasions when humans impersonate spirits; so too is absolute – or what we would today refer to as 'private' – property. In such societies, there turns out to be a profound formal similarity between the notion of private property and the notion of the sacred. Both are, essentially, structures of exclusion.

Much of this is implicit – if never clearly stated or developed – in Émile Durkheim's classic definition of 'the sacred' as that which is 'set apart': removed from the world, and placed on a pedestal, at some times literally and at other times figuratively, because of its imperceptible connection with a higher force or being. Durkheim argued that the clearest expression of the sacred was the Polynesian term *tabu*, meaning 'not to be touched'. But when we speak of absolute, private property, are we not talking about something very similar – almost identical in fact, in its underlying logic and social effects?

As British legal theorists like to put it, individual property rights are held, notionally at least, 'against the whole world'. If you own a car, you have the right to prevent anyone in the entire world from entering or using it. (If you think about it, this is the only right you have in your car that's really absolute. Almost anything else you can do with a car is strictly regulated: where and how you can drive it, park it, and so forth. But you can keep absolutely anyone else in the world from getting inside it.) In this case the object is set apart, fenced about by invisible or visible barriers – not because it is tied to some supernatural being, but because it's sacred to a specific, living human individual. In other respects, the logic is much the same.

To recognize the close parallels between private property and notions of the sacred is also to recognize what is so historically odd about European social thought. Which is that – quite unlike free societies – we take this absolute, sacred quality in private property as a paradigm for *all* human rights and freedoms. This is what the political scientist C. B. Macpherson meant by 'possessive individualism'. Just as every man's home is his castle, so your right not to be killed, tortured or arbitrarily imprisoned rests on the idea that you *own* your own body, just as you own your chattels and possessions, and legally have the right to exclude others from your land, or house, or car, and so on.[53] As we've seen, those who did not share this particular European conception of the sacred could indeed be killed, tortured or arbitrarily imprisoned – and, from Amazonia to Oceania, they often were.[54]

*

For most Native American societies, this kind of attitude was profoundly alien. If it applied anywhere at all, then it was only with regard to sacred objects, or what the anthropologist Robert Lowie termed 'sacra' when he pointed out long ago that many of the most important forms of indigenous property were immaterial or incorporeal: magic formulae, stories, medical knowledge, the right to perform a certain dance, or stitch a certain pattern on one's mantle. It was often the case that weapons, tools and even territories used to hunt game were freely shared – but the esoteric powers to safeguard the reproduction of game from one season to the next, or ensure luck in the chase, were individually owned and jealously guarded.[55]

Quite often, sacra have both material *and* immaterial elements; as among the Kwakiutl, where ownership of an heirloom wooden feast-dish also conveyed the right to gather berries on a certain stretch of land with which to fill it; which in turn afforded its owner the right to present those berries while singing a certain song at a certain feast, and so forth.[56] Such forms of sacred property are endlessly complex and variable. Among Plains societies of North America, for instance, sacred bundles (which normally included not only physical objects but accompanying dances, rituals and songs) were often the only objects in that society to be treated as private property: not just owned exclusively by individuals, but also inherited, bought and sold.[57]

Often, the true 'owners' of land or other natural resources were said to be gods or spirits; mortal humans are merely squatters, poachers, or at best caretakers. People variously adopted a predatory attitude to resources – as with hunters, who appropriate what really belongs to the gods – or that of a caretaker (where one is only the 'owner' or 'master' of a village, or men's house, or stretch of territory if one is ultimately responsible for maintaining and looking after it). Sometimes these attitudes coexist, as in Amazonia, where the paradigm for ownership (or 'mastery' – it's always the same word) involves capturing wild animals and then adopting them as pets; that is, precisely the point where violent appropriation of the natural world turns into nurture or 'taking care'.[58]

It is not unusual for ethnographers working with indigenous Amazonian societies to discover that almost everything around them has an owner, or could potentially be owned, from lakes and mountains to cultivars, liana groves and animals. As ethnographers also note, such ownership always carries a double meaning of domination and care. To be without an owner is to be exposed, unprotected.[59] In what anthropologists refer to as totemic systems, of the kind we discussed for Australia and North America, the responsibility of care takes on a particularly extreme form. Each human clan is said to 'own' a certain species of animal – thus making them the 'Bear clan', 'Elk clan', 'Eagle clan' and so forth – but what this means is precisely that members of that clan cannot hunt, kill, harm or otherwise consume animals of that species. In fact, they are expected to take part in rituals that promote its existence and make it flourish.

What makes the Roman Law conception of property – the basis of almost all legal systems today – unique is that the responsibility to care and share is reduced to a minimum, or even eliminated entirely. In Roman Law there are three basic rights relating to possession: *usus* (the right to use), *fructus* (the right to enjoy the products of a property, for instance the fruit of a tree), and *abusus* (the right to damage or destroy). If one has only the first two rights this is referred to as *usufruct*, and is not considered true possession under the law. The defining feature of true legal property, then, is that one has the option of *not* taking care of it, or even destroying it at will.

We are now, finally, approaching a general conclusion about the coming of private property, which can be illustrated by one last and especially striking example: the famous initiation rituals of the Australian Western Desert. Here adult males of each clan act as guardians or custodians of particular territories. There are certain sacra, known as *churinga* or *tsurinja* by the Aranda, which are relics of ancestors who effectively created each clan's territory in ancient times. Mostly, they are smoothed pieces of wood or stone inscribed with a totemic emblem. The same objects could also embody legal title to those lands. Émile Durkheim considered them the very archetype of the sacred: things set apart from the ordinary world and accorded pious devotion; effectively, the 'Holy Ark of the clan'.[60]

During periodic rites of initiation, new cohorts of male Aranda youths are taught about the history of the land and the nature of its resources. They are also charged with the responsibility of caring for it, which in particular means the duty to maintain *churinga* and the sacred sites associated with them, which only the initiated should properly know about in the first place. As observed by T. G. H. Strehlow – an anthropologist and the son of a Lutheran missionary, who spent many years among the Aranda in the early twentieth century, becoming the foremost non-Aranda authority on this topic – the weight of duty is conveyed through terror, torture and mutilation:

> One or two months after the novice has submitted to circumcision, there follows the second principal initiation rite, that of sub-incision . . . The novice has now undergone all the requisite physical operations which have been designed to make him worthy of a man's estate, and he has learned to obey the commands of the old men implicitly. His newly-found blind obedience stands in striking contrast to the unbridled insolence and general unruliness of temper which characterized his behaviour in the days of his childhood. Native children are usually spoiled by their parents. Mothers gratify every whim of their offspring, and fathers do not bother about any disciplinary measures. The deliberate cruelty with which the traditional initiation rites are carried out at a later age is carefully calculated to punish insolent and lawless boys for their past impudence and to train them into obedient, dutiful 'citizens' who will obey their elders without a murmur, and be fit heirs to the ancient sacred traditions of their clan.[61]

Here is another, painfully clear example of how behaviour observed in ritual contexts takes exactly the opposite form to the free and equal relations that prevail in ordinary life. It is only within such contexts that exclusive (sacred) forms of property exist, strict and top-down hierarchies are enforced, and where orders given are dutifully obeyed.[62]

Looking back again to prehistory, it is – as we've already noted – impossible to know precisely which forms of property or ownership existed at places like Göbekli Tepe, Poverty Point, Sannai Maruyama or Stonehenge, any more than we can know if regalia buried with the 'princes' of the Upper Palaeolithic were their personal possessions. What we can now suggest, in light of these wider considerations, is that such carefully co-ordinated ritual theatres, often laid out with geometrical precision, were exactly the kinds of places where exclusive claims to rights over property – together with strict demands for unquestioning obedience – were likely to be made, among otherwise free people. If private property has an 'origin', it is as old as the idea of the sacred, which is likely as old as humanity itself. The pertinent question to ask is not so much when this happened, as how it eventually came to order so many other aspects of human affairs.

5
Many Seasons Ago

Why Canadian foragers kept slaves and their Californian neighbours didn't; or, the problem with 'modes of production'

Our world as it existed just before the dawn of agriculture was anything but a world of roving hunter-gatherer bands. It was marked, in many places, by sedentary villages and towns, some by then already ancient, as well as monumental sanctuaries and stockpiled wealth, much of it the work of ritual specialists, highly skilled artisans and architects.

When considering the broad sweep of history, most scholars either completely ignore this pre-agricultural world or write it off as some kind of strange anomaly: a false start to civilization. Palaeolithic hunters and Mesolithic fisherfolk may have buried their dead like aristocracy, but the 'origins' of class stratification are still sought in much later periods. Louisiana's Poverty Point may have had the dimensions and at least some functions of an ancient city, but it is absent from most histories of North American urbanism, let alone urbanism in general; just as 10,000 years of Japanese civilization is sometimes written off as a prelude to the coming of rice-farming and metallurgy. Even the Calusa of Florida Keys are often referred to as an 'incipient chiefdom'. What's deemed important is not what they were, but the fact that they could be on the brink of turning into something else: a 'proper' kingdom, presumably, whose subjects paid tribute in crops.

This peculiar habit of thought requires us to treat whole populations of 'complex hunter-gatherers' either as deviants, who took some kind of diversion from the evolutionary highway, or as lingering on the cusp of an 'Agricultural Revolution' that never quite took place. It's bad enough when this is applied to a people like the Calusa, who were after all relatively small in number, living in complicated historical circumstances. Yet the same logic is regularly applied to the history of entire indigenous populations along the Pacific Coast of North America, in a territory running from present-day greater Los Angeles to the surroundings of Vancouver.

When Christopher Columbus set sail from Palos de la Frontera in 1492, these lands were home to hundreds of thousands, perhaps even millions of inhabitants.[1] They were foragers, but about as different from the Hadza, Mbuti or !Kung as one can imagine. Living in an unusually bounteous environment, often occupying villages year-round, the indigenous peoples of California, for example, were notorious for their industry and, in many cases, near-obsession with the accumulation of wealth. Archaeologists often characterize their techniques of land management as a kind of incipient agriculture; some even use Aboriginal California as a model for what the prehistoric inhabitants of the Fertile Crescent – who first began domesticating wheat and barley 10,000 years ago in the Middle East – might have been like.

To be fair to the archaeologists, it's an obvious comparison, since ecologically California – with its 'Mediterranean' climate, exceptionally fertile soils and tight juxtaposition of micro-environments (deserts, forests, valleys, coastlands and mountains) – is remarkably similar to the western flank of the Middle East (the area, say, from modern Gaza or Amman north to Beirut and Damascus). On the other hand, a comparison with the inventors of farming makes little sense from the perspective of indigenous Californians, who could hardly have failed to notice the nearby presence – particularly among their Southwest neighbours – of tropical crops, including maize corn, which first arrived there from Mesoamerica around 4,000 years ago.[2] While the free peoples of North America's eastern seaboard nearly all adopted at least some food crops, those of the West Coast uniformly rejected them. Indigenous peoples of California were not pre-agricultural. If anything, they were anti-agricultural.

IN WHICH WE FIRST CONSIDER THE QUESTION OF CULTURAL DIFFERENTIATION

The systematic nature of this rejection of agriculture is a fascinating phenomenon in itself. Most who attempt to explain it nowadays appeal almost entirely to environmental factors: relying on acorns or pine nuts as one's staple in California, or aquatic resources further north, was simply ecologically more efficient than the maize agriculture adopted in other parts of North America. No doubt this was true on the whole, but in an area spanning several thousand miles and a wide variety of different ecosystems, it seems unlikely that there was not a single region where maize cultivation would have been advantageous. And if efficiency was the only consideration, one would have to imagine there were some cultigens – beans, squash, pumpkins, watermelons, any one of an endless variety of leafy vegetables – that someone, somewhere along the coast might have found worth adopting.

The systematic rejection of *all* domesticated foodstuffs is even more striking when one realizes that many Californians and Northwest Coast peoples did plant and grow tobacco, as well as other plants – such as springbank clover and Pacific silverweed – which they used for ritual purposes, or as luxuries consumed only at special feasts.[3] In other words, they were perfectly familiar with the techniques for planting and tending to cultigens. Yet they comprehensively rejected the idea of planting everyday foodstuffs or treating crops as staples.

One reason this rejection is significant is that it offers a clue as to how one might answer the much broader question we posed – but then left dangling – at the beginning of Chapter Four: what is it that causes human beings to spend so much effort trying to demonstrate that they are different from their neighbours? Recall how, after the end of the last Ice Age, the archaeological record is increasingly characterized by 'culture areas': that is, localized populations with their own characteristic styles of clothing, cooking and architecture; and no doubt also their own stories about the origin of the universe, rules for the marriage of cousins, and so forth. Ever since Mesolithic times, the broad tendency has been for human beings to further subdivide, coming up with endless new ways to distinguish themselves from their neighbours.

It is curious how little anthropologists speculate about why this whole process of subdivision ever happened. It's usually treated as self-evident, an inescapable fact of human existence. If any explanation is offered, it's assumed to be an effect of language. Tribes or nations are regularly referred to as 'ethno-linguistic' groups; that is, what is really important about them is the fact they share the same language. Those who share the same language are presumed, all other things being equal, also to share the same customs, sensibilities and traditions of family life. Languages, in turn, are generally assumed to branch off from one another by something like a natural process.

In this line of reasoning, a key breakthrough was the realization – usually attributed to Sir William Jones, a British colonial official stationed in Bengal towards the end of the eighteenth century – that Greek, Latin and Sanskrit all seem to derive from a common root. Before long, linguists had determined that Celtic, Germanic and Slavic languages – as well as Persian, Armenian, Kurdish and more – all belonged to the same 'Indo-European' family. Others, for instance Semitic, Turkic and East Asian languages, did not. Studying relationships among these various linguistic groups eventually led to the science of glottochronology: how distinct languages diverge from a common source. Since all languages are continually changing, and since that change appears to occur at a relatively steady pace, it became possible to reconstruct how and when Turkic languages began to separate from Mongolic, or the relative temporal distance between Spanish and French, Finnish and Estonian, Hawaiian and Malagasy, and so on. All this led to the construction of a series of linguistic family trees, and eventually an attempt – still highly controversial – to trace virtually all Eurasian languages to a single hypothetical ancestor called 'Nostratic'. Nostratic was believed to have existed sometime during the later Palaeolithic, or even to have been the original phylum from which every human language sprang.

It might seem strange to imagine linguistic drift causing a single idiom to evolve into languages as different as English, Chinese and Apache; but, given the extraordinarily long periods of time being considered here, even an accretion of tiny generational changes can, it seems, eventually transform the vocabulary and sound-structure, even the grammar of a language completely.

If cultural differences largely correspond to what happens in language, then distinct human cultures, more generally, would have to be the product of a similar process of gradual drift. As populations migrated or became otherwise isolated from one another, they formed not only their own characteristic languages but their own traditional customs as well. All this involves any number of largely unexamined assumptions – for instance, why is it that languages are always changing to begin with? – but the main point is this. Even if we take such an explanation as a given, it doesn't really explain what we actually observe on the ground.

Consider an ethno-linguistic map of northern California in the early twentieth century, set into a larger map of North American 'culture areas' as defined by ethnologists at that time:

What we are presented with here is a collection of people with broadly similar cultural practices, but speaking a jumble of languages, many drawn from entirely different language families – as distant from one another as, say, Arabic, Tamil and Portuguese. All these groups shared broad similarities: in terms of how they went about gathering and processing foodstuffs; in their most important religious rituals; in the organization of their political life, and so on. But there were also subtle or not-so-subtle differences between them, so that members of each group saw themselves as distinct kinds of people: Yurok, Hupa, Karok and so forth.

These local identities did map on to linguistic differences. However, neighbouring peoples speaking languages drawn from different families (Athabascan, Na-Dene, Uto-Aztecan and so on) actually had far more in common with each other, in almost every other way, than they did with speakers of languages from the same linguistic family living in other parts of North America. The same can be said of the First Nations of the Canadian Northwest Coast, who also speak a variety of unrelated languages, but in other ways resemble one another far more closely than they do speakers of the same languages from outside the Northwest Coast, including in California.

Of course, European colonization had a profound and catastrophic impact on the distribution of Native American peoples, but what we are seeing here also reflects a deeper continuity of culture-historical development, a process that tended to occur at various points in human history, when modern nation states were not around to order populations into neat ethno-linguistic groups. Arguably, the very idea that the world is divided into such homogeneous units, each with its own history, is largely a product of the modern nation state, and the

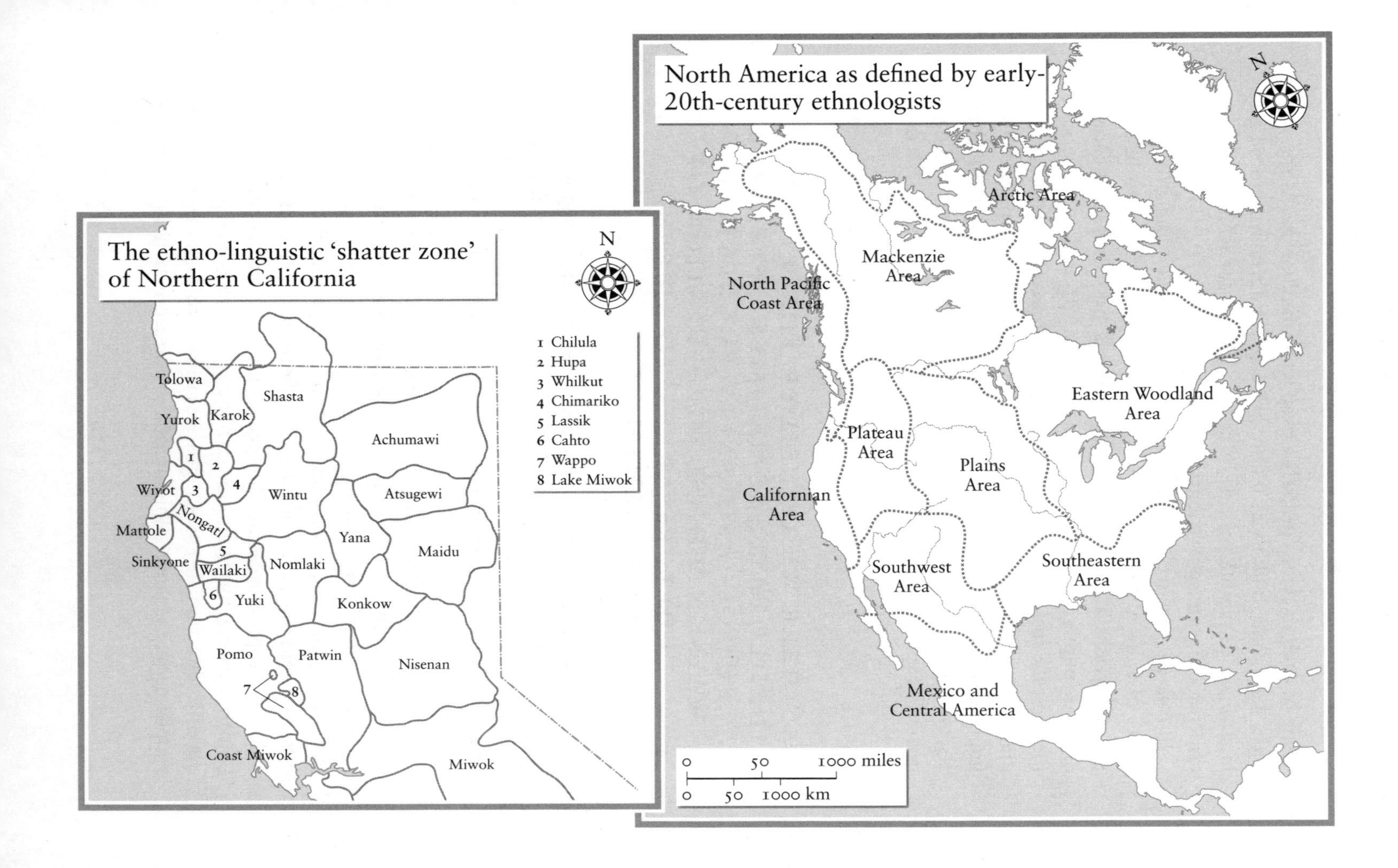

The ethno-linguistic 'shatter zone' of Northern California
N
1 Chilula
2 Hupa
3 Whilkut
4 Chimariko
5 Lassik
6 Cahto
7 Wappo
8 Lake Miwok
Tolowa
Yurok
Karok
Shasta
Achumawi
Wiyot
Wintu
Atsugewi
Nongatl
Mattole
Yana
Maidu
Sinkyone
Wailaki
Nomlaki
Yuki
Konkow
Pomo
Patwin
Nisenan
Coast Miwok
Miwok
North America as defined by early-20th-century ethnologists
N
Arctic Area
Mackenzie Area
North Pacific Coast Area
Eastern Woodland Area
Plateau Area
Plains Area
Californian Area
Southwest Area
Southeastern Area
Mexico and Central America
0 50 1000 miles
0 50 1000 km

desire of each to claim for itself a deep territorial lineage. At the very least, we should think twice before projecting such uniformities back in time, on to remote periods of human history for which no direct evidence of language distributions even exists.

In this chapter, we want to explore what actually *did* drive processes of cultural subdivision for the greater part of human history. Such processes are crucial to understanding how human freedoms, once taken for granted, eventually came to be lost. In doing this, we'll focus on the history of those non-agricultural peoples who inhabited the western coast of North America. As their refusal of agriculture implies, these processes were likely far more self-conscious than scholars usually imagine. In some cases, as we'll see, they appear to have involved explicit reflection and argument about the nature of freedom itself.

WHERE WE CONSIDER THE WILDLY INADEQUATE, SOMETIMES OFFENSIVE BUT OCCASIONALLY SUGGESTIVE WAYS IN WHICH THE QUESTION OF 'CULTURE AREAS' HAS BEEN BROACHED BEFORE

How did earlier generations of scholars describe these regional clusters of societies? The term most commonly used, up until the middle of the twentieth century, was 'culture areas' (or 'culture circles'), a concept which, nowadays, has either been forgotten or fallen into disrepute.

The notion of 'culture areas' first emerged in the last decades of the nineteenth century and the first of the twentieth. Since the Renaissance, human history had been seen largely as the story of great migrations: humans, having fallen from grace, wandering ever more distant from the Garden of Eden. Family trees showing the dispersal of Indo-European or Semitic languages did nothing to discourage this kind of thinking. But the notion of human progress pulled in the exact opposite direction: it encouraged researchers to imagine 'primitive' peoples as tiny, isolated communities, cut off from each other and the larger world. This, of course, is what made it possible to treat them as specimens of earlier stages of human development in the first place: if everyone were in regular contact with each other, this sort of evolutionist analysis wouldn't really work.[4]

The notion of 'culture areas', by contrast, came largely out of museums, and particularly in North America. Curators organizing art and artefacts had to decide whether to arrange their material so as to illustrate theories about the different stages of human adaptation (Lower Savagery, Upper Savagery, Lower Barbarism and so on); or so as to trace the history of ancient migrations, whether real or imagined (in the American context this would mean organizing them by language family, then assumed, for no particularly good reason, to correspond with 'racial' stocks); or whether to simply organize them into regional clusters.[5] Though the last of these seemed most arbitrary, it proved to be the one that really worked best. Art and technology from different Eastern Woodlands tribes, for instance, appeared to have much more in common than material from, say, all speakers of Athabascan languages; or all people who relied mainly on fishing, or cultivated maize. This method turned out to work quite well for archaeological material too, with prehistorians like the Australian V. Gordon Childe observing similar patterns among Neolithic villages stretching across central Europe, forming regional clusters of evidence relating to domestic life, art and ritual.

At first, the most prominent exponent of the culture area approach was Franz Boas. Boas, it will be recalled, was a transplanted German ethnologist[6] who in 1899 landed a chair in anthropology at New York's Columbia University. He also gained a position in charge of ethnographic collections at the American Museum of Natural History, where his halls dedicated to the Eastern Woodlands and Northwest Coast still remain popular attractions over a century later. Boas's student and successor at the museum, Clark Wissler, tried to systematize his ideas by dividing the Americas as a whole, from Newfoundland to Tierra del Fuego, into fifteen different regional systems, each with its own characteristic customs, aesthetic styles, ways of obtaining and preparing food, and forms of social organization. Before long, other ethnologists were undertaking similar projects, mapping out regions from Europe to Oceania.

Boas was a staunch anti-racist. As a German Jew, he was particularly troubled by the way the American obsession with race and eugenics was being taken up in his own mother country.[7] When Wissler began to embrace certain eugenicist ideas, the pair had a bitter falling-out. But the original impetus for the culture area concept was precisely to find a way of talking about human history which avoided ranking populations into higher or lower on *any* grounds, whether claiming some were of superior genetic stock or had reached a more advanced level of moral and technological evolution. Instead, Boas and his students proposed that anthropologists reconstruct the diffusion of what were then referred to as 'culture traits' (ceramics, sweat lodges, the organization of young men into competing warrior societies), and try to understand why, as Wissler put it, tribes of a certain region came to share 'the same mesh of culture traits'.[8]

This resulted in a peculiar fascination with reconstructing the historical movement, or 'diffusion', of specific customs and ideas. Flipping through anthropological journals from the turn of the twentieth century, you find that the majority of the essays in a given number are of this type. They paid special attention to contemporary games and musical instruments used, say, in various different parts of Africa, or of Oceania – perhaps because, of all culture traits, these seemed least affected by practical considerations or constraints, and their distribution might therefore shed light on historical patterns of contact and influence. One especially lively area of debate concerned the string-figure game known as cat's cradles. During the Torres Straits expedition of 1898, Professors Alfred Haddon and W. H. R. Rivers, then leading figures in British anthropology, developed a uniform method of diagramming string figures used in children's play, which made it possible for systematic comparisons to be made. Before long, rival theories concerning the origins and diffusion of particular patterns of string figures (the Palm Tree, the Bagobo Diamond . . .) among different societies were being hotly contested in the pages of the *Journal of the Royal Anthropological Society* and similar erudite venues.[9]

The obvious questions, then, were: why culture traits cluster as they do; and how they come to be 'enmeshed' in regional patterns to begin with. Boas himself was convinced that while geography might have defined the circulation of ideas within particular regions (mountains and deserts forming natural barriers), what happened inside those regions was, effectively, down to historical accident. Others hypothesized about the predominating ethos or form of organization within a given region; or dreamed of creating a kind of natural science that might one day explain or even predict the ebb and flow of styles, habits and social forms. Almost no one reads this literature any more. Like the cat's cradles, today it's considered at best an amusing token of the discipline's childhood.

Still, important issues were raised here: issues which no one to this day has really been able to address. For example, why are the peoples of California so similar to one another, and so different from neighbouring peoples of the American Southwest, or the Canadian Northwest Coast? Perhaps the most insightful contribution came from Marcel Mauss, who tackled the notion of 'culture areas' in a series of essays on nationalism and civilization written between 1910 and 1930.[10] Mauss thought the idea of cultural 'diffusion' was mostly nonsense; not for the reasons most anthropologists do now (that it's pointless and uninteresting),[11] but because he felt it was based on a false assumption: that the movement of people, technologies and ideas was somehow unusual.

The exact opposite was true, Mauss argued. People in past times, he wrote, appear to have travelled a great deal – more than they do today – and it's simply impossible to imagine that anyone back then would have been unaware of the existence of basketry, feather pillows, or the wheel if such objects were regularly employed a month or two's journey away; the same could presumably be said of ancestor cults or syncopated drum rhythms. Mauss went further. He was convinced the entire Pacific Rim had once been a single realm of cultural exchange, with voyagers criss-crossing it at regular intervals. He too was interested in the distribution of games across the entire region. Once, he taught a college course called 'On the greasy pole, the ball play, and other games on the periphery of the Pacific Ocean', his premise being that, at least when it came to games, all lands bordering the Pacific – from Japan to New Zealand to California – could be treated, effectively, as a single culture area.[12] Legend has it that when Mauss, visiting New York's American Museum of Natural History, was shown the famous Kwakiutl war canoe in Boas's Northwest Coast wing, his first reaction was to say that now he knew precisely what ancient China must have looked like.

Though Mauss overstated his case, his exaggeration nonetheless led him to reframe the entire question of 'culture areas' in an intriguing way.[13] For if everyone was broadly aware of what surrounding people were up to, and if knowledge of foreign customs, arts and technologies was widespread, or at least easily available, then the question becomes not why certain culture traits spread, but why other culture traits didn't. The answer, Mauss felt, is that this is precisely how cultures define themselves against their neighbours. Cultures were, effectively, structures of refusal. Chinese are people who use chopsticks, but not knives and forks; Thai are people who use spoons, but not chopsticks, and so forth. It's easy enough to see how this could be true of aesthetics – styles of art, music or table manners – but surprisingly, Mauss found, it extended even to technologies which held obvious adaptive or utilitarian benefits. He was intrigued, for example, by the fact that Athabascans in Alaska steadfastly refused to adopt Inuit kayaks, despite these being self-evidently more suited to the environment than their own boats. Inuit, for their part, refused to adopt Athabascan snowshoes.

What was true of particular cultures was equally true of culture areas; or, as Mauss preferred, 'civilizations'. Since almost any existing style, form or technique has always been potentially available to almost anyone, these too must always have come about through some such combination of borrowing *and* refusal. Crucially, Mauss noted, this process tends to be quite self-conscious. He especially liked to evoke the example of debates in Chinese courts about the adoption of foreign styles and customs, such as the remarkable argument put forward by a king of the Zhou Dynasty to his advisors and great feudal vassals, who were refusing to wear the Hunnish (Manchu) dress and to ride horses instead of driving chariots: he painstakingly tried to show them the difference between rites and customs, between the arts and fashion. 'Societies', wrote Mauss, 'live by borrowing from each other, but they define themselves rather by the refusal of borrowing than by its acceptance.'[14]

Nor are such reflections limited to what historians think of as 'high' (that is, literate) civilizations. Inuit did not simply react with instinctual revulsion when they first encountered someone wearing snowshoes, and then refused to change their minds. They reflected on what adopting, or not adopting, snowshoes might say about the kind of people they considered themselves to be. In fact, Mauss concluded, it is precisely in comparing themselves with their neighbours that people come to think of themselves as distinct groups.

Framed in this way, the question of how 'culture areas' formed is necessarily a political one. It raises the possibility that decisions such as whether or not to adopt agriculture weren't just calculations of caloric advantage or matters of random cultural taste, but also reflected questions about values, about what humans really are (and consider themselves to be), and how they should properly relate to one another. Just the kinds of issues, in fact, which our own post-Enlightenment intellectual tradition tends to express through terms like freedom, responsibility, authority, equality, solidarity and justice.

IN WHICH WE APPLY MAUSS'S INSIGHT TO THE PACIFIC COAST AND CONSIDER WHY WALTER GOLDSCHMIDT'S DESCRIPTION OF ABORIGINAL CALIFORNIANS AS 'PROTESTANT FORAGERS', WHILE IN MANY WAYS ABSURD, STILL HAS SOMETHING TO TELL US

Let us return, then, to the Pacific. Since around the start of the twentieth century, anthropologists have divided the indigenous inhabitants of North America's western littoral into two broad culture areas: 'California' and the 'Northwest Coast'. Before the nineteenth century, when the effects of the fur trade and then the Gold Rush wreaked havoc on indigenous groups and many were exterminated, these populations formed a continuous chain of foraging societies extending through much of the West Coast: at that time, perhaps the largest continuous distribution of foraging peoples in the world. If nothing else, it was a highly efficient way of life; both the Northwest Coast peoples and those of California maintained higher densities of population than, say, maize, beans and squash farmers of the nearby Great Basin and American Southwest.

In other ways, the northern and southern zones were profoundly different, both ecologically and culturally. The peoples of the Canadian Northwest Coast relied heavily on fishing, and particularly the harvesting of anadromous fish such as salmon and eulachon, which migrate upriver from the sea to spawn; as well as a variety of marine mammals, terrestrial plants and game resources. As we saw a couple of chapters ago, these groups divided their year between very large coastal winter villages, holding ceremonies of great complexity, and, in spring and summer, smaller social units that were more pragmatically focused on the provision of food. Expert woodworkers, they also transformed the local conifers (fir, spruce, redwood, yew and cedar) into a dazzling material culture of carved and painted masks, containers, tribal crests, totem poles, richly decorated houses and canoes which ranks among the world's most striking artistic traditions.

Aboriginal societies in California, to the south, occupied one of the world's most diverse habitats. They made use of a staggering variety of terrestrial resources, which they managed by careful techniques of burning, clearing and pruning. The region's 'Mediterranean' climate and tightly compressed topography of mountains, deserts, foothills, river valleys and coastlines made for a rich assortment of local flora and fauna, exchanged at inter-tribal trade fairs. Most Californians were proficient fishers and hunters, but many also followed an ancient reliance on tree crops – especially nuts and acorns – as staple foods. Their artistic traditions differed from those of the Northwest Coast. House exteriors were generally plain and simple. There was almost nothing similar to the Northwest Coast masks or monumental sculptures that so delight museum curators; rather, aesthetic activity focused on the weaving of highly patterned baskets used for storing and serving food.[15]

There was a further important difference between these two extensive groupings of societies, one that for some reason is far less remarked on by scholars today. From the Klamath River northwards, there existed societies dominated by warrior aristocracies engaged in frequent inter-group raiding and in which, traditionally, a significant portion of the population had consisted of chattel slaves. This apparently had been true as long as anyone living there could remember. But none of this was the case further south. How exactly did this happen? How did a boundary emerge between one extended 'family' of foraging societies that habitually raided each other for slaves, and another that did not keep slaves at all?

You might think there would be a lively debate about this among scholars, but in fact there isn't. Instead, most treat the differences as insignificant, preferring to lump all Californian and Northwest Coast societies together in a single category of 'affluent foragers' or 'complex hunter-gatherers'.[16] If differences between them are considered at all, they are usually understood as mechanical responses to their contrasting modes of subsistence: aquatic (fish-based) economies, it's argued, simply tended to foster warlike societies, just as terrestrial (acorn-based) foraging economies somehow did not.[17] We will shortly consider the merits and limitations of such recent arguments, but first it is useful to turn back to some of the ethnographic work undertaken by earlier generations.

Some of the most striking research about the indigenous peoples of California was done by the twentieth-century anthropologist Walter Goldschmidt. One of his key writings, unobtrusively entitled 'an ethnological contribution to the sociology of knowledge', was concerned with the Yurok and other related groups who inhabited the northwestern corner of California, just south of the mountain ranges where Oregon begins.[18] For Goldschmidt and members of his anthropological circle, the Yurok were famous for the central role that money – which took the form of white dentalium shells arranged on strings, and headbands made of bright red woodpecker scalps – played in every aspect of their social lives.

It's worth mentioning here that settlers in different parts of North America referred to a whole variety of things as 'Indian money'. Often these were shell beads or actual shells. But in almost every case, the term is largely a projection of European categories on to objects that look like money, but really aren't. Probably the most famous of these, *wampum*, did eventually come to be used as a trade currency in transactions between settlers and indigenous peoples of the Northeast, and was even accepted as currency in several American states for transactions between settlers (in Massachusetts and New York, for instance, *wampum* was legal tender in shops). In dealings between indigenous people, however, it was almost never used to buy or sell anything. Rather, it was employed to pay fines, and as a way of forming and remembering compacts and agreements. This was true in California as well. But in California, unusually, money also seems to have been used in more or less the way we expect money to have been used: for purchases, rentals and loans. In California in general, and its northwest corner in particular, the central role of money in indigenous societies was combined with a cultural emphasis on thrift and simplicity, a disapproval of wasteful pleasures, and a glorification of work that – according to Goldschmidt – bore an uncanny resemblance to the Puritan attitudes described by Max Weber in his famous 1905 essay, *The Protestant Ethic and the Spirit of Capitalism.*

This analogy might seem a bit of a stretch, and in many ways it was. But it's important to understand the comparison that Goldschmidt was actually making. Weber's essay, familiar to just about anyone who's ever taken a social science course, is often misunderstood. Weber was trying to answer a very specific question: why capitalism emerged in western Europe, and not elsewhere. Capitalism, as he defined it, was itself a kind of moral imperative. Almost everywhere in the world, he noted, and certainly in China, India and the Islamic world, one found commerce, wealthy merchants and people who might justly be referred to as 'capitalists'. But almost everywhere, anyone who acquired an enormous fortune would eventually cash in their chips. They would either buy themselves a palace and enjoy life, or come under enormous moral pressure from their community to spend their profits on religious or public works, or boozy popular festivities (usually they did a bit of both).

Capitalism, on the other hand, involved constant reinvestment, turning one's wealth into an engine for creating ever more wealth, increasing production, expanding operations, and so forth. But imagine, Weber suggested, being the very first person in one's community to act this way. To do so would have meant defying all social expectations, to be utterly despised by almost all your neighbours – who would, increasingly, also become your employees. Anyone capable of acting in such a defiantly single-minded manner, Weber observed, would 'have to be some sort of hero'. This, he said, is the reason why it took a Puritanical strain of Christianity, like Calvinism, to make capitalism possible. Puritans not only believed almost anything they could spend their profits on was sinful; but also, joining a Puritan congregation meant one had a moral community whose support would allow one to endure the hostility of one's hell-bound neighbours.

Obviously, none of this was true in an eighteenth-century Yurok village. Aboriginal Californians did not hire one another as wage labourers, lend money at interest, or invest the profits of commercial ventures to expand production. There were no 'capitalists' in the literal sense. What there was, however, was a remarkable cultural emphasis on private property. As Goldschmidt notes, all property, whether natural resources, money or items of wealth, was 'privately (and for the most part individually) owned', including fishing, hunting and gathering grounds. Individual ownership was complete, with full rights of alienation. Such a highly developed concept of property, Goldschmidt observed, requires the use of money, such that in Northwest California 'money buys everything – wealth, resources, food, honor and wives.'[19]

This very unusual property regime corresponded to a broad ethos, which Goldschmidt compared to Weber's 'spirit' of capitalism (though, one might object, it corresponds more to how capitalists like to imagine the world than to how capitalism actually works). The Yurok were what we've called 'possessive individualists'. They took it for granted that we are all born equal, and that it is up to each of us to make something of ourselves through self-discipline, self-denial and hard work. What's more, this ethos appears to have been largely applied in practice.

As we've seen, the indigenous peoples of the Northwest Coast were just as industrious as those of California, and in both cases those who accumulated wealth were expected to give much of it away by sponsoring collective festivals. The underlying ethos, however, could not have been more different. Where the wealthy Yurok were expected to be modest, Kwakiutl chiefs were boastful and vainglorious; so much so that one anthropologist compared them to paranoid schizophrenics. Where wealthy Yurok made little of their ancestry, Northwest

Coast households had much in common with the noble houses and dynastic estates of medieval Europe, in which a class of nobles jockeyed for position within ranks of hereditary privilege, staging dazzling banquets to enhance their reputations and secure their claims to honorific titles and heirloom treasures stretching back to the beginning of time.[20]

It's hard to imagine that the existence of such striking cultural differences between neighbouring populations could be completely coincidental, but it's also extremely difficult to find any studies that even begin to address the question of how this contrast came about.[21] Is it possible to see indigenous Californians and peoples of the Northwest Coast as defining themselves against each other, rather in the manner that Californians and New Yorkers do today? If so, then how much of their way of life can we really explain as being motivated by a desire to be unlike other groups of people? Here, we need to bring back our earlier discussion of schismogenesis, which we introduced to help make sense of the intellectual encounter between seventeenth-century French colonists and the Wendat people of North America's Eastern Woodlands.

Schismogenesis, you'll recall, describes how societies in contact with each other end up joined within a common system of differences, even as they attempt to distinguish themselves from one another. Perhaps the classic historical example (in both senses of the term 'classic') would be the ancient Greek city-states of Athens and Sparta, in the fifth century BC. As Marshall Sahlins puts it:

> Dynamically interconnected, they were then reciprocally constituted . . . Athens was to Sparta as sea to land, cosmopolitan to xenophobic, commercial to autarkic, luxurious to frugal, democratic to oligarchic, urban to villageois, autochthonous to immigrant, logomanic to laconic: one cannot finish enumerating the dichotomies . . . Athens and Sparta were antitypes.[22]

Each society performs a mirror image of the other. In doing so, it becomes an indispensable alter ego, the necessary and ever-present example of what one should never wish to be. Might a similar logic apply to the history of foraging societies in California and on the Northwest Coast?

WHERE WE MAKE A CASE FOR SCHISMOGENESIS BETWEEN 'PROTESTANT FORAGERS' AND 'FISHER KINGS'

Let's look more closely at what might be described, in Weber's sense, as the 'spirit' of northern Californian foragers. At root, it was a series of ethical imperatives, in Goldschmidt's words: 'the moral demand to work and by extension pursuit of gain; the moral demand of self-denial; and the individuation of moral responsibility'.[23] Bound up in this was a passion for individual autonomy as absolute as that of any Kalahari Bushman – even if it took a strikingly different form. Yurok men scrupulously avoided being placed in a situation of debt or ongoing obligation to anyone else. Even the collective management of resources was frowned upon; foraging grounds were individually owned and could be rented out in times of shortfall.

Property was sacred, and not only in the legal sense that poachers could be shot. It also had a spiritual value. Yurok men would often spend long hours meditating on money, while the highest objects of wealth – precious hides and obsidian blades displayed only at festivals – were the ultimate sacra. Yurok struck outsiders as puritanical in a literal sense as well: as Goldschmidt reports, ambitious Yurok men were 'exhorted to abstain from any kind of indulgence – eating, sexual gratification, play or sloth'. Big eaters were considered 'vulgar'. Young men and women were lectured on the need to eat slowly and modestly, to keep their bodies slim and lithe. Wealthy Yurok men would gather every day in sweat lodges, where an almost daily test of these ascetic values was the need to crawl headfirst through a tiny aperture that no overweight body could possibly enter. Repasts were kept bland and spartan, decoration simple, dancing modest and restrained. There were no inherited ranks or titles. Even those who did inherit wealth continued to emphasize their personal hard work, frugality and achievement; and while the rich were expected to be generous towards the less fortunate and look after their own lands and possessions, responsibilities for sharing and caring were modest in comparison with foraging societies almost anywhere else.

Northwest Coast societies, in contrast, became notorious among outside observers for the delight they took in displays of excess. They were best known to European ethnologists for the festivals called *potlatch*, usually held by aristocrats acceding to some new noble title (nobles would often accumulate many of these over the course of a lifetime). In these feasts they sought to display their grandeur and contempt for ordinary worldly possessions by performing magnificent feats of generosity, overwhelming their rivals with gallons of candlefish oil, berries and quantities of fatty and greasy fish. Such feasts were scenes of dramatic contests, sometimes culminating in the ostentatious destruction of heirloom copper shields and other treasures, just as in the early period of colonial contact, around the turn of the nineteenth century, they sometimes culminated in the sacrificial killing of slaves. Each treasure was unique; there was nothing that resembled money. *Potlatch* was an occasion for gluttony and indulgence, 'grease feasts' designed to leave the body shiny and fat. Nobles often compared themselves to mountains, with the gifts they bestowed rolling off them like boulders, to flatten and crush their rivals.

The Northwest Coast group we know best are the Kwakwaka'wakw (Kwakiutl), among whom Boas conducted fieldwork. They became famous for the exuberant ornamentation of their art – their love of masks within masks – and the theatrical stage effects employed in their rituals, including fake blood, trap doors and violent clown-police. All the surrounding societies – including the Nootka, Haida and Tsimshian – appear to have shared the same broad ethos: similarly dazzling material cultures and performances could be found all the way from Alaska south to the area of Washington State. They also shared the same basic social structure, with hereditary ranks of nobles, commoners and slaves. Throughout this entire region, a 1,500-mile strip of land from the Copper River delta to Cape Mendocino, inter-group raiding for slaves was endemic, and had been for as long as anyone could recall.

In all these societies of the Northwest Coast, nobles alone enjoyed the ritual prerogative to engage with guardian spirits, who conferred access to aristocratic titles, and the right to keep the slaves captured in raids. Commoners, including brilliant artists and craftspeople, were largely free to decide which noble house they wished to align themselves with; chiefs vied for their allegiance by sponsoring feasts, entertainment and vicarious participation in their heroic adventures. 'Take good care of your people,' went the elder's advice to a young Nuu-chah-nulth (Nootka) chief. 'If your people don't like you, you're nothing.'[24]

In many ways, the behaviour of Northwest Coast aristocrats resembles that of Mafia dons, with their strict codes of honour and patronage relations; or what sociologists speak of as 'court societies' – the sort of arrangement one might expect in, say, feudal Sicily, from which the Mafia derived many of its cultural codes.[25] But this is emphatically not what we are taught to expect among foragers. Granted, the followers of any one of these 'fisher-kings' rarely numbered more than 100 or 200 people, not much larger than the size of a Californian village; in neither the Northwest Coast nor the Californian culture area were there overarching political, economic or religious organizations of any kind. But within the tiny communities that did exist, entirely different principles of social life applied.

All this begins to make the anthropologists' habit of lumping Yurok notables and Kwakiutl artists together as 'affluent foragers' or 'complex hunter-gatherers' seem rather silly: the equivalent of saying a Texas oil executive and a medieval Egyptian poet were both 'complex agriculturalists' because they ate a lot of wheat.

But how do we explain the differences between these two culture areas? Do we start from the institutional structure (the rank system and importance of *potlatch* in the Northwest Coast, the role of money and private property in California), then try to understand how the prevailing ethos of each society emerges from it? Or did the ethos come first – a certain conception of the nature of humanity and its role in the cosmos – and did the institutional structures emerge from that? Or are both simply effects of a different technological adaptation to the environment?

These are fundamental questions about the nature of society. Theorists have been batting them about for centuries, and probably will be for centuries to come. To put the matter more technically, we might ask what ultimately determines the shape a society takes: economic factors, organizational imperatives or cultural meanings and ideas?

Following in the footsteps of Mauss, we might also suggest a fourth possibility. Are societies in effect self-determining, building and reproducing themselves primarily with reference to each other?

There's a lot riding on the answer we give in this particular case. The indigenous history of the Pacific Coast might not provide a very good model for what the first 'proto-farmers' in the Fertile Crescent were like, 10,000 years ago. But it does shed unique light on other kinds of cultural processes, which – as we explored above – have been going on for just as long, if not longer: whereby certain foraging peoples, in particular times and places, came to accept permanent inequalities, structures of domination and the loss of freedoms.

Let's now go through the possible explanations, one by one.

The most striking difference between the indigenous societies of California and the Northwest Coast is the absence, in California, of formal ranks and the institution of *potlatch*. The second really follows from the first. In California there were feasts and festivals, to be sure, but since there was no title system, these lacked almost all the distinctive features of *potlatch*: the division between 'high' and 'low' forms of cuisine, the use of ranked seating orders and serving equipment, obligatory eating of oily foods, competitive gifting, self-aggrandizing speeches, or any other public manifestations of rivalry between nobles fighting over titular privilege.[26]

In many ways, the seasonal gatherings of Californian tribes seem exactly to reverse the principles of *potlatch*. Staple rather than luxury foods were consumed; ritual dances were playful rather than regimented or menacing, often involving the humorous transgression of social boundaries between men and women, children and elders (they seem to be one of the few occasions when the otherwise staid Yurok allowed themselves to have a bit of fun). Valuables such as obsidian blades and deer skins were never sacrificed or gifted to enemies as a challenge or insult, but carefully unwrapped and passed into the trust of temporary 'dance leaders', as if to underline how much their owners wished to avoid drawing undue attention to themselves.[27]

Local headmen in California certainly did benefit by hosting such occasions: social connections were made, and an enhanced reputation could often mean later opportunities to make money.[28] But insofar as feast sponsors could be seen as self-aggrandizing, they themselves went to great lengths to downplay their roles, and anyway, attributing a secret desire for profit to them seems reductive in the extreme, even rather insulting, considering the actual redistribution of resources that went on in Californian trade feasts and 'deerskin dances', and their well-documented importance in promoting solidarity between groups from neighbouring hamlets.[29]

So are we talking about the same basic institution (a 'redistributive feast') carried out in an entirely different spirit, or two entirely different institutions, or even, *potlatch* and anti-*potlatch*? How are we to tell? Clearly the issue is much broader, and touches on the very nature of 'culture areas' and what actually constitutes a threshold or boundary between them. We are looking for a key to this problem. It lies in the institution of slavery, which, as we've noted, was endemic on the Northwest Coast but correspondingly absent south of the Klamath River in California.

Slaves on the Northwest Coast were hewers of wood and drawers of water, but they were especially involved in the mass harvesting, cleaning and processing of salmon and other anadromous fish. There's no consensus, however, on how far back the indigenous practice of slavery actually went there. The first European accounts of the region in the late eighteenth century speak of slaves, and express mild surprise in doing so, since full-fledged chattel slavery was quite unusual in other parts of aboriginal North America. These accounts suggest that perhaps a quarter of the indigenous Northwest Coast population lived in bondage – which is about equivalent to proportions found in the Roman Empire, or classical Athens, or indeed the cotton plantations of the American South. What's more, slavery on the Northwest Coast was a hereditary status: if you were a slave, your children were also fated to be so.[30]

Given the limitations of our sources, it's always possible that these European accounts were describing what was, at the time, a recent innovation. Current archaeological and ethno-historical research, though, suggests that the institution of slavery goes back a very long way indeed on the Northwest Coast, many centuries before European ships began docking at Nootka Sound to trade in otter pelts and blankets.

CONCERNING THE NATURE OF SLAVERY AND 'MODES OF PRODUCTION' MORE GENERALLY

It's fiendishly difficult to 'find slavery' in the archaeological record, unaided by written records; but on the West Coast we can at least observe how many of the elements that later came together in the institution of slavery emerged at roughly the same time, starting around 1850 BC, in what's called the Middle Pacific period. This is when we first observe the bulk harvesting of anadromous fish, an incredibly bounteous resource – later travellers recounted salmon runs so massive one could not see the water for the fish – but one that involved a dramatic intensification of labour demands. It's presumably no coincidence that around this same time, we see also the first signs of warfare and the building of defensive fortifications, and expanding trade networks.[31] There are also some other pointers.

Cemeteries of Middle Pacific age, between 1850 BC and AD 200, reveal extreme disparities in treatments of the dead, something not seen in earlier times. At the 'top end', the most privileged burials exhibit formal systems of body ornamentation, and the somewhat macabre staging of corpses in seated, reclining or other fixed positions, presumably referencing a strict hierarchy of ritual postures and manners among the living. At the 'bottom' we see quite the other extreme: mutilation of certain individuals' bodies, recycling of human bone to make tools and containers, and the 'offering' of people as grave goods (i.e. human sacrifice). The overall impression is of a wide spectrum of formalized statuses, ranging from high rank to people whose lives and deaths appear to have mattered little.[32]

Turning now to California, one thing we can note straight away is the absence of all these features in correspondingly early periods. South of Cape Mendocino we seem to be dealing with a different kind of Middle Pacific – a more 'pacific' one, in fact. But we can't put these differences down to a lack of contact between the two groups. On the contrary, archaeological and linguistic evidence demonstrates extensive movement of people and goods along much of the West Coast. A vibrant, canoe-borne maritime commerce already linked coastal and island societies, conveying valuables such as shell beads, copper, obsidian and a host of organic commodities across the diverse ecologies of the Pacific littoral. Various lines of evidence also point to the movement of human captives as a feature of inter-group warfare and trade. As early as 1500 BC, some parts of the shoreline around the Salish Sea were already equipped with fortifications and shelters, in apparent anticipation of raids.[33]

So far, we have been talking about slavery without really defining the term. This is a little unwise, because Amerindian slavery had certain specific features that make it very different from ancient Greek or Roman household slavery, let alone European plantation slavery in the Caribbean or in America's Deep South. While slavery of any sort was a fairly unusual institution among indigenous peoples of the Americas, some of these distinctively Amerindian features were shared, at least in their broad outlines, across much of the continent, including the tropics, where the earliest Spanish sources document local forms of slavery back to the fifteenth century AD. The Brazilian anthropologist Fernando Santos-Granero has coined a term for Amerindian societies that possessed these features. He calls them 'capturing societies'.[34]

Before exploring what he means, let's define slavery itself. What makes a slave different from a serf, a peon, captive or inmate is their lack of social ties. In legal terms, at least, a slave has no family, no kin, no community; they can make no promises and forge no ongoing connections with other human beings. This is why the English word 'free' is actually derived from a root meaning 'friend'. Slaves could not have friends because they could not make commitments to others, since they were entirely under someone else's power and their only obligation was to do exactly what their master said. If a Roman legionary was captured in battle and enslaved, then managed to escape and return home, he had to go through an elaborate process of restoring all his social relationships, including remarrying his wife, since the act of enslaving him was considered to have severed all previous relationships. The West Indian sociologist Orlando Patterson has referred to this as a condition of 'social death'.[35]

Unsurprisingly, the archetypical slaves are usually war captives, who are typically far from home amid people who owe them nothing. There is another practical reason for turning war captives into slaves. A slave's master has a responsibility to keep them alive in a fit state to work. Most human beings need a good deal of care and resources, and can usually be considered a net economic loss until they are twelve or sometimes fifteen years old. It rarely makes economic sense to breed slaves – which is why, globally, slaves have so often been the product of military aggression (though many were also products of debt traps, punitive judicial decisions or banditry). Seen one way, a slave-raider is stealing the years of caring labour another society invested to create a work-capable human being.[36]

What, then, do Amerindian 'capturing societies' have in common which makes them distinctive from other kinds of slave-holding societies? On the face of it, not much. And least of all their modes of subsistence, which were about as diverse as could be imagined. As Santos-Granero points out, in Northwest Amazonia the dominant peoples were sedentary horticulturalists and fishermen living along the largest rivers, who raided the nomadic hunting-gathering bands of the hinterland. By contrast, in the Paraguay River basin it was semi-itinerant hunter-gatherers who raided or subjugated village agriculturalists. In southern Florida the hegemonic groups (Calusa, in this case) were fishermen-gatherers who lived in large, permanent villages but moved seasonally to fishing and gathering sites, raiding both fishing and farming communities.[37]

Classifying these groups according to how much they farmed, fished or hunted tells us little of their actual histories. What really mattered, in terms of the ebb and flow of power and resources, was their use of organized violence to 'feed off' other populations. Sometimes the foraging peoples – such as the Guaicurú of the Paraguay palm savannah, or the Calusa of Florida Keys – had the upper hand militarily over their agricultural neighbours. In such cases, taking slaves and exacting tribute exempted a portion of the dominant society from basic subsistence chores, and supported the existence of leisured elites. It also supported the training of specialized warrior castes, which in turn created the means for further appropriation and further tribute.

Here, again, the idea of classifying human societies by 'modes of subsistence' looks decidedly naive. How, for instance, would we propose to classify foragers who consume quantities of *domestic* crops, exacted as tribute from nearby farming populations? Marxists, who refer to 'modes of production', do sometimes allow for a 'Tributary Mode,' but this has always been linked to the growth of agrarian states and empires, back to Book III of Marx's *Capital*.[38] What really needs to be theorized here is not just the mode of production practised by *victims* of predation, but also that of the non-producers who prey on them. Now wait. A non-productive mode of production? This sounds like a contradiction in terms. But it's only so if we limit the meaning of 'production' strictly to the creation of food or goods. And maybe we shouldn't.

'Capturing societies' in the Americas considered slave-taking as a mode of subsistence in its own right, but not in the usual sense of producing calories. Raiders almost invariably insisted that slaves were captured for their life force or 'vitality' – vitality which was consumed by the conquering group.[39] Now, you might say this is literally true: if you exploit another human being for their labour, either directly or indirectly, you are living off their energies or life force; and if they are providing you with food, you are in fact eating it. But there is slightly more going on here.

Let's recall Amazonian ideas of ownership. You appropriate something from nature, killing or uprooting it, but then this initial act of violence is transformed into a relation of caring, as you maintain and tend what is captured. Slave-raiding was talked about in similar terms, as hunting (traditionally men's work), and captives were likened to vanquished prey. Experiencing social death, they would come to be regarded as something more like 'pets'. While being re-socialized in their captors' households they had to be nurtured, cooked for, fed and instructed in the proper ways of civilization; in short, domesticated (these tasks were usually women's work). If the socialization was completed, the captive ceased to be a slave. However, captives could sometimes be kept suspended in social death, as part of a permanent pool of victims awaiting their actual, physical death. Typically they would be killed at collective feasts (akin to the Northwest Coast *potlatch*) presided over by ritual specialists, and this would sometimes result in the eating of enemy flesh.[40]

All this may seem exotic. However, it echoes the way exploited people everywhere and throughout history tend to feel about their situation: their bosses, or landlords, or superiors are blood-sucking vampires, and they are treated at best as pets and at worst as cattle. It's just that in the Americas, a handful of societies enacted those relationships in a quite literal fashion. The more important point, concerning 'modes of production' or 'modes of subsistence', is that this kind of exploitation often took the form of ongoing relations *between* societies. Slavery almost always tends to do this, since imposing 'social death' on people whose biological relatives speak the same language as you and can easily travel to where you live will always create problems.

Let's recall how some of the first European travellers to the Americas compared 'savage' males to noblemen back home – because, like these noblemen, they dedicated almost all their time to politics, hunting, raiding and waging war on neighbouring groups. A German observer in 1548 spoke of Arawakan villagers of the Grand Chaco in Paraguay as serfs of Guaicurú foragers, 'in the same way as German rustics are with respect to their lords'. The implication was that little really separates a Guaicurú warrior from a Swabian feudal baron, who likely spoke French at home, feasted regularly on wild game and lived off the labour of German-speaking peasants, even though he had never touched a plough. At what point, we might then ask, were the Guaicurú, who lived amid piles of maize, manioc (cassava) and other agricultural products delivered as tribute, as well as slaves secured in raids on societies even further distant, no longer simply 'hunter-gatherers' (especially if they were also hunting and gathering other humans)?

True, crops were sent as tribute from nearby conquered villages, but tributary villages also sent servants, and raids on villages further out tended to concentrate on enslaving women, who could serve as concubines, nursemaids and domestics – allowing Guaicurú 'princesses', their bodies often completely covered with intricate tattoos and spiral designs painted on daily by their domestics, to devote their days to leisure. Early Spanish commentators always remarked that Guaicurú treated their slaves with care and even tenderness, almost exactly as they did their pet parrots and dogs,[41] but what was really going on here? If slavery is the theft of labour that other societies invest in bringing up children, and the main purpose to which slaves were put was caring for children, or attending to and grooming a leisure class, then, paradoxically, the main objective of slave-taking for the 'capturing society' seems to have been to increase its internal capacity for caring labour. What was ultimately being produced here, within Guaicurú society, were certain kinds of people: nobles, princesses, warriors, commoners, servants, and so on.[42]

What needs emphasizing – since it will become extremely important as our story unfolds – is the profound ambivalence, or perhaps we might better say double-edged-ness, of these caring relationships. Amerindian societies typically referred to themselves by some term that can be roughly translated as 'human beings' – most of the tribal names traditionally applied to them by Europeans are derogatory terms used by their neighbours ('Eskimo', for example, means 'people who don't cook their fish', and 'Iroquois' is derived from an Algonkian term meaning 'vicious killers'). Almost all these societies took pride in their ability to adopt children or captives – even from among those whom they considered the most benighted of their neighbours – and, through care and education, turn them into what they considered to be proper human beings. Slaves, it follows, were an anomaly: people who were neither killed nor adopted, but who hovered somewhere in between; abruptly and violently suspended in the midpoint of a process that should normally lead from prey to pet to family. As such, the captive as slave becomes trapped in the role of 'caring for others', a non-person whose work is largely directed towards enabling those others to become persons, warriors, princesses, 'human beings' of a particularly valued and special kind.

As these examples show, if we want to understand the origins of violent domination in human societies, this is precisely where we need to look. Mere acts of violence are passing; acts of violence transformed into caring relations have a tendency to endure. Now that we have a clearer idea of what Amerindian slavery actually involved, let us return to the Pacific Coast of North America and try to understand some of the specific conditions that made chattel slavery so prevalent on the Northwest Coast, and so unusual in California. We'll start with a piece of oral history, an old story.

IN WHICH WE CONSIDER 'THE STORY OF THE WOGIES' – AN INDIGENOUS CAUTIONARY TALE ABOUT THE DANGERS OF TRYING TO GET RICH QUICK BY ENSLAVING OTHERS (AND INDULGE OURSELVES IN AN ASIDE ON 'GUNS, GERMS AND STEEL')

The story we're about to recount is first attested in 1873 by the geographer A. W. Chase. Chase claims it was related to him by people of the Chetco Nation of Oregon. It concerns the origins of the word 'Wogie' (pronounced 'Wâgeh'), which across much of the coastal region was an indigenous term for white settlers. The story didn't really register among scholars; it was repeated a couple of times in the following half-century or so, but otherwise that was it. Yet this long-overlooked story contains some precious gems of information, especially about indigenous attitudes to slavery, at precisely the interface between California and the Northwest Coast that we've been exploring.

Barely a handful of Chetco exist today. Originally dominating the southern shoreline of Oregon, they were largely wiped out in genocidal massacres carried out by invading settlers in the mid nineteenth century. By the 1870s, a small number of survivors were living in the Siletz Reservation, now in Lincoln County. This is what their ancestors told Chase about their origins and where they had come from:

> The Chetkos say that, many seasons ago, their ancestors came in canoes from the far north, and landed at the river's mouth. They found two tribes in possession, one a warlike race, resembling themselves; these they soon conquered and exterminated. The other was a diminutive people, of an exceedingly mild disposition, and white. These called themselves, or were called by the new-comers, 'Wogies.' They were skillful in the manufacture of baskets, robes, and canoes, and had many methods of taking game and fish which were unknown to the invaders. Refusing to fight, the Wogies were made slaves of, and kept at work to provide food and shelter and articles of use for the more warlike race,

> who waxed very fat and lazy. One night, however, after a grand feast, the Wogies packed up and fled, and were never more seen. When the first white men appeared, the Chetkos supposed that they were the Wogies returned. They soon found out their mistake, however, but retained among themselves the appellation for the white men, who are known as Wogies by all the coast tribes in the vicinity.[43]

The tale might seem unassuming, but there's a lot packed into it. That the survivors of a forager group on the Oregon coast should narrate Euro-American colonization as an act of historical vengeance is unsurprising.[44] Neither is there anything implausible about an indigenous slave-holding society migrating south by sea into new territory, at some remote time, and either subjugating or killing the autochthonous inhabitants.[45]

Similarly to the Guaicurú, the aggressors appear to have made a point of subduing people with skills they themselves lacked. What the 'proto-Chetco' acquired was not just physical brawn ('Wogie labour') or even care, but the accumulated *savoir-faire* of a hunter-fisher-forager people not entirely unlike themselves and, according to the story at least, in many respects more capable.

Another intriguing feature of this story is its setting. The Chetco lived in the intermediate zone between our two major culture regions, precisely where one would imagine the institution of slavery to be most explicitly debated and contested. And indeed, the story has a distinctly ethical flavour, as if it were a cautionary tale aimed at anyone tempted to render others slaves, or acquire wealth and leisure through raiding. Having forced their victims into servitude, growing 'fat and lazy' on the proceeds, it's the Chetcos' newfound sloth that makes them unable to pursue the fleeing Wogies. The Wogies come out of the whole affair on top by virtue of their pacifism, industriousness, craft skills and capacity for innovation; indeed, they get to make a lethal return – in spirit, at least – as Euro-American settlers equipped with 'guns, germs, and steel'.[46]

Taking this into account, the tale of the Wogies points to some intriguing possibilities. Most importantly, it indicates that the rejection of slavery among groups in the region between California and the Northwest Coast had strong ethical and political dimensions. And indeed, once one starts looking, it's not hard to find further evidence for this. The Yurok, for example, did hold a small number of slaves, mainly debt peons or captives not yet ransomed by their relatives. But their legends evince a strong disapproval. To take one example, a heroic protagonist makes his fame by defeating a maritime adventurer named Le'mekwelolmei, who would pillage and enslave passing travellers. After defeating him in combat, our hero rejects his appeal to join forces:

> 'No, I do not want to be like you, summoning boats to the shore, seizing them and their cargo, and making people slaves. As long as you live you will never be tyrannous again, but like other men.'
>
> 'I will do so,' said Le'mekwelolmei.
>
> 'If you return to your former ways, I will kill you. Perhaps I should take you for a slave now, but I will not. Stay in your home and keep what is yours and leave people alone.' To the slaves who stood about nearly filling the river bank, he said, 'Go to your homes. You are free now.'
>
> The people who had been enslaved surrounded him, weeping and thanking him and wanting to drag his boat back to the water. 'No, I will drag it myself,' he said, and then with one hand he lifted it to the river. So the freed people all scattered, some down-river and some upriver to their homes.[47]

Northwest Coast-style maritime raiding was in no sense celebrated, to say the least.

Still, one might ask: might there not be a more straightforward explanation for the prevalence of slavery on the Northwest Coast, and its absence further south? It's easy to express moral disapproval of a practice if there's not much economic incentive to practise it anyway. An ecological determinist would almost necessarily argue this, and in fact there is a body of literature that makes just such an argument for the Pacific Coast – and it's about the only literature that does actually take on the question of why different coastal societies looked so different in the first place. This is a branch of behavioural ecology called 'optimal foraging theory'. Its proponents make some interesting points. Before proceeding, then, let us consider them.

IN WHICH WE ASK: WOULD YOU RATHER FISH, OR GATHER ACORNS?

Optimal foraging theory is a style of predictive modelling that originates in the study of non-human species such as starlings, honeybees or fish. Applied to humans, it typically frames behaviour in terms of economic rationality, i.e.: 'foragers will design their hunting and collecting strategies with the intention of obtaining a maximum return in calories, for a minimum outlay of labour.' This is what behavioural ecologists call a 'cost-benefit' calculation. First you figure out how foragers ought to act, if they are trying to be as efficient as possible. Then you examine how they do in fact act. If it doesn't correspond to the optimum foraging strategy, something else must be going on.

From this perspective, the behaviour of indigenous Californians was far from optimal. As we've noted, they relied primarily on gathering acorns and pine nuts as staples. In a region as bounteous as California, there's no obvious reason to do this. Acorns and pine nuts offer tiny individual food packages and require a great deal of labour to process. To render them edible, most varieties require the back-breaking work of leaching and grinding to be carried out, to remove toxins and release nutrients. Nut yields can vary dramatically from one season to the next, a risky pattern of boom and bust. At the same time, fish are found in abundance from the Pacific Coast inland at least as far as the confluence of the Sacramento and San Joaquin Rivers. Fish are both more nutritious and more reliable than nuts. Despite this, salmon and other aquatic foods generally came second to tree crops in Californian diets, and this seems to have been the case long before the arrival of Europeans.[48]

In terms of 'optimal foraging theory', then, the behaviour of Californians simply makes no sense. Salmon can be harvested and processed in great quantities on an annual basis, and they provide oil and fats as well as protein. In terms of cost-benefit calculations, the peoples of the Northwest Coast are eminently more sensible than Californians, and have been for hundreds or even thousands of years.[49] Granted, they also had little choice, since nut-gathering was never a serious option on the Northwest Coast (the main forest species there are conifers). It's also true that Northwest Coast peoples enjoyed a greater range of fish than Californians, including eulachon (candlefish), intensively exploited for its oil, which was both a staple food and a core ingredient in 'grease feasts,' where nobles ladled great quantities of this stuff on to the burning hearth, and occasionally on to one another. But the Californians *did* have a choice.

California, then, is an ecological puzzle. Most of its indigenous inhabitants appear to have prided themselves on their hard work, clear-sighted practicality and prudence in monetary affairs – quite unlike the wild and excessive self-image of Northwest Coast chiefs, who liked to boast that they 'didn't care about anything' – but as it turns out, the Californians were the ones basing their entire regional economy on apparently irrational choices. Why did they choose to intensify the use of oak groves and pinion stands when so many rich fisheries were available?

Ecological determinists sometimes try to solve the puzzle by appealing to food security. Brigands like Le'mekwelolmei might have been seen as villains, at least in some quarters, but brigands, they argue, will always exist. And what is more attractive to thieves and raiders than stockpiles of already processed, easy-to-transport food? But dead fish, for reasons that should be obvious to all of us, cannot be left lying around. They must be either eaten immediately or cleaned, filleted, dried and smoked to prevent infestation. On the Northwest Coast these tasks were completed like clockwork in the spring and summer, because they were critical for the group's physical survival, and also its social survival in the competitive feasting exploits of the winter season.[50]

In the technical language of behavioural ecology, fish are 'front-loaded'. You have to do most of the work of preparation right away. As a result, one could argue that a decision to rely heavily on fish – while undoubtedly sensible in purely nutritional terms – is also weaving a noose for one's own neck. It meant investing in the creation of a storable surplus of processed and packaged foods (not just preserved meat, but also fats and oils), which also meant creating an irresistible temptation for plunderers.[51] Acorns and nuts, on the other hand, present neither such risks nor such temptations. They are 'back-loaded'. Harvesting them was a simple and fairly leisurely affair,[52] and, crucially, there was no need for processing prior to storage. Instead most of the hard work took place only just before consumption: leaching and grinding to make porridges, cakes and biscuits. (This is the very opposite of smoked fish, which you don't even have to cook if you don't want to.)

So there was little point in raiding a store of raw acorns. As a result, there was also no real incentive to develop organized ways of defending these stores against potential raiders. One can begin to see the logic here. Salmon-fishing and acorn-gathering simply have very different practical affordances, which over the long term might be expected to produce very different sorts of societies: one warlike and prone to raiding (and after you have made off with the food, it's not much of a leap to begin carrying off prisoners as well), the other essentially peaceful.[53] Northwest Coast societies, then, were warlike because they simply didn't have the option of relying on a war-proof staple food.

It's certainly an elegant theory, quite clever and satisfying in its own way.[54] The problem is it just doesn't seem to match up to historical reality. The first and most obvious difficulty is that the capture of dried fish, or foodstuffs of any kind, was never a significant aim of Northwest Coast inter-group raiding. To put it bluntly, there's only so many smoked fish one can pile up in a war canoe. And carrying bulk products overland was even more difficult: pack animals being entirely absent in this part of the Americas, everything had to be carried by human beings, and on a long trip a slave is likely to eat about as much as they can carry. The main aim of raids was always to capture people, never food.[55] But this was also one of the most densely populated regions of North America. Where, then, did this hunger for people come from? These are precisely the kind of questions that 'optimal foraging theory' and other 'rational-choice' approaches seem utterly unable to answer.

In fact, the ultimate causes of slavery didn't lie in environmental or demographic conditions, but in Northwest Coast concepts of the proper ordering of society; and these, in turn, were the result of political jockeying by different sectors of the population who, as everywhere, had somewhat different perspectives on what a proper society should be. The simple reality is that there was no shortage of working hands in Northwest Coast households. But a good proportion of those hands belonged to aristocratic title holders who felt strongly that they should be exempted from menial work. They might hunt manatees or killer whales, but it was inconceivable for them to be seen building weirs or gutting fish. First-hand accounts show this often became an issue in the spring and summer, when the only limits on fish-harvesting were the number of hands available to process and preserve the catch. Rules of decorum prevented nobles from joining in, while low-ranking commoners ('perpetual transients', as one ethnographer called them)[56] would instantly defect to a rival household if pressed too hard or called upon too often.

In other words, aristocrats probably did feel that commoners should be working like slaves for them, but commoners had other opinions. Many were happy to devote long hours to art, but considered fish runs quite another matter. Indeed, the relation between title-holding nobles and their dependants seems to have been under constant negotiation. Sometimes it was not entirely clear who was serving whom:

> High rank was a birthright but a noble could not rest on his laurels. He had to 'keep up' his name through generous feasting, potlatching, and general open-handedness. Otherwise he ran the risk not only of losing face but in extreme cases actually losing his position, or even his life. Swadesh tells of a despotic [Nootka] chief who was murdered for 'robbing' his commoners by demanding all of his fishermen's catch, rather than the usual tributary portion. His successor outdid himself in generosity, saying when he caught a whale, 'You people cut it up and everyone take one chunk; just leave the little dorsal fin for me.'[57]

The result, from the nobles' point of view, was a perennial shortage, not of labour as such but of *controllable* labour at key times of year. This was the problem to which slavery addressed itself. And such were the immediate causes, which made 'harvesting people' from neighbouring clans no less essential to the aboriginal economy of the Northwest Coast than constructing weirs, clam gardens or terraced root plots.[58]

So we must conclude that ecology does not explain the presence of slavery on the Northwest Coast. Freedom does. Title-holding aristocrats, locked in rivalry with one another, simply lacked the means to compel their own subjects to support their endless games of magnificence. They were forced to look abroad.

What, then, of California?

Picking up where we left off, with the 'tale of the Wogies', a logical place to start is precisely the boundary zone between these two culture areas. As it turns out, the Yurok and other 'Protestant foragers' of northern California were, even by Californian standards, unusual, and it behoves us to understand why.

IN WHICH WE TURN TO THE CULTIVATION OF DIFFERENCE IN THE PACIFIC 'SHATTER ZONE'

Alfred Kroeber, who pioneered the ethnographic study of California's indigenous population, described its northwest section as a 'shatter zone', an area of unusual diversity, bridging the two great culture areas of the Pacific littoral. Here the distribution of ethnic and language groups – Yurok, Karuk, Hupa, Tolowa, and as many as a dozen even smaller societies – compressed like the bellows of an accordion. Some of these micro-nations spoke languages of the Athabascan family; others, in their domestic arrangements and architecture, retained traces of aristocracy that point clearly to their origins somewhere up on the Northwest Coast. Still, with very few exceptions, none practised chattel slavery.[59]

To underscore the contrast, we should note that in any true Northwest Coast settlement hereditary slaves might have constituted up to a quarter of the population. These figures are striking. As we noted earlier, they rival the demographic balance in the colonial South at the height of the cotton boom and are in line with estimates for household slavery in classical Athens.[60] If so, these were full-blown 'slave societies' where unfree labour underpinned the domestic economy and sustained the prosperity of nobles and commoners alike. Assuming that many groups came south from the Northwest Coast, as linguistic and other evidence suggests, and that at least some of this movement took place after about 1800 BC (when slavery was most likely institutionalized), the question becomes: when and how did foragers in the 'shatter zone' come to lose the habit of keeping slaves?

The 'when' part of this question is really a matter for future research. The 'how' part is more accessible. In many of these societies one can observe customs that seem explicitly designed to head off the danger of captive status becoming permanent. Consider, for example, the Yurok requirement for victors in battle to pay compensation for each life taken, at the same rate one would pay if one were guilty of murder. This seems a highly efficient way of making inter-group raiding both fiscally pointless and morally bankrupt. In monetary terms, military advantage became a liability to the winning side. As Kroeber put it, 'The *vae victis* of civilization might well have been replaced among the Yurok, in a monetary sense at least, by the dictum: "Woe to the victors."'[61]

The Chetcos' cautionary tale of the Wogies offers some further pointers. It suggests that populations directly adjacent to the Californian 'shatter zone' were aware of their northern neighbours and saw them as warlike, and as disposed to a life of luxury based on exploiting the labour of those they subdued. It implies they recognized such exploitation as a possibility in their own societies yet rejected it, since keeping slaves would undermine important social values (they would become 'fat and lazy'). Turning south, to the California shatter zone itself, we find evidence that, in many key areas of social life, the foragers of this region were indeed building their communities, in good schismogenetic fashion, as a kind of mirror image; a conscious inversion of those on the Northwest Coast. Some examples are in order.

Clues emerge from the simplest and most apparently pragmatic details. Let us cite just one or two. No free member of a Northwest Coast household would ever be seen chopping or carrying wood.[62] To do so was to undermine one's own status, effectively making oneself the equivalent of a slave. Californian chiefs, by contrast, seem to have elevated these exact same activities into a solemn public duty, incorporating them into the core rituals of the sweat lodge. As Goldschmidt observed:

> All men, particularly the youths, were exhorted to gather wood for use in sweating. This was not exploitation of child labor, but an important religious act, freighted with significance. Special wood was brought

> from the mountain ridges; it was used for an important purification ritual. The gathering itself was a religious act, for it was a means of acquiring 'luck.' It had to be done with the proper psychological attitude of which restrained demeanor and constant thinking about the acquisition of riches were the chief elements. The job became a moral end rather than a means to an end, with both religious and economic involvements.[63]

Similarly, the ritual sweating that ensued – by purging the Californian male's body of surplus fluid – inverts the excessive consumption of fat, blubber and grease that signified masculine status on the Northwest Coast. To enhance his status and impress his ancestors, the nobleman of the Northwest Coast ladled candlefish oil into the fire at the tournament fields of the *potlatch*; the Californian chief, by contrast, burned calories in the closed seclusion of his sweat lodge.

Native Californians seem to have been well aware of the kinds of values they were rejecting. They even institutionalized them in the figure of the clown,[64] whose public antics of sloth, gluttony and megalomania – while offering a platform from which to sound off about local problems and discontents – also seem to parody the most coveted values of a proximate civilization. Further inversions occur in the domains of spiritual and aesthetic life. Artistic traditions of the Northwest Coast are all about spectacle and deception: the theatrical trickery of masks that flicker open and shut, of surface figures pulling the gaze in sharply opposed directions. The native word for 'ritual' in most Northwest Coast languages actually translates as 'fraud' or 'illusion'.[65] Californian spirituality provides an almost perfect antithesis. What mattered was cultivation of the *inner* self through discipline, earnest training, and hard work. Californian art entirely avoids the use of masks.

Moreover, Californian songs and poetry show that disciplined training and work were ways of connecting with what is authentic in life. So, while Northwest Coast groups were not averse to adopting Europeans in lavish naming ceremonies, would-be Californians – like Robert Frank, adopted by the Yurok in the late nineteenth century – were more likely to find themselves hauling wood from the mountains, weeping with each footfall, as they earned their place among the 'real people'.[66]

If we accept that what we call 'society' refers to the mutual creation of human beings, and that 'value' refers to the most conscious aspects of that process, then it really is hard to see the Northwest Coast and California as anything but opposites. People in both regions engaged in extravagant expenditures of labour, but the forms and functions of that labour could not have differed more. In the Northwest Coast, the exuberant multiplication of furniture, crests, poles, masks, mantles and boxes was consistent with the extravagance and theatricality of *potlatch*. The ultimate purpose of all this work and ritual creativity, however, was to 'fasten on' names and titles to aristocratic contenders – to fashion specific sorts of persons. The result, among other things, is that Northwest Coast artistic traditions are still widely considered among the most dazzling the world has ever seen; immediately recognizable for their strong focus on the theme of exteriority – a world of masks, illusions and façades.[67]

Societies in the Californian shatter zone were equally extravagant in their own way. But if they were '*potlatch*ing' anything, then surely it was labour itself. As one ethnographer wrote of another Yurok neighbour, the Atsugewi: 'The ideal individual was both wealthy and industrious. In the first grey haze of dawn he arose to begin his day's work, never ceasing activity until late at night. Early rising and the ability to go without sleep were great virtues. It was extremely complimentary to say "he doesn't know how to sleep."'[68] Wealthy men – and it should be noted that all these societies were decidedly patriarchal – were typically seen as providers for poorer dependants, improvident folk and foolish drifters, by virtue of their own self-discipline and labour and that of their wives.

With its 'Protestant' emphasis on interiority and introspection, Californian spirituality offers a perfect counterpoint to the smoke and mirrors of Northwest Coast ceremonials. Among the Yurok, work properly performed became a way of connecting with a true reality, of which prized objects like *dentalia* and hummingbird scalps were mere outward manifestations. A contemporary ethnographer explains:

> As he 'accumulates' himself and becomes cleaner, the person in training sees himself as more and more 'real' and thus the world as more and more 'beautiful': a real place in experience rather than merely a setting

> for a 'story,' for intellectual knowledge . . . In 1865, Captain Spott, for instance, trained for many weeks as he helped the medicine man prepare for the First Salmon ceremony at the mouth of the Klamath River . . . 'the old [medicine] man sent him to bring down sweathouse wood. On the way he cried with nearly every step because now he was seeing with his own eyes how it was done.' . . . Tears, crying, are of crucial importance in Yurok spiritual training as manifestations of personal yearning, sincerity, humility, and openness.[69]

Through such exertions one discovered one's true vocation and purpose; and when 'someone else's purpose in life is to interfere with you,' the same ethnographer was told, 'he must be stopped, lest you become his slave, his "pet".'

The Yurok, with their puritanical manners and extraordinary cultural emphasis on work and money, might seem an odd choice to celebrate as anti-slavery heroes (though many Calvinist Abolitionists were not so very different). But of course we're not introducing them as heroes, any more than we wish to represent their Northwest Coast neighbours as the villains of the piece. We are introducing them as a way to illustrate how the process by which cultures define themselves against one another is always, at root, political, since it involves self-conscious arguments about the proper way to live. Revealingly, the arguments appear to have been most intense precisely in this border zone between anthropological 'culture areas'.

As we mentioned, the Yurok and their immediate neighbours were somewhat unusual, even by Californian standards. Yet they are unusual in contradictory ways. On the one hand, they actually did hold slaves, if few in number. Almost all the peoples of central and southern California, the Maidu, Wintu, Pomo and so on, rejected the institution entirely.[70] There appear to have been at least two reasons for this. First, almost everywhere except in the northwest, a man or woman's money and other wealth was ritually burned at death – and as a result, the institution served as an effective levelling mechanism.[71] The Yurok-Karuk-Hupa area was one of the few places where dentalium could actually be inherited. Combine this with the fact that quarrels did lead to war much more frequently here than anywhere else, and you have a kind of shrunken, diminished version of the Northwest Coast ranking system, in this case a tripartite division between wealthy families, ordinary Yurok and paupers.[72]

Captives were not slaves, all sources insist they were redeemed quickly, and all killers had to pay compensation; but all this required money. This meant the important men who often instigated wars could profit handsomely from the affair by lending to those unable to pay, and the latter were thus either reduced to debt peons, or retreated to live ignominiously in isolated homesteads in the woods.[73] One might see the intense focus on obtaining money, and resultant puritanism, and also the strong moral opposition to slave-raiding as a result of tensions created by living in this unstable and chaotic buffer zone between the two regions. Elsewhere in California, formal chiefs or headmen existed, and though they wielded no power of compulsion they settled conflicts by raising funds for compensation collectively, and the focus of cultural life was less on the accumulation of property than on organizing annual rites of world renewal.

Here one might say things have turned full circle. The ostensible purpose of the *potlatch* and spectacular competitions over wealth and heirloom titles on the Northwest Coast was, ultimately, to win prized roles in the great midwinter masquerades that were, similarly, intended to revive the forces of nature. California chiefs too were ultimately concerned with winter masquerades – being Californians, they did not employ literal masks, but, as in the Kwakiutl midwinter ceremonial, gods came down to earth and were embodied in costumed dancers – designed to regenerate the world and save it from imminent destruction. The difference, of course, was that in the absence of a servile labour force or any system of hereditary titles, Californian Pomo or Maidu chiefs had to go about organizing such rituals in an entirely different way.

SOME CONCLUSIONS

Environmental determinists have an unfortunate tendency to treat humans as little more than automata, living out some economist's fantasy of rational calculation. To be fair, they don't deny that human beings are quirky and imaginative creatures – they just seem to reason that, in the long run, this fact makes very little difference. Those who don't follow an optimal pathway for the use of resources are destined for the ash heap of history. Anthropologists who object to this kind of determinism will typically appeal to culture, but ultimately this comes down to little more than insisting that explanation is impossible: English people act the way they do because they are English, Yurok act the way they do because they're Yurok; why they are English or Yurok is not really ours to say. Humans – from this other perspective, which is just as extreme in its own way – are at best an arbitrary constellation of cultural elements, perhaps assembled according to some prevailing spirit, code or ethos, and which society ends up with which ethos is treated as beyond explanation, little more than a random roll of the dice.

Putting matters in such stark terms does not mean there is no truth to either position. The intersection of environment and technology does make a difference, often a huge difference, and to some degree, cultural difference really is just an arbitrary roll of the dice: there's no 'explanation' for why Chinese is a tonal language and Finnish an agglutinative one; that's just the way things happened to turn out. Still, if one treats the arbitrariness of linguistic difference as the foundation of all social theory – which is basically what structuralism did, and post-structuralism continues to do – the result is just as mechanically deterministic as the most extreme form of environmental determination. 'Language speaks us.' We are doomed to endlessly enact patterns of behaviour not of our own creation; not of anyone's creation really, until some seismic shift in the cultural equivalent of tectonic plates lands us somehow in a new, equally inexplicable arrangement.

In other words, both approaches presume that we are already, effectively, stuck. This is why we ourselves place so much emphasis on the notion of self-determination. Just as it is reasonable to assume that Pleistocene mammoth hunters, moving back and forth between different seasonal forms of organization, must have developed a degree of political self-consciousness – to have thought about the relative merits of different ways of living with one another – so too the intricate webs of cultural difference that came to characterize human societies after the end of the last Ice Age must surely have involved a degree of political introspection. Once again, our intention is simply to treat those who created these forms of culture as intelligent adults, capable of reflecting on the social worlds they were building or rejecting.

Obviously, this approach, like any other, can be taken to ridiculous extremes. Returning momentarily to Weber's *Protestant Ethic*, it is popular in certain circles to claim that 'nations make choices', that some have chosen to be Protestant and others Catholic, and that this is the main reason so many people in the United States or Germany are rich, and so many in Brazil or Italy are poor. This makes about as much sense as arguing that since everyone is free to make their own decisions, the fact that some people end up as financial consultants and others as security guards is entirely their own doing (indeed, it's usually the same sort of people who make both sorts of argument). Perhaps Marx put it best: we make our own history, but not under conditions of our own choosing.

In fact, one reason social theorists will always be debating this issue is that we can't really know how much difference 'human agency' – the preferred term, currently, for what used to be called 'free will' – really makes. Historical events by definition happen only once, and there's no real way to know if they 'might' have turned out otherwise (might Spain have never conquered Mexico? Could the steam engine have been invented in Ptolemaic Egypt, leading to an ancient industrial revolution?), or what the point of asking is even supposed to be. It seems part of the human condition that while we cannot predict future events, as soon as those events do happen we find it hard to see them as anything but inevitable. There's no way to know. So precisely where one wishes to set the dial between freedom and determinism is largely a matter of taste.

Since this book is mainly about freedom, it seems appropriate to set the dial a bit further to the left than usual, and to explore the possibility that human beings have more collective say over their own destiny than we ordinarily assume. Rather than defining the indigenous inhabitants of the Pacific Coast of North America as 'incipient' farmers or as examples of 'emerging' complexity – which is really just an updated way of saying they were all 'rushing headlong for their chains' – we have explored the possibility that they might have been proceeding with (more or less) open eyes, and found plenty of evidence to support it.

Slavery, we've argued, became commonplace on the Northwest Coast largely because an ambitious aristocracy found itself unable to reduce its free subjects to a dependable workforce. The ensuing violence seems to have spread until those in what we've been calling the 'shatter zone' of northern California gradually found themselves obliged to create institutions capable of insulating them from it, or at least its worst extremes. A schismogenetic process ensued, whereby coastal peoples came to define themselves increasingly against each other. This was by no means just an argument about slavery; it appears to have affected everything from the configuration of households, law, ritual and art to conceptions of what it meant to be an admirable human being, and was most evident in contrasting attitudes to work, food and material wealth.[74]

All this played a crucial role in shaping what outsiders came to see as the predominant sensibility of each resulting 'culture area' – the flamboyant extravagance of one, the austere simplicity of the other. But it also resulted in the overwhelming rejection of the practice of slavery, and the class system it entailed, throughout every part of California except for its northwesternmost corner; and even there it remained sharply limited.

What does this tell us about the emergence of similar forms of domination in earlier phases of human history? Nothing for certain, of course. It is difficult to know for sure whether Mesolithic societies of the Baltic or Breton coast that remind us, superficially, of indigenous societies on the Northwest Coast of Canada were, in fact, organized on similar principles. 'Complexity' – as reflected in the co-ordination of labour or elaborate ritual systems – need not mean domination. But it seems likely that similar arrangements were, indeed, emerging in some parts of the world, in some times and places, and that when they did they did not go uncontested. Regional processes of cultural differentiation, of the kind one begins to see more evidence for after the end of the last Ice Age, were probably every bit as political as those of later ages, including the ones we have considered in this chapter.

Second, we can now see more clearly that domination begins at home. The fact that these arrangements became subjects of political contestation does not mean they were political in origin. Slavery finds its origins in war. But everywhere we encounter it slavery is also, at first, a domestic institution. Hierarchy and property may derive from notions of the sacred, but the most brutal forms of exploitation have their origins in the most intimate of social relations: as perversions of nurture, love and caring. Certainly, those origins are not to be found in government. Northwest Coast societies lacked anything that could be remotely described as an overarching polity; the closest they came were the organizing committees of annual masquerades. Instead, one finds an endless succession of great wooden houses, tiny courts each centring on a title-holding family, the commoners attached to them, and their personal slaves. Even the rank system referred to divisions within the household. It seems very likely this was true in non-agricultural societies elsewhere as well.

Finally, all this suggests that, historically speaking, hierarchy and equality tend to emerge together, as complements to one another. Tlingit or Haida commoners on the Northwest Coast were effectively equals in that they were all equally excluded from the ranks of title holders and therefore, in comparison to the aristocrats – with their unique identities – formed a kind of undifferentiated mass. Insofar as Californian societies rejected that entire arrangement, they could be described as self-consciously egalitarian, but in a quite different sense. Odd as it may seem, this comes through most clearly in their enthusiastic embrace of money, and again comparisons with their northern neighbours are instructive. For Northwest Coast societies, wealth, which was sacred in every sense of the term, consisted above all of heirloom treasures, whose value was based on the fact that each was unique and there was nothing in the world like it. Equality between title holders was simply inconceivable, much though they might have argued about who ultimately outranked whom. In California, the most important forms of wealth consisted of currencies whose value lay in the degree to which each string of dentalium or band of woodpecker scalps was exactly the same, and could therefore be counted – and, generally speaking, such wealth was not inherited but destroyed on the owner's death.

As our story continues, we will encounter this dynamic repeatedly.

We might refer to it, perhaps, as 'inequality from below'. Domination first appears on the most intimate, domestic level. Self-consciously egalitarian politics emerge to prevent such relations from extending beyond those small worlds into the public sphere (which often comes to be imagined, in the process, as an exclusive sphere for adult men). These are the kind of dynamics that culminated in phenomena like ancient Athenian democracy. But their roots probably extend much further back in time, to well before the advent of farming and agricultural societies.

6
Gardens of Adonis

The revolution that never happened: how Neolithic peoples avoided agriculture

Let us turn, then, to the origins of farming.

PLATONIC PREJUDICES, AND HOW THEY CLOUD OUR IDEAS ABOUT THE INVENTION OF FARMING

'Tell me this,' writes Plato:

> Would a serious and intelligent farmer, with seeds he cared about and wished to grow to fruition, sow them in summer in the gardens of Adonis and rejoice as he watched them become beautiful in a matter of eight days; or if he did it at all, would he do this for fun and festivity? For things he really was serious about, would he not use his farmer's craft, plant them in a suitable environment, and be content if everything he planted came to maturity in the eighth month?[1]

The gardens of Adonis, to which Plato is referring here, were a sort of festive speed farming which produced no food. For the philosopher, they offered a convenient simile for all things precocious, alluring, but ultimately sterile. In the dog days of summer, when nothing can grow, the women of ancient Athens fashioned these little gardens in baskets and pots. Each held a mix of quick-sprouting grain and herbs. The makeshift seedbeds were carried up ladders on to the flat roofs of private houses and left to wilt in the sun: a botanical re-enactment of the premature death of Adonis, the fallen hunter, slain in his prime by a wild boar. Then, beyond the public gaze of men and civic authority, began the rooftop rites. Open to women from all classes of Athenian society, including prostitutes, these were rites of grieving but also wanton drunkenness, and no doubt other forms of ecstatic behaviour as well.

Historians agree that the roots of this women's cult lie in Mesopotamian fertility rites of Dumuzi/Tammuz, the shepherd-god and personification of plant life, mourned on his death each summer. Most likely the worship of Adonis, his ancient Greek incarnation, spread westwards to Greece from Phoenicia in the wake of Assyrian expansion, in the seventh century BC. Nowadays, some scholars see the whole thing as a riotous subversion of patriarchal values: an antithesis to the staid and proper state-sponsored Thesmophoria (the autumn festival of the Greek fertility goddess, Demeter), celebrated by the wives of Athenian citizens and dedicated to the serious farming on which the life of the city depended. Others read the story of Adonis the other way round, as a requiem for the primeval drama of serious hunting, cast into shadow by the advent of agriculture, but not forgotten – an echo of lost masculinity.[2]

All well and good, you may say, but what does any of this have to do with the origins of farming? What have the gardens of Adonis got to do with the first Neolithic stirrings of agriculture some 8,000 years before Plato? Well, in a sense, everything. Because these scholarly debates encapsulate just the sort of problems that surround any modern investigation of this crucial topic. Was farming from the very beginning about the serious business of producing more food to supply growing populations? Most scholars assume, as a matter of course, that this had to be the principal reason for its invention. But maybe farming began as a more playful or even subversive kind of process – or perhaps even as a side effect of other concerns, such as the desire to spend longer in particular kinds of locations, where hunting and trading were the real priorities. Which of these two ideas really embodies the spirit of the first agriculturalists; is it the stately and pragmatic Thesmophoria, or the playful and self-indulgent gardens of Adonis?

No doubt the peoples of the Neolithic – the world's first farmers – themselves spent a good deal of time debating similar questions. To get a sense of why we say this, let's consider what is probably the most famous Neolithic site in the world, Çatalhöyük.

IN WHICH WE DISCUSS HOW ÇATALHÖYÜK, THE WORLD'S OLDEST TOWN, GOT A NEW HISTORY

Located on the Konya Plain of central Turkey, Çatalhöyük was first settled around 7400 BC, and continued to be populated for some 1,500 years (for the purposes of mental calibration, roughly the same period of time that separates us from Amalafrida, Queen of the Vandals, who reached the height of her influence around AD 523). The site's renown derives partly from its surprising scale. At thirteen hectares, it was more town than village, with a population of some 5,000. Yet it was a town with no apparent centre or communal facilities, or even streets: just a dense agglomeration of one household after another, all of similar sizes and layout, each accessed by ladder from the roof.

If the overall plan of Çatalhöyük suggests an ethos of dreary uniformity, a maze of identical mud walls, the internal life of its buildings points in exactly the opposite direction. In fact, another reason for the site's fame is its inhabitants' distinctly macabre sense of interior design. If you've ever glimpsed the inside of a Çatalhöyük house you will never forget it: central living rooms, no more than sixteen feet across, with the skulls and horns of cattle and other creatures projecting inwards from the walls, and sometimes outwards from the fittings and furnishings. Many rooms also had vivid wall paintings and figurative mouldings, and contained platforms under which resided some portion of the household dead – remains of between six and sixty individuals in any given house – propping up the living. We can't help recalling Maurice Sendak's vision of a magical house where 'the walls became the world all around'.[3]

Generations of archaeologists have wanted to see Çatalhöyük as a monument to the 'origins of farming'. Certainly, it's easy to understand why this should be. It is among the first large settlements we know of whose inhabitants practised agriculture, and who got most of their nutrition from domesticated cereals, pulses, sheep and goats. It seems reasonable to see them, then, as the very engineers of what has been referred to since the time of V. Gordon Childe – prehistorian and author of *Man Makes Himself* (1936) and *What Happened in History* (1942) – as the 'Agricultural Revolution', and when first excavated in the 1960s Çatalhöyük's remarkable material culture was interpreted in this way. Clay figurines of seated women, including a famous example flanked by felines, were understood as depictions of a Mother Goddess, presiding over the fertility of women and crops. The wall-mounted ox-skulls ('*bucrania*') were assumed to be those of domestic cattle, dedicated to a taurine deity responsible for the protection and reproduction of herds. Certain buildings were identified as 'shrines'. All this ritual life was assumed to refer to serious farming – a Neolithic pageant play, more in the spirit of Demeter than Adonis.[4]

But more recent excavations suggest we have been too quick to write off Adonis.[5] Since the 1990s, new methods of fieldwork at Çatalhöyük produced a string of surprises, which oblige us to revise both the history of the world's oldest town and also how we think about the origins of farming in general. The cattle, it turns out, were not domestic: those impressive skulls belonged to fierce, wild aurochs. The shrines were not shrines, but houses in which people engaged in such everyday tasks as cooking, eating and crafts – just like anywhere else, except they happened to contain a larger density of ritual paraphernalia. Even the Mother Goddess has been cast into shadow. It is not so much that corpulent female figurines stopped turning up entirely in the excavations, but that the new finds tended to appear, not in shrines or on thrones, but in trash dumps outside houses with the heads broken off and didn't really seem to have been treated as objects of religious veneration.[6]

Today, most archaeologists consider it deeply unsound to interpret prehistoric images of corpulent women as 'fertility goddesses'. The very idea that they should be is the result of long-outmoded Victorian fantasies about 'primitive matriarchy'. In the nineteenth century, it's true, matriarchy was considered the default mode of political organization for Neolithic societies (as opposed to the oppressive patriarchy of the ensuing Bronze Age). As a result, almost every image of a fertile-looking woman was interpreted as a goddess. Nowadays, archaeologists are more likely to point out that many figurines could just as easily have been the local equivalents of Barbie dolls (the kind of Barbie dolls one might have in a society with very different standards of female beauty); or that different figurines might have served entirely different purposes (no doubt correct); or to dismiss the entire debate by insisting we simply have no idea why people created so many female images and never will, so any interpretations on offer are more likely to be projections of our own assumptions about women, gender or fertility than anything that would have made sense to an inhabitant of Neolithic Anatolia.

All of which might seem a bit pedantic, but in this hair-splitting, as we'll see, there's a great deal at stake.

IN WHICH WE ENTER SOMETHING OF AN ACADEMIC NO-GO ZONE, AND DISCUSS THE POSSIBILITY OF NEOLITHIC MATRIARCHIES

It's not just the idea of 'primitive matriarchy' that's become such a bugaboo today: even to suggest that women had unusually prominent positions in early farming communities is to invite academic censure. Perhaps it's not entirely surprising. In the same way that social rebels, since the 1960s, tended to idealize hunter-gatherer bands, earlier generations of poets, anarchists and bohemians had tended to idealize the Neolithic as an imaginary, beneficent theocracy ruled over by priestesses of the Great Goddess, the all-powerful distant ancestor of Inanna, Ishtar, Astarte and Demeter herself – that is, until such societies were overwhelmed by violent, patriarchal Indo-European-speaking horsemen descending from the steppes, or, in the case of the Middle East, Semitic-speaking nomads from the deserts. How people saw this imagined confrontation became the source of a major political divide in the late nineteenth and early twentieth centuries.

To give you a flavour of this, let's look at Matilda Joslyn Gage (1826–98), considered in her lifetime one of the most prominent American feminists. Gage was also an anti-Christian, attracted to the Haudenosaunee 'matriarchate', which she believed to be one of the few surviving examples of Neolithic social organization, and a staunch defender of indigenous rights, so much so that she was eventually adopted as a Mohawk clan mother. (She spent the last years of her life in the home of her devoted son-in-law, L. Frank Baum, author of the *Oz* books – a series of a dozen volumes in which, as many have pointed out, there are queens, good witches and princesses, but not a single legitimate male figure of authority.) In *Woman, Church, and State* (1893), Gage posited the universal existence of an early form of society 'known as the Matriarchate or Mother-rule', where institutions of government and religion were modelled on the relationship of mother to child in the household.

Or consider one of Sigmund Freud's two favourite students: Otto Gross, an anarchist who in the years before the First World War developed a theory that the superego was in fact patriarchy and needed to be destroyed so as to unleash the benevolent, matriarchal collective unconscious, which he saw as the hidden but still-living residue of the Neolithic. (This he set out to accomplish largely through the use of drugs and polyamorous sexual relationships; Gross's work is now largely remembered for its influence on Freud's other favourite student, Carl Jung, who kept the idea of the collective unconscious but rejected Gross's political conclusions.) After the Great War, Nazis began to take up the same story of the 'Aryan' invasions from the exact opposite perspective, representing the imagined, patriarchal invaders as the ancestors of their master race.

With such intense politicization of what were obviously fanciful readings of prehistory, it's hardly surprising that the topic of 'primitive matriarchy' became something of an embarrassment – the intellectual equivalent of a no-go zone – for subsequent generations. But it's hard to avoid the impression something else is going on here. The degree of erasure has been extraordinary, and far more than is warranted by mere suspicion of an overstated or outdated theory. Among academics today, belief in primitive matriarchy is treated as a kind of intellectual offence, almost on a par with 'scientific racism', and its exponents have been written out of history: Gage from the history of feminism, Gross from that of psychology (despite inventing such concepts as introversion and extroversion, and having worked closely with everyone from Franz Kafka and the Berlin Dadaists to Max Weber).

This is odd. After all, a century or so does seem more than enough time for the dust to settle. Why is the matter still so shrouded in taboo?

Much of this present-day sensitivity stems from a backlash against the legacy of a Lithuanian-American archaeologist named Marija Gimbutas. In the 1960s and 1970s, Gimbutas was a leading authority on the later prehistory of eastern Europe. Nowadays, she is often represented as just as much of an oddball as psychiatric rebels like Otto Gross, accused of having attempted to revive the most ridiculous of old Victorian fantasies in modern guise. This is not only untrue (very few of those who dismiss her work seem to have actually read any of it), but it has created a situation where scholars find it difficult even to speculate as to how hierarchy and exploitation came to take root in the domestic sphere – unless one wants to return to Rousseau, and the simplistic notion that settled farming somehow automatically generated the power of husbands over wives and fathers over children.

In fact, if you read the books of Gimbutas – such as *The Goddesses and Gods of Old Europe* (1982) – you quickly realize that their author was attempting to do something which, until then, only men had been allowed to do: to craft a grand narrative for the origins of Eurasian civilization. She did so taking as her building blocks the very kind of 'culture areas' we discussed in the last chapter and using them to argue that, in some ways (though certainly not all), the old Victorian story about goddess-worshipping farmers and Aryan invaders was actually true.

Gimbutas was largely concerned with trying to understand the broad contours of a cultural tradition she referred to as 'Old Europe', a world of settled Neolithic villages centring on the Balkans and eastern Mediterranean (but also extending further north), in which, as Gimbutas saw it, men and women were equally valued, and differences of wealth and status were sharply circumscribed. Old Europe, by her estimation, endured from roughly 7000 BC to 3500 BC – which is, again, quite a respectable period of time. She believed these societies to be essentially peaceful, and argued that they shared a common pantheon under the tutelage of a supreme goddess, whose cult is attested in many hundreds of female figurines – some depicted with masks – found in Neolithic settlements, from the Middle East to the Balkans.[7]

According to Gimbutas, 'Old Europe' came to a catastrophic end in the third millennium BC, when the Balkans were overrun by a migration of cattle-keeping peoples – the so-called '*kurgan*' folk – originating on the Pontic steppe, north of the Black Sea. *Kurgan* refers to the most archaeologically recognizable feature of these groups: earthen tumuli heaped over the graves of (typically male) warriors, buried with weapons and ornaments of gold, and with extravagant sacrifices of animals and occasionally also human 'retainers'. All these features attested values antithetical to the communitarian ethos of Old Europe. The incoming groups were aristocratic and 'androcratic' (i.e., patriarchal), and were extremely warlike. Gimbutas considered them responsible for the westward spread of Indo-European languages, the establishment of new kinds of societies based on the radical subordination of women, and the elevation of warriors to a ruling caste.

As we've noted, all this bore a certain resemblance to the old Victorian fantasies – but there were key differences. The older version was rooted in an evolutionary anthropology that assumed matriarchy was the original condition of humankind because, at first, people supposedly didn't understand physiological paternity and assumed women were single-handedly responsible for producing babies. This meant, of course, that hunter-gatherer communities before them should be just as matrilineal and matriarchal, if not more so, than early farmers – something many did indeed argue from first principles, despite a complete lack of any sort of evidence. Gimbutas, though, was not proposing anything of this sort: she was arguing for women's autonomy and ritual priority in the Middle Eastern and European Neolithic. Yet by the 1990s many of her ideas had become a charter for ecofeminists, New Age religions and a host of other social movements; in turn, they inspired a slew of popular books, ranging from the philosophical to the ridiculous – and in the process became entangled with some of the more extravagant older Victorian ideas.

Given all this, many archaeologists and historians concluded that Gimbutas was muddying the waters between scientific research and pop literature. Before long, she was being accused of just about everything the academy could think to throw at her: from cherry-picking evidence to failing to keep up with methodological advances; accusations of reverse sexism; or that she was indulging in 'myth-making'. She was even subject to the supreme insult of public psychoanalysis, as leading academic journals published articles suggesting her theories about the displacement of Old Europe were basically phantasmagorical projections of her own tumultuous life experience, Gimbutas having fled her mother country, Lithuania, at the close of the Second World War in the wake of foreign invasions.[8]

Mercifully, perhaps, Gimbutas herself, who died in 1994, was not around to see most of this. But that also meant she was never able to respond. Some, maybe most of these criticisms had truth in them – though similar criticisms could no doubt be made of pretty much any archaeologist who makes a sweeping historical argument. Gimbutas's arguments involved myth-making of a sort, which in part explains this wholesale takedown of her work by the academic community. But when male scholars engage in similar myth-making – and, as we have seen, they frequently do – they not only go unchallenged but often win prestigious literary prizes and have honorary lectures created in their name. Arguably Gimbutas was seen as meddling in, and quite consciously subverting, a genre of grand narrative that had been (and still is) entirely dominated by male writers such as ourselves. Yet her reward was not a literary prize, or even a place among the revered ancestors of archaeology; it was near-universal posthumous vilification, or, even worse, becoming an object of dismissive contempt.

At least, until quite recently.

Over the last few years, the analysis of ancient DNA – unavailable in Gimbutas's time – has led a number of leading archaeologists to concede that at least one significant part of her reconstruction was probably right. If these new arguments, put forward on the basis of population genetics, are even broadly correct, then there really was an expansion of herding peoples from the grasslands north of the Black Sea around the time Gimbutas believed it to have happened: the third millennium BC. Some scholars are even arguing that massive migrations took place out of the Eurasian steppe at that time, leading to population replacement and perhaps the spread of Indo-European languages across large swathes of central Europe, just as Gimbutas envisaged. Others are far more cautious; but either way, after decades of virtual silence, people are suddenly talking about such issues, and hence about Gimbutas's work, again.[9]

So what about the other half of Gimbutas's argument, that Early

Neolithic societies were relatively free of ranks and hierarchies? Before even beginning to answer this question, we need to clear up a few misconceptions. Gimbutas in fact never argued outright for the existence of Neolithic matriarchies. Indeed, the term seems to mean very different things to different authors. Insofar as 'matriarchy' describes a society where women hold a preponderance of formal political positions, one can indeed say this is exceedingly rare in human history. There are plenty of examples of individual women wielding real executive power, leading armies or creating laws, but few if any societies in which *only* women are normally expected to wield executive power or lead armies or create laws. Even strong queens like Elizabeth I of England, the Dowager Empress of China or Ranavalona I of Madagascar did not primarily appoint other women to be their chief advisors, commanders, judges and officials.

In any case, another term – 'gynarchy', or 'gynaecocracy' – describes the political rule of women. The word 'matriarchy' means something rather different. There is a certain logic here: 'patriarchy', after all, refers not primarily to the fact that men wield public office, but first and foremost to the authority of patriarchs, that is, male heads of household – an authority which then acts as a symbolic model for, and economic basis of, male power in other fields of social life. Matriarchy might refer to an equivalent situation, in which the role of mothers in the household similarly becomes a model for, and economic basis of, female authority in other aspects of life (which doesn't necessarily imply dominance in a violent or exclusionary sense), where women as a result hold a preponderance of overall day-to-day power.

Looked at this way, matriarchies are real enough. Kandiaronk himself arguably lived in one. In his day, Iroquoian-speaking groups such as the Wendat lived in towns that were made up of longhouses of five or six families. Each longhouse was run by a council of women – the men who lived there did not have a parallel council of their own – whose members controlled all the key stockpiles of clothing, tools and food. The political sphere in which Kandiaronk himself moved was perhaps the only one in Wendat society where women did not predominate, and even so there existed women's councils which held veto power over any decision of the male councils. On this definition, the Pueblo nations such as Hopi and Zuñi might also qualify as matriarchies, while the Minangkabau, a Muslim people of Sumatra, describe themselves as matriarchal for exactly the same reasons.[10]

True, such matriarchal arrangements are somewhat unusual – at least in the ethnographic record, which covers roughly the last 200 years. But once it's clear that such arrangements can exist, we have no particular reason to exclude the possibility that they were more common in Neolithic times, or to assume that Gimbutas – by searching for them there – was doing something inherently fanciful or misguided. As with any hypothesis, it's more a matter of weighing up the evidence.

Which takes us back to Çatalhöyük.

IN WHICH WE CONSIDER WHAT LIFE IN THE WORLD'S MOST FAMOUS NEOLITHIC TOWN MIGHT HAVE ACTUALLY BEEN LIKE

Recently, a number of discoveries among the miniature art of Çatalhöyük appear to show that the female form was a special focus of ritual attention, skilled artisanship and symbolic reflection on life and death. One is a clay figure with typically corpulent female front, transitioning at the back to a carefully modelled skeleton via arms that look emaciated. Its head, now lost, was fixed into a hole at the top. Another female figurine has a tiny cavity in the centre of her back, into which a single seed from a wild plant had been placed. And within a domestic platform of the sort used for burials, excavators found one particularly revealing and exquisitely carved limestone figure of a woman. Its detailed rendering clarifies an aspect of the more common figures made in clay: the sagging breasts, drooping belly and rolls of fat appear to signify not pregnancy, as once was believed, but age.[11]

Such findings suggest that the more ubiquitous female figurines, while clearly not all objects of worship, weren't necessarily all dolls or toys either. Goddesses? Probably not. But quite possibly matriarchs of some sort, their forms revealing an interest in female elders. And no equivalent representations of male elders have been found. Of course, this doesn't mean we should ignore the many other Neolithic figurines that have possible phallic attributes, or mixed male-female attributes, or that are so schematic we shouldn't really try to identify them as male or female, or even as clearly human. Similarly, the occasional links between Neolithic figurines and masking – attested both in the Middle East and eastern Europe[12] – may relate to occasions or performances where such categorical distinctions were deliberately blurred, or even inverted (not unlike, say, the masquerades of the Pacific Coast of North America, where the deities and those impersonating them were almost invariably male).

There is no evidence that Çatalhöyük's female inhabitants enjoyed better standards of living than its male ones. Detailed studies of human teeth and skeletons reveal a basic parity of diet and health, as does the ritual treatment of male and female bodies in death.[13] Yet the point remains that there exist no similarly elaborate or highly crafted depictions of male forms in the portable art of Çatalhöyük. Wall decoration is another matter. Where coherent scenes emerge from the surviving murals, they are mainly concerned with the hunting and teasing of game animals such as boar, deer, bear and bulls. The participants are men and boys, apparently depicted in different stages of life, or perhaps entering those stages through the initiatory trials of the chase. Some of these spritely figures wear leopard skins; in one deer-baiting scene, all have beards.

One thing to emerge clearly from the newer investigations at Çatalhöyük is the way in which household organization permeates almost every aspect of social life. Despite the considerable size and density of the built-up area, there is no evidence for central authority. Each household appears more or less a world unto itself – a discrete locus of storage, production and consumption. Each also seems to have held a significant degree of control over its own rituals, especially where treatment of the dead was concerned, although ritual experts may of course have moved between them. While it's unclear what social rules and habits were responsible for maintaining the autonomy of households, what seems evident is that these rules were learned mainly within the household itself; not just through its ceremonies, but also its micro-routines of cooking, cleaning floors, resurfacing walls with plaster, and so on.[14] All this is vaguely reminiscent of the Northwest Coast, where society was a collection of great houses, except that the inhabitants of these Neolithic houses show no sign of being divided into ranks.

The residents of Çatalhöyük seem to have placed great value on routine. We see this most clearly in the fastidious reproduction of domestic layouts over time. Individual houses were typically in use for between fifty and 100 years, after which they were carefully dismantled and filled in to make foundations for superseding houses. Clay wall went up on clay wall, in the same location, for century after century, over periods reaching up to a full millennium. Still more astonishing, smaller features such as mud-built hearths, ovens, storage bins and platforms often follow the same repetitive patterns of construction, over similarly long periods. Even particular images and ritual installations come back, again and again, in different renderings but the same locations, often widely separated in time.

Was Çatalhöyük, then, an 'egalitarian society'? There is no sign of any self-conscious egalitarian ideal in the sense of, say, a concern with uniformity in the art, architecture or material culture; but neither are there many explicit signs of rank. Nonetheless, as individual houses built up histories, they also appear to have acquired a degree of cumulative prestige. This is reflected in a certain density of hunting trophies, burial platforms and obsidian – a dark volcanic glass, obtained from sources in the highlands of Cappadocia, some 125 miles north. The authority of long-lived houses seems consistent with the idea that elders, and perhaps elder women in particular, held positions of influence. But the more prestigious households are distributed among the less, and do not coalesce into elite neighbourhoods. In terms of gender relations, we can acknowledge a degree of symmetry, or at least complementarity. In pictorial art, masculine themes do not encompass the feminine, nor vice versa. If anything, the two domains seem to be kept apart, in different sectors of dwellings.

What were the underlying realities of social life and labour at Çatalhöyük? Perhaps the most striking thing about all this art and ritual is that it makes almost no reference to agriculture. As we've noted, domestic cereals (wheat and barley) and livestock (sheep and goats) were far more important than wild resources in terms of nutrition. We know this because of organic remains recovered in quantity from every house. Yet for 1,000 years the cultural life of the community remained stubbornly oriented around the worlds of hunting and foraging. At this point, one has to ask how complete our picture of life at Çatalhöyük really is and where the largest gaps may lie.

HOW THE SEASONALITY OF SOCIAL LIFE IN EARLY FARMING COMMUNITIES MIGHT HAVE WORKED

Only something like 5 per cent of Neolithic Çatalhöyük has been excavated.[15] Soundings and surveys offer no particular reason to believe that other parts of the town were substantially different, but it's a reminder of how little we really know, and that we also have to think about what is missing from the archaeological record. For instance, it is clear that house floors were regularly swept clean, so the distribution of artefacts around them is far from a straightforward representation of past activities, which can only be reliably tracked through tiny fragments and residue embedded in the plaster.[16] Traces have also been found of reed mats that covered living surfaces and furnishings, further disturbing the picture. We don't necessarily know everything that was happening in the houses, or perhaps even half of it – or, indeed, how much time was actually spent living in these cramped and peculiar structures at all.

In considering this, it's worth taking a broader look at the site of Çatalhöyük in relation to its ancient surroundings, which archaeological science allows us to reconstruct, at least in outline. Çatalhöyük was situated in an area of wetlands (whence all the mud and clay) seasonally flooded by the Çarşamba River, which split its course as it entered the Konya Plain. Swamps would have surrounded the site for much of the year, interspersed with raised areas of dry land. Winters were cold and damp, summers oppressively hot. From spring to autumn, sheep and goats would have been moved between areas of pasture within the plain, and sometimes further into the highlands. Arable crops were most likely sown late in the spring on the receding floodplain of the Çarşamba, where they could ripen in as little as three months, with harvesting and processing in the late summer: fast-growing grains, in the season of Adonis.[17]

While all these tasks may have taken place quite close to the town, they will inevitably have involved a periodic dispersal and reconfiguration of working arrangements and of general social affairs. And, as the rites of Adonis remind us, another kind of social life altogether may have existed on the rooftops. It is in fact quite likely that what we are seeing in the surviving remains of Çatalhöyük's built environment are largely the social arrangements prevalent in winter, with their intense and distinctive ceremonialism focused upon hunting and the veneration of the dead. At that time of year, with the harvest in, the organization required for agricultural labour would have given way to a different type of social reality as the community's life shrank back towards its houses, just as its herds of sheep and goats shrank back into the confines of their pens.

Seasonal variations of social structure[18] were alive and well at Çatalhöyük, and these carefully balanced alternations seem central to understanding why the town endured. An impressive degree of material equality prevailed in the everyday exchanges of family life, within and between houses. Yet at the same time, hierarchy developed to slower rhythms, played out in rituals that joined the living to the dead. Shepherding and cultivation surely involved a strict division of labour, to safeguard the annual crop and protect the herds – but if so it found little space in the ceremonial life of the household, which drew its energy from older sources, more Adonis than Demeter.

A certain controversy has arisen, however, concerning just where the people of Çatalhöyük planted their crops. At first, microscopic studies of cereal remains suggested a dry-land location. Given the known extent of ancient swamps in the Konya basin, this would imply that arable fields were located at least eight miles from the town, which hardly seems plausible in the absence of donkeys or ox carts (remember, cattle were not yet domesticated in this region, let alone harnessed to anything). Subsequent analyses support a more local setting, on the alluvial soils of the Çarşamba floodplain.[19] The distinction is important for a variety of reasons, not just ecological but also historical, even political, because how we picture its practical realities has direct implications for how we view the social consequences of Neolithic farming.

We must take an even broader perspective to see exactly why.

ON BREAKING APART THE FERTILE CRESCENT

When Çatalhöyük was first investigated, in the 1960s, the striking discovery of houses lined with cattle skulls led many to assume, quite reasonably, that the plain of Konya was an early cradle of animal domestication. These days it is known that cattle (and boar) were first domesticated 1,000 years before Çatalhöyük was founded, and in another location altogether: around the upper reaches of the Tigris and Euphrates valleys, which lie further east into Asia, within the area known as the Fertile Crescent. It was from that general direction that the founders of Çatalhöyük obtained the basis of their farming economy, including domestic cereals, pulses, sheep and goats. But they didn't adopt domestic cattle or pigs. Why not?

Since no environmental obstacles were present, one has to assume an element of cultural refusal here. The best contender for an explanation is also the most obvious. As Çatalhöyük's art and ritual suggests, wild cattle and boar were highly valued as prey, and probably had been for as long as anyone could remember. In terms of prestige, there was much to be lost, perhaps especially for men, by the prospect of surrounding these dangerous animals with more docile, domestic varieties. Allowing cattle to remain exclusively in their ancient wild form – a big beast, but also lean, fast and highly impressive – also meant keeping intact a certain sort of human society. Accordingly, cattle remained wild and glamorous until around 6000 BC.[20]

So, what exactly *is*, or was, the Fertile Crescent? First, it's important to note that this is a completely modern concept, the origins of which are as much geopolitical as environmental. The term Fertile Crescent was invented in the nineteenth century, when Europe's imperial powers were carving up the Middle East according to their own strategic interests. Partly because of the close ties between archaeology, ancient history and the modern institutions of empire, the term became widely adopted among researchers to describe an area from the eastern shores of the Mediterranean (modern Palestine, Israel and Lebanon) to the foothills of the Zagros Mountains (roughly the Iran–Iraq border), crossing parts of Syria, Turkey and Iraq on the way.

Now it is only prehistorians who still use it, to indicate the region where farming began: a roughly crescent-shaped belt of arable lands bounded by deserts and mountains.[21]

Yet in ecological terms, it's really not one crescent but two – or no doubt even more, depending how closely one chooses to look. At the end of the last glacial period, around 10,000 BC, this region developed in two clearly distinct directions. Going with the topography, we can discern an 'upland crescent' and a 'lowland crescent'. The upland crescent follows the foothills of the Taurus and Zagros Mountains, running north of the modern border between Syria and Turkey. For foragers at the end of the last Ice Age, it would have been something of an open frontier; an expanding belt of oak-pistachio forest and game-rich prairie intersected by river valleys.[22] The lowland crescent to the south was characterized by *Pistacia* woodlands, as well as tracts of fertile terrain bound tightly to river systems or to the shores of lakes and artesian springs, beyond which lay deserts and barren plateaus.[23]

Between 10,000 and 8000 BC, foraging societies in the 'upland' and 'lowland' sectors of the Fertile Crescent underwent marked transformations, but in quite different directions. The differences cannot easily be expressed in terms of modes of subsistence or habitation. In both regions, in fact, we find a complex mosaic of human settlement: villages, hamlets, seasonal camps and centres of ritual and ceremonial activity marked out by impressive public buildings. Both regions, too, have produced varying degrees of evidence for plant cultivation and livestock management, within a broader spectrum of hunting and foraging activities. Yet there are also cultural differences, some so striking as to suggest a process of schismogenesis, of the sort we described in the previous chapter. It might even be argued that, after the last Ice Age, the ecological frontier between 'lowland' and 'upland' Fertile Crescent also became a cultural frontier with zones of relative uniformity on either side, distinguished almost as sharply as the 'Protestant foragers' and 'fisher kings' of the Pacific Coast.

In the uplands, there was a striking turn towards hierarchy among settled hunter-foragers, most dramatically attested at the megalithic centre of Göbekli Tepe and at nearby sites like that recently discovered at Karahan Tepe. In the lowlands of the Euphrates and Jordan valleys, by contrast, such megalithic monuments are absent, and Neolithic societies followed a distinct but equally precocious path of change, which we will shortly describe. What's more, these two adjacent families of societies – let's call them 'lowlanders' and 'uplanders' – were well acquainted. We know this because they traded durable materials with each other over long distances, among them the same materials, in fact, that we found circulating as valuables on the West Coast of North America: obsidian and minerals from the mountains, and mollusc shells from the coasts. Obsidian from the Turkish highlands flowed south, and shells (perhaps used as currency) flowed north from the shores of the Red Sea, ensuring that uplanders and lowlanders stayed in touch.[24]

The routes of this prehistoric trade circuit contracted as they progressed southwards into less evenly populated areas, starting at the Syrian bend of the Euphrates, winding through the Damascus basin and down into the Jordan valley. This route formed the so-called 'Levantine Corridor'. And the lowlanders who lived here were devoted craft specialists and traders. Each hamlet seems to have developed its own expertise (stone-grinding, bead-carving, shell-processing and so on), and industries were often associated with special 'cult buildings' or seasonal lodges, pointing to the control of such skills by guilds or secret societies. By the ninth millennium BC, larger settlements had developed along the principal trade routes. Lowland foragers occupied fertile pockets of land among the drainages of the Jordan valley, using trade wealth to support increasingly large, settled populations. Sites of impressive scale sprang up in such propitious locations, some, such as Jericho and Basta, approaching ten hectares in size.[25]

To understand the importance of trade in this process is to appreciate that the lowland crescent was a landscape of intimate contrasts and conjunctures (very similar, in this respect, to California). There were constant opportunities for foragers to exchange complementary products – which included foods, medicines, drugs and cosmetics – since the local growth cycles of wild resources were staggered by sharp differences in climate and topography.[26] Farming itself seems to have started in precisely this way, as one of so many 'niche' activities or local forms of specialization. The founder crops of early agriculture – among them emmer wheat, einkorn, barley and rye – were not

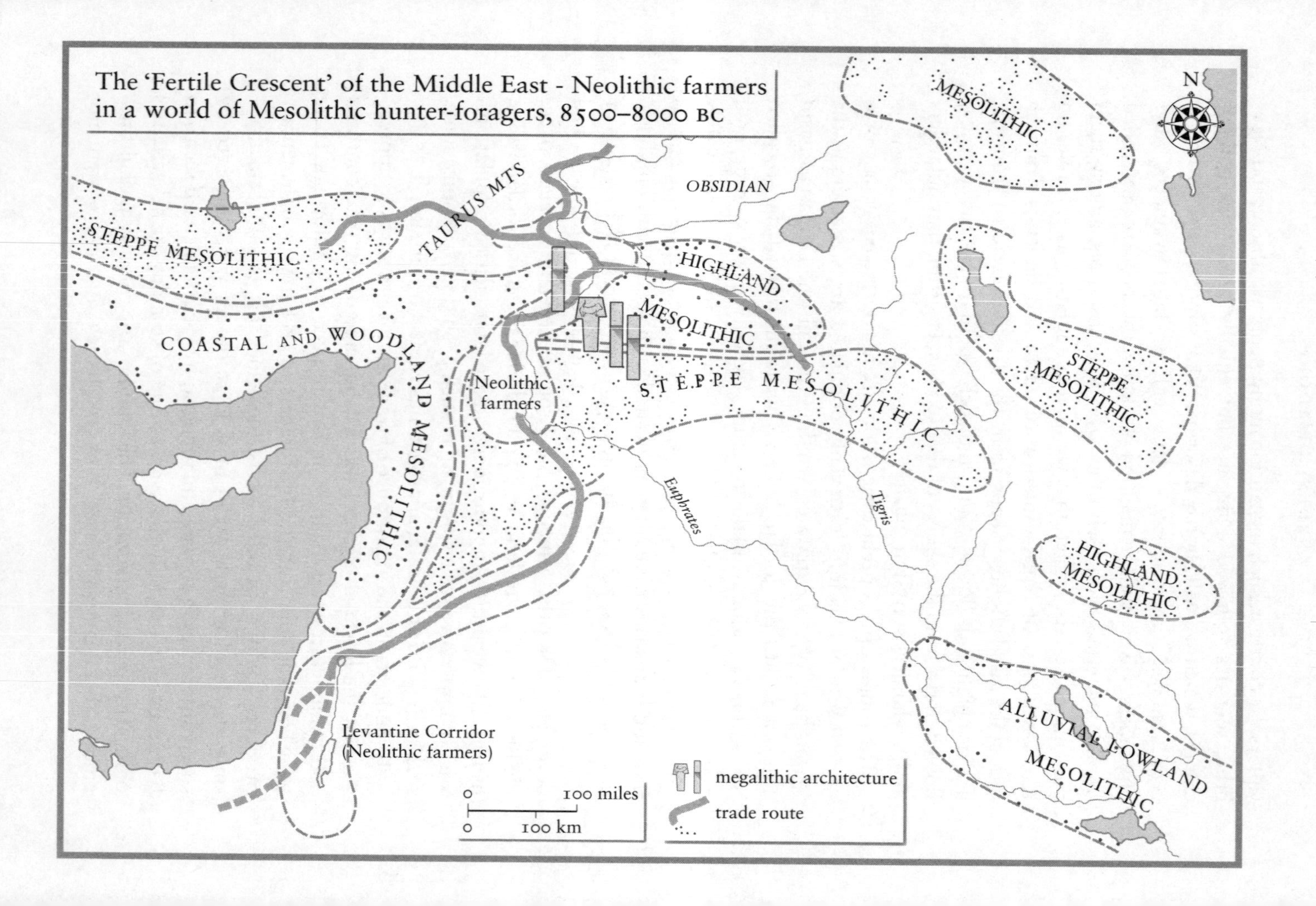
The 'Fertile Crescent' of the Middle East - Neolithic farmers in a world of Mesolithic hunter-foragers, 8500–8000 BC
N
MESOLITHIC
OBSIDIAN
TAURUS MTS
STEPPE MESOLITHIC
HIGHLAND
MESOLITHIC
STEPPE MESOLITHIC
COASTAL AND WOODLAND MESOLITHIC
Neolithic farmers
STEPPE MESOLITHIC
Euphrates
Tigris
HIGHLAND MESOLITHIC
ALLUVIAL LOWLAND MESOLITHIC
Levantine Corridor (Neolithic farmers)
0
100 miles
0
100 km
megalithic architecture
trade route

domesticated in a single 'core' area (as once supposed), but at different stops along the Levantine Corridor, scattered from the Jordan valley to the Syrian Euphrates, and perhaps further north as well.[27]

At higher altitudes, in the upland crescent, we find some of the earliest evidence for the management of livestock (sheep and goats in western Iran, cattle too in eastern Anatolia), incorporated into seasonal rounds of hunting and foraging.[28] Cereal cultivation began in a similar way, as a fairly minor supplement to economies based mainly on wild resources: nuts, berries, legumes and other readily accessible foodstuffs. Cultivation, however, is rarely just about calories. Cereal production also brought people together in new ways to perform communal tasks, mostly repetitive, labour-intensive and no doubt freighted with symbolic meaning; and the resulting foods were incorporated into their ceremonial lives. At the site of Jerf el-Ahmar, on the banks of the Syrian Euphrates – where upland and lowland sectors of the Fertile Crescent converge – the storage and processing of grain was associated less with ordinary dwellings than with subterranean lodges, entered from an opening in the roof and suffused with ritual associations.[29]

Before exploring some further contrasts between lowlanders and uplanders, it seems important to consider in a little more detail what these very earliest kinds of farming were actually like. To do this, we have to go deeper into the process of domestication.

ON SLOW WHEAT, AND POP THEORIES OF HOW WE BECAME FARMERS

In crops, domestication is what happens when plants under cultivation lose features that allow them to reproduce in the wild. Among the most important is the facility to disperse seed without human assistance. In wheat, seeds growing on the stalk are contained by tiny aerodynamic capsules known as spikelets. As wild wheat ripens, the connection between spikelet and stem (an element called the rachis) shatters. The spikelets free themselves and fall to the ground. Their spiky ends penetrate the soil, deep enough for at least some seed to survive and grow (the other ends project upwards, equipped with bristle-like awns to deter birds, rodents and browsing animals).

In domestic varieties, these aids to survival are lost. A genetic mutation takes place, switching off the mechanism for spontaneous seed dispersal and turning wheat from a hardy survivor into a hopeless dependant. Unable to separate from its mother plant, the rachis becomes a locus of attachment. Instead of spreading out to take on the big bad world, the spikelets stay rigidly fixed to the top part of the stem (the 'ear'). And there they remain, until someone comes along to harvest them, or until they rot, or are eaten by animals. So how did these genetic and behavioural changes in crops come about, how long did it take, and what had to happen in human societies to make them possible? Historians sometimes like to turn this question on its head. It is wheat, they remind us, that has domesticated people, just as much as people ever domesticated wheat.

Yuval Harari waxes eloquent on this point, asking us to think 'for a moment about the Agricultural Revolution from the viewpoint of wheat'. Ten thousand years ago, he points out, wheat was just another form of wild grass, of no special significance; but within the space of a few millennia it was growing over large parts of the planet. How did it happen? The answer, according to Harari, is that wheat did it by manipulating *Homo sapiens* to its advantage. 'This ape', he writes, 'had been living a fairly comfortable life hunting and gathering until about 10,000 years ago, but then began to invest more and more effort in cultivating wheat.' If wheat didn't like stones, humans had to clear them from their fields; if wheat didn't want to share its space with other plants, people were obliged to labour under the hot sun weeding them out; if wheat craved water, people had to lug it from one place to another, and so on.[30]

There's something ineluctable about all this. But only if we accept the premise that it does in fact make sense to look at the whole process 'from the viewpoint of wheat'. On reflection, why should we? Humans are very large-brained and intelligent primates and wheat is, well . . . a sort of grass. Of course, there are non-human species that have, in a sense, domesticated themselves – the house mouse and sparrow are among them, and so too probably the dog, all found, incidentally, in Early Neolithic villages of the Middle East. It's also undoubtedly true that, over the long term, ours is a species that has become enslaved to its crops: wheat, rice, millet and corn feed the world, and it's hard to envisage modern life without them.

But to make sense of the beginnings of Neolithic farming, we surely need to try and see it from the perspective of the Palaeolithic, not of the present, and still less from the viewpoint of some imaginary race of bourgeois ape-men. Of course, this is harder to do, but the alternative is to slip back into the realms of myth-making: retelling the past as a 'just-so' story, which makes our present situation seem somehow inevitable or preordained. Harari's retelling is appealing, we suggest, not because it's based on any evidence, but because we've heard it a thousand times before, just with a different cast of characters. In fact, many of us have been hearing it from infancy. Once again, we're back in the Garden of Eden. Except now, it's not a wily serpent who tricks humanity into sampling the forbidden fruit of knowledge. It's the fruit itself (i.e. the cereal grains).

We already know how this one goes. Humans were once living a 'fairly comfortable life', subsisting from the blessings of Nature, but then we made our most fatal mistake. Lured by the prospect of a still easier life – of surplus and luxury, of living like gods – we had to go and tamper with that harmonious State of Nature, and thus unwittingly turned ourselves into slaves.

What happens if we put aside this fable and consider what botanists, geneticists and archaeologists have found out in the past few decades? Let's focus on wheat and barley.

After the last Ice Age, these particular crops were among the first to be domesticated, along with lentils, flax, peas, chickpeas and bitter vetch. As we've noted, this process occurred in various different parts of the Fertile Crescent, rather than a single centre. Wild varieties of some of these crops grow there today, giving researchers the chance to make direct observations about how those plants behave, and even to reconstruct certain aspects of the technical process that led, 10,000 years ago, to domestication. Armed with such knowledge, they can also examine actual remains of ancient seeds and other plant remains, recovered in the many hundreds from archaeological sites in the same region. Scientists can then compare the biological process of domestication (reproduced under technological conditions similar to those of Neolithic cultivation) with the actual process that took place in prehistoric times and see how they match up.

Once cultivation became widespread in Neolithic societies, we might expect to find evidence of a relatively quick or at least continuous transition from wild to domestic forms of cereals (which is exactly what terms like the 'Agricultural Revolution' lead us to think), but in fact this is not at all what the results of archaeological science show. And despite the Middle Eastern setting, those findings do not add up to anything remotely resembling a Garden of Eden-type story about how humans haplessly stumbled their way into a Faustian pact with wheat. Just how far we are (or should be) from that kind of story was already clear to researchers some decades ago, once they began comparing actual prehistoric rates of crop domestication to those achieved under experimental conditions.

Experiments of this kind with wild wheat were first undertaken in the 1980s.[31] What they showed was that the key genetic mutation leading to crop domestication could be achieved in as little as twenty to thirty years, or at most 200 years, using simple harvesting techniques like reaping with flint sickles or uprooting by hand. All it would have taken, then, is for humans to follow the cues provided by the crops themselves. That meant harvesting after they began to ripen, doing it in ways that left the grain on the stem (e.g. cutting or pulling, as opposed to beating grain straight off the ear with a paddle), sowing new seed on virgin soil (away from wild competitors), learning from errors, and repeating the winning formula next year. For foragers seasoned in the harvesting of wild crops, these changes need not have posed major logistical or conceptual challenges. And there may also have been other good reasons to harvest wild cereals in this manner, besides obtaining food.

Harvesting by sickle yields straw as well as grain. Today we consider straw a by-product of cereal-farming, the primary purpose being to produce food. But archaeological evidence suggests things started the other way round.[32] Human populations in the Middle East began settling in permanent villages long before cereals became a major component of their diets.[33] In doing so, they found new uses for the stalks of wild grasses; these included fuel for lighting fires, and the temper that transformed mud and clay from so much friable matter into a vital tectonic resource, used to build houses, ovens, storage bins and other fixed structures. Straw could also be used to make baskets, clothing, matting and thatch. As people intensified the harvesting of wild grasses for straw (either by sickle or simply uprooting), they also produced one of the key conditions for some of these grasses to lose their natural mechanisms of seed dispersal.

Now here's the key point: if crops, rather than humans, had been setting the pace, these two processes would have gone hand in hand, leading to the domestication of large-seeded grasses within a few decades. Wheat would have gained its human handmaidens, and humans would have gained a plant resource that could be efficiently harvested with little loss of seed and that was eminently storable, but that also required much greater outlays of labour in the form of land management and the post-harvesting work of threshing and winnowing (a process which occurs naturally in wild cereals). Within a few human generations, the Faustian pact between people and crops would have been sealed. But here again, the evidence flatly contradicts these expectations.

In fact, the latest research shows that the process of plant domestication in the Fertile Crescent was not fully completed until much later: as much as 3,000 years after the cultivation of wild cereals first began.[34] (Once again, to get a sense of the scale here, think: the time between the putative Trojan War and today.) And while some modern historians may allow themselves the luxury of disposing with 'a few short millennia' here or there, we can hardly extend this attitude to the prehistoric actors whose lives we are trying to understand. At this point, you might reasonably ask what we mean by 'cultivation', and how we can possibly know when it began, if it didn't lead to clear changes in the reproductive behaviour of wild plants? The answers lie in weeds (and in research methods dreamed up in an inventive subbranch of archaeology, known as 'archaeobotany').

WHY NEOLITHIC FARMING TOOK SO LONG TO EVOLVE, AND DID NOT, AS ROUSSEAU IMAGINED, INVOLVE THE ENCLOSURE OF FIXED FIELDS

Since the early 2000s, archaeobotanists have been studying a phenomenon known as 'pre-domestication cultivation'. Cultivation in general refers to the work done by humans to improve the life chances of favoured crops, whether these be wild or domestic. This usually involves, at minimum, clearing and tilling the soil. Soil preparation induces changes in the size and shape of wild cereal grains, though such changes need not lead to domestication (basically they just get bigger). It also attracts other flora that flourish in disturbed soils, including arable weeds such as clover, fenugreek, gromwell and indeed members of the colourful crowfoot family (genus *Adonis*!), quick to flower and just as quick to die.

Since the 1980s, researchers have accumulated statistical evidence from prehistoric sites in the Middle East, analysing this evidence for changes over time in grain size and proportions of arable weed flora. Samples now number in the many tens of thousands. What they show is that, in certain parts of the region such as northern Syria, the cultivation of wild cereals dates back at least to 10,000 BC.[35] Yet in these same regions, the biological process of crop domestication (including the crucial switch-over from brittle rachis to tough) was not completed until closer to 7000 BC – that is roughly ten times as long as it need have taken – if, that is, humans really had stumbled blindly into the whole process, following the trajectory dictated by changes in their crops.[36] To be clear: that's 3,000 years of human history, far too long to constitute an 'Agricultural Revolution' or even to be considered some kind of transitional state on the road to farming.

To us, with our Platonic prejudices, all this looks like a very long and unnecessary delay, but clearly it was not experienced that way by people in Neolithic times. We need to understand this 3,000-year period as an important phase of human history in its own right. It's a phase marked by foragers moving in and out of cultivation – and as we've seen, there's nothing unusual or anomalous about this flirting and tinkering with the possibilities of farming, in just the ways Plato would have despised – but in no way enslaving themselves to the needs of their crops or herds. So long as it didn't become too onerous, cultivation was just one of many ways in which early settled communities managed their environments. Separating wild and domestic plant populations need not have been a major concern for them, even if it appears that way to us.[37]

On reflection, this approach makes perfectly good sense. Cultivating domestic cereals, as the 'affluent' foragers of the Pacific Coast knew well, is enormously hard work.[38] Serious farming meant serious soil maintenance and weed clearance. It meant threshing and winnowing after harvest. All these activities would have got in the way of hunting, wild food collection, craft production, marriages and any number of other things, not to mention storytelling, gambling, travelling and organizing masquerades. Indeed, to balance out their dietary needs and labour costs, early cultivators may even have strategically chosen practices that worked against the morphological changes which signal the onset of domestication in plants.[39]

This balancing act involved a special kind of cultivation, which brings us back full circle to Çatalhöyük and its wetland location. Called 'flood retreat', 'flood recession' or *décrue* farming, it takes place on the margins of seasonally flooding lakes or rivers. Flood-retreat farming is a distinctly lackadaisical way to raise crops. The work of soil preparation is given over mostly to nature. Seasonal flooding does the work of tillage, annually sifting and refreshing the soil. As the waters recede they leave behind a fertile bed of alluvial earth, where seed can be broadcast. This was garden cultivation on a small scale with no need for deforestation, weeding or irrigation, except perhaps the construction of small stone or earthen barriers ('bunds') to nudge the distribution of water this way or that. Areas of high groundwater, such as the edges of artesian springs, could also be exploited in this way.[40]

In terms of labour, flood-retreat farming is not only pretty light, it also requires little central management. Critically, such systems have a kind of inbuilt resistance to the enclosure and measurement of land. Any given parcel of territory might be fertile one year, and then either flooded or dried out the next, so there is little incentive for long-term ownership or enclosure of fixed plots. It makes little sense to set up boundary stones when the ground itself is shifting underneath you. No form of human ecology is 'innately' egalitarian, but much as Rousseau and his epigones would have been surprised to hear it, these early cultivation systems did not lend themselves to the development of private property. If anything, flood-retreat farming was practically oriented towards the collective holding of land, or at least flexible systems of field reallocation.[41]

Flood-retreat farming was an especially important feature of Early

Neolithic economies in the more arid, lowland sectors of the Fertile Crescent, and particularly the Levantine Corridor, where important sites often developed on the margins of springs or lakes (e.g. Jericho, Tell Aswad) or on riverbanks (e.g. Abu Hureyra, Jerf el-Ahmar). Because the densest stands of wild grain crops actually lay in upland areas with higher rainfall, the inhabitants of such lowland sites had opportunities to isolate cultivated from wild stock, setting in motion a process of divergence and domestication by gathering grains from the highlands and broadcasting them in lowland, flood-retreat areas. This makes the extremely long timescale of cereal domestication more striking still. Early cultivators, it seems, were doing the minimum amount of subsistence work needed to stay in their given locations, which they occupied for reasons other than farming: hunting, foraging, fishing, trading and more.

ON WOMAN, THE SCIENTIST

Rejecting a Garden of Eden-type narrative for the origins of farming also means rejecting, or at least questioning, the gendered assumptions lurking behind that narrative.[42] Apart from being a story about the loss of primordial innocence, the Book of Genesis is also one of history's most enduring charters for the hatred of women, rivalled only (in the Western tradition) by the prejudices of Greek authors like Hesiod, or for that matter Plato. It is Eve, after all, who proves too weak to resist the exhortations of the crafty serpent and is first to bite the forbidden fruit, because she is the one who desires knowledge and wisdom. Her punishment (and that of all women following her) is to bear children in severe pain and live under the rule of her husband, whose own destiny is to subsist by the sweat of his brow.

When today's writers speculate about 'wheat domesticating humans' (as opposed to 'humans domesticating wheat'), what they are really doing is replacing a question about concrete scientific (human) achievements with something rather more mystical. In this view, we're not asking questions about who might actually have been doing all the intellectual and practical work of manipulating wild plants: exploring their properties in different soils and water regimes; experimenting with harvesting techniques, accumulating observations about the effects these all have on growth, reproduction and nutrition; debating the social implications. Instead, we find ourselves waxing lyrical about the temptations of forbidden fruits and musing on the unforeseen consequences of adopting a technology (agriculture) that Jared Diamond has characterized – again, with biblical overtones – as 'the worst mistake in the history of the human race'.[43]

Consciously or not, it is the contributions of women that get written out of such accounts. Harvesting wild plants and turning them into food, medicine and complex structures like baskets or clothing is almost everywhere a female activity, and may be gendered female even when practised by men. This is not quite an anthropological universal, but it's about as close to one as you are ever likely to get.[44] Hypothetically, of course, it is possible that things haven't always been so. It's even conceivable that the current situation is really the result of some great global switch-around of gender roles and language structures that took place in the last few thousand years – but one would imagine that such an epochal change would have left other traces, and no one has so much as suggested what such traces might be. True, archaeological evidence of any kind is hard to come by, because aside from charred seeds, very little of what was done culturally with plants survives from prehistoric times. But where evidence exists, it points to strong associations between women and plant-based knowledge as far back as one can trace such things.[45]

By plant-based knowledge we don't just mean new ways of working with wild flora to produce food, spices, medicines, pigments or poisons. We also mean the development of fibre-based crafts and industries, and the more abstract forms of knowledge these tend to generate about properties of time, space and structure. Textiles, basketry, network, matting and cordage were most likely always developed in parallel with the cultivation of edible plants, which also implies the development of mathematical and geometrical knowledge that is (quite literally) intertwined with the practice of these crafts.[46] Women's association with such knowledge extends back to some of the earliest surviving depictions of the human form: the ubiquitous sculpted female figurines of the last Ice Age with their woven headgear, string skirts and belts made of cord.[47]

There is a peculiar tendency among (male) scholars to skip over the gendered aspects of this kind of knowledge or veil it in abstractions. Consider Claude Lévi-Strauss's famous comments on the 'savage mind', those 'Neolithic scientists' he imagined as having created a parallel route of discovery to modern science, but one that started from concrete interactions with the natural world rather than generalizing laws and theorems. The former method of experimentation proceeds 'from the angle of sensible qualities', and according to Lévi-Strauss it flowered in the Neolithic period, giving us the basis of agriculture, animal husbandry, pottery, weaving, conservation and preparation of food, etc.; while the latter mode of discovery, starting from the definition of formal properties and theories, only came to fruition much more recently, with the advent of modern scientific procedures.[48]

Nowhere in *The Savage Mind* – a book ostensibly dedicated to understanding that other sort of knowledge, the Neolithic 'science of the concrete' – does Lévi-Strauss even mention the possibility that those responsible for its 'flowering' might, very often, have been women.

If we take these kinds of considerations (instead of some imaginary State of Nature) as our starting point, then entirely different sorts of questions arise about the invention of Neolithic farming. In fact, a whole new language becomes necessary to describe it, since part of the problem with conventional approaches lies in the very terms 'agriculture' and 'domestication'. Agriculture is essentially about the production of food, which was just one (quite limited) aspect of the Neolithic relationship between people and plants. Domestication usually implies some form of domination or control over the unruly forces of 'wild nature'. Feminist critiques have already done much to unpack the gendered assumptions behind both concepts, neither of which seems appropriate to describe the ecology of early cultivators.[49]

What if we shifted the emphasis away from agriculture and domestication to, say, botany or even gardening? At once we find ourselves closer to the realities of Neolithic ecology, which seem little concerned with taming wild nature or squeezing as many calories as possible from a handful of seed grasses. What it really seems to have been about is creating garden plots – artificial, often temporary habitats – in which the ecological scales were tipped in favour of preferred species. Those species included plants that modern botanists separate out into competing classes of 'weeds', 'drugs', 'herbs' and 'food crops', but which Neolithic botanists (schooled by hands-on experience, not textbooks) preferred to grow side by side.

Instead of fixed fields, they exploited alluvial soils on the margins of lakes and springs, which shifted location from year to year. Instead of hewing wood, tilling fields and carrying water, they found ways of 'persuading' nature to do much of this labour for them. Theirs was not a science of domination and classification, but one of bending and coaxing, nurturing and cajoling, or even tricking the forces of nature, to increase the likelihood of securing a favourable outcome.[50] Their 'laboratory' was the real world of plants and animals, whose innate tendencies they exploited through close observation and experimentation. This Neolithic mode of cultivation was, moreover, highly successful.

In lowland regions of the Fertile Crescent, such as the Jordan and Euphrates valleys, ecological systems of this kind fostered the incremental growth of settlements and populations for three millennia. Pretending it was all just some kind of very extended transition or rehearsal for the advent of 'serious' agriculture is to miss the real point. It's also to ignore what to many has long seemed an obvious connection between Neolithic ecology and the visibility of women in contemporary art and ritual. Whether one calls these figures 'goddesses' or 'scientists' is perhaps less important than recognizing how their very appearance signals a new awareness of women's status, which was surely based on their concrete achievements in binding together these new forms of society.

Part of the difficulty with studying scientific innovation in prehistory is that we have to imagine a world without laboratories; or rather, a world in which laboratories are potentially everywhere and anywhere. Here Lévi-Strauss is much more on the ball:

> . . . there are two distinct modes of scientific thought. These are certainly not a function of different stages of development of the human

> mind but rather of two strategic levels at which nature is accessible to scientific enquiry: one roughly adapted to that of perception and the imagination: the other at a remove from it. It is as if the necessary connections which are the object of all science, Neolithic or modern, could be arrived at by two different routes, one very close to, and the other more remote from, sensible intuition.[51]

Lévi-Strauss, as we noted, called the first route to discovery a 'science of the concrete'. And it's important to recall that most of humanity's greatest scientific discoveries – the invention of farming, pottery, weaving, metallurgy, systems of maritime navigation, monumental architecture, the classification and indeed domestication of plants and animals, and so on – were made under precisely those other (Neolithic) sorts of conditions. Judged by its results, then, this concrete approach was undeniably science. But what does 'science of the concrete' actually look like, in the archaeological record? How can we hope to see it at work, when so many thousands of years stand between us and the processes of innovation we are trying to understand? The answer here lies precisely in its 'concreteness'. Invention in one domain finds echoes and analogies across a whole range of others, which might otherwise seem completely unrelated.

We can see this clearly in Early Neolithic cereal cultivation. Recall that flood-retreat farming required people to establish durable settlements in mud-based environments, like swamps and lake margins. Doing so meant becoming intimate with the properties of soils and clays, carefully observing their fertility under different conditions, but also experimenting with them as tectonic materials, or even as vehicles of abstract thought. As well as supporting new forms of cultivation, soil and clay – mixed with wheat and chaff – became basic materials of construction: essential in building the first permanent houses; used to make ovens, furniture and insulation – almost everything, in fact, except pottery, a later invention in this part of the world.

But clay was also used, in the same times and places, to (literally) model relationships of utterly different kinds, between men and women, people and animals. People started using its plastic qualities to figure out mental problems, making small geometric tokens that many see as direct precursors to later systems of mathematical notation. Archaeologists find these tiny numerical devices in direct association with figurines of herd animals and full-bodied women: the kind of miniatures that stimulate so much modern speculation about Neolithic spirituality, and which find later echoes in myths about the demiurgic, life-giving properties of clay.[52] As we'll soon see, earth and clay even came to redefine relationships between the living and the dead.

Seen this way, the 'origins of farming' start to look less like an economic transition and more like a media revolution, which was also a social revolution, encompassing everything from horticulture to architecture, mathematics to thermodynamics, and from religion to the remodelling of gender roles. And while we can't know exactly who was doing what in this brave new world, it's abundantly clear that women's work and knowledge were central to its creation; that the whole process was a fairly leisurely, even playful one, not forced by any environmental catastrophe or demographic tipping point and unmarked by major violent conflict. What's more, it was all carried out in ways that made radical inequality an extremely unlikely outcome.

All this applies most clearly to the development of Early Neolithic societies in lowland parts of the Fertile Crescent, and especially along the valleys of the Jordan and Euphrates Rivers. But these communities did not develop in isolation. For almost the entire period we've been discussing, the upland crescent – following the foothills of the Taurus and Zagros Mountains and the adjoining steppe – was also home to settled populations, adept in managing a variety of wild plant and animal resources. They too were often village dwellers, who adopted strategies of cultivation and herding as they saw fit, while still deriving the bulk of their diet from non-domesticated species. But in other ways they are clearly marked out from their lowland neighbours, their construction of megalithic architecture, including the famous structures of Göbekli Tepe, being just the most obvious. Some of these groups lived in proximity to lowland Neolithic societies, especially along the upper reaches of the Euphrates, but their art and ritual suggest a radically different orientation to the world, as sharply distinguished from the latter as Northwest Coast foragers were from their Californian neighbours.

TO FARM OR NOT TO FARM: IT'S ALL IN YOUR HEAD (WHERE WE RETURN TO GÖBEKLI TEPE)

At the frontier between the upland and lowland sectors of the Fertile Crescent stands Göbekli Tepe itself. It is actually one of a series of megalithic centres that sprang up around the Urfa valley, near the modern border of Syria and Turkey, in the ninth millennium BC.[53] Most are still not excavated. Only the tops of their great T-shaped pillars can be seen projecting from the deep valley soils. While direct evidence is still lacking, this style of stone architecture probably marks the apex of a building tradition that began in timber. Wooden prototypes may also lie behind Göbekli Tepe's tradition of sculptural art, which evokes a world of fearsome images, far removed from the visual arts of the lowlands, with their humble figurines of women and domestic animals, and hamlets of clay.

In both medium and message, Göbekli Tepe could hardly be more different from the world of early farming communities. Its standing remains were wrought from stone, a material little used for construction in the Euphrates and Jordan valleys. Carved on these stone pillars is an imagery dominated by wild and venomous animals; scavengers and predators, almost exclusively sexed male. On a limestone pillar a lion rears up in high relief, teeth gnashing, claws outstretched, penis and scrotum on show. Elsewhere lurks a malevolent boar, its male sex also displayed. The most often repeated images depict raptors taking human heads. One remarkable sculpture, resembling a totem pole, comprises superimposed pairings of victims and predators: disembodied skulls and sharp-eyed birds of prey. Elsewhere, flesh-eating birds and other carnivores are shown grasping, tossing about or otherwise playing with their catch of human crania; carved below one such figure on a monumental pillar is the image of a headless man with an erect penis (conceivably this depicts the kind of immediate post-mortem erection or 'priapism' that occurs in victims of hanging or beheading as a result of massive trauma to the spinal cord).[54]

What are these images telling us? Could the taking of trophy heads among upland populations of the steppe-forest zone be part of the picture? At the settlement of Nevalı Çori – also in Urfa province, and with similar monuments to Göbekli Tepe – burials with detached skulls were found, including one of a young woman with a flint dagger still lodged under her jaw; while from Jerf el-Ahmar – on the Upper Euphrates, where the lowland crescent approaches the uplands – comes the startling find of a splayed skeleton (again, a young woman) still lying inside a burnt-down building, prone and missing her head.[55] At Göbekli Tepe itself, the chopping of human heads was mimicked in statuary: anthropomorphic sculptures were made, only to have their tops smashed off and the stone heads buried adjacent to pillars within the shrines.[56] For all this, archaeologists remain rightly cautious about linking such practices to conflict or predation; so far, there is only limited evidence for interpersonal violence, let alone warfare at this time.[57]

Here we might also consider evidence from Çayönü Tepesi, in the Ergani plain. This was the site of a large prehistoric settlement comprising substantial houses built on stone foundations, as well as public buildings. It lay on a tributary of the Tigris in the hill country of Diyarbakır, not far north of Göbekli Tepe, and was established around the same time by a community of hunter-foragers and sometime herders.[58] Near the centre of the settlement stood a long-lived structure that archaeologists call the 'House of Skulls', for the simple reason that it was found to hold the remains of over 450 people, including headless corpses and over ninety crania, all crammed into small compartments. Cervical vertebrae were attached to some skulls, indicating they were severed from fleshed (but not necessarily living) bodies. Most of the heads were taken from young adults or adolescents, individuals in the prime of life, and ten from children. If any of these were trophy skulls, claimed from victims or enemies, then they were chosen for their vitality. The skulls were left bare, with no trace of decoration.[59]

Human remains in the House of Skulls were stored together with those of large prey animals, and a wild cattle skull was mounted on an outside wall. In its later stages of use, the building was furnished with a polished stone table, erected near the entrance in an open plaza that could have hosted large gatherings. Studies of blood residues from the surface, and from associated objects, led researchers to identify this as an altar on which public sacrifice and processing of bodies took place, the victims both animal and human. Whether or not the detail of this reconstruction is correct, the association of vanquished animals and human remains is suggestive. The House of Skulls met its end in a violent conflagration, after which the people of Çayönü covered the whole complex under a deep blanket of pebbles and soil.

Perhaps what we're detecting in the House of Skulls, but in a rather different form, is a complex of ideas already familiar from Amazonia and elsewhere: hunting as predation, shifting subtly from a mode of subsistence to a way of modelling and enacting dominance over other human beings. After all, even feudal lords in Europe tended to identify themselves with lions, hawks and predatory beasts (they were also fond of the symbolism of putting heads on poles; 'off with his head!' is still the most popular phrase identified with the British monarchy).[60] But what about Göbekli Tepe itself? If the display of trophy heads really was an important aspect of the site's function, surely some direct trace would remain, other than just some suggestive stone carvings.

Human remains are so far rare at Göbekli Tepe. Which makes it even more remarkable that – of the few hundred scraps of prehistoric human bone so far recovered from the site – some two-thirds are indeed segments of skulls or facial bones, some retaining signs of defleshing, and even decapitation. Among them were found remnants of three skulls, recovered from the area of the stone shrines, which bear evidence for more elaborate types of cultural modification in the form of deep incisions and drill-holes, allowing the skull to dangle from a string or be mounted on a pole.[61]

In earlier chapters, we've explored why farming was much less of a rupture in human affairs than we tend to assume. Now we're finally in a position to bring the various strands of this chapter together and say something about why this matters. Let's recap.

Neolithic farming began in Southwest Asia as a series of local specializations in crop-raising and animal-herding, scattered across various parts of the region, with no epicentre. These local strategies were pursued, it seems, in order to sustain access to trade partnerships and optimal locations for hunting and gathering, which continued unabated alongside cultivation. As we discussed back in Chapter One, this 'trade' might well have had more to do with sociability, romance or adventure than material advantage as we'd normally conceive it.

Still, whatever the reasons, over thousands of years such local innovations – everything from non-shattering wheat to docile sheep – were exchanged between villages, producing a degree of uniformity among a coalition of societies across the Middle East. A standard 'package' of mixed farming emerged, from the Iranian Zagros to the eastern shores of the Mediterranean, and then spread beyond it, albeit, as we'll see, with very mixed success.

But from its earliest beginnings, farming was much more than a new economy. It also saw the creation of patterns of life and ritual that remain doggedly with us millennia later, and have since become fixtures of social existence among a broad sector of humanity: everything from harvest festivals to habits of sitting on benches, putting cheese on bread, entering and exiting via doorways, or looking at the world through windows. Originally, as we've seen, much of this Neolithic lifestyle developed alongside an alternative cultural pattern in the steppe and upland zones of the Fertile Crescent, most clearly distinguished by the building of grand monuments in stone, and by a symbolism of male virility and predation that largely excluded female concerns. By contrast, the art and ritual of lowland settlements in the Euphrates and Jordan valleys presents women as co-creators of a distinct form of society – learned through the productive routines of cultivation, herding and village life – and celebrated by modelling and binding soft materials, such as clay or fibres, into symbolic forms.[62]

Of course, we could put these cultural oppositions down to coincidence, or perhaps even environmental factors. But considering the close proximity of the two cultural patterns, and how the groups responsible for them exchanged goods and were keenly aware of each other's existence, it is equally possible, and perhaps more plausible, to see what happened as the result of mutual and self-conscious differentiation, or schismogenesis, akin to what we traced in the last chapter among the recent foraging societies of America's West Coast. The more that uplanders came to organize their artistic and ceremonial lives around the theme of predatory male violence, the more lowlanders tended to organize theirs around female knowledge and symbolism – and vice versa. With no written sources to guide us, the clearest evidence we can find for such mutual oppositions is when things get (quite literally, in our case) turned on their head, as when one group of people seems to make a great display of going against some highly characteristic behaviour of their neighbours.

Such evidence is not at all hard to find, since lowland villagers, like their upland neighbours, also attached great ritual significance to human heads, but chose to treat them in ways that would have been utterly foreign to the uplanders. Let us briefly illustrate what we mean.

Perhaps the most recognizable – and definitely the most macabre – objects found in Early Neolithic villages of the Levantine Corridor (Israel, Palestine, Jordan, Lebanon and the Syrian Euphrates) are 'skull portraits'. These are heads that were removed from burials of women, men and occasionally children in a secondary process, after the corpse had decomposed. Once separated from the body, they were cleaned and carefully modelled over with clay, then coated with layers of plaster to become something altogether different. Shells were often fixed into the eye sockets, just as clay and plaster filled in for the flesh and skin. Red and white paint added further life. Skull portraits appear to have been treasured heirlooms, carefully stored and repaired over generations. They reached their height of popularity in the eighth millennium BC, as Göbekli Tepe fell into decline, when the practice spread as far as Çatalhöyük; there, one such modelled head was found in an intimate situation, clutched to the chest of a female burial.[63]

Ever since these intriguing objects first came to light at Jericho in the early twentieth century, archaeologists have puzzled over their meaning. Many scholars see them as expressions of care and reverence for ancestors. But there are literally countless ways one might show respect or grief for ancestors without systematically removing crania from their places of rest and modelling life into them by adding layers of clay, plaster, shell, fibre and pigment. Even in the lowland parts of the Fertile Crescent, this treatment was reserved for a minority of individuals. More often, human crania removed from burials were left bare, while others had complex histories as ritual objects, such as a group of skulls from Tell Qarassa in southern Syria, found to have been deliberately mutilated around the face in what appears to have been an act of post-mortem desecration.[64]

In the Jordan and Euphrates valleys and adjacent coastlands, the practice of curating human crania has an even longer history, extending back to Natufian hunter-gatherers, before the onset of the

Neolithic period; but longevity need not imply an entirely local context for later ritual innovations, such as the addition of decorative materials to make skull portraits. Perhaps making skull portraits in this particular way was not just about reconnecting with the dead, but also negating the logic of stripping, cutting, piercing and accumulating heads as trophies. At the very least, it offers a further indication that lowland and upland populations in the Fertile Crescent were following quite different – and in some ways, mutually opposed – cultural trajectories throughout the centuries when plants and animals were first domesticated.[65]

ON SEMANTIC SNARES AND METAPHYSICAL MIRAGES

Back in the 1970s, a brilliant Cambridge archaeologist called David Clarke predicted that, with modern research, almost every aspect of the old edifice of human evolution, 'the explanations of the development of modern man, domestication, metallurgy, urbanization and civilization – may in perspective emerge as semantic snares and metaphysical mirages.'[66] It is beginning to seem like he was right.

Let's recap a little further. A founding block in that old edifice of human social evolution was the allocation of a specific place in history to foraging societies, which was to be the prelude to an 'Agricultural Revolution' that supposedly changed everything about the course of history. The job of foragers in this conventional narrative is to be all that farming is not (and thus also to explain, by implication, what farming is). If farmers are sedentary, foragers must be mobile; if farmers actively produce food, foragers must merely collect it; if farmers have private property, foragers must renounce it; and if farming societies are unequal, this is by contrast with the 'innate' egalitarianism of foragers. Finally, if a particular group of foragers should happen to possess any such features in common with farmers, the dominant narrative demands that these can only be 'incipient', 'emergent' or 'deviant' in nature, so that the destiny of foragers is either to 'evolve' into farmers, or eventually to wither and die.

It will by now be increasingly obvious to any reader that almost nothing about this established narrative matches the available evidence. In the Fertile Crescent of the Middle East, long regarded as the cradle of the 'Agricultural Revolution', there was in fact no 'switch' from Palaeolithic forager to Neolithic farmer. The transition from living mainly on wild resources to a life based on food production took something in the order of 3,000 years. And while agriculture allowed for the *possibility* of more unequal concentrations of wealth, in most cases this only began to happen millennia after its inception. In the centuries between, people were effectively trying farming on for size, 'play farming' if you will, switching between modes of production, much as they switched their social structures back and forth.

Clearly, it no longer makes any sense to use phrases like 'the Agricultural Revolution' when dealing with processes of such inordinate length and complexity. And since there was no Eden-like state from which the first farmers could take their first steps on the road to inequality, it makes even less sense to talk about agriculture as marking the origins of social rank, inequality or private property. In the Fertile Crescent, it is – if anything – among upland groups, furthest removed from a dependence on agriculture, that we find stratification and violence becoming entrenched; while their lowland counterparts, who linked the production of crops to important social rituals, come out looking decidedly more egalitarian; and much of this egalitarianism relates to an increase in the economic and social visibility of women, reflected in their art and ritual. In that sense, the work of Gimbutas – while painted with brush strokes that were broad, sometimes to the point of caricature – was not entirely wide of the mark.

All this raises an obvious question: if the adoption of farming actually set humanity, or some small part of it, on a course *away* from violent domination, what went wrong?

7
The Ecology of Freedom

How farming first hopped, stumbled and bluffed its way around the world

In a way, the Fertile Crescent of the Middle East is unusual precisely because we know so much about what happened there. Long recognized as a crucible of plant and animal domestication, it has been more intensively studied by archaeologists than almost any other region outside Europe. This accumulation of evidence allows us to begin to tease out some of the social changes that accompanied the first steps to crop and animal domestication, even to rely to a certain extent on negative evidence. It is difficult, for instance, to make any sort of convincing argument that warfare was a significant feature of early farming societies in the Middle East, as by now one would expect some evidence for it to have shown up in the record. On the other hand, there is abundant evidence for the proliferation of trade and specialized crafts, and for the importance of female figures in art and ritual.

For the same reasons, we're able to draw comparisons between the lowland part of the Fertile Crescent (especially the Levantine Corridor passing via the Jordan valley) and its upland sector (the plains and foothills of eastern Turkey), where equally precocious developments in village life and local industries were associated with the raising of stone monuments adorned with masculine symbolism and an imagery of predatory violence.[1] Some scholars have tried to see all these developments as somehow part of a single process, heading in the same general direction, towards the 'birth of agriculture'. But the first farmers were reluctant farmers who seem to have understood the logistical implications of agriculture and avoided any major commitment to it. Their upland neighbours, also living settled lives in areas with diverse wild resources, had even less incentive to tie their existence to a narrow range of crops and livestock.

If the situation in just one cradle of early farming was that complicated, then surely it no longer makes sense to ask, 'what were the social implications of the transition to farming?' – as if there was necessarily just one transition, and one set of implications. Certainly, it's wrong to assume that planting seeds or tending sheep means one is *necessarily* obliged to accept more unequal social arrangements, simply to avert a 'tragedy of the commons'. There is a paradox here. Most general works on the course of human history do actually assume something like this; but almost nobody, if pressed, would seriously defend such a point because it's an obvious straw man. Any student of agrarian societies knows that people inclined to expand agriculture sustainably, without privatizing land or surrendering its management to a class of overseers, have always found ways to do so.

Communal tenure, 'open-field' principles, periodic redistribution of plots and co-operative management of pasture are not particularly exceptional and were often practised for centuries in the same locations.[2] The Russian *mir* is a famous example, but similar systems of land redistribution once existed all over Europe, from the Highlands of Scotland to the Balkans, occasionally into very recent times. The Anglo-Saxon term was *run-rig* or *rundale*. Of course, the rules of redistribution varied from one case to the next – in some, it was made *per stirpes*, in others according to the number of people in a family. Most often, the precise location of each strip was determined by lottery, with each family receiving one strip per land tract of differing quality, so that nobody was obliged to travel much further than anyone else to his fields or to work soil of consistently lower quality.[3]

Of course, it wasn't just in Europe that such things happened. In his 1875 *Lectures on the Early History of Institutions*, Henry Sumner Maine – who held the first chair of historical and comparative jurisprudence at Oxford – was already discussing cases of periodic land redistribution and *rundale*-type institutions from India to Ireland, noting that almost up until his own day, 'cases were frequent in which the arable land was divided into farms which shifted among the tenant-families periodically, and sometimes annually.' And that in pre-industrial Germany, where land tenure was apportioned between 'mark associations', each tenant would receive lots divided among the three main qualities of soil. Importantly, he notes, these were not so much forms of property as 'modes of occupation', not unlike the rights of access found in many forager groups.[4] We could go on piling up the examples (the Palestinian *mash'a* system, for instance, or Balinese *subak*).[5]

In short, there is simply no reason to assume that the adoption of agriculture in more remote periods also meant the inception of private land ownership, territoriality, or an irreversible departure from forager egalitarianism. It may have happened that way sometimes, but this can no longer be treated as a default assumption. As we saw in the last chapter, exactly the opposite seems true in the Fertile Crescent of the Middle East, at least for the first few thousand years after the appearance of farming. If the situation in just one cradle of early farming was so different from our evolutionary expectations, then we can only wonder what other stories remain to be told, in other places where farming emerged. Indeed, these other locations are multiplying in light of new evidence, genetic and botanical, as well as archaeological. It turns out the process was far messier, and far less unidirectional, than anyone had guessed; and so we have to consider a broader range of possibilities than once assumed. In this chapter, we'll show just how much the picture is changing and point towards some of the surprising new patterns that are starting to emerge.

Geographers and historians used to believe that plants and animals were first domesticated in just a few 'nuclear' zones: the same areas in which large-scale, politically centralized societies later appeared. In the Middle East there was wheat and barley, as well as sheep, goats, pigs and cattle; in China there was rice (*japonica*), soybeans and a different variety of pig; potatoes, quinoa and llamas were brought under domestication in the Peruvian Andes; and maize, avocado and chilli in Mesoamerica. Such neat geographical alignments between early centres of crop domestication and the rise of centralized states invited speculation that the former led to the latter: that food production was responsible for the emergence of cities, writing, and centralized political organization, providing a surplus of calories to support large populations and elite classes of administrators, warriors and politicians. Invent agriculture – or so the story once went – and you set yourself on a course that will eventually lead to Assyrian charioteers, Confucian bureaucrats, Inca sun-kings or Aztec priests carrying away a significant chunk of your grain. Domination – and most often violent, ugly domination – was sure to follow; it was just a matter of time.

Archaeological science has changed all this. Experts now identify between fifteen and twenty independent centres of domestication, many of which followed very different paths of development to China, Peru, Mesoamerica or Mesopotamia (which themselves all followed quite different paths, as we'll see in later chapters). To those centres of early farming must now be added, among others, the Indian subcontinent (where browntop millet, mungbeans, horse gram, *indica* rice and humped zebu cattle were domesticated); the grasslands of West Africa (pearl millet); the central highlands of New Guinea (bananas, taroes and yams); the tropical forests of South America (manioc and peanuts); and the Eastern Woodlands of North America, where a distinct suite of local seed crops – goosefoot, sunflower and sumpweed – was raised, long before the introduction of maize from Mesoamerica.[6]

We know much less about the prehistory of these other regions than we do about the Fertile Crescent. None followed a linear trajectory from food production to state formation. Nor is there any reason to assume a rapid spread of farming beyond them to neighbouring areas. Food production did not always present itself to foragers, fishers and hunters as an obviously beneficial thing. Historians painting with a broad brush sometimes write as if it did, or as if the only barriers to the 'spread of farming' were natural ones, such as climate and topography. This sets up something of a paradox, because even foragers living in highly suitable environments, and clearly aware of the possibilities of cereal-farming, often chose not to adopt it. Take Jared Diamond:

> Just as some regions proved much more suitable than others for origins of food production, the ease of its spread also varied greatly around the world. Some areas that are ecologically very suitable for food production never acquired it in prehistoric times at all, even though areas of prehistoric food production existed nearby. The most conspicuous such

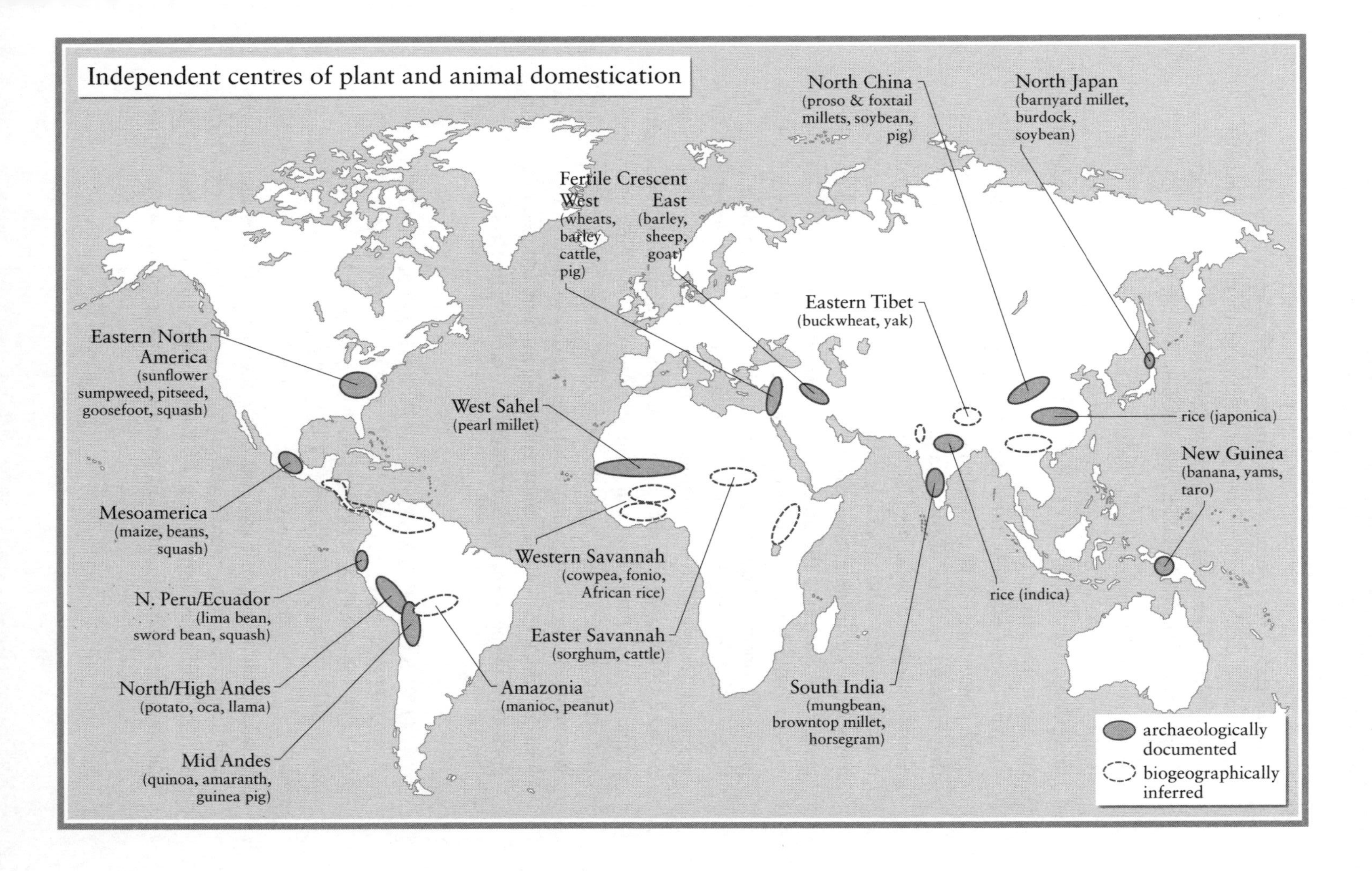

Independent centres of plant and animal domestication
North China
(proso & foxtail millets, soybean, pig)
North Japan
(barnyard millet, burdock, soybean)
Fertile Crescent
West
(wheats, barley cattle, pig)
East
(barley, sheep, goat)
Eastern North America
(sunflower sumpweed, pitseed, goosefoot, squash)
Eastern Tibet
(buckwheat, yak)
West Sahel
(pearl millet)
rice (japonica)
New Guinea
(banana, yams, taro)
Mesoamerica
(maize, beans, squash)
Western Savannah
(cowpea, fonio, African rice)
N. Peru/Ecuador
(lima bean, sword bean, squash)
rice (indica)
Easter Savannah
(sorghum, cattle)
North/High Andes
(potato, oca, llama)
Amazonia
(manioc, peanut)
South India
(mungbean, browntop millet, horsegram)
Mid Andes
(quinoa, amaranth, guinea pig)
archaeologically documented
biogeographically inferred

> examples are the failure of both farming and herding to reach Native American California from the U.S. Southwest or to reach Australia from New Guinea and Indonesia, and the failure of farming to spread from South Africa's Natal Province to South Africa's Cape.[7]

As we saw in Chapter Five, the failure of farming to 'reach' California is not a particularly compelling way to frame the problem. This is just an updated version of the old diffusionist approach, which identifies culture traits (cat's cradles, musical instruments, agriculture and so on) and maps out how they migrate across the globe, and why in some places they fail to do so. In reality, there's every reason to believe that farming 'reached' California just as soon as it reached anywhere else in North America. It's just that (despite a work ethic that valorized strenuous labour, and a regional exchange system that would have allowed information about innovations to spread rapidly) people there rejected the practice as definitively as they did slavery.

Even in the American Southwest, the overall trend for 500 years or so before Europeans arrived was the gradual abandonment of maize and beans, which people had been growing in some cases for thousands of years, and a return to a foraging way of life. If anything, during this period Californians were the ones doing the spreading, with populations originally from the east of the state bringing new foraging techniques, and replacing previously agricultural peoples, as far away as Utah and Wyoming. By the time Spaniards arrived in the Southwest, the Pueblo societies which had once dominated the region were reduced to isolated pockets of farmers, entirely surrounded by hunter-gatherers.[8]

ON SOME ISSUES OF TERMINOLOGY WHEN DISCUSSING THE MOVEMENT OF DOMESTIC CROPS AND ANIMALS AROUND THE GLOBE

In books on world history, you often encounter phrases like 'crops and livestock spread rapidly through Eurasia', or 'the plant package of the Fertile Crescent launched food production from Ireland to the Indus', or 'maize diffused northwards at a snail's pace.' How appropriate is such language when describing the expansion of Neolithic economies many thousands of years ago?

If anything, it seems to reflect the experience of the last few centuries, when Old World domesticates did indeed conquer the environments of the Americas and Oceania. In those more recent times, crops and livestock were able to 'spread' like wildfire, transforming existing habitats in ways that often rendered them unrecognizable within a few generations. But this has less to do with the nature of seed cultivation itself than with imperial and commercial expansion: seeds can spread very quickly if those carrying them have an army and are driven by the need endlessly to expand their enterprises to maintain profits. The Neolithic situation was altogether different. Especially for the first several thousand years after the end of the last Ice Age, most people were still not farmers, and farmers' crops had to compete with a whole panoply of wild predators and parasites, most of which have since been eliminated from agricultural landscapes.

To begin with, domestic plants and animals could not 'spread' beyond their original ecological limits without significant effort on the part of their human planters and keepers. Suitable environments not only had to be found but also modified by weeding, manuring, terracing, and so on. The landscape modifications involved may seem small-scale – little more than ecological tinkering – to our eyes, but they were onerous enough by local standards, and crucial in extending the range of domestic species.[9] Of course, there were always paths of least resistance, topographical features and climatic regimes conducive or less conducive to the Neolithic economy. The east–west axis of Eurasia discussed by Jared Diamond in his *Guns, Germs and Steel* (1997) or the 'lucky latitudes' of Ian Morris's *Why the West Rules – For Now* (2010) are ecological corridors of this sort.

Eurasia, as these authors point out, has few equivalents to the sharp climatic variations of the Americas, or indeed of Africa. Terrestrial species can travel across the breadth of the Eurasian continent without crossing boundaries between tropical and temperate zones. Continents whose extremities tilt north to south are a different proposition, and perhaps less amenable to such ecological transfers. The basic geographical point is surely sound, at least for the last 10,000 years of history. It explains why cereals of Fertile Crescent origin are successfully grown today in such distant locations as Ireland and Japan. It may also explain, to some extent, why many thousands of years elapsed before American crops – such as maize or squash (first domesticated in the tropics) – were accepted in the temperate northern part of the American continent, by contrast with the relatively rapid adoption of Eurasian crops outside their areas of origin.

To what extent can such observations help to make sense of human history on a larger scale? How far can geography go in explaining history, rather than simply informing it?

Back in the 1970s and 1980s, a geographer called Alfred W. Crosby came up with a number of important theories about how ecology shaped the course of history. Among other things, he was the first to draw attention to the 'Columbian exchange', the remarkable crossover of non-human species set in motion by Europeans' arrival in the Americas after 1492, and its transformative effect on the global configuration of culture, economy and cuisine. Tobacco, peppers, potatoes and turkeys flowed into Eurasia; maize, rubber and chickens entered Africa; and citrus fruits, coffee, horses, donkeys and livestock travelled to the Americas. Crosby went on to argue that the global ascendance of European economies since the sixteenth century could be accounted for by a process he called 'ecological imperialism'.[10]

The temperate zones of North America and Oceania, as Crosby pointed out, were ideally suited to Eurasian crops and livestock; not only because of their climate, but because they possessed few native competitors and no local parasites, such as the various funguses, insects or field mice that have developed to specialize in sharing human-grown wheat. Unleashed on such fresh environments, Old World domesticates went into reproductive overdrive, even going feral again in some cases. Outgrowing and out-grazing local flora and fauna, they began to turn native ecosystems on their heads, creating 'Neo-Europes' – carbon copies of European environments, of the sort one sees today when driving through the countryside of New Zealand's North Island, for example; or much of New England. The ecological assault on native habitats also included infectious diseases, such as smallpox and measles, which originated in Old World environments where humans and cattle cohabited. While European plants thrived in the absence of pests, diseases brought with domestic animals (or by humans accustomed to living alongside them) wreaked havoc on indigenous populations, creating casualty rates as high as 95 per cent, even in places where settlers were not enslaving or actively massacring the indigenous population – which, of course, they often were.

Viewed in this light, the success of modern European imperialism owed more to 'the Old World Neolithic Revolution' – with its roots in the Fertile Crescent – than to the specific achievements of Columbus, Magellan, Cook and all the rest. And in a sense this is true. But the story of agricultural expansion *before* the sixteenth century is very far from being a one-way street; in fact, it is full of false starts, hiccups and reversals. This becomes truer the further back we go in time. To appreciate why, we will have to look beyond the Middle East to consider how the earliest farming populations fared in some other parts of the world after the end of the last Ice Age. But first there is a more basic point to address: why is our discussion of these issues confined only to the last 10,000 or so years of human history? Given that humans have been around for upwards of 200,000 years, why didn't farming develop much earlier?

WHY AGRICULTURE DID NOT DEVELOP SOONER

Since our species came into existence, there have been only two sustained periods of warm climate of the kind that might support an agricultural economy for long enough to leave some trace in the archaeological record.[11] The first was the Eemian interglacial, which took place around 130,000 years ago. Global temperatures stabilized at slightly above their present-day levels, sustaining the spread of boreal forests as far north as Alaska and Finland. Hippos basked on the banks of the Thames and the Rhine. But the impact on human populations was limited by our then restricted geographical range. The second is the one we are living in now. When it began, around 12,000 years ago, people were already present on all the world's continents, and in many different kinds of environment. Geologists call this period the Holocene, from Greek *holos* (entire), *kainos* (new).

Many earth scientists now consider the Holocene over and done. For at least the last two centuries we have been entering a new geological epoch, the Anthropocene, in which for the first time in history human activities are the main drivers of global climate change. Where exactly the Anthropocene begins is a scientific bone of contention. Most experts point to the Industrial Revolution, but some put its origins earlier, in the late 1500s and early 1600s. At that time, a global drop in surface air temperatures occurred – part of the 'Little Ice Age' – which natural forces can't explain. Quite likely, European expansion in the Americas played a role. With perhaps 90 per cent of the indigenous population eliminated by the effects of conquest and infectious disease, forests reclaimed regions in which terraced agriculture and irrigation had been practised for centuries. In Mesoamerica, Amazonia and the Andes, some 50 million hectares of cultivated land may have reverted to wilderness. Carbon uptake from vegetation increased on a scale sufficient to change the Earth System and bring about a human-driven phase of global cooling.[12]

Wherever one starts it, the Anthropocene is what we have done with the legacy of a Holocene Age, which in some ways had been a 'clean sheet' for humanity. At its onset, many things really were new. As the ice receded, flora and fauna – once confined to small refuge zones – spread out to new vistas. People followed, helping favoured species on their way by setting fires and clearing land. The effect of global warming on the world's shorelines was more complex, as coastal shelves formerly under ice sprang back to the surface, while others sank below rising seawaters, fed from glacial melt.[13] For many historians, the onset of the Holocene is significant because it created conditions for the origins of agriculture. Yet in many parts of the world, as we've already seen, it was also a Golden Age for foragers, and it's important to remember that this forager paradise was the context in which the first farmers set up shop.

The most vigorous expansion of foraging populations was in coastal environments, freshly exposed by glacial retreat. Such locations offered a bonanza of wild resources. Saltwater fish and sea birds, whales and dolphins, seals and otters, crabs, shrimps, oysters, periwinkles and more besides. Freshwater rivers and lagoons, fed by mountain glaciers, now teemed with pike and bream, attracting migratory waterfowl. Around estuaries, deltas and lake margins, annual rounds of fishing and foraging took place at increasingly close range, leading to sustained patterns of human aggregation quite unlike those of the glacial period, when long seasonal migrations of mammoth and other large game structured much of social life.[14]

Scrub and forest replaced open steppe and tundra across much of this postglacial world. As in earlier times, foragers used various techniques of land management to stimulate the growth of desired species, such as fruit and nut-bearing trees. By 8000 BC, their efforts had contributed to the extinction of roughly two-thirds of the world's megafauna, which were ill suited to the warmer and more enclosed habitats of the Holocene.[15] Expanding woodlands offered a superabundance of nutritious and storable foods: wild nuts, berries, fruits, leaves and fungi, processed with a new suite of composite ('micro-lithic') tools. Where forest took over from steppe, human hunting techniques shifted from the seasonal co-ordination of mass kills to more opportunistic and versatile strategies, focused on smaller mammals with more limited home ranges, among them elk, deer, boar and wild cattle.[16]

What is easy to forget, with hindsight, is that farmers entered into this whole new world very much as the cultural underdogs. Their earliest expansions were about as far removed as one could imagine from the *missions civilisatrices* of modern agrarian empires. Mostly, as we'll see, they filled in the territorial gaps left behind by foragers: geographical spaces either too remote, inaccessible or simply undesirable to attract the sustained attention of hunters, fishers and gatherers. Even in such locations, these outlier economies of the Holocene would have decidedly mixed fortunes. Nowhere is this more dramatically illustrated than in the Early Neolithic period of central Europe, where farming endured one of its first and most conspicuous failures. To better understand the reasons why this failure occurred, we will then consider some more successful expansions of early farming populations in Africa, Oceania and the tropical lowlands of South America.

Historically speaking, there is no direct connection among these cases; but what they show, collectively, is how the fate of early farming societies often hinged less on 'ecological imperialism' than on what we might call – to adapt a phrase from the pioneer of social ecology, Murray Bookchin – an 'ecology of freedom'.[17] By this we mean something quite specific. If peasants are people 'existentially involved in cultivation',[18] then the ecology of freedom ('play farming', in short) is precisely the opposite condition. The ecology of freedom describes the proclivity of human societies to move (freely) in and out of farming; to farm without fully becoming farmers; raise crops and animals without surrendering too much of one's existence to the logistical rigours of agriculture; and retain a food web sufficiently broad as to prevent cultivation from becoming a matter of life and death. It is just this sort of ecological flexibility that tends to be excluded from conventional narratives of world history, which present the planting of a single seed as a point of no return.

Moving freely in and out of farming in this way, or hovering on its threshold, turns out to be something our species has done successfully for a large part of its past.[19] Such fluid ecological arrangements – combining garden cultivation, flood-retreat farming on the margins of lakes or springs, small-scale landscape management (e.g. by burning, pruning and terracing) and the corralling or keeping of animals in semi-wild states, combined with a spectrum of hunting, fishing and collecting activities – were once typical of human societies in many parts of the world. Often these activities were sustained for thousands of years, and not infrequently supported large populations. As we'll see, they may also have been crucial to the survival of those first human populations to incorporate domesticated plants and animals. Biodiversity – not bio-power – was the initial key to the growth of Neolithic food production.

IN WHICH WE CONSIDER A NEOLITHIC CAUTIONARY TALE: THE GRISLY AND SURPRISING FATE OF CENTRAL EUROPE'S FIRST FARMERS

Kilianstädten, Talheim, Schletz and Herxheim are all names of Early Neolithic sites on the loess plains of Austria and Germany. Collectively, they tell a very unfamiliar story of early agriculture.

In these places, starting around 5500 BC, villages of a similar cultural outlook – known as the 'Linear Pottery' tradition – were established. They are among the villages of central Europe's first farmers. But, unlike most other early farming settlements, each ended its life in a period of turmoil, marked by the digging and filling of mass graves. The contents of these graves attest to the annihilation, or attempted annihilation, of an entire community: crudely dug trenches or reused ditches containing chaotic jumbles of human remains, including adults and children of both sexes, disposed of like so much refuse. Their bones show the telltale marks of torture, mutilation and violent death – the breaking of limbs, taking of scalps, butchering for cannibalism. At Kilianstädten and Asparn, younger women were missing from the assemblage, suggesting their appropriation as captives.[20]

The Neolithic farming economy had arrived in central Europe, carried by migrants from the southeast, and with ultimately catastrophic consequences for some of those whose ancestors brought it there.[21] The earliest settlements of these newcomers to the central European plains suggest a relatively free society, with few indicators of status difference either within or between communities. Their basic family units – timber longhouses – were all approximately the same size; but around 5000 BC, disparities began to appear between them, as also in the kind of goods placed with their dead. People enclosed their settlements within large ditches, which yield evidence of warfare in the forms of arrows, axe heads and human remains. In some cases, when the sites were overrun, these ditches were turned into mass graves for the residents they had failed to defend.[22]

Such is the quality and quantity of accurately dated material that researchers are able to model demographic trends accompanying these changes. Their reconstructions have come as something of a surprise. The arrival of farming in central Europe was associated with an initial and quite massive upsurge in population – which is of course exactly what one would expect. But what followed was not the anticipated 'up and up' pattern of demographic growth. Instead came a disastrous downturn, a boom and bust, between 5000 and 4500 BC, and something approaching a regional collapse.[23] These Early Neolithic groups arrived, they settled, and then in many (but, we should emphasize, not all) areas their numbers dwindled into obscurity, while in others they were bolstered through intermarriage with more established forager populations. Only after a hiatus of roughly 1,000 years did extensive cereal-farming take off again in central and northern Europe.[24]

Older narratives of prehistory tended simply to assume that Neolithic colonists held the upper hand over native foraging populations, demographically and socially; that they either replaced them, or converted them to a superior way of life through trade and intermarriage. The boom-and-bust pattern of early farming now documented in temperate Europe contradicts this picture and raises wider questions about the viability of Neolithic economies in a world of foragers. To address these questions, we need to know a bit more about the foraging populations themselves, and how they developed their Pleistocene traditions after the Ice Age and into the Holocene.

Much of what we know about postglacial (Mesolithic) forager populations in Europe derives from findings along the Baltic and Atlantic coasts. Much more is lost to the sea. We learn a great deal about these Holocene hunter-foragers from their funerary customs. From northern Russia through Scandinavia, and down to the Breton coast, they are illuminated by finds of prehistoric cemeteries. Quite often, the burials were richly adorned. In the Baltic and Iberian regions they include copious amounts of amber. Corpses lie in striking postures – sitting or leaning, even flipped on their heads – suggesting complex and now largely unfathomable codes of hierarchy. On the fringes of northern Eurasia, peat bogs and waterlogged sites preserve glimpses of a wood-carving tradition that produced decorated ski runners, sledges, canoes and monuments resembling the totem poles of the Pacific Northwest Coast.[25] Staffs topped with elk and reindeer effigies, reminiscent of Pleistocene rock art depictions, appear over broad areas: a stable symbolism of authority, crossing the boundaries of local foraging groups.[26]

How did Europe's deep interior, where incoming farmers settled, look from the vantage point of these established Mesolithic populations? Most probably like an ecological dead end, lacking the obvious advantages of coastal environments. It may have been precisely this that allowed Linear Pottery colonists to spread freely west and north on the loess plains to begin with: they were moving into areas with little or no prior occupation. Whether that reflects a conscious policy of avoiding local foragers is unclear. What's clearer is that this wave of advance began to break as the new farming groups approached more densely populated coastlands. What exactly this might have meant in practice is often ambiguous. For example, human remains of coastal foragers, found on Mesolithic sites in Brittany, show anomalous levels of terrestrial protein in the diet of many young females, contrasting with the general prevalence of marine foods among the rest of the population. It seems that women of inland origin (who until then had been eating largely meat, not fish) were joining coastal groups.[27]

What does this tell us? It may indicate that women had been captured and transported in raids, conceivably including raids by foragers on farming communities.[28] This can only be speculative; we cannot know for sure that women moved involuntarily, or even that they moved at the behest of men. And while raiding and warfare were clearly part of the picture, it would be simplistic to attribute the initial failure of Neolithic farming in Europe to such factors alone. We'll consider some broader explanations in due course. First, though, we should take a reprieve from Europe and examine some of the success stories of early farming. We will start with Africa, then move on to Oceania, and lastly the rather different but instructive case of Amazonia.

SOME VERY DIFFERENT PLACES WHERE NEOLITHIC FARMING FOUND ITS FEET: THE TRANSFORMATION OF THE NILE VALLEY (*c.*5000–4000 BC) AND THE COLONIZATION OF ISLAND OCEANIA (*c.*1600–500 BC)

Around the time that Linear Pottery settlements were established in central Europe, the Neolithic farming economy made its first appearance in Africa. The African variant had the same ultimate origin, in Southwest Asia. It comprised the same basic suite of crops (emmer wheat and einkorn) and animals (domestic sheep, goats and cattle – with perhaps some admixture of local African aurochs). Yet the African reception of this Neolithic 'package' could not have been more different. It is almost as if the first African farmers opened up the package, threw out some of the contents, then rewrapped it in such strikingly distinct ways that one could easily mistake it for a completely local invention. As, in many ways, it was.

The place where much of this happened was a region largely ignored by foragers until then, but soon to become a major axis of demographic and political change: the Nile valley of Egypt and Sudan. By 3000 BC, the political integration of its lower reaches with the Nile delta would produce the first territorial kingdom of ancient Egypt, facing the Mediterranean. However, the cultural roots of this and all later Nilotic civilizations lay in much earlier transformations, linked to the adoption of farming between 5000 and 4000 BC, with their centre of gravity more firmly in Africa. These first African farmers reinvented the Neolithic in their own image. Cereal cultivation was relegated to a minor pursuit (regaining its status only centuries later), and the idea that one's social identity was represented by hearth and home was largely thrown out too. In their place came a quite different Neolithic: supple, vibrant and travelling on the hoof.[29]

This new form of Neolithic economy relied heavily on livestock-herding, combined with annual rounds of fishing, hunting and foraging on the rich floodplain of the Nile, and in the oases and seasonal streams (*wadis*) of what are now the neighbouring deserts, which were then still watered by annual rains. Herders moved periodically in and out of this 'Green Sahara', both west and east to the Red Sea coast. Complex systems of bodily display developed. New forms of personal adornment employed cosmetic pigments and minerals, prospected from the adjacent deserts, and a dazzling array of beadwork, combs, bangles and other ornaments made of ivory and bone, all richly attested in Neolithic cemeteries running the length of the Nile valley, from Central Sudan to Middle Egypt.[30]

What survives today of this amazing gear now graces the shelves of museum displays the world over, reminding us that – before there were pharaohs – almost anyone could hope to be buried like a king, queen, prince or princess.

*

Another of the world's great Neolithic expansions took place in island Oceania. Its origins lay at the other end of Asia, in the rice- and millet-growing cultures of Taiwan and the Philippines (the deeper roots are in China). Around 1600 BC a striking dispersal of farming groups took place, starting here and ending over 5,000 miles to the east in Polynesia.

Known as the 'Lapita horizon' (after the site in New Caledonia where its decorated pottery was first identified), this precocious expansion – which called into being the world's first deep-ocean outrigger canoes – is often connected to the spread of Austronesian languages. Rice and millet, poorly suited to tropical climates, were jettisoned in its early stages of dispersal. But as the Lapita horizon advanced, their place was taken by a rich admixture of tubers and fruit crops encountered along the way, together with a growing menagerie of animal domesticates (pigs, joined by dogs and chickens; rats too hitched along for the ride). These species travelled with Lapita colonists to previously uninhabited islands – among them Fiji, Tonga and Samoa – where they put down roots (quite literally, in the case of taro and other tubers).[31]

Like the Linear Pottery farmers of central Europe, Lapita groups seem to have avoided established centres of population. They gave a wide berth to the forager stronghold of Australia, and skirted largely clear of Papua New Guinea, where a local form of farming was already well established in the uplands around the Wahgi valley.[32] On virgin islands and beside vacant lagoons they founded their villages, comprising houses perched on stilts. With stone adzes, a mainstay of their travelling toolkit, they cleared patches of forest to make gardens for their crops – taroes, yams and bananas – which they supplemented with animal domesticates and a rich diet of fish, shellfish and marine turtles, wild birds and fruit bats.[33]

Unlike Europe's first farmers, the carriers of the Lapita horizon diversified their economy continuously as they spread. And this was not just true of their crops and animals. Voyaging eastwards, Lapita peoples left a trail of distinctive pottery, their most consistent signal in the archaeological record. Along the way they also encountered many new materials. The most valued, such as particular types of shell, were crafted into multi-media ornaments – arm rings, necklaces, pendants – which left a trace on Melanesian and Polynesian island culture that was still visible many centuries later, when Captain Cook (unwittingly retracing the steps of Lapita) caught sight of New Caledonia in 1774 and wrote that it reminded him of Scotland.

Lapita prestige items also included bird-feather headdresses (depicted on the pottery), fine pandanus leaf mats and obsidian. Obsidian blades, circulating thousands of miles away from their sources in the Bismarck Archipelago, were used in tattooing and scarification to apply pigment and plant matter to the skin. While the tattoos themselves do not survive, the impressed decoration of Lapita pots gives some hint of their underlying schema, transferred from skin to ceramic. More recent traditions of Polynesian tattooing and body art – 'wrapping the body in images', as a famous anthropological study puts it – remind us how little we really know of the vibrant conceptual worlds of earlier times, and those who first carried such practices across remote Pacific island-scapes.[34]

ON THE CASE OF AMAZONIA, AND THE POSSIBILITIES OF 'PLAY FARMING'

On first inspection, these three variations on 'the Neolithic' – European, African, Oceanic – might seem to have almost nothing in common. However, all share two important features. First, each involved a serious commitment to farming. Of the three, the Linear Pottery culture of Europe enmeshed itself most deeply in the raising of cereals and livestock. The Nile valley was fully wedded to its herds, as was the Lapita to its pigs and yams. In every case, the species in question was fully domesticated, reliant on human intervention for its survival, and was no longer able to reproduce unassisted in the wild. For their part, the people in question had oriented their lives around the needs of certain plants and animals; enclosing, protecting and breeding those species was a perennial feature of their existence and a cornerstone of their diets. All of them had become 'serious' farmers.

Second, all three cases involved a targeted spread of farming to lands largely uninhabited by existing populations. The highly mobile

Neolithic of the Nile valley extended seasonally into the adjacent steppe-desert, but avoided regions that were already densely settled, such as the Nile delta, the Sudanese *gezira* and the major oases (including the Fayum, where lakeside fisher-foragers prevailed, adopting and abandoning farming practices largely as it suited them).[35] Similarly, the Linear Pottery culture of Europe took root in niches left open by Mesolithic foragers, such as patches of loess soil and unused river levees. The Lapita horizon, too, was a relatively closed system, interacting with others when necessary, but otherwise enfolding new resources into its own pattern of life. Serious farmers tended to form societies with hard boundaries, ethnic and, in some cases, also linguistic.[36]

But not all early farming expansions were of this 'serious' variety. In the lowland tropics of South America, archaeological research has uncovered a distinctly more playful tradition of Holocene food production. Similar practices were still widely in evidence in Amazonia until recently, such as we found among the Nambikwara of Brazil's Mato Grosso region. Well into the twentieth century, they spent the rainy season in riverside villages, clearing gardens and orchards to grow a panoply of crops including sweet and bitter manioc, maize, tobacco, beans, cotton, groundnuts, gourds and more besides. Cultivation was a relaxed affair, with little effort spent on keeping different species apart. And as the dry season commenced, these tangled house gardens were abandoned altogether. The entire group dispersed into small nomadic bands to hunt and forage, only to begin the whole process again the following year, often in a different location.

In Greater Amazonia, such seasonal moves in and out of farming are documented among a wide range of indigenous societies and are of considerable antiquity.[37] So is the habit of keeping pets. It is often stated that Amazonia has no indigenous animal domesticates, and from a biological standpoint this is true. From a cultural perspective, things look more complicated. Many rainforest groups carry with them what can only be described as a small zoo comprising tamed forest creatures: monkeys, parrots, collared peccaries, and so on. These pets are often the orphaned offspring of animals hunted and killed for food. Taken in by human foster-parents, fed and nurtured through infancy, they become utterly dependent on their masters. This subservience lasts into maturity. Pets are not eaten. Nor are their keepers interested in breeding them. They live as individual members of the community, who treat them much like children, as subjects of affection and sources of amusement.[38]

Amazonian societies blur our conventional distinction between 'wild' and 'domestic' in other ways. Animals they routinely hunt and capture for food include peccary, agouti and others we would classify as 'wild'. Yet these same species are locally considered as already domesticated, at least in the sense of being subjects of supernatural 'masters of the animals' who protect them and to whom they are bound. 'Master' or 'Mistress of the Animals' figures are actually very common in hunting societies; sometimes they take the form of a huge or perfect specimen of a certain type of beast, a kind of embodiment of the species, but at the same time they appear as human or human-like owners of the species, to whom the souls of all deer, or seals, or caribou must be returned after hunters take them. In Amazonia, what this means in practice is that people avoid intervening in the reproduction of those particular species lest they usurp the role of spirits.

In other words, there was no obvious cultural route, in Amazonia, that might lead humans to become both the primary carers for and consumers of other species; relationships were either too remote (in the case of game) or too intimate (in the case of pets). We are dealing here with people who possess all the requisite ecological skills to raise crops and livestock, but who nevertheless pull back from the threshold, maintaining a careful balancing act between forager (or better, perhaps, forester) and farmer.[39]

Amazonia shows how this 'in-and-out-of-farming' game could be far more than a transient affair. It seems to have played out over thousands of years, since during that time there is evidence of plant domestication and land management, but little commitment to agriculture.[40] From 500 BC, this neotropical mode of food production expanded from its heartlands on the Orinoco and Rio Negro, tracking river systems through the rainforest, and ultimately becoming established all the way from Bolivia to the Antilles. Its legacy is clearest in the distribution of living and historical groups speaking languages of the Arawak family.[41]

Arawak-speaking groups were famed in recent centuries as master blenders of culture – traders and diplomats, forging diverse alliances, often for commercial advantage. Over 2,000 years ago, a similar process of strategic cultural mixing (quite unlike the avoidance strategies of more 'serious' farmers) seems to have brought about the convergence of the Amazon basin into a regional system. Arawak languages and their derivatives are spoken all along the *várzea* (alluvial terraces), from the mouths of the Orinoco and Amazon to their Peruvian headwaters. But their users have little in the way of shared genetic ancestry. The various dialects are structurally closer to those of their non-Arawak neighbours than to each other, or to any putative *Ur*-language.

The impression is not at all of a uniform spread, but a targeted interweaving of groups along the main routes of canoe-borne transport and trade. The result was an interlaced network of cultural exchange, lacking clear boundaries or a centre. The latticework schema on Amazonian pottery, cotton fabrics and skin painting – recurring in strikingly similar styles from one edge of the rainforest to the other – seem to model these connective principles, entangling human bodies in a complex cartography of relations.[42]

Until quite recently, Amazonia was regarded as a timeless refuge of solitary tribes, about as close to Rousseau or Hobbes's State of Nature as one could possibly get. As we've seen, such romantic notions persisted in anthropology well into the 1980s, through studies that cast groups like the Yanomami in the role of 'contemporary ancestors', windows on to our evolutionary past. Research in the fields of archaeology and ethnohistory is now overturning this picture.

We now know that, by the beginning of the Christian era, the Amazonian landscape was already studded with towns, terraces, monuments and roadways, reaching all the way from the highland kingdoms of Peru to the Caribbean. The first Europeans to arrive there in the sixteenth century described lively floodplain settlements governed by paramount chiefs who dominated their neighbours. It is tempting to dismiss these accounts as adventurers' hyperbole, designed to impress the sponsors at home – but, as archaeological science brings the outlines of this rainforest civilization into view, it is increasingly difficult to do so. Partly this new understanding is the result of controlled research; partly a consequence of industrial deforestation, which in the Upper Amazon basin (looking west to the Andes) has exposed from the canopy a tradition of monumental earthworks, executed to precise geometrical plans and linked by road systems.[43]

What exactly was the reason for this ancient Amazonian efflorescence? Up until a few decades ago, all these developments were explained as the result of yet another 'Agricultural Revolution'. It was supposed that, in the first millennium BC, intensified manioc-farming raised Amazonian population levels, generating a wave of human expansion throughout the lowland tropics. The basis for this hypothesis lay in finds of domesticated manioc, dating back as early as 7000 BC; more recently, in southern Amazonia, the cultivation of maize and squash has been traced back to similarly early periods.[44] Yet there is little evidence for widespread farming of these crops in the key period of cultural convergence, beginning around 500 BC. In fact, manioc only seems to have become a staple crop *after* European contact. All this implies that at least some early inhabitants of Amazonia were well aware of plant domestication but did not select it as the basis of their economy, opting instead for a more flexible kind of agroforestry.[45]

Modern rainforest agriculture relies on slash-and-burn techniques, labour-intensive methods geared to the extensive cultivation of a small number of crops. The more ancient mode, which we've been describing, allowed for a much wider range of cultivars, grown in doorstep gardens or small forest clearings close to settlements. Such ancient plant nurseries rested on special soils (or, more strictly, 'anthrosols'), which are locally called *terra preta de índio* ('black earth of the Indians') and *terra mulata* ('brown earth'): dark earths with carrying capacities well in excess of ordinary tropical soils. The dark earths owe their fertility to absorption of organic by-products such as food residues, excrement and charcoal from everyday village life (forming *terras pretas*) and/or earlier episodes of localized burning and cultivation (*terras mulatas*).[46] Soil enrichment in ancient Amazonia was a slow and ongoing process, not an annual task.

'Play farming' of this sort, in the Amazon as elsewhere, has had its recent advantages for indigenous peoples. Elaborate and unpredictable subsistence routines are an excellent deterrent against the colonial State: an ecology of freedom in the literal sense. It is difficult to tax and monitor a group that refuses to stay in one location, obtaining its livelihood without making long-term commitments to fixed resources, or growing much of its food invisibly underground (as with tubers and other root vegetables).[47] While this may be so, the deeper history of the American tropics shows that similarly loose and flexible patterns of food production sustained civilizational growth on a continent-wide scale, long before Europeans arrived.

In fact, farming of this particular sort ('low-level food production' is the more technical term) has characterized a very wide range of Holocene societies, including the earliest cultivators of the Fertile Crescent and Mesoamerica.[48] In Mexico, domestic forms of squash and maize existed by 7000 BC.[49] Yet these crops only became staple foods around 5,000 years later. Similarly, in the Eastern Woodlands of North America local seed crops were cultivated by 3000 BC, but there was no 'serious farming' until around AD 1000.[50] China follows a similar pattern. Millet-farming began on a small scale around 8000 BC, on the northern plains, as a seasonal complement to foraging and dog-assisted hunting. It remained so for 3,000 years, until the introduction of cultivated millets into the basin of the Yellow River. Similarly, on the lower and middle reaches of the Yangtze, fully domesticated rice strains only appear fifteen centuries after the first cultivation of wild rice in paddy fields. It might have even taken longer were it not for a snap of global cooling around 5000 BC, which depleted wild rice stands and nut harvests.[51]

In both parts of China, long after their domestication pigs still came second to wild boar and deer in terms of dietary significance. This was also the case in the wooded uplands of the Fertile Crescent, where Çayönü with its House of Skulls is located, and where human–pig relations long remained more a matter of flirtation than full domestication.[52] So while it's tempting to hold Amazonia up as a 'New World' alternative to the 'Old World Neolithic', the truth is that Holocene developments in both hemispheres are starting to look increasingly similar, at least in terms of the overall pace of change. And in both cases, they look increasingly un-revolutionary. In the beginning, many of the world's farming societies were Amazonian in spirit. They hovered at the threshold of agriculture while remaining wedded to the cultural values of hunting and foraging. The 'smiling fields' of Rousseau's *Discourse* still lay far off in the future.

It may be that further research reveals demographic fluctuations among early farming (or forester-farmer) populations in Amazonia, Oceania or even among the first herding peoples of the Nile valley, similar to those now observed for central Europe. Indeed, some sort of decline, or at least major reconfiguration of settlement, took place in the Fertile Crescent itself during the seventh millennium BC.[53] At any rate, we shouldn't be too categorical about the contrasts among these various regions, given the different amounts of evidence available for each. Still, based on what is currently known, we can at least reframe our initial question and ask: why did Neolithic farmers in certain parts of Europe initially suffer population collapse on a scale currently unknown, or undetected elsewhere?

Clues lie in the tiniest of details.

Cereal-farming, as it turns out, underwent some important changes during its transfer from Southwest Asia to central Europe via the Balkans. Originally there were three kinds of wheat (einkorn, emmer and free threshing) and two kinds of barley (hulled and naked) under cultivation, but also five different pulses (pea, lentil, bitter vetch, chickpea and grass pea). By contrast, the majority of Linear Pottery sites contain just glume wheats (emmer and einkorn) and one or two kinds of pulse. The Neolithic economy had become increasingly narrow and uniform, a diminished subset of the Middle Eastern original. Furthermore, the loess landscapes of central Europe offered little topographical variability and few opportunities to add new resources, while dense forager populations limited expansion towards the coasts.[54]

Almost everything came to revolve around a single food web for Europe's earliest farmers. Cereal-farming fed the community. Its by-products – chaff and straw – provided fuel, fodder for their animals, as well as basic materials for construction, including temper for pottery and daub for houses. Livestock supplied occasional meat, dairy and wool, as well as manure for gardens.[55] With their wattle-and-daub longhouses and sparse material culture, these first European farming settlements bear a peculiar resemblance to the rural peasant societies of much later eras. Most likely, they were also subject to some of the same weaknesses – not just periodic raiding from the outside, but also internal labour crunches, soil exhaustion, disease and harvest failures across a whole string of like-for-like communities, with little scope for mutual aid.

Neolithic farming was an experiment that could fail – and, on occasion, did.

BUT WHY DOES IT ALL MATTER? (A QUICK REPRISE ON THE DANGERS OF TELEOLOGICAL REASONING)

In this chapter we have tracked the fate of some of the world's first farmers as they hopped, stumbled and bluffed their way around the globe, with mixed success. But what does this tell us about the overall course of human history? Surely, the sceptical reader might object, what matters in the wider scheme of things are not the first faltering steps towards agriculture, but its long-term effects. After all, by no later than 2000 BC agriculture was supporting great cities, from China to the Mediterranean; and by 500 BC food-producing societies of one sort or another had colonized pretty much all of Eurasia, with the exception of southern Africa, the sub-Arctic region and a handful of subtropical islands.

A sceptic might continue: agriculture alone could unlock the carrying capacity of lands that foragers were either unable or unwilling to exploit to anything like the same degree. So long as people were willing to give up their mobility and settle, even small parcels of arable soil could be made to yield food surpluses, especially once ploughs and irrigation were introduced. Even if there were temporary downturns, or even catastrophic failures, over the long term the odds were surely always stacked in favour of those who could intensify land use to sustain ever larger and denser populations. And let's face it, the same sceptic might conclude, the world's population could only grow from perhaps 5 million at the start of the Holocene to 900 million by AD 1800, and now to billions, because of agriculture.

How too, for that matter, could such large populations be fed, without chains of command to organize the masses, formal offices of leadership; full-time administrators, soldiers, police, and other non-food-producers, who in turn could only be supported by the surpluses that agriculture provides? These seem like reasonable questions to ask, and those who make the first point almost invariably make the second. But in doing so, they risk parting company with history. You can't simply jump from the beginning of the story to the end, and then just assume you know what happened in the middle. Well, you can, but then you are slipping back into the very fairy tales we've been dealing with throughout this book. So instead, let's recap very briefly what we've learned about the origins and spread of farming, and then turn to examine some of the more dramatic things that actually did happen to human societies over the last 5,000 or so years.

Farming, as we can now see, often started out as an economy of deprivation: you only invented it when there was nothing else to be done, which is why it tended to happen first in areas where wild resources were thinnest on the ground. It was the odd one out in the strategies of the Early Holocene, but it had explosive growth potential, especially after domestic livestock were added to cereal crops. Even so, it was the new kid on the block. Since the first farmers made more rubbish, and often built houses of baked mud, they are also more visible to archaeologists. That's one reason why imaginative in-filling is necessary if we want to avoid missing the action going on in much richer environments at the same time, among populations still largely reliant on wild resources.

Seasonally erected monuments like those of Göbekli Tepe or Lake Shigirskoe are as clear a signal as one could wish for that big things were afoot among Holocene hunter-fisher-gatherers. But what were all the non-farming people doing, and where were they living, for the rest of the time? Upland forested areas, like the uplands of eastern Turkey or the foothills of the Urals, are one candidate, but since most construction was in wood, very little of this habitation survives. Most likely, the largest communities were concentrated around lakes, rivers and coastlands, and especially at their junctures: delta environments – such as those of southern Mesopotamia, the lower reaches of the Nile and the Indus – where many of the world's first cities arose, and to which we must now turn in order to find out exactly what living in large and densely populated settlements really did (and did not) imply for the development of human societies.

8

Imaginary Cities

Eurasia's first urbanites – in Mesopotamia, the Indus valley, Ukraine and China – and how they built cities without kings

Cities begin in the mind.

Or so proposed Elias Canetti, a novelist and social philosopher often written off as one of those offbeat mid-century central European thinkers no one knows quite what to do with. Canetti speculated that Palaeolithic hunter-gatherers living in small communities must, inevitably, have spent time wondering what larger ones would be like. Proof, he felt, was on the walls of caves, where they faithfully depicted herd animals that moved together in uncountable masses. How could they not have wondered what human herds might be like, in all their terrible glory? No doubt they also considered the dead, outnumbering the living by orders of magnitude. What if everyone who'd ever died were all in one place? What would that be like? These 'invisible crowds', Canetti proposed, were in a sense the first human cities, even if they existed only in the imagination.

All this might seem idle speculation (in fact, speculation *about* speculation), but current advances in the study of human cognition suggest that Canetti had put his finger on something important, something almost everyone else had overlooked. Very large social units are always, in a sense, imaginary. Or, to put it in a slightly different way: there is always a fundamental distinction between the way one relates to friends, family, neighbourhood, people and places that we actually know directly, and the way one relates to empires, nations and metropolises, phenomena that exist largely, or at least most of the time, in our heads. Much of social theory can be seen as an attempt to square these two dimensions of our experience.

*

In the standard, textbook version of human history, scale is crucial. The tiny bands of foragers in which humans were thought to have spent most of their evolutionary history could be relatively democratic and egalitarian precisely because they were small. It's common to assume – and is often stated as self-evident fact – that our social sensibilities, even our capacity to keep track of names and faces, are largely determined by the fact that we spent 95 per cent of our evolutionary history in tiny groups of at best a few dozen individuals. We're designed to work in small teams. As a result, large agglomerations of people are often treated as if they were by definition somewhat unnatural, and humans as psychologically ill equipped to handle life inside them. This is the reason, the argument often goes, that we require such elaborate 'scaffolding' to make larger communities work: such things as urban planners, social workers, tax auditors and police.[1]

If so, it would make perfect sense that the appearance of the first cities, the first truly large concentrations of people permanently settled in one place, would also correspond to the rise of states. For a long time, the archaeological evidence – from Egypt, Mesopotamia, China, Central America and elsewhere – did appear to confirm this. If you put enough people in one place, the evidence seemed to show, they would almost inevitably develop writing or something like it, together with administrators, storage and redistribution facilities, workshops and overseers. Before long, they would also start dividing themselves into social classes. 'Civilization' came as a package. It meant misery and suffering for some (since some would inevitably be reduced to serfs, slaves or debt peons), but also allowed for the possibility of philosophy, art and the accumulation of scientific knowledge.

The evidence no longer suggests anything of the sort. In fact, much of what we have come to learn in the last forty or fifty years has thrown conventional wisdom into disarray. In some regions, we now know, cities governed themselves for centuries without any sign of the temples and palaces that would only emerge later; in others, temples and palaces never emerged at all. In many early cities, there is simply no evidence of either a class of administrators or any other sort of ruling stratum. In others, centralized power seems to appear and then disappear. It would seem that the mere fact of urban life does not, necessarily, imply any particular form of political organization, and never did.

This has all sorts of important implications: for one thing, it suggests a much less pessimistic assessment of human possibilities, since the mere fact that much of the world's population now live in cities may not determine *how* we live, to anything like the extent you might assume – but before even starting to think about that, we need to ask how we got things so extraordinarily wrong to begin with.

IN WHICH WE FIRST TAKE ON THE NOTORIOUS ISSUE OF 'SCALE'

'Common sense' is a peculiar expression. Sometimes it means exactly what it seems to mean: practical wisdom born of real-life experience, avoiding stupid, obvious pitfalls. This is what we mean when we say that a cartoon villain who puts a clearly marked 'self-destruct' button on his doomsday device, or who fails to block the ventilation passages in his secret headquarters, is lacking common sense. On the other hand, it occasionally turns out that things which seem like simple common sense are, in fact, not.

For a long time, it was considered almost universal common sense that women make poor soldiers. After all, it was noted, women tend to be smaller and have less upper-body strength. Then various military forces made the experiment and discovered that women also tend to be much better shots. Similarly, it is almost universal common sense that it's relatively easy for a small group to treat each other as equals and come to decisions democratically, but that the larger the number of people involved, the more difficult this becomes. If you think about it, this isn't really as commonsensical as it seems, since it clearly isn't true of groups that endure. Over time, any group of intimate friends, let alone a family, will eventually develop a complicated history that makes coming to agreement on almost anything difficult; whereas the larger the group, the less likely it is to contain a significant proportion of people you specifically detest. But for various reasons, the problem of scale has now become a matter of simple common sense not only to scholars, but to almost everyone else.

Since the problem is typically seen as a result of our evolutionary inheritance, it might be helpful for a moment to return to the source and consider how evolutionary psychologists like Robin Dunbar have typically framed the question. Most begin by observing that the social organization of hunter-gatherers – both ancient and modern – operates at different tiers or levels, 'nested' inside one another like Russian dolls. The most basic social unit is the pair-bonded family, with shared investment in offspring. To provide for themselves and dependants, these nuclear units are obliged (or so the argument goes) to cluster together in 'bands' made up of five or six closely related families. On ritual occasions, or when game is particularly abundant, such bands coalesce to form 'residential groups' (or 'clans') of roughly 150 persons, which – according to Dunbar – is also around the upper limit of stable, trusting relationships we are cognitively able to keep track of in our heads. And this, he suggests, is no coincidence. Beyond 150 (which has come to be known as 'Dunbar's Number') larger groups such as 'tribes' may form – but, Dunbar asserts, these larger groups will inevitably lack the solidarity of smaller, kin-based ones, and so conflicts will tend to arise within them.[2]

Dunbar considers such 'nested' arrangements to be among the factors which shaped human cognition in deep evolutionary time, such that even today a whole plethora of institutions that require high levels of social commitment, from military brigades to church congregations, still tend to gravitate around the original figure of 150 relationships. It's a fascinating hypothesis. As formulated by evolutionary psychologists, it hinges on the idea that living hunter-gatherers do actually provide evidence for this supposedly ancient way of scaling social relationships upwards from core family units to bands and residential groups, with each larger group reproducing that same sense of loyalty to one's natal kin, just on a greater scale, all the way up to things like 'brothers' – or indeed 'sisters' – in arms. But here comes the worm in the bud.

There is an obvious objection to evolutionary models which assume that our strongest social ties are based on close biological kinship: many humans just don't like their families very much. And this appears to be just as true of present-day hunter-gatherers as anybody else. Many seem to find the prospect of living their entire lives surrounded by close relatives so unpleasant that they will travel very long distances just to get away from them. New work on the demography of modern hunter-gatherers – drawing statistical comparisons from a global sample of cases, ranging from the Hadza in Tanzania to the Australian Martu[3] – shows that residential groups turn out not to be made up of biological kin at all; and the burgeoning field of human genomics is beginning to suggest a similar picture for ancient hunter-gatherers as well, all the way back to the Pleistocene.[4]

While modern Martu, for instance, might speak of themselves as if they were all descended from some common totemic ancestor, it turns out that primary biological kin actually make up less than 10 per cent of the total membership of any given residential group. Most participants are drawn from a much wider pool who do not share close genetic relationships, whose origins are scattered over very large territories, and who may not even have grown up speaking the same languages. Anyone recognized to be Martu is a potential member of any Martu band, and the same turns out to be true of the Hadza, BaYaka, !Kung San, and so on. The truly adventurous, meanwhile, can often contrive to abandon their own larger group entirely. This is all the more surprising in places like Australia, where there tend to be very elaborate kinship systems in which almost all social arrangements are ostensibly organized around genealogical descent from totemic ancestors.

It would seem, then, that kinship in such cases is really a kind of metaphor for social attachments, in much the same way we'd say 'all men are brothers' when trying to express internationalism (even if we can't stand our actual brother and haven't spoken to him for years). What's more, the shared metaphor often extended over very long distances, as we've seen with the way that Turtle or Bear clans once existed across North America, or moiety systems across Australia. This made it a relatively simple matter for anyone disenchanted with their immediate biological kin to travel very long distances and still find a welcome.

It is as though modern forager societies exist simultaneously at two radically different scales: one small and intimate, the other spanning vast territories, even continents. This might seem odd, but from the perspective of cognitive science it makes perfect sense. It's precisely this capacity to shift between scales that most obviously separates human social cognition from that of other primates.[5] Apes may vie for affection or dominance, but any victory is temporary and open to being renegotiated. Nothing is imagined as eternal. Nothing is really imagined at all. Humans tend to live simultaneously with the 150-odd people they know personally, and inside imaginary structures shared by perhaps millions or even billions of other humans. Sometimes, as in the case of modern nations, these are imagined as being based on kin ties; sometimes they are not.[6]

In this, at least, modern foragers are no different from modern city dwellers or ancient hunter-gatherers. We all have the capacity to feel bound to people we will probably never meet; to take part in a macro-society which exists most of the time as 'virtual reality', a world of possible relationships with its own rules, roles and structures that are held in the mind and recalled through the cognitive work of image-making and ritual. Foragers may sometimes exist in small groups, but they do not – and probably have not ever – lived in small-scale *societies*.[7]

None of which is to say that scale – in the sense of absolute population size – makes no difference at all. What it means is that these things do not necessarily matter in the seemingly common-sense sort of way we tend to assume. On this particular point, at least, Canetti had it right. Mass society exists in the mind before it becomes physical reality. And crucially, it also exists in the mind *after* it becomes physical reality.

At this point we can return to cities.

Cities are tangible things. Certain elements of their physical infrastructure – walls, roads, parks, sewers – might remain fixed for hundreds or even thousands of years; but in human terms they are never stable. People are constantly moving in and out of them, whether permanently, or seasonally for holidays and festivals, to visit relatives, trade, raid, tour around, and so on; or just in the course of their daily rounds. Yet cities have a life that transcends all this. This is not because of the permanence of stone or brick or adobe; neither is it because most people in a city actually meet one another. It is because they will often think and act as people who *belong* to the city – as Londoners or Muscovites or Calcuttans. As the urban sociologist Claude Fischer put it:

> Most city dwellers lead sensible, circumscribed lives, rarely go downtown, hardly know areas of the city they neither live nor work in, and see (in any sociologically meaningful way) only a tiny fraction of the city's population. Certainly, they may on occasion – during rush hours, football games, etc. – be in the presence of thousands of strangers, but that does not necessarily have any direct effect on their personal lives ... urbanites live in small social worlds that touch but do not interpenetrate.[8]

All this applies in equal measure to ancient cities. Aristotle, for example, insisted that Babylon was so large that, two or three days after it had been captured by a foreign army, some parts of the city still hadn't heard the news. In other words, from the perspective of someone living in an ancient city, the city itself was not so entirely different from earlier landscapes of clans or moieties that extended across hundreds of miles. It was a structure raised primarily in the human imagination, which allowed for the possibility of amicable relations with people they had never met.

In Chapter Four we suggested that for much of human history, the geographical range in which most human beings were operating was actually shrinking. Palaeolithic 'culture areas' spanned continents. Mesolithic and Neolithic culture zones still covered much wider areas than the home territory of most contemporary ethno-linguistic groups (what anthropologists refer to as 'cultures'). Cities were part of that process of contraction, since urbanites could, and many did, spend almost their entire lives within a few miles' radius – something that would hardly have been conceivable for people of an earlier age. One way to think about this would be to imagine a vast regional system, of the kind that once spanned much of Australia or North America, being squeezed into a single urban space – while still maintaining its virtual quality. If that is even roughly what happened when the earliest cities formed, then there's no reason to assume there were any special cognitive challenges involved. Living in unbounded, eternal, largely imaginary groups is effectively what humans had been doing all along.

So what was really new here? Let's go back to the archaeological evidence. Settlements inhabited by tens of thousands of people make their first appearance in human history around 6000 years ago, on almost every continent, at first in isolation. Then they multiply. One of the things that makes it so difficult to fit what we now know about them into an old-fashioned evolutionary sequence, where cities, states, bureaucracies and social classes all emerge together,[9] is just how different these cities are. It's not just that some early cities lack class divisions, wealth monopolies, or hierarchies of administration. They exhibit such extreme variability as to imply, from the very beginning, a conscious experimentation in urban form.

Contemporary archaeology shows, among other things, that surprisingly few of these early cities contain signs of authoritarian rule. It also shows that their ecology was far more diverse than once believed: cities do not necessarily depend on a rural hinterland in which serfs or peasants engage in back-breaking labour, hauling in cartloads of grain for consumption by urban dwellers. Certainly, that situation became increasingly typical in later ages, but in the first cities small-scale gardening and animal-keeping were often at least as important; so too were the resources of rivers and seas, and for that matter the continued hunting and collecting of wild seasonal foods in forests or in marshes. The particular mix depended largely on where in the world the cities happened to be, but it's becoming increasingly apparent that history's first city dwellers did not always leave a harsh footprint on the environment, or on each other.

What were these early cities like to live in?

In what follows we'll mainly describe what happened in Eurasia, before moving over to Mesoamerica in the next chapter. Of course, the whole story could be told from other geographical perspectives (that of sub-Saharan Africa, for instance, where local trajectories of urban development in the Middle Niger delta stretch back long before the spread of Islam), but there is only so much one can cover in a single volume without doing excessive violence to the subject.[10] Each region we consider presents a distinct range of source material for the archaeologist or historian to sift and weigh. In most cases, written evidence is either lacking or extremely limited in scope. (We are still talking here, for the most part, about very early periods of human history, and cultural traditions very different from our own.)

We may never be able to reconstruct in any detail the unwritten constitutions of the world's first cities, or the upheavals that appear to have periodically changed them. Still, what evidence does exist is robust enough, not just to upend the conventional narrative but to open our eyes to possibilities we would otherwise never have considered. Before looking at specific cases, we should at least briefly consider why cities ever appeared in the first place. Did the sort of temporary, seasonal aggregation sites we discussed in earlier chapters gradually become permanent, year-round settlements? That would be a gratifyingly simple story. Unfortunately, it doesn't seem to be what happened. The reality is more complex and, as usual, a good deal more interesting.

IN WHICH WE SET THE SCENE BROADLY FOR A WORLD OF CITIES, AND SPECULATE AS TO WHY THEY FIRST AROSE

Wherever cities emerged, they defined a new phase of world history.[11] Let's call it the 'early urban world', an admittedly bland term for what was in many ways a strange phase of the human past. Perhaps it is one of the hardest for us now to grasp, since it was simultaneously so familiar and so alien. We will consider the familiar parts first.

Almost everywhere, in these early cities, we find grand, self-conscious statements of civic unity, the arrangement of built spaces in harmonious and often beautiful patterns, clearly reflecting some kind of planning at the municipal scale. Where we do have written sources (ancient Mesopotamia, for example), we find large groups of citizens referring to themselves, not in the idiom of kinship or ethnic ties, but simply as 'the people' of a given city (or often its 'sons and daughters'), united by devotion to its founding ancestors, its gods or heroes, its civic infrastructure and ritual calendar, which always involves at least some occasions for popular festivity.[12] Civic festivals were moments when the imaginary structures to which people deferred in their daily lives, but which couldn't normally be seen, temporarily took on tangible, material form.

Where there is evidence to be had, we also find differences. People who lived in cities often came from far away. The great city of Teotihuacan in the Valley of Mexico was already attracting residents from such distant areas as Yucatán and the Gulf Coast in the third or fourth century AD; migrants settled there in their own neighbourhoods, including a possible Maya district. Immigrants from across the great floodplains of the Indus buried their loved ones in the cemeteries of Harappa. Typically, ancient cities divided themselves into quarters, which often developed enduring rivalries, and this seems to have been true of the very first cities. Marked out by walls, gates or ditches, consolidated neighbourhoods of this sort were probably not different in any fundamental respect from their modern counterparts.[13]

What makes these cities strange, at least to us, is largely what isn't there. This is especially true of technology, whether advanced metallurgy, intensive agriculture, social technologies like administrative records, or even the wheel. Any one of these things may, or may not, have been present, depending where in this early urban world we cast our gaze. Here it's worth recalling that in most of the Americas, before the European invasion, there were neither metal tools nor horses, donkeys, camels or oxen. All movement of people and things was either by foot, canoe or travois. But the scale of pre-Columbian capitals like Teotihuacan or Tenochtitlan dwarfs that of the earliest cities in China and Mesopotamia, and makes the 'city-states' of Bronze Age Greece (like Tiryns and Mycenae) seem little more than fortified hamlets.

In point of fact, the largest early cities, those with the greatest populations, did not appear in Eurasia – with its many technical and logistical advantages – but in Mesoamerica, which had no wheeled vehicles or sailing ships, no animal-powered traction or transport, and much less in the way of metallurgy or literate bureaucracy. This raises an obvious question: why did so many end up living in the same place to begin with? The conventional story looks for the ultimate causes in technological factors: cities were a delayed, but inevitable, effect of the 'Agricultural Revolution', which started populations on an upward trajectory and set off a chain of other developments, for instance in transport and administration, which made it possible to support large populations living in one place. These large populations then required states to administer them. As we've seen, neither part of this story seems to be borne out by the facts.

Indeed, it's hard to find a single story. Teotihuacan, for instance, appears to have become such a large city, peaking at perhaps 100,000 souls, mainly because a series of volcanic eruptions and related natural disasters drove entire populations out of their homelands to settle there.[14] Ecological factors often played a role in the formation of cities, but in this particular case these would appear to be only obliquely related to the intensification of agriculture. Still, there are hints of a pattern. Across many parts of Eurasia, and in a few parts of the Americas, the appearance of cities follows quite closely on a secondary, post-Ice Age shuffling of the ecological pack which started around 5000 BC. At least two environmental changes were at work here.

The first concerns rivers. At the beginning of the Holocene, the world's great rivers were mostly still wild and unpredictable. Then, around 7,000 years ago, flood regimes started changing, giving way to more settled routines. This is what created wide and highly fertile floodplains along the Yellow River, the Indus, the Tigris and other rivers that we associate with the first urban civilizations. Parallel to this, the melting of polar glaciers slowed down in the Middle Holocene to a point that allowed sea levels the world over to stabilize, at least to a greater degree than they ever had before. The combined effect of these two processes was dramatic; especially where great rivers met the open waters, depositing their seasonal loads of fertile silt faster than seawaters could push them back. This was the origin of those great fan-like deltas we see today at the head of the Mississippi, the Nile or the Euphrates, for instance.[15]

Comprising well-watered soils, annually sifted by river action, and rich wetland and waterside habitats favoured by migratory game and waterfowl, such deltaic environments were major attractors for human populations. Neolithic farmers gravitated to them, along with their crops and livestock. Hardly surprising, considering these were effectively scaled-up versions of the kind of river, spring and lakeside environments in which Neolithic horticulture first began, but with one other major difference: just over the horizon lay the open sea, and before it expansive marshlands supplying aquatic resources to buffer the risks of farming, as well as a perennial source of organic materials (reeds, fibres, silt) to support construction and manufacturing.[16]

All this, combined with the fertility of alluvial soils further inland, promoted the growth of more specialized forms of farming in Eurasia, including the use of animal-drawn ploughs (also adopted in Egypt by 3000 BC), and the breeding of sheep for wool. Extensive agriculture may thus have been an outcome, not a cause, of urbanization.[17] Choices about which crops and animals to farm often had less to do with brute subsistence than the burgeoning industries of early cities, notably textile production, as well as popular forms of urban cuisine such as alcoholic drinks, leavened bread and dairy products. Hunters and foragers, fishers and fowlers were no less important to these new urban economies than farmers and shepherds.[18] Peasantries, on the other hand, were a later, secondary development.

Wetlands and floodplains are no friends to archaeological survival. Often, these earliest phases of urban occupation lie beneath later deposits of silt, or the remains of cities grown over them. In many parts of the world, the first available evidence relates to an already mature phase of urban expansion: by the time the picture comes into focus, we already see a marsh metropolis, or network of centres, outscaling all previous known settlements by a factor of ten to one. Some of these cities in former wetlands have only emerged very recently into historical view – virgin births from the bulrushes. The results are often striking, and their implications still unclear.

We now know, for instance, that in China's Shandong province, on the lower reaches of the Yellow River, settlements of 300 hectares or more – such as Liangchengzhen and Yaowangcheng – were present by no later than 2500 BC, which is over 1,000 years before the earliest royal dynasties developed on the Central Chinese plains. On the other side of the Pacific, around the same time, ceremonial centres of great magnitude developed in the valley of Peru's Rio Supe, notably at the site of Caral, where archaeologists have uncovered sunken plazas and monumental platforms four millennia older than the Inca Empire.[19] The extent of human habitation around these great centres is still to be determined.

These new findings show that archaeologists still have much to find out about the distribution of the world's first cities. They also indicate how much older those cities may be than the systems of authoritarian government and literate administration that were once assumed necessary for their foundation. Similar revelations are emerging from the Maya lowlands, where ceremonial centres of truly enormous size – and, so far, presenting no evidence of monarchy or stratification – can now be dated back as far as 1000 BC: more than 1,000 years before the rise of Classic Maya kings, whose royal cities were notably smaller in scale.[20] This, in turn, raises a fascinating but difficult question. What held the earliest experiments in urbanization together, other than reeds, fibres and clay? What was their social glue? It is high time for some examples but, before we examine the great valley civilizations of the Tigris, Indus and Yellow Rivers, we will first visit the interior grasslands of eastern Europe.

ON 'MEGA-SITES', AND HOW ARCHAEOLOGICAL FINDINGS IN UKRAINE ARE OVERTURNING CONVENTIONAL WISDOM ON THE ORIGINS OF CITIES

The remote history of the countries around the Black Sea is awash with gold. At least, any casual visitor to the major museums of Sofia, Kiev or Tbilisi could be forgiven for leaving with this impression. Ever since the days of Herodotus, outsiders to the region have come home full of lurid tales about the lavish funerals of warrior-kings, and the mass slaughter of horses and retainers that accompanied them. Over 1,000 years later, in the tenth century AD, the traveller Ibn Fadlan was telling almost identical stories to impress and titillate his Arab readers.

As a result, in these lands the term 'prehistory' (or sometimes 'proto-history') has always evoked the legacy of aristocratic tribes and lavish tombs crammed with treasure. Such tombs are, certainly, there to be found. On the region's western flank, in Bulgaria, they begin with the gold-soaked cemetery of Varna, oddly placed in what regional archaeologists refer to as the Copper Age, corresponding to the fifth millennium BC. To the east, in southernmost Russia, a tradition of extravagant funeral rites began shortly after, associated with burial mounds known as *kurgans*, which do indeed mark the resting places of warrior princes of one sort or another.[21]

But it turns out this wasn't the whole story. In fact, magnificent warrior tombs might not even be the most interesting aspect of the region's prehistory. There were also cities. Archaeologists in Ukraine and Moldova got their first inkling of them in the 1970s, when they began to detect the existence of human settlements older and much larger than anything they had previously encountered.[22] Further research showed that these settlements, often referred to as 'mega-sites' – with their modern names of Taljanky, Maidenetske, Nebelivka and so on – dated to the early and middle centuries of the fourth millennium BC, which meant that some existed even before the earliest known cities in Mesopotamia. They were also larger in area.

Yet, even now, in scholarly discussions about the origins of urbanism, these Ukrainian sites almost never come up. Indeed, the very use of the term 'mega-site' is a kind of euphemism, signalling to a wider audience that these should not be thought of as proper cities but as something more like villages that for some reason had expanded inordinately in size. Some archaeologists even refer to them outright as 'overgrown villages'. How do we account for this reluctance to welcome the Ukrainian mega-sites into the charmed circle of urban origins? Why has anyone with even a passing interest in the origin of cities heard of Uruk or Mohenjo-daro, but almost no one of Taljanky?

The answer is largely political. Some of it concerns simple geopolitics: much of the initial work of discovery was carried out by Eastern Bloc scholars during the Cold War, which not only slowed down the reception of their findings in Western academic circles but tended to tinge any news of surprising discoveries with at least a tiny bit of scepticism. Even more, perhaps, it had to do with the internal political life of the prehistoric settlements themselves. That is, according to conventional views of politics, there didn't seem to be any. No evidence was unearthed of centralized government or administration – or indeed, any form of ruling class. In other words, these enormous settlements had all the hallmarks of what evolutionists would call a 'simple', not a 'complex' society.

It's hard here not to recall Ursula Le Guin's famous short story 'The Ones Who Walk Away from Omelas', about the imaginary city of Omelas, a city which also made do without kings, wars, slaves or secret police. We have a tendency, Le Guin notes, to write off such a community as 'simple', but in fact these citizens of Omelas were 'not simple folk, not dulcet shepherds, noble savages, bland utopians. They were not less complex than us.' The trouble is just that 'we have a bad habit, encouraged by pedants and sophisticates, of considering happiness as something rather stupid.'

Le Guin has a point. Obviously, we have no idea how relatively happy the inhabitants of Ukrainian mega-sites like Maidenetske or Nebelivka were, compared to the lords who constructed *kurgan* burials, or even the retainers ritually sacrificed at their funerals; or the bonded labourers who provided wheat and barley to the inhabitants of later Greek colonies along the Black Sea coast (though we can guess), and as anyone who has read the story knows, Omelas had some problems too. But the point remains: why do we assume that people who have figured out a way for a large population to govern and support itself without temples, palaces and military fortifications – that is, without overt displays of arrogance, self-abasement and cruelty – are somehow less complex than those who have not?

Why would we hesitate to dignify such a place with the name of 'city'?

The mega-sites of Ukraine and adjoining regions were inhabited from roughly 4100 to 3300 BC, that is, for something in the order of eight centuries, which is considerably longer than most subsequent urban traditions. Why were they there at all? Like the cities of Mesopotamia and the Indus valley, they appear to have been born of ecological opportunism in the middle phase of the Holocene. Not floodplain dynamics, in this case, but processes of soil formation on the flatlands north of the Black Sea. These black earths (Russian: *chernozem*) are legendary for their fertility; for the empires of later antiquity, they made the lands between the Southern Bug and Dniepr Rivers a breadbasket (which is why Greek city-states established colonies in the region and enslaved or made serfs of the local populations to begin with: ancient Athens was largely fed by Black Sea grain).

By 4500 BC, *chernozem* was widely distributed between the Carpathian and the Ural Mountains, where a mosaic landscape of open prairie and woodland emerged capable of supporting dense human habitation.[23] The Neolithic people who settled there had travelled east from the lower reaches of the Danube, passing through the Carpathian Mountains. We do not know why, but we do know that – throughout their peregrinations in river valleys and mountain passes – they retained a cohesive social identity. Their villages, often small in scale, shared similar cultural practices, reflected in the forms taken by their dwellings, female figurines and ways of making and serving food. The archaeological name given to this particular 'design for life' is the Cucuteni-Tripolye culture, after the sites where it was first recorded.[24]

So the Ukrainian and Moldovan mega-sites did not come out of thin air. They were the physical realization of an extended community that already existed long before its constituent units coalesced into large settlements. Some tens of these settlements have now been documented. The biggest currently known – Taljanky – extends over an area of 300 hectares, outspanning the earliest phases of the city of Uruk in southern Mesopotamia. It presents no evidence of central administration or communal storage facilities. Nor have any government buildings, fortifications or monumental architecture been found. There is no acropolis or civic centre; no equivalent to Uruk's raised public district called Eanna ('House of Heaven') or the Great Bath of Mohenjo-daro.

What we do find are houses; well over 1,000 in the case of Taljanky. Rectangular houses, sixteen or so feet wide and twice as long, built of wattle and daub on timber frames, with stone foundations. With their attached gardens, these houses form such neat circular patterns that from a bird's-eye view, any mega-site resembles the inside of a tree trunk: great rings, with concentric spaces between. The innermost ring frames a big gap in the middle of the settlement, where early excavators at first expected to find something dramatic, whether magnificent buildings or grand burials. But in every known case, the central area is simply empty; guesses for its function range from popular assemblies to ceremonies or the seasonal penning of animals – or possibly all three.[25] In consequence, the standard archaeological plan of a Ukrainian mega-site is all flesh, no core.

Just as surprising as their scale is the distribution of these massive

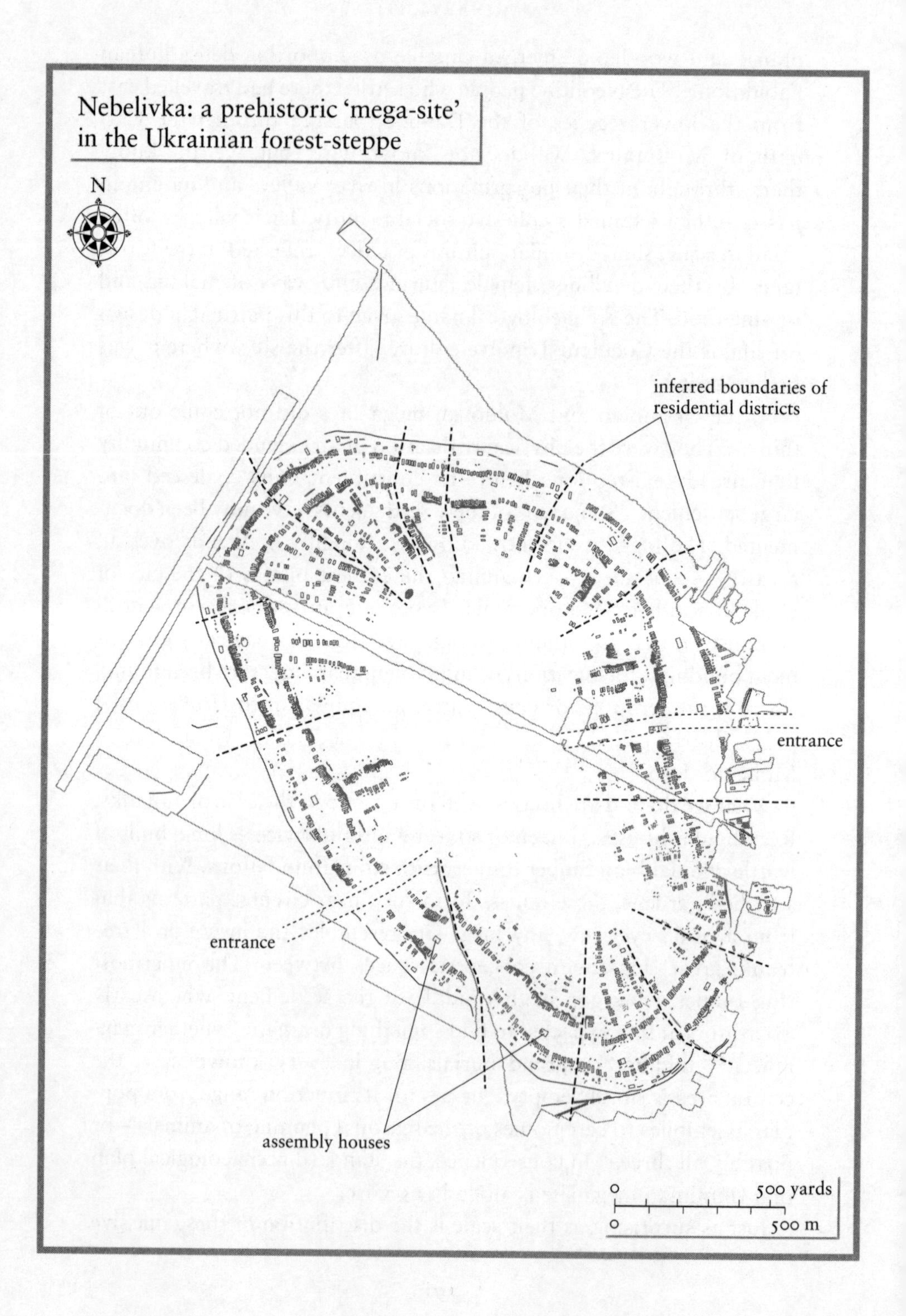
Nebelivka: a prehistoric 'mega-site' in the Ukrainian forest-steppe
N
inferred boundaries of residential districts
entrance
entrance
assembly houses
0
500 yards
0
500 m

settlements, which are all quite close to each other, at most six to nine miles apart.[26] Their total population – estimated in the many thousands per mega-site, and probably well over 10,000 in some cases – would therefore have had to draw resources from a common hinterland. Yet their ecological footprint appears to have been surprisingly light.[27] There are a number of possible explanations. Some have suggested the mega-sites were only occupied part of the year, even for just a season,[28] making them urban-scale versions of the kind of temporary aggregation sites we discussed in Chapter Three. This is difficult to reconcile with the substantial nature of their houses (consider the effort expended in felling trees, laying foundations, making good walls etc.). More probably, the mega-sites were much like most other cities, neither permanently inhabited nor strictly seasonal, but somewhere in between.[29]

We should also consider if the inhabitants of the mega-sites consciously managed their ecosystem to avoid large-scale deforestation. This is consistent with archaeological studies of their economy, which suggest a pattern of small-scale gardening, often taking place within the bounds of the settlement, combined with the keeping of livestock, cultivation of orchards, and a wide spectrum of hunting and foraging activities. The diversity is actually remarkable, as is its sustainability. As well as wheat, barley and pulses, the citizens' plant diet included apples, pears, cherries, sloes, acorns, hazelnuts and apricots. Mega-site dwellers were hunters of red deer, roe deer and wild boar as well as farmers and foresters. It was 'play farming' on a grand scale: an urban populus supporting itself through small-scale cultivation and herding, combined with an extraordinary array of wild foods.[30]

This way of life was by no means 'simple'. As well as managing orchards, gardens, livestock and woodlands, the inhabitants of these cities imported salt in bulk from springs in the eastern Carpathians and the Black Sea littoral. Flint extraction by the ton took place in the Dniestr valley, furnishing material for tools. A household potting industry flourished, its products considered among the finest ceramics of the prehistoric world; and regular supplies of copper flowed in from the Balkans.[31] There is no firm consensus among archaeologists about what sort of social arrangements all this required, but most would agree the logistical challenges were daunting. A surplus was definitely produced, and with it ample potential for some to seize control of the stocks and supplies, to lord it over others or battle for the spoils; but over eight centuries we find little evidence for warfare or the rise of social elites. The true complexity of the mega-sites lies in the strategies they adopted to prevent such things.

How did it all work? In the absence of written records (or a time machine), there are serious limits to what we can say about kinship and inheritance, or how people in these cities went about making collective decisions.[32] Still, some clues exist, beginning at the level of individual households. Each of these had a roughly common plan, but each was also, in its own way, unique. From one dwelling to the next there is constant innovation, even playfulness, in the rules of commensality. Each family unit invented its own slight variations on domestic rituals, reflected in its unique assemblage of serving and eating vessels, painted with polychrome designs of often mesmerizing intensity and made in a dazzling variety of forms. It's as if every household was an artists' collective which invented its own unique aesthetic style.

Some of this household pottery evokes the bodies of women; and among the other items most commonly found within the remains of houses are female figurines of clay. Model houses and tiny replicas of furniture and eating equipment also survive – miniature representations of lost social worlds, again, affirming the prominent role of women within them.[33] All this tells us a little about the cultural atmosphere of these households (and one can easily see why Marija Gimbutas, whose syntheses of Eurasian prehistory we discussed earlier, considered the Cucuteni-Tripolye culture to be part of 'Old Europe', with its cultural roots in the early farming societies of Anatolia and the Middle East). But how did these households come together in such numbers to form the great concentric arrangements which give the Ukrainian mega-sites their distinctive plan?

The first impression of these sites is one of rigid uniformity, a closed circuit of social interaction, but closer study reveals constant deviation from the norm. Individual households would sometimes opt to cluster together in groups of between three and ten families. Ditches or pits marked their boundaries. At some sites these groups coalesce into neighbourhoods, radiating out from the centre to the perimeter of the city, and even forming larger residential districts or quarters. Each had access to at least one assembly house, a structure larger than an ordinary dwelling where a wider sector of the population might gather periodically for activities we can only guess at (political meetings? legal proceedings? seasonal festivities?).[34]

Careful analysis by archaeologists shows how the apparent uniformity of the Ukrainian mega-sites arose from the bottom up, through processes of local decision-making.[35] This would have to mean that members of individual households – or at least, their neighbourhood representatives – shared a conceptual framework for the settlement as a whole. We can also safely infer that this framework was based on the image of a circle and its properties of transformation. To understand how the citizens put this mental image into effect, translating it into a workable social reality at such enormous scales, we cannot rely on archaeology alone. Fortunately, the burgeoning field of ethno-mathematics shows exactly how such a system might have worked in practice. The most informative case we know of is that of traditional Basque settlements in the highlands of the Pyrénées-Atlantiques.

These modern Basque societies – tucked down in the southwest corner of France – also imagine their communities in circular form, just as they imagine themselves as being surrounded by a circle of mountains. They do so as a way of emphasizing the ideal equality of households and family units. Now, obviously, the social arrangements of these existing communities are unlikely to be quite the same as those of ancient Ukraine. Nonetheless, they provide an excellent illustration of how such circular arrangements can form part of self-conscious egalitarian projects, in which 'everyone has neighbours to the left and neighbours to the right. No one is first, and no one is last.'[36]

In the commune of Sainte-Engrâce, for instance, the circular template of the village is also a dynamic model used as a counting device, to ensure the seasonal rotation of essential tasks and duties. Each Sunday, one household will bless two loaves at the local church, eat one, then present the other to its 'first neighbour' (the house to their right); the next week that neighbour will do the same to the next house to its right, and so on in a clockwise direction, so that in a community of 100 households it would take about two years to complete a full cycle.[37]

As so often with such matters, there is an entire cosmology, a theory of the human condition, baked in, as it were: the loaves are spoken of as 'semen', as something that gives life; meanwhile, care for the dead and dying travels in the opposite, counter-clockwise direction. But the system is also the basis for economic co-operation. If any one household is for any reason unable to fulfil its obligations when it is time to do so, a careful system of substitution comes into play, so neighbours at first, second and sometimes third remove can temporarily take their place. This in turn provides the model for virtually all forms of co-operation. The same system of 'first neighbours' and substitution, the same serial model of reciprocity, is used to call up anything that requires more hands than a single family can provide: from planting and harvesting to cheese-making and slaughtering pigs. It follows that households cannot simply schedule their daily labour in line with their own needs. They also have to consider their obligations to other households, which in turn have their own obligations to other, different households, and so on. Factoring in that some tasks – such as moving flocks to highland pastures, or the demands of milking, shearing and guarding herds – may require the combined efforts of ten different households, and that households have to balance the scheduling of numerous different sorts of commitment, we begin to get a sense of the complexities involved.

In other words, such 'simple' economies are rarely all that simple. They often involve logistical challenges of striking complexity, resolved on a basis of intricate systems of mutual aid, all without any need of centralized control or administration. Basque villagers in this region are self-conscious egalitarians, in the sense that they insist each household is ultimately the same and has the same responsibilities as any others; yet rather than governing themselves through communal assemblies (which earlier generations of Basque townsfolk famously created in places like Guernica), they rely on mathematical principles such as rotation, serial replacement and alternation. But the end result is the same, and the system flexible enough that changes in the number of households or the capacities of their individual members can be continually taken into account, ensuring relations of equality are preserved over the long term, with an almost complete absence of internal conflict.

There is no reason to assume that such a system would only work on a small scale: a village of 100 households is already way beyond Dunbar's proposed cognitive threshold of 150 people (the number of stable, trusting relationships we are able to keep track of in our minds, before – according, that is, to Dunbar – we are obliged to start putting chiefs and administrators in charge of social affairs); and Basque villages and towns used to be far larger than this. One can at least begin to see how – in a different context – such egalitarian systems might scale up to communities of many hundreds or even thousands of households. Returning to the Ukrainian mega-sites, we must admit that much remains unknown. Around the middle of the fourth millennium BC, most of them were basically abandoned. We still don't know why. What they offer us, in the meantime, is significant: proof that highly egalitarian organization has been possible on an urban scale.[38] With this in mind, we can look with fresh eyes at some better-known cases from other parts of Eurasia. Let's start with Mesopotamia.

ON MESOPOTAMIA, AND 'NOT-SO-PRIMITIVE' DEMOCRACY

'Mesopotamia' means 'land between the two rivers'. Archaeologists sometimes also call this region the 'heartland of cities'.[39] Its floodplains cross the otherwise arid landscape of southern Iraq, turning to marshland as they near the head of the Persian Gulf.[40] Urban life here goes back at least to 3500 BC. In the more northerly lands between the Tigris and Euphrates, where the rivers flow through rain-fed plains, the history of cities may go even further back, beyond 4000 BC.[41]

Unlike the Ukrainian mega-sites, or the Bronze Age cities of the Indus valley to which we'll turn shortly, Mesopotamia was already part of modern memory before any archaeologist put a spade into one of its ancient mounds.[42] Anyone who had read the Bible knew about the kingdoms of Babylonia and Assyria; and in the Victorian era of high empire, biblical scholars and Orientalists began excavating sites with scriptural associations, like Nineveh and Nimrud, hoping to uncover cities ruled by figures of legend such as Nebuchadnezzar,

Sennacherib or Tiglath-Pileser. They did find these; but in those places and elsewhere they discovered other things that were even more spectacular, like a basalt stela bearing the law code of Hammurabi, ruler of Babylon in the eighteenth century BC, unearthed at Susa in western Iran; clay tablets from Nineveh bearing copies of the *Epic of Gilgamesh*, fabled ruler of Uruk; and the Royal Tombs of Ur in southern Iraq, where kings and queens unknown to the Bible were interred with startling riches and the remains of sacrificed retainers around 2500 BC.

There were even bigger surprises. The oldest remains of cities and kingdoms – including the Royal Tombs of Ur – belonged to a culture previously unknown, and not mentioned in scripture: the Sumerians, who used a language unrelated to the Semitic family from which Hebrew and Arabic derive.[43] (In fact, as in the case of Basque, there's no consensus on what language family Sumerian does belong to.) But in general, the first decades of archaeological work in the region, from the late nineteenth to the early twentieth century, confirmed an expected association of ancient Mesopotamia with empire and monarchy. The Sumerians, at least on first sighting, seemed no exception.[44] In fact, they set the tone. Such was public interest in the findings from Ur that in the 1920s the *Illustrated London News* (England's 'window on the world') devoted no less than thirty feature articles to Leonard Woolley's excavation of the Royal Tombs.

All this reinforced a popular picture of Mesopotamia as a civilization of cities, monarchy and aristocracy, all tinged with the excitement of uncovering the 'truth' behind biblical scripture ('Ur of the Chaldees,' as well as being a Sumerian city, appears in the Hebrew Bible as the birthplace of the patriarch Abraham). But one of the major accomplishments of modern archaeology and epigraphy has been to redraw this picture entirely: to show that Mesopotamia was never, in fact, an eternal 'land of kings'. The real story is far more complicated.

The earliest Mesopotamian cities – those of the fourth and early third millennia BC – present no clear evidence for monarchy at all. Now, you might object, it's difficult to prove for certain that something *isn't* there. However, we know what evidence for monarchy in such cities would be like, because half a millennium later (from around 2800 BC onwards) monarchy starts popping up everywhere: palaces, aristocratic burials and royal inscriptions, along with defensive walls for cities and organized militia to guard them. But the birth of cities, and with it the basic elements of Mesopotamian civic life – the ancient building blocks of its urban society – begin considerably before this 'Early Dynastic' period.

These original urban elements include some which have been wrongly characterized as inventions of royal statecraft, such as the institution that historians call by the French term corvée. This refers to obligatory labour on civic projects exacted from free citizens on a seasonal basis, and it has always been assumed to be a form of tax extracted by powerful rulers: taxes paid not in goods, but in services. From a Mesopotamian perspective, though, corvée was already very ancient. As old as humanity itself. The flood-myth *Atrahasis* – the prototype for the Old Testament story of Noah – tells how the gods first created people to perform corvée on their behalf. Mesopotamian gods were unusually hands-on, and had originally worked themselves. Eventually tiring of digging irrigation canals, they created minor deities to do the work, but they too rebelled, and – receiving a much more favourable hearing than Lucifer would in Heaven – the gods conceded to their demands and created people.[45]

Everyone had to do corvée. Even the most powerful Mesopotamian rulers of later periods had to heave a basket of clay to the construction site of an important temple. The Sumerian word for corvée (*dubsig*) refers to this basket of earth, written with a pictogram showing a person lifting it on to their head, like kings do on monuments such as the Plaque of Ur-Nanshe, carved around 2500 BC. Free citizens performed *dubsig* for weeks or even months. When they did, high-ranking clerics and administrators worked alongside artisans, shepherds and cereal farmers. Later kings could grant exemptions, allowing the rich to pay tax in lieu, or employ others to do the work for them. Still, all contributed in some way.[46]

Royal hymns describe the 'happy faces' and 'joyous hearts' of corvée workers. No doubt there's an element of propaganda here, but it's clear that, even in periods of monarchy and empire, these seasonal projects were undertaken in a festive spirit, labourers receiving copious rewards of bread, beer, dates, cheese and meat. There was also something of the carnival about them. They were occasions when the moral order of the city spun on its axis, and distinctions between citizens dissolved away. The 'Hymns of Gudea' – the governor (*ensi*) of the city-state of Lagash – convey something of the atmosphere in which they took place. Dating from around the end of the third millennium BC, they eulogize the restoration of a temple called *Eninnu*, the House of Ningirsu, patron deity of the city:

> Women did not carry baskets,
> only the top warriors did the building
> for him; the whip did not strike;
> mother did not hit her (disobedient) child;
> The general,
> The colonel,
> The captain,
> (and) the conscript,
> they (all) shared the work equally;
> the supervision indeed was (like)
> soft wool in their hands.[47]

More lasting benefits for the citizenry at large included debt cancellation by the governor.[48] Times of labour mobilization were thus seen as moments of absolute equality before the gods – when even slaves might be placed on an equal footing to their masters – as well as times when the imaginary city became real, as its inhabitants shed their day-to-day identities as bakers or tavern keepers or inhabitants of such and such a neighbourhood, or later generals or slaves, and briefly assembled to become 'the people' of Lagash, or Kish, Eridu, or Larsa as they built or rebuilt some part of the city or the network of irrigation canals that sustained it.

If this is at least partly how cities were built, it's hard to write such festivals off as pure symbolic display. What's more, there were other institutions, also said to originate in the Predynastic age, which ensured that ordinary citizens had a significant hand in government. Even the most autocratic rulers of later city-states were answerable to a panoply of town councils, neighbourhood wards and assemblies – in all of which women often participated alongside men.[49] The 'sons and daughters' of a city could make their voices heard, influencing everything from taxation to foreign policy. These urban assemblies might not have been so powerful as those of ancient Greece – but, on the other hand, slavery was not nearly so developed in Mesopotamia, and women were not excluded from politics to anything like the same degree.[50] In diplomatic correspondence, we also catch occasional glimpses of corporate bodies rising up against unpopular rulers or policies, often successfully.

The term used by modern scholars for this general state of affairs is 'primitive democracy'. It's not a very good term, since there's no particular reason to think any of these institutions were in any way crude or unsophisticated. Arguably, the continued use of this odd term by researchers has inhibited wider discussion, which remains mostly confined to the specialist field of Assyriology: the study of ancient Mesopotamia and its written legacy in the cuneiform script. Let's take a closer look at the argument, and some of its implications.

The idea that Mesopotamia possessed a 'primitive democracy' was first advanced in the 1940s by Thorkild Jacobsen, the Danish historian and Assyriologist.[51] Today, scholars in that field have extended his idea even further. District councils and assemblies of elders – representing the interests of urban publics – were not just a feature of the earliest Mesopotamian cities, as Jacobsen thought; there is evidence for them in all later periods of Mesopotamian history too, right down to the time of the Assyrian, Babylonian and Persian Empires, whose memory lived on through biblical scripture.

Popular councils and citizen assemblies (Sumerian: *ukkin*; Akkadian: *puhrum*) were stable features of government, not just in Mesopotamian cities, but also their colonial offshoots (like the Old Assyrian *karum* of Kanesh, in Anatolia), and in the urban societies of neighbouring peoples such as the Hittites, Phoenicians, Philistines and Israelites.[52] In fact, it is almost impossible to find a city anywhere in the ancient Near East that did not have some equivalent to a popular assembly – or often several assemblies (for instance, different ones representing the interests of 'the young' and 'the old'). This was the case even in areas such as the Syrian steppe and northern Mesopotamia, where traditions of monarchy ran deep.[53] Still, we know very little about how these assemblies functioned, their composition, or often even where they met.[54] Likely as not, an ancient Greek observer might have described some of them as democratic, others oligarchic, still others as a mix of democratic, oligarchic and monarchic principles. But for the most part, experts can only guess.

Some of the clearest evidence comes from between the ninth and seventh centuries BC. Assyrian emperors like Sennacherib and Ashurbanipal have been famous since biblical times for their brutality, creating monuments that boasted of the bloody vengeance they carried out against rebels. But when dealing with loyal subjects they were strikingly hands-off, often granting near-total autonomy to citizen bodies that made decisions collectively.[55] We know this because governors stationed far from the Assyrian court, in southern Mesopotamia's major cities – Babylon, Nippur, Uruk, Ur, and so on – sent letters to their overlords. Many of these were letters recovered by archaeologists during the excavation of royal archives at the ancient imperial capital of Nineveh. In them, city governors relay information to the Assyrian court about decisions made by civic councils. We learn the 'will of the people' on matters ranging from foreign policy to the election of governors; also, that citizen bodies sometimes took matters into their own hands, raising soldiers or taxes to support civic projects, and playing their overlords off against each other.

Neighbourhood wards (Akkadian: *bābtum*, after the word for 'gate') were active in local administration, and sometimes appear to have replicated certain aspects of village or tribal governance in an urban setting.[56] Murder trials, divorce and property disputes seem to have been mostly in the hands of town councils. Texts found at Nippur give unusual details about the composition of one such assembly, summoned to act as a jury for a homicide case. Among those sitting we find one bird catcher, one potter, two gardeners and a soldier in the service of a temple. The Trinidadian intellectual C. L. R. James once said of fifth-century Athens that 'every cook can govern'. In Mesopotamia, or at least in many parts of it, it seems this was literally true: being a manual labourer did not exclude one from direct participation in law and politics.[57]

Participatory government in ancient Mesopotamian cities was organized at multiple levels, from wards – sometimes defined on ethnic lines or in terms of professional affiliations – up to larger urban districts, and ultimately the city as a whole. The interests of individual citizens might be represented at every tier, but the surviving written evidence contains frustratingly few details about how this system of urban government worked in practice. Historians attribute this lack of information to the key role of assemblies, operating at various scales, and conducting their deliberations (about local property disputes, divorce and inheritance cases, accusations of theft or murder, and so on) in ways that were largely independent of central government and did not require its written authorization.[58]

Archaeologists find themselves in general agreement with the historians, although one might reasonably ask how archaeology can shed independent light on such political matters. One answer comes from the site of Mashkan-shapir, an important centre under the kings of Larsa, around 2000 BC. As with most Mesopotamian cities, the urban landscape of Mashkan-shapir was dominated by its main temple – in this case, the sanctuary of Nergal, god of the underworld – raised up high on a ziggurat platform; but intensive archaeological survey of the city's harbour, gateways and residential districts revealed a strikingly even distribution of wealth, craft production and administrative tools across the five main districts, with no obvious centre of commercial or political power.[59] In terms of day-to-day affairs, city dwellers (even under monarchies) largely governed themselves, presumably much as they had before kings appeared on the scene to begin with.

Things could work the other way round. Sometimes, the arrival of an authoritarian ruler from outside the city sent urban life into reverse. Such was the case with the Amorite Dynasty of the Lims – Yaggid-Lim, Yahudun-Lim and Zimri-Lim – which conquered much of the Syrian Euphrates around the same time Mashkan-shapir was thriving far to the south. The Lims decided to set up their centre of operations in the ancient city called Mari (modern Tell Hariri, on the Syrian Euphrates), and occupied government buildings in its heart. Their arrival seems to have precipitated a mass exodus of Mari's urban population, who left to join up with smaller townships or tent-dwelling herders scattered across the Syrian steppe. Before the sack of Mari by Hammurabi of Babylon in 1761 BC, the last 'city' of the Amorite kings comprised little more than the royal residence, harem, attached temples and a handful of other official buildings.[60]

Written correspondence of this period offers direct evidence of antipathy between arriviste monarchy of this kind and the established power of urban assemblies. Letters to Zimri-Lim from Terru – lord of the ancient Hurrian capital of Urkesh (modern Tell Mozan) – convey his impotence in the face of the city's councils and assemblies. On one occasion, Terru tells Zimri-Lim: 'Because I am submitted to my lord's pleasure, the inhabitants of my town despise me, and two and three times I have snatched my head back from death by their hand.' To which the Mari king responds: 'I did not realize that the inhabitants of your town despised you on account of me. You belong to me even if the town of Urkesh belongs to someone else.' All this came to a head when Terru confessed he had to flee from public opinion ('the mouth of Urkesh'), taking refuge in a nearby town.[61]

So, far from needing rulers to manage urban life, it seems most Mesopotamian urbanites were organized into autonomous self-governing units, which might react to offensive overlords either by driving them out or by abandoning the city entirely. None of this necessarily answers the question, 'what was the nature of government in Mesopotamian cities *before* the appearance of kingship?' (though it's certainly suggestive). Instead, the answers depend to a slightly alarming degree on discoveries from a single site: the city of Uruk – modern Warka, biblical Erech – whose later mythology inspired Jacobsen's original search for 'primitive democracy'.[62]

IN WHICH WE DESCRIBE HOW (WRITTEN) HISTORY, AND PROBABLY (ORAL) EPIC TOO, BEGAN: WITH BIG COUNCILS IN THE CITIES, AND SMALL KINGDOMS IN THE HILLS

At 3300 BC, Uruk was a city of around 200 hectares, dwarfing her neighbours on the southern Mesopotamian floodplain. Estimates of Uruk's population at this time range widely, between 20,000 and 50,000. The first residential quarters are built over by later urban settlement, which continued down to the time of Alexander the Great in the fourth century BC.[63] Cuneiform script may well have been invented at Uruk, around 3300 BC, and we can see its early stages of development in numerical tablets and other forms of administrative notation. Bookkeeping in the city's temples was writing's main function at that point.[64] Thousands of years later, it was also in the temples of Uruk that cuneiform script finally passed into obsolescence, by which time it had been elaborated to record, among other things, the world's earliest written literature and law codes.

What do we know about the original city of Uruk? By the late fourth millennium BC it had a high acropolis, much of which was taken up by the raised public district called Eanna, 'House of Heaven', dedicated to the Goddess Inanna. On its summit stood nine monumental buildings, of which only the foundations of imported limestone survive, together with bits of stairwells and fragments of columned halls decorated with coloured mosaic. The roofs of these broad civic structures must originally have been constructed of exotic timbers, brought by river barge from the 'Cedar Forest' of Syria, which form the backdrop to the Mesopotamian *Epic of Gilgamesh*.

For the urban historian, Uruk remains something of a strange fruit. A bit like a Ukrainian mega-site in reverse, its oldest known architectural layout is all core with no surrounding flesh, since we know almost nothing of the residential districts beyond the Eanna precinct, which were ignored by early excavators at the site. In other words, we glimpse something of the city's public sector, but as yet we have no private sector against which to define it. Still, let's press on with what we do know.

Most of these public buildings seem to have been great communal assembly halls, clearly modelled on the plan of ordinary households, but constructed as houses of the gods.[65] There was also a Great Court comprising an enormous sunken plaza, 165 feet across, entirely surrounded by two tiers of benches and equipped with water channels to feed trees and gardens, which offered much-needed shade for open-air gatherings. This sort of arrangement – a series of magnificent, open temples accompanied by a congenial space for public meetings – is exactly what one might expect were Uruk to have been governed by a popular assembly; and, as Jacobsen emphasized, the Gilgamesh epic (which begins in Predynastic Uruk) does speak of such assemblies, including one reserved for the young men of the city.

To draw an obvious parallel: the Athenian agora in the time of

Pericles (the fifth century BC) was also full of public temples, but the actual democratic assemblies took place in an open space called the Pnyx, a low hill equipped with seating for the Council of Five Hundred citizens, appointed – by sortition, with rotating membership – to run the everyday affairs of the city (all other citizens were expected to stand). Meetings at the Pnyx could involve anywhere between 6,000 and 12,000 people, groups comprising free adult males drawn from perhaps 20 per cent of the city's total population. The Great Court at Uruk is considerably larger, and while we have little idea what the total population of Uruk was in, say, 3500 BC, it's hard to imagine it was anywhere near that of classical Athens. This suggests a wider range of participation, which would make sense if women were not entirely excluded and if early Uruk did not, like later Athens, define some 30 per cent of its population as resident aliens with no voting rights, and up to 40 per cent as slaves.

Much of this remains speculative, but what's clear is that in later periods things change. Around 3200 BC the original public buildings of the Eanna sanctuary were razed and covered with debris, and its sacred landscape redesigned around a series of gated courts and ziggurats. By 2900 BC, we have evidence for local kings of rival city-states battling it out for supremacy over Uruk, in response to which a five-and-a-half-mile fortification wall (whose building was later attributed to Gilgamesh) went up around the city's perimeter. Within a few centuries, city rulers were setting themselves up as neighbours of the goddesses and the gods, building their own palatial houses on the doorstep to the House of Heaven and stamping their names into its sacred brickwork.[66]

Once again, while evidence of democratic self-governance is always a bit ambiguous (would anyone guess what was really going on in fifth-century Athens, from archaeological evidence alone?), evidence for royal rule, when it appears, is entirely unmistakable.

What Uruk is really famous for is writing. It is the first city for which we have extensive written records, and some of these documents do date back to the period before royal rule. Unfortunately, while they can be read, they are also extremely difficult to interpret.

Most are cuneiform tablets recovered from trash dumps dug into the foundations of the acropolis, and they appear to provide only a very narrow window on to city life. The great majority are bureaucratic receipts, recording transactions of goods and services. There are also 'school texts' which comprise sign-lists, copied out by scribes in training to familiarize them with the standard administrative lexicon of the time. The historical value of the latter is unclear, because scribes may have had to learn to write all kinds of cuneiform signs – which were executed by pressing a reed stylus into moist clay – that had little application in practice. Such learning may well have been part of what was considered a proper literate training at the time.[67]

Still, the mere existence of a college of scribes administering complicated relations between people, animals and things shows us there was much more going on in the large 'houses of the gods' than just ritual gatherings. There were goods and industries to be administered, and a body of citizens who developed pedagogical techniques that quickly became so essential to this particular form of urban life that they remain with us to this day. To get a sense of how pervasive some of these innovations were, consider that just about anyone reading this book is likely to have first learned to read in classrooms, sitting in rows opposite a teacher, who follows a standard curriculum. This rather stern way of learning was itself a Sumerian invention, one now to be found in virtually every corner of our world.[68]

So what do we know about these houses of the gods? For one thing, it is clear that in many ways they resembled factories more than churches. Even the earliest for which there is evidence had considerable amounts of human labour at their disposal, along with workshops and stockpiles of raw materials. Some details of the way these Sumerian temples organized themselves are still with us, including the quantification of human labour into standard workloads and units of time. Sumerian officials counted all sorts of things – including days, months and years – using a sexagesimal (base-60) system from which ultimately derives (via many and varied pathways of transmission) our own system of time-reckoning.[69] In their bookkeeping records we find the ancient seedbeds of modern industrialism, finance and bureaucracy.

It is often hard to determine exactly who these temple labourers were, or even what sort of people were being organized in this way, allotted meals and having their outputs inventoried – were they permanently attached to the temple, or just ordinary citizens fulfilling their annual corvée duty? – but the presence of children in the lists suggests at least some may have lived there. If so, then this was most likely because they had nowhere else to go. If later Sumerian temples are anything to go by, this workforce will have comprised a whole assortment of the urban needy: widows, orphans and others rendered vulnerable by debt, crime, conflict, poverty, disease or disability, who found in the temple a place of refuge and support.[70]

For the time being, though, let us just emphasize the remarkable number of industries that developed in these temple workshops, as documented in the cuneiform accounts. Among them we find the first large-scale dairy and wool production; also the manufacture of leavened bread, beer and wine, including facilities for standardized packaging. Some eighty varieties of fish – fresh- and saltwater – appear in the administrative accounts along with their associated oil and food products, preserved and stored in temple repositories. From this we can deduce that a primary economic function of this temple sector was to co-ordinate labour at key times of year, and to provide quality control for processed goods that differed from those made in ordinary households.[71]

This particular kind of work, unlike the maintenance of irrigation dykes and building of roads and embankments, was routinely carried out under central administrative control. In other words, in the early phases of Mesopotamian urban life, what we would ordinarily imagine as the state sector (e.g. public works, international relations) was managed largely by local or city-wide assemblies; while top-down bureaucratic procedures were limited to what we would now think of as the economic or commodity sphere.[72]

Of course, Uruk's inhabitants didn't have an explicit concept of 'the economy' – no one did until very recent times. For Sumerians, the ultimate purpose of all these factories and workshops was to provide the gods and goddesses of the city with an illustrious residence where they would receive offerings of food, fine clothing and care, which also meant servicing their cult and organizing their festivals. The latter activity was probably depicted on the Uruk Vase, one of the few surviving examples of narrative art from this early period, whose carved decoration shows a number of identical nude males parading behind one larger male figure to the temple precinct of the Goddess Inanna with their yield of field, orchard and flock.[73]

It's not entirely clear who the larger, leading male figure – or 'Uruk man', as he's sometimes called in the literature – is supposed to be. According to the much later story of Gilgamesh, which is set in Uruk, one of the leaders of the youth assembly did manage to catapult himself into the position of *lugal*, or king – but if anything like this happened it left no trace in the written records of the fourth millennium BC, since lists of Uruk office holders have been found, dating to that time, and *lugal* is not among them. (The term only shows up much later, around 2600 BC, at a time when there are also palaces and other clear signs of royalty.) There is no reason to think that monarchy – ceremonial or otherwise – played any significant role in the earliest cities of southern Mesopotamia. Quite the opposite, in fact.[74]

Yet it's also clear that early inscriptions prise open only a very narrow window on urban life. We know something about the mass production of woollen garments and other commodities in temples; we can also infer that – somehow or other – these woollens and other temple manufactures were being traded for wood, metal and precious stone that were not available in the river valleys, but abounded in the surrounding high country. We know little about how this trade was organized in its earliest days, but we do know from archaeological evidence that Uruk was establishing colonies, tiny versions of itself, at many strategic points along the trade routes. Uruk colonies appear to have been both commercial outposts and religious centres, and traces of them are found as far north as the Taurus Mountains and as far east as the Iranian Zagros.[75]

'Uruk expansion', as it is called in the archaeological literature, is puzzling. There's no real evidence of violent conquest, no weapons or fortifications, yet at the same time there seems to have been an effort to transform – in effect, to colonize – the lives of nearby peoples, to disseminate the new habits of urban life. In this, the emissaries of Uruk seem to have proceeded with an almost missionary zeal. Temples were established, and with them new sorts of clothing, new dairy products, wines and woollens were disseminated to local populations. While these products might not have been entirely novel, what the temples introduced was the principle of standardization: urban temple-factories were literally outputting products in uniform packages, with the houses of the gods guaranteeing purity and quality control.[76]

The entire process was, in a sense, colonial, and it did not go unopposed. As it turns out, we cannot really understand the rise of what we have come to call 'the state' – and specifically of aristocracies and monarchies – except in the larger context of that counter-reaction.

Perhaps the most revealing site, in this respect, is called Arslantepe – the 'Hill of the Lion', in the Malatya Plain of eastern Turkey. Around the same time Uruk was becoming a large city, Arslantepe was coming into its own as a regional centre of some significance, where the upper reaches of the Euphrates arc towards the Anti-Taurus Mountains, with their rich sources of metal and timber. The site may have started life as some kind of seasonal trade fair; at nearly 3,300 feet above sea level, it was likely snowed in over the winter months. Even at its peak it was never larger than five hectares, and there were probably never more than a few hundred people actually living on the spot. Within those five hectares, however, archaeologists have unearthed evidence for a remarkable sequence of political developments.[77]

The story of Arslantepe begins around 3300 BC, when a temple was built on the site. This temple resembled those of Uruk and her colonies, with storage areas for food and carefully arranged archives of administrative seals, just as in any temple of the Mesopotamian floodplain. But within a few generations the temple was dismantled, and in its place was built a massive private structure enclosing a grand audience chamber and living quarters, as well as storage areas, including an armoury. An assemblage of swords and spearheads – finely crafted of arsenic-rich copper and quite unlike anything found in public buildings of the lowlands at this time – signals not only control over, but a celebration of the means to enact violence: a new aesthetics of personal combat and killing. The excavators have labelled this building the world's 'earliest known palace'.

From 3100 BC, across the hilly country of what's now eastern Turkey, and then in other places on the edge of urban civilization, we see evidence for the rise of a warrior aristocracy, heavily armed with metal spears and swords, living in what appear to be hill forts or small palaces. All traces of bureaucracy disappear. In their place we find not just aristocratic households – reminiscent of Beowulf's mead hall, or indeed the Pacific Northwest Coast in the nineteenth century – but for the first time also tombs of men who, in life, were clearly considered heroic individuals of some sort, accompanied to the afterlife by prodigious quantities of metal weaponry, treasures, elaborate textiles and drinking gear.[78]

Everything about these tombs and their makers, living on the frontiers of urban life, bespeaks a spirit of extravagance. Copious amounts of fine food, drink and personal jewellery were deposited. There are signs that such funerals could spiral into spectacles of competitive one-upmanship, as what must have been priceless trophies, heirlooms and prizes of unparalleled magnificence were offered up or even intentionally destroyed; some, too, are accompanied by subsidiary burials of those apparently slaughtered at the graveside as offerings.[79] Unlike the isolated 'princes' and 'princesses' of the Ice Age, there are whole cemeteries full of such burials – for example at Başur Höyük, on the way to Lake Van, while at Arslantepe we see exactly the kind of physical infrastructure (forts, storehouses) we might expect from a society dominated by some sort of warrior aristocracy.

Here we have the very beginnings of an aristocratic ethos with a long afterlife and some wide ramifications in the history of Eurasia (something we touched on earlier, when alluding to Herodotus' account of the Scythians, and Ibn Fadlan's later observations on the 'barbarian' Germanic tribes of the Volga). We are witnessing the first known emergence of what Hector Munro Chadwick famously called 'heroic societies' and, moreover, these societies all seem to have emerged just where his analysis tells us to expect them: on the margins of bureaucratically ordered cities.

Writing in the 1920s, Chadwick – Professor of Anglo-Saxon at Cambridge, at much the same time J. R. R. Tolkien held that post at Oxford – was initially concerned with why great traditions of epic poetry (Nordic sagas, the works of Homer, the *Ramayana*) always seemed to emerge among people in contact with and often employed by the urban civilizations of their day, but who ultimately rejected the values of those same civilizations. For a long time, his notion of 'heroic societies' fell into a certain disfavour: there was a widespread assumption that such societies did not really exist but were, like the society represented in Homer's *Iliad*, retroactively reconstructed in epic literature.

But as archaeologists have more recently discovered, there is a very real pattern of heroic burials, indicating in turn an emerging cultural emphasis on feasting, drinking, the beauty and fame of the individual male warrior.[80] And it appears time and again around the fringes of urban life, often in strikingly similar forms, over the course of the Eurasian Bronze Age. In searching for the common features of such 'heroic societies', we can find a fairly consistent list in precisely the traditions of epic poetry that Chadwick compared (in each region, the first written versions being much later in date than the heroic burials themselves, but shedding light on earlier customs). It's a list which applies just as well, in most of its features, to the *potlatch* societies of the Northwest Coast or, for that matter, the Māori of New Zealand.

All these cultures were aristocracies, without any centralized authority or principle of sovereignty (or, maybe, some largely symbolic, formal one). Instead of a single centre, we find numerous heroic figures competing fiercely with one another for retainers and slaves. 'Politics', in such societies, was composed of a history of personal debts of loyalty or vengeance between heroic individuals; all, moreover, focus on game-like contests as the primary business of ritual, indeed political, life.[81] Often, massive amounts of loot or wealth were squandered, sacrificed or given away in such theatrical performances. Moreover, all such groups explicitly resisted certain features of nearby urban civilizations: above all, writing, for which they tended to substitute poets or priests who engaged in rote memorization or elaborate techniques of oral composition. Inside their own societies, at least, they also rejected commerce. Hence standardized currency, either in physical or credit forms, tended to be eschewed, with the focus instead on unique material treasures.

It goes without saying that we cannot possibly hope to trace all these various tendencies back into periods for which no written testimony exists. But it is equally clear that, insofar as modern archaeology allows us to identify an ultimate origin for 'heroic societies' of this sort, it is to be found precisely on the spatial and cultural margins of the world's first great urban expansion (indeed, some of the earliest aristocratic tombs in the Turkish highlands were dug directly into the ruins of abandoned Uruk colonies).[82] Aristocracies, perhaps monarchy itself, first emerged in opposition to the egalitarian cities of the Mesopotamian plains, for which they likely had much the same mixed but ultimately hostile and murderous feelings as Alaric the Goth would later have towards Rome and everything it stood for, Genghis Khan towards Samarkand or Merv, or Timur towards Delhi.

IN WHICH WE CONSIDER WHETHER THE INDUS CIVILIZATION WAS AN EXAMPLE OF CASTE BEFORE KINGSHIP

Fast-forward now 1,000 years from the Uruk expansion to around 2600 BC. On the banks of the Indus River, in what is today the Pakistani province of Sindh, a city was founded on virgin soil: Mohenjo-daro. It remained there for 700 years.[83] The city is considered the greatest expression of a new form of society that flourished in the valley of the Indus at the time; a form of society which archaeologists have come to know simply as the 'Indus' or 'Harappan' civilization. It was South Asia's first urban culture. Here we will find further evidence that Bronze Age cities – the world's first large-scale, planned human settlements – could emerge in the absence of ruling classes and managerial elites; but those of the Indus valley also present some uniquely puzzling features, which archaeologists have debated for more than a century.[84] Let's introduce both the problem and its key locus – the site of Mohenjo-daro – in a little more detail.

On first inspection, Mohenjo-daro bears out its reputation as the most completely preserved city of the Bronze Age world. There's something staggering about it all: a brazen modernity, which was not lost on the first excavators of the site, who didn't hesitate to designate certain areas 'high streets', 'police barracks' and so on (though much of this initial interpretation, as it turned out, was fantasy). Most of the city consists of the brick-built houses of the Lower Town, with its grid-like arrangement of streets, long boulevards and sophisticated drainage and sanitation systems (terracotta sewage pipes, private and public toilets and bathrooms were ubiquitous). Above these surprisingly comfortable arrangements loomed the Upper Citadel, a raised civic centre, also known (for reasons we'll explain) as the Mound of the Great Bath. Though both parts of the city stood on massive artificial foundations of heaped earth, lifting them above the floodplain, the Upper Citadel was also encased in a wall of baked bricks made to standard dimensions which extended all the way round it, affording further protection when the Indus broke its banks.[85]

In the wider ambit of Indus civilization, there is only one rival to Mohenjo-daro: the site of Harappa (whence the alternative term 'Harappan civilization'). Of similar magnitude, it lies about 370 miles upstream on the Ravi River, a tributary of the Indus. Many other sites of the same date and cultural family exist, ranging from large towns to hamlets. They extend over most of the area of modern-day Pakistan, and well beyond the floodplain of the Indus, into northern India. For instance, perched on an island amid the salt flats of the Great Rann of Kutch lie the striking remains of Dholavira, a town equipped with over fifteen brick-built reservoirs to capture rainwater and run-off from local streams. The Indus civilization had colonial outposts as far as the Oxus River in northern Afghanistan, where the site of Shortugai presents a miniature replica of its urban mother-culture: ideally placed to tap the rich mineral sources of the Central Asian highlands (lapis, tin and other gemstones and metals). Such materials were prized by lowland artisans and their commercial partners as far away as Iran, Arabia and Mesopotamia. At Lothal, on Gujarat's Gulf of Khambhat, lie remains of a well-appointed port town facing the Arabian Sea, presumably built by Indus engineers to service maritime trade.[86]

The Indus civilization had its own script, which appeared and vanished together with its cities. It has not been deciphered. What survives to us are mainly short captions, stamped or incised on storage jars, copper tools and the remnants of a lonely piece of street signage from Dholavira. Short inscriptions also feature on tiny stone amulets, captioning pictorial vignettes or miniature animal figures, carved with striking precision. Most of these are realistic depictions of water buffalo, elephant, rhinoceros, tiger and other local fauna, but they also include fantastic beasts, most often unicorns. Debate surrounds the amulets' function: were they worn as personal identifiers, for passage through the city's gated quarters and walled compounds, or perhaps to gain entry to ceremonial occasions? Or were they used for administration, to impress identifying signs on commodities passing among unknown parties: a Bronze Age origin of product-branding? Could they be all of these things?[87]

Aside from our inability to make sense of the Indus script, there are many puzzling aspects of Harappa and Mohenjo-daro. Both were excavated in the early twentieth century, when archaeology was a large-scale and broad-brush affair, with sometimes thousands of workers digging simultaneously. Rapid work on this scale produced striking spatial exposures of street plans, residential neighbourhoods and entire ceremonial precincts. But it largely neglected to chart the site's development over time, a process that can only be disentangled with more careful methods. For instance, early excavators recorded just the baked-brick foundations of buildings. The superstructures were of softer mud-brick, often missed or unwittingly destroyed in the course of rapid digging; while the upper storeys of large civic structures were originally of fine timber, rotted or removed for reuse in antiquity. What seems in plan to be a single phase of urban construction is, in reality, a false composite made up of different elements from various periods of the city's history – a city inhabited for over 500 years.[88]

All of which leaves us with plenty of known unknowns, including the city's size or population (recent estimates suggest up to 40,000 residents, but really we can only guess).[89] It's not even clear where to draw the city boundaries. Some scholars include only the immediately visible areas of the planned Lower Town and the Upper Citadel as part of the city proper, yielding a total area of 100 hectares. Others note scattered evidence for the city's extension over a far greater area, maybe three times this size – we'd have to call them 'Lower, Lower Towns' – long since submerged by floodplain soils: a poignant illustration of that conspiracy between nature and culture which so often makes us forget that shanty dwellers even exist.

But it's this last point that leads us in more promising directions. Despite all its problems, Mohenjo-daro and its sister sites in the

Punjab do offer some insights into the nature of civic life in the first cities of South Asia, and into the wider question that we posed at the start of this chapter: is there a causal relationship between scale and inequality in human societies?

Let's consider, for a moment, what archaeology tells us about wealth distribution at Mohenjo-daro. Contrary to what we might expect, there is no concentration of material wealth on the Upper Citadel. Quite the opposite, in fact. Metals, gemstones and worked shell – for example – were widely available to households of the Lower Town; archaeologists have recovered such goods from caches beneath house floors, and bundles of them are scattered over every quarter of the site.[90] The same goes for little terracotta figures of people wearing bangles, diadems and other flashy personal adornment. Not so the Upper Citadel.

Writing, and also standard weights and measures, were also widely distributed across the Lower Town; so too evidence for craft occupations and industries from metalworking and potting to the manufacture of beads. All flourished down there, in the Lower Town, but are absent from the city's Upper Citadel, where the main civic structures stood.[91] Objects made for personal display had little place, it seems, in the most elevated quarters of the city. Instead, what defines the Upper Citadel are buildings like the Great Bath – a large sunken pool measuring roughly forty feet long and over six feet deep, lined with carefully executed brickwork, sealed with plaster and bitumen and entered on either side via steps with timber treads – all constructed to the finest architectural standards, yet unmarked by monuments dedicated to particular rulers, or indeed any other signs of personal aggrandizement.

Because of its lack of royal sculpture, or indeed other forms of monumental depiction, the Indus valley has been termed a 'faceless civilization'.[92] At Mohenjo-daro, it seems, the focus of civic life was not a palace or cenotaph, but a public facility for purifying the body. Brick-made bathing floors and platforms also were a standard fixture in most dwellings of the Lower Town. Citizens seem to have been familiar with very specific notions of cleanliness, with daily ablutions apparently forming part of their domestic routine. The Great Bath was, at one level, an outsized version of these residential washing facilities. On another level, though, life on the Upper Citadel seems to negate that of the Lower Town.

So long as the Great Bath was in use – and it was for some centuries – we find no evidence of industrial activities nearby. The narrowing lanes on the acropolis effectively prevented the use of ox-drawn carts and similar commercial traffic. Here, it was the Bath itself – and the act of bathing – that became the focus of social life and labour. Barracks and storerooms adjacent to the Bath housed a staff (whether in attached or rotating service, we cannot know) and their essential supplies. The Upper Citadel was a special sort of 'city within the city', in which ordinary principles of household organization went into reverse.[93]

All this is redolent of the inequality of the caste system, with its hierarchical division of social functions, organized on an ascending scale of purity.[94] But the earliest recorded reference to caste in South Asia comes only 1,000 years later, in the *Rig Veda* – an anthology of sacrificial hymns, probably composed around 1200 BC. The system, as described in later Sanskrit epics, consisted of four hereditary ranks or *varnas*: priests (*brahmins*), warriors or nobles (*kshatriyas*), farmers and traders (*vaishyas*) and labourers (*shudras*); and also those so lowly as to be excluded from the *varnas* entirely. The very top ranks belong to world-renouncers, whose abstention from trappings of personal status raises them to a higher spiritual plane. Commerce, industry and status rivalries may all thrive, but the wealth, power or prosperity being fought over is always seen as of lesser value – in the great scheme of things – than the purity of priestly caste.

The *varna* system is about as 'unequal' as any social system can possibly be, yet where one ranks within it has less to do with how many material goods one can pile up or lay claim to than with one's relation to certain (polluting) substances – physical dirt and waste, but also bodily matter linked to birth, death and menstruation – and the people who handle them. All this creates serious problems for any contemporary scholar seeking to apply Gini coefficients or any other property-based measure of 'inequality' to the society in question. On the other hand, and despite the great gaps in time between our sources, it might allow us to make sense of some of Mohenjo-daro's otherwise puzzling features, such as the fact that those residential buildings most closely resembling palaces are not located on the Upper Citadel but crammed into the streets of the Lower Town – that bit closer to the mud, sewage pipes and paddy fields, where such jostling for worldly status seems to have properly belonged.[95]

Clearly, we can't just project the social world evoked in Sanskrit literature indiscriminately on to the much earlier Indus civilization. If the first South Asian cities were indeed organized on caste-like principles, then we would immediately have to acknowledge a major difference from the system of ranks described over a millennium later in Sanskrit texts, where second-highest status (just below *brahmins*) is reserved for the warrior caste known as *kshatriyas*. In the Bronze Age Indus valley there is no evidence of anything like a *kshatriya* class of warrior-nobles, nor of the kind of aggrandizing behaviour associated with such groups in later epic tales such as the *Mahabharata* or *Ramayana*. Even the largest cities, like Harappa and Mohenjo-daro, yield no evidence of spectacular sacrifices or feasts, no pictorial narratives of military prowess or celebrations of famous deeds, no sign of tournaments in which anyone vied over titles and treasures, no aristocratic burials. And if such things were going on in the Indus cities at the time, there would be ways to know.

Indus civilization wasn't some kind of commercial or spiritual arcadia; nor was it an entirely peaceful society.[96] But neither does it contain any evidence for charismatic authority figures: war leaders, lawgivers and the like. A small, cloaked sculpture made of yellow limestone from Mohenjo-daro, known in the literature as the 'priest-king', is often presented as such. But, in fact, there's no particular reason to believe the figure really is a priest-king or an authority figure of any sort. It's simply a limestone image of an urbane Bronze Age man with a beard. The fact that past generations of scholars have insisted on referring to him as 'priest-king' is testimony more to their own assumptions about what they think must have been happening in early Asian cities than anything the evidence implies.

Over time, experts have largely come to agree that there's no evidence for priest-kings, warrior nobility, or anything like what we would recognize as a 'state' in the urban civilization of the Indus valley. Can we speak, then, of 'egalitarian cities' here as well, and if so, in what sense? If the Upper Citadel at Mohenjo-daro really was dominated by some sort of ascetic order, literally 'higher' than everyone else, and the area around the citadel by wealthy merchants, then there was a clear hierarchy between groups. Yet this doesn't necessarily mean that the groups themselves were hierarchical in their internal organization, or that ascetics and merchants had a greater say than anyone else when it came to matters of day-to-day governance.

Now, you might at this point be objecting: 'well, yes, technically that may be true, but honestly, what's the chance that they weren't hierarchical, or that the pure or the wealthy did not have greater say in running the city's affairs?' In fact, it seems very difficult for most of us even to imagine how self-conscious egalitarianism on a large scale would work. But this again simply serves to demonstrate how automatically we have come to accept an evolutionary narrative in which authoritarian rule is somehow the natural outcome whenever a large enough group of people are brought together (and, by implication, that something called 'democracy' emerges only much later, as a conceptual breakthrough – and most likely just once, in ancient Greece).

Scholars tend to demand clear and irrefutable evidence for the existence of democratic institutions of any sort in the distant past. It's striking how they never demand comparably rigorous proof for top-down structures of authority. These latter are usually treated as a default mode of history: the kind of social structures you would simply expect to see in the absence of evidence for anything else.[97] We could speculate about where this habit of thought comes from, but it wouldn't help us to decide if the everyday governance of early Indus cities could have proceeded on egalitarian lines, alongside the existence of ascetic social orders. It is more useful, we suggest, to level the interpretive playing field by asking if there are cases of such things happening in later, better-documented periods of South Asian history.

In fact, such cases are not difficult to find. Consider the social milieu from which Buddhist monasteries, or *sangha*, arose. The word *sangha* was actually first used for the popular assemblies that governed many South Asian cities in the Buddha's lifetime – roughly the fifth century BC – and early Buddhist texts insist that the Buddha was himself inspired by the example of these republics, and in particular the importance they accorded to convening full and frequent public assemblies. Early Buddhist *sanghas* were meticulous in their demands for all monks to gather together in order to reach unanimous decisions on matters of general concern, resorting to majority vote only when consensus broke down.[98] All this remains true of *sangha* to this day. Over the course of time, Buddhist monasteries have varied a great deal in governance – many have been extremely hierarchical in practice. But the important thing here is that even 2,000 years ago it was not considered in any way unusual for members of ascetic orders to make decisions in much the same way as, for example, contemporary anti-authoritarian activists do in Europe or Latin America (by consensus process, with a fallback on majority vote); that these forms of governance were based on an ideal of equality; and that there were entire cities governed in what was seen to be exactly the same way.[99]

We might go further still and ask: are there any known examples of societies with formal caste hierarchies, in which practical governance nonetheless takes place on egalitarian lines? It may seem paradoxical but the answer, again, is yes: there is plenty of evidence for such arrangements, some of which continue to this day. Perhaps best documented is the *seka* system on the island of Bali, whose population adopted Hinduism in the Middle Ages. Balinese are not only divided by caste: their society is conceived as a total hierarchy in which not just every group but every individual knows (or at least, should know) their exact position in relation to everyone else. In principle, then, there are no equals, and most Balinese would argue that in the greater cosmic scheme of things, this must always be so.

At the same time, however, practical affairs such as the management of communities, temples and agricultural life are organized according to the *seka* system, in which everyone is expected to participate on equal terms and come to decisions by consensus. For instance, if a neighbourhood association meets to discuss repairing the roofs of public buildings, or what to serve for food during an upcoming dance contest, those who consider themselves particularly high and mighty, offended by the prospect of having to sit in a circle on the ground with lowly neighbours, may choose not to attend; but in that case they are obliged to pay fines for non-attendance – fines which are then used to pay for the feast or the repairs.[100] We currently have no way of knowing if such a system prevailed in the Indus valley over 4,000 years ago. The example merely serves to underscore that there is no necessary correspondence between overarching concepts of social hierarchy and the practical mechanics of local governance.

The same is, incidentally, true of kingdoms and empires. One very common theory held that these tended to first appear in river valleys, because agriculture there involved the maintenance of complex irrigation systems, which in turn required some form of administrative co-ordination and control. Bali again provides the perfect counter-example. For most of its history Bali was divided into a series of kingdoms, endlessly squabbling over this or that. It is also famous as a rather small volcanic island which manages to support one of the densest populations on earth by a complex system of irrigated wet-rice agriculture. Yet the kingdoms seem to have had no role whatsoever in the management of the irrigation system. This was governed by a series of 'water-temples', through which the distribution of water was managed by an even more complex system of consensual decision-making, according to egalitarian principles, by the farmers themselves.[101]

CONCERNING AN APPARENT CASE OF 'URBAN REVOLUTION' IN CHINESE PREHISTORY

So far in this chapter we've looked at what happened when cities first appeared in three distinct parts of Eurasia. In each case, we noted the absence of monarchs or any evidence of a warrior elite, and the corresponding likelihood that each had instead developed institutions of communal self-governance. Within those broad parameters, each regional tradition was very different. Contrasts between the expansion of Uruk and the Ukrainian mega-sites illustrate this point with particular clarity. Both appear to have developed an ethos of explicit egalitarianism – but it took strikingly different forms in each.

It is possible to express these differences at a purely formal level. A self-conscious ethos of egalitarianism, at any point in history, might take either of two diametrically opposing forms. We can insist that everyone is, or should be, precisely the same (at least in the ways that we consider important); or alternatively, we can insist that everyone is so utterly different from each other that there are simply no criteria for comparison (for example, we are all unique individuals, and so there is no basis upon which any one of us can be considered better than another). Real-life egalitarianism will normally tend to involve a bit of both.

Yet it could be argued that Mesopotamia – with its standardized household products, allocation of uniform payments to temple employees, and public assemblies – seems to have largely embraced the first version. Ukrainian mega-sites, in which each household seems to have developed its own unique artistic style and, presumably, idiosyncratic domestic rituals, embraced the second.[102] The Indus valley appears – if our interpretation is broadly correct – to represent yet a third possibility, where rigorous equality in certain areas (even the bricks were all precisely the same size) was complemented by explicit hierarchy in others.

It's important to stress that we are not arguing that the very first cities to appear in any region of the world were invariably founded on egalitarian principles (in fact, we will shortly see a perfect counter-example). What we are saying is that archaeological evidence shows this to have been a surprisingly common pattern, which goes against conventional evolutionary assumptions about the effects of scale on human society. In each of the cases we've considered so far – Ukrainian mega-sites, Uruk Mesopotamia, the Indus valley – a dramatic increase in the scale of organized human settlement took place with no resulting concentration of wealth or power in the hands of ruling elites. In short, archaeological research has shifted the burden of proof on to those theorists who claim causal connections between the origins of cities and the rise of stratified states, and whose claims now look increasingly hollow.

So far we've been providing what are effectively a series of snapshot views of cities that, in most cases, were occupied for centuries. It seems unlikely that they did not have their own share of upheavals, transformations and constitutional crises. In some cases we can be certain they did. At Mohenjo-daro, for instance, we know that roughly 200 years before the city's demise, the Great Bath had already fallen into disrepair. Industrial facilities and ordinary residences crept beyond the Lower Town, on to the Upper Citadel, and even the site of the Bath itself. Within the Lower Town, we now find buildings of truly palatial dimensions with attached craft workshops.[103] This 'other' Mohenjo-daro existed for generations, and seems to represent a self-conscious project of transforming the city's (by then centuries-old) hierarchy into something else – though archaeologists have yet to fathom quite what that other thing was supposed to be.

Like the Ukrainian cities, those of the Indus were eventually abandoned entirely, to be replaced by societies of much smaller scale where heroic aristocrats held sway. In Mesopotamian cities palaces eventually appear. Overall, one might be forgiven for thinking that history was progressing uniformly in an authoritarian direction. And in the very long run it was; at least, by the time we have written histories, lords and kings and would-be world emperors have popped up almost everywhere (though civic institutions and independent cities never entirely go away).[104] Still, rushing to this conclusion would be unwise. Dramatic reversals have sometimes taken place in the other direction – for instance in China.

In China, archaeology has opened a yawning chasm between the birth of cities and the appearance of the earliest named royal dynasty, the Shang. Since the early twentieth century discovery of inscribed oracle bones at Anyang in the north-central province of Henan, political history in China has started with the Shang rulers, who came to power around 1200 BC.[105] Until quite recently, Shang civilization was thought to be a fusion of earlier urban ('Erligang' and 'Erlitou') and aristocratic or 'nomadic' elements, the latter taking the form of bronze casting techniques, new types of weaponry, and horse-drawn chariots first developed on the Inner Asian steppe, home to a series of powerful and highly mobile societies who played so much havoc with later Chinese history.[106]

Before the Shang, nothing particularly interesting was supposed to have happened – just a few decades ago, textbooks on early China simply presented a long series of 'Neolithic' cultures receding into the distant past, defined by technological trends in farming and stylistic changes in regional traditions of pottery and the design of ritual jades. The underlying assumption was that these were pretty much the same as Neolithic farmers were imagined to be anywhere else: living in villages, developing embryonic forms of social inequality, preparing the way for the sudden leap that would bring the rise of cities and, with cities, the first dynastic states and empires. But we now know this is not what happened at all.

Today, archaeologists in China speak of a 'Late Neolithic' or 'Longshan' period marked by what can be described, without equivocation, as cities. Already by 2600 BC we find a spread of settlements surrounded by rammed earth walls across the entire valley of the Yellow River, from the coastal margins of Shandong to the mountains of southern Shanxi. They range in size from centres of more than 300 hectares to tiny principalities, little more than villages but still fortified.[107] The major demographic hubs lay far away, on the lower reaches of the Yellow River to the east; also to the west of Henan, in the Fen River valley of Shanxi province; and in the Liangzhu culture of southern Jiangsu and northern Zhejiang.[108]

Many of the largest Neolithic cities contain cemeteries, where individual burials hold tens or even hundreds of carved ritual jades. These may be badges of office, or perhaps a form of ritual currency: in ancestral rites, the stacking and combination of such jades, often in great number, allowed differences of rank to be measured along a common scale of value, spanning the living and the dead. Accommodating such finds in the annals of written Chinese history proved an uncomfortable task, since we are speaking of a long and apparently tumultuous epoch that just wasn't supposed to have happened.[109]

The problem is not merely one of time, but also of space. Astonishingly, some of the most striking 'Neolithic' leaps towards urban life are now known to have taken place in the far north, on the frontier with Mongolia. From the perspective of later Chinese empires (and the historians who described them), these regions were already halfway to 'nomad-barbarian' and would eventually end up beyond the Great Wall. Nobody expected archaeologists to find there, of all places, a 4,000-year-old city, extending over 400 hectares, with a great stone wall enclosing palaces and a step-pyramid, lording it over a subservient rural hinterland nearly 1,000 years pre-Shang.

The excavations at Shimao, on the Tuwei River, have revealed all this, along with abundant evidence for sophisticated crafts – including bone-working and bronze-casting – and warfare, including the mass killing and burial of captives, in around 2000 BC.[110] Here we sense a much livelier political scene than was ever imagined in the annals of later courtly tradition. Some of it had a grisly aspect, including the decapitation of captured foes, and the burial of some thousands of ancestral jade axes and sceptres in cracks between great stone blocks of the city wall, not to be found or seen again until the prying eyes of archaeologists uncovered them over four millennia later. The likely intention of all this was to disrupt, demoralize and delegitimize rival lineages ('all in all, you're just another jade in the wall').

At the site of Taosi – contemporary with Shimao, but located far to the south in the Jinnan basin – we find a rather different story. Between 2300 and 1800 BC, Taosi went through three phases of expansion. First, a fortified town of sixty hectares arose on the ruins of a village, expanding subsequently to a city of 300 hectares. In these early and middle periods, Taosi presents evidence for social stratification almost as dramatic as what we see at Shimao, or indeed what we might expect of a later imperial Chinese capital. There were massive enclosure walls, road systems and large, protected storage areas; also rigid segregation between commoner and elite quarters, with craft workshops and a calendrical monument clustered around what was most likely some sort of palace.

Burials in the early town cemetery of Taosi fell into clearly distinct social classes. Commoner tombs were modest; elite tombs were full of hundreds of lacquered vessels, ceremonial jade axes and remains of extravagant pork feasts. Then suddenly, around 2000 BC, everything seems to change. As the excavator describes it:

> The city wall was razed flat, and . . . the original functional divisions destroyed, resulting in a lack of spatial regulation. Commoners' residential areas now covered almost the entire site, even reaching beyond the boundaries of the middle-period large city wall. The size of the city became even larger, reaching a total area of 300 hectares. In addition,

> the ritual area in the south was abandoned. The former palace area now included a poor-quality rammed-earth foundation of about 2,000 square metres, surrounded by trash pits used by relatively low-status people. Stone tool workshops occupied what had been the lower-level elite residential area. The city clearly had lost its status as a capital, and was in a state of anarchy.[111]

What's more, there are clues that this was a conscious process of transformation, most likely involving a significant degree of violence. Commoner graves burst in on the elite cemetery, and in the palace district a mass burial, with signs of torture and grotesque violations of the corpses, appears to be evidence for what the excavator describes as an 'act of political retribution'.[112]

Now, it is considered bad form to question an excavator's first-hand judgement about a site, but we cannot resist a couple of observations. First, the ostensible 'state of anarchy' (elsewhere described as 'collapse and chaos')[113] lasted for a considerable period of time, between two and three centuries. Second, the overall size of Taosi during the latter period actually grew from 280 to 300 hectares. This sounds a lot less like collapse than an age of widespread prosperity, following the abolition of a rigid class system. It suggests that after the destruction of the palace, people did not fall into a Hobbesian 'war of all against all' but simply got on with their lives – presumably under what they considered a more equitable system of local self-governance.

Here, on the banks of the Fen River, we might conceivably be in the presence of evidence for the world's first documented social revolution, or at least the first in an urban setting. Other interpretations are no doubt possible. But at the very least, the case of Taosi invites us to consider the world's earliest cities as places of self-conscious social experimentation, where very different visions of what a city could be like might clash – sometimes peacefully, sometimes erupting in bursts of extraordinary violence. Increasing the number of people living in one place may vastly increase the range of social possibilities, but in no sense does it predetermine which of those possibilities will ultimately be realized.

As we'll see in the next chapter, the history of central Mexico suggests that the kinds of revolution we've been talking about – urban revolutions of the political kind – may well be a lot more common in human history than we tend to think. Again, we may never be able fully to reconstruct the unwritten constitutions of the earliest cities to appear in various parts of the world, or the reforms undergone in their first centuries, but we can no longer doubt that these existed.

9
Hiding in Plain Sight

The indigenous origins of social housing and democracy in the Americas

Sometime around AD 1150, a people called the Mexica migrated south from a place called Aztlán – its location is now unknown – to take up a new home in the heart of the Valley of Mexico, which now bears their name.[1] There they were eventually to carve out an empire, the Aztec Triple Alliance,[2] and build its capital at Tenochtitlan, an island-city in Lake Texcoco – one link in a chain of great lakes and lake-cities, and part of an urban landscape ringed by mountains. Lacking an urban tradition of their own, the Mexica modelled the layout of Tenochtitlan on that of another city they found, lying in ruins and virtually abandoned, in a valley about one day's journey distant. They called that other city Teotihuacan, the 'Place of Gods'.

It had been some time since anyone lived in Teotihuacan. By the twelfth century, when the Mexica arrived, nobody even seems to have remembered the city's original name. Still, the new arrivals clearly found the city – with its two colossal pyramids set against the Cerro Gordo – both alien and alluring, and far too large simply to ignore. Their response, aside from using it as a model for their own great city, was to veil Teotihuacan in myth, and cage its standing remains in a dense forest of names and symbols. As a result, we still see Teotihuacan largely through Aztec (Culhua-Mexica) eyes.[3]

Written references to Teotihuacan from the time it was still inhabited comprise a few tantalizing inscriptions from far to the east in the Maya lowlands, which call it 'the place of cattail reeds', corresponding to the Nahuatl word 'Tollan' and evoking a primordial, perfect city by the water.[4] Otherwise all we have are sixteenth-century transcriptions of chronicles, set down in Spanish and Nahuatl, which describe Teotihuacan as a place full of mountain pools and primal voids, from which the planets sprang at the beginning of time. The planets were followed by gods, and the gods by a mysterious race of fish-men, whose world had to be destroyed to make way for our own.

In historical terms, such sources are not very useful, especially since we have no way of knowing if these myths were ever told in the city when it was actually inhabited, or whether they were just invented by the Aztecs. Still, the legacy of those stories continues. It was the Aztecs, for instance, who made up the names 'Pyramid of the Sun', 'Pyramid of the Moon' and 'Way of the Dead', which archaeologists and tourists alike use to this day when describing the city's most visible monuments and the road that links them all.[5]

For all their facility with astronomical calculation, the builders of Tenochtitlan either didn't know or didn't find it important to know when, precisely, Teotihuacan had last been inhabited. Here, at least, archaeology has been able to fill in the gaps. We know now that the city of Teotihuacan had its heyday eight centuries before the coming of the Mexica, and more than 1,000 years before the arrival of the Spanish. Its foundation dates to around 100 BC, and its decline to around AD 600. We also know that, in the course of those centuries, Teotihuacan became a city of such grandeur and sophistication that it could easily be put on a par with Rome at the height of its imperial power.

We don't actually know if Teotihuacan was, like Rome, the centre of a great empire, but even conservative estimates place its population at around 100,000[6] (perhaps as much as five times the likely population of Mohenjo-daro, Uruk or any of the other early Eurasian cities we discussed in the last chapter). At its zenith, there were probably at least a million people distributed across the Valley of Mexico and surrounding lands, many of whom had only visited the great city once, or perhaps only knew someone who had, but nonetheless considered Teotihuacan the most important place in the entire world.

This much is broadly accepted by virtually every scholar and historian of ancient Mexico. More controversial is the question of what sort of city Teotihuacan was, and how it was governed. Pose this question to a specialist in the study of Mesoamerican history or archaeology (as we often have done), and you'll likely get the same reaction: a roll of the eyes and a resigned acknowledgement that there's just something 'weird' about the place. Not merely because of its exceptional size, but because of its stubborn refusal to conform to expectations of how an early Mesoamerican city *should* have functioned.

At this point, the reader can probably guess what's coming. All the evidence suggests that Teotihuacan had, at its height of its power, found a way to govern itself without overlords – as did the much earlier cities of prehistoric Ukraine, Uruk-period Mesopotamia and Bronze Age Pakistan. Yet it did so with a very different technological foundation, and on an even larger scale.

But first some background.

As we've seen, when kings appear in the historical record, they tend to leave unmistakeable traces. We can expect to find palaces, rich burials and monuments celebrating their conquests. All this is true in Mesoamerica as well.

In the wider region, the paradigm is set by a series of dynastic polities, located far from the Valley of Mexico in the Yucatán Peninsula and adjacent highlands. Today's historians know these polities as the Classic Maya (*c.*AD 150–900 – the term 'classic' is also applied to their ancient written language and to the chronological period in question). Cities like Tikal, Calakmul or Palenque were dominated by royal temples, ball-courts (settings for competitive, sometimes lethal games), images of war and humiliated captives (often publicly killed after ball games), complex calendrical rituals celebrating royal ancestors, and records of the deeds and biographies of living kings. In the modern imagination this has become the 'standard package' of Mesoamerican kingship, associated with ancient cities throughout the region from Monte Alban (in Oaxaca, *c.*AD 500–800) to Tula (in central Mexico, *c.*AD 850–1150), and arguably reaching as far north as Cahokia (near what's now East St Louis, *c.*AD 800–1200).

In Teotihuacan, all this seems to have been strikingly absent. Unlike in the Mayan cities, there are few written inscriptions in general.[7] (For this reason, we don't know what language was spoken by the majority of Teotihuacan's inhabitants, although we know the city was sufficiently cosmopolitan to include among its population both Maya and Zapotec minorities familiar with the use of writing.)[8] However, there remains plenty of pictorial art. Teotihuacan's citizens were prolific craft specialists and makers of images, leaving behind everything from monumental stone sculptures to diminutive terracotta figures that could be held in the palm, as well as vivid wall paintings bustling with human activity (picture something like the carnivalesque feel of a Bruegel street scene and you are not too far off). Still, nowhere among some thousands of such images do we find even a single representation of a ruler striking, binding or otherwise dominating a subordinate – unlike in the contemporary arts of the Maya and Zapotec, where this is a constant theme. Today scholars pore over Teotihuacan imagery, searching for anything that might be construed as a kingly figure, but largely they fail. In many cases the artists seem to have deliberately frustrated such efforts, for instance by making all the figures in a given scene exactly the same size.

Another key element of royal display in the ancient kingdoms of Mesoamerica, the ceremonial ball-court is also conspicuous by its absence at Teotihuacan.[9] Neither has there been found any equivalent to the great tombs of Sihyaj Chan K'awiil at Tikal or K'inich Janaab Pakal in Palenque. And not for lack of trying. Archaeologists have combed through the ancient tunnels around the Pyramids of the Sun and the Moon and under the Temple of the Feathered Serpent, only to discover that the passages do not lead to royal tombs, or even robbed-out tomb chambers, but to chthonic labyrinths and mineral-crusted shrines: evocations of other worlds, no doubt, but not the graves of sacred rulers.[10]

Some have suggested that the self-conscious rejection of outside convention at Teotihuacan runs even deeper. For instance, the city's artists appear to have been aware of formal and compositional principles found among their Mesoamerican neighbours, and to have set about deliberately inverting them. Where Maya and Zapotec art draws on a tradition of relief carving derived from the earlier Olmec kings of Veracruz, favouring curves and flowing forms, the sculpture of Teotihuacan shows humans and humanoid figures as flat composites, tightly fitted to angular blocks. Some decades ago, these contrasts led Esther Pasztory – a Hungarian-American art historian who spent much of her career studying Teotihuacan's art and imagery – to a radical conclusion. What we have, she argued, with highland Teotihuacan and the lowland Maya, is nothing less than a case of conscious cultural inversion – or what we've been calling schismogenesis – but this time on the scale of urban civilizations.[11]

Teotihuacan, in Pasztory's view, created a new tradition of art to express the ways in which its society was different from that of its contemporaries elsewhere in Mesoamerica. In doing so it rejected both the specific visual trope of ruler and captive and the glorification of aristocratic individuals in general. In this it was strikingly different from both the earlier cultural tradition of the Olmec, and from contemporary Maya polities. If the visual arts of Teotihuacan celebrated anything, Pasztory insisted, then it was the community as a whole and its collective values, which – over a period of some centuries – successfully prevented the emergence of 'dynastic personality cults'.[12]

According to Pasztory, Teotihuacan was not just 'anti-dynastic' in spirit, it was itself a utopian experiment in urban life. Those who created it thought of themselves as creating a new and different kind of city, a Tollan for the people, without overlords or kings. Following in Pasztory's footsteps, other scholars, eliminating virtually every other possibility, arrived at similar conclusions. In its early years, they concluded, Teotihuacan had gone some way down the road to authoritarian rule, but then around AD 300 suddenly reversed course: possibly there was a revolution of sorts, followed by a more equal distribution of the city's resources and the establishment of a kind of 'collective governance'.[13]

The general consensus among those who know the site best is that Teotihuacan was, in fact, a city organized along some sort of self-consciously egalitarian lines. And, as we've seen, in world-historical terms all this is not nearly as weird or anomalous as scholars – or anyone else, for that matter – tend to assume. It is equally true if we simply try to understand Teotihuacan within its Mesoamerican context. The city didn't come out of nowhere. While there might be a recognizable 'package' of Mesoamerican kingship, there also appears to have been a very different, dare we say republican, tradition as well.

What we propose to do in this chapter, then, is bring to the surface this neglected strand of Mesoamerican social history: one of urban republics, large-scale projects of social welfare, and indigenous forms of democracy that can be followed down to the time of the Spanish conquest and beyond.

IN WHICH WE FIRST CONSIDER AN EXAMPLE OF STRANGER-KINGS IN THE MAYA LOWLANDS, AND THEIR AFFILIATION WITH TEOTIHUACAN

Let us start by leaving behind the city itself, and the valleys and plateaus of central Mexico, for the tropical forest kingdoms of the Classic Maya, whose ruins lie to the east: in Mexico's Yucatán Peninsula, and within the modern countries of Guatemala, Belize, Honduras and El Salvador. In the fifth century AD, something remarkable happens in the art and writing of some of these Maya city-states, including the largest and most prominent among them, Tikal.

Finely carved scenes on Mayan monuments of this period show figures seated on thrones, and wearing what can be instantly recognized as foreign, Teotihuacan-style dress and weaponry (the spear-throwers called *atlatls*, feathered shields, and so on), clearly distinct from the garb and finery of local rulers. Archaeologists working in western Honduras, near the border with Guatemala, have even unearthed what, judging by the grave goods, appear to be the actual burials of these stranger-kings at the base level of a temple at the site of Copán, which went through seven further phases of construction. Here, glyphic inscriptions describe at least some of these individuals as actually coming from the Land of the Cattail Reeds.[14]

Two things (at least) are very hard to explain here. First, why are there images of what appear to be Teotihuacano lords on thrones in Tikal, when there are no similar images of lords sitting on thrones at Teotihuacan itself? Second, how could Teotihuacan ever have mounted a successful military expedition against a kingdom over 600 miles away? Most experts assume the latter was simply impossible on logistical grounds, and they are probably correct to do so (although we should keep an open mind; after all, who could have predicted on logistical grounds that a motley crew of Spaniards would bring down a Mesoamerican empire of many millions?). The first question certainly requires more careful consideration. Were the individuals depicted as seated kings really from central Mexico at all?

It's possible we are just dealing here with local lords who had a taste for exotica. We know from art and inscriptions that Maya grandees sometimes enjoyed dressing up in Teotihuacan warrior gear, sometimes beheld visions of Teotihuacan spirits after ritual bloodletting, and generally liked to style themselves 'Lords and Ladies of the West'. The city was certainly far enough away for the Maya to see it as a place of exotic fantasies, some kind of distant Shangri-La. But there are reasons to suspect it was more than just that. For one thing, people did regularly move back and forth. Obsidian from Teotihuacan adorned the Maya gods, and Teotihuacan's deities wore green quetzal feathers from the Maya lowlands. Mercenaries and traders went both ways, pilgrimages and diplomatic visits followed; immigrants from Teotihuacan built temples in Maya cities, and there was even a Maya neighbourhood, replete with murals, at Teotihuacan itself.[15]

How do we resolve the puzzle of this Mayan depiction of Teotihuacan kings? Well, first of all, if history teaches us anything about long-distance trade routes, it's that they are likely to be full of unscrupulous characters of various sorts: bandits, runaways, grifters, smugglers, religious visionaries, spies – or figures who may be any combination of these at a given time. This was no less true in Mesoamerica than anywhere else. The Aztecs, for instance, employed orders of heavily armed warrior-merchants called *pochteca*, who also gathered intelligence on the cities where they traded.

History is also full of stories of adventurous travellers who either find themselves taken into some alien society and miraculously transformed there into kings or embodiments of sacred power: 'stranger-kings' like Captain James Cook, who – on casting anchor in Hawaii in 1779 – was accorded the status of an ancient Polynesian fertility god called Lono; or others who, like Hernán Cortés, did their best to convince local people that they should be welcomed as such.[16] Worldwide, a remarkably large percentage of dynastic histories begin precisely this way, with a man (it's almost always a man) who mysteriously appears from somewhere far away. It is easy to see how an adventurous traveller from a famous city might have taken advantage of such notions. Could something like this have happened in the Maya lowlands in the fifth century AD?

From inscriptions at Tikal, we do know the names of some of these particular stranger-kings and their close associates, or at least the names they adopted as Maya nobles. One, called Sihyaj K'ahk' ('Born of Fire'), seems never himself to have ruled but helped install a series of Teotihuacano 'princes' on Maya thrones, including the throne of Tikal. We also know that these princes married local women of high rank, and that their offspring became Maya rulers, who also celebrated their ancestral connection to Teotihuacan: the 'Tollan of the West'.

From examination of burials at Copán we also know that, before their elevation to royal status, at least some of these adventurous individuals led extremely colourful lives, fighting and travelling and fighting again, and that they may not originally have come either from Copán or from Teotihuacan but somewhere else entirely.[17] Taking all lines of evidence into account, it seems likely that these progenitors of Maya dynasties were originally members of groups that specialized in long-distance travel – traders, soldiers of fortune, missionaries or perhaps even spies – who, perhaps quite suddenly, found themselves elevated to royalty.[18]

There is a remarkable analogy for this process closer to our times. Many centuries later, when the focus of Maya culture – and most of its largest cities – had shifted to Yucatán in the north, a similar wave of central Mexican influence occurred, most dramatically evident in the city of Chichén Itzá, whose Temple of the Warriors seems to be directly modelled on the Toltec capital of Tula (a later Tollan). Again, we don't really know what happened, but later chronicles, written secretly under Spanish rule, described the Itzá in almost exactly these terms: as a band of uprooted warriors, 'stuttering foreigners' from the west, who managed to seize control of a series of cities in Yucatán and ended up in a prolonged rivalry with another dynasty of Toltec exiles – or at least, exiles who insisted they were originally Toltec – called the Xiu.[19] These chronicles are full of accounts of the exiles' wanderings in the wilderness, temporary periods of glory, accusations of oppression, and sombre prophecies of future tribulation. Once again, we seem to be dealing with a feeling among the Maya that kings really *should* come from somewhere far away, and with the willingness of at least a few unscrupulous foreigners to take advantage of this idea.

All this is only guesswork. Still, it's clear the images and records from places like Tikal tell us more about Maya concepts of royal power than they do about Teotihuacan itself, where not a shred of compelling evidence for the institution of kingship has yet been found. The 'Mexican' princes of the Maya lowlands, bedecked in regalia and seated on thrones, were engaging in exactly the sort of grandiose political gestures that had no place in their putative homeland. If not a monarchy, then, what was Teotihuacan? There is, we suggest, no one answer to this question – and over a period of five centuries there is no particular reason why there ought to be.

Let's look at a central portion of the standard architectural plan of Teotihuacan, pieced together from the most exhaustive survey of an urban landscape ever undertaken by archaeologists.[20] Having gone to the lengths of recording a built environment on that scale – all eight square miles of it – archaeologists naturally want to see it all at once, in a single gasp. Modern archaeology often presents to us something like the chronologically collapsed plan of Mohenjo-daro and other 'first cities' with centuries or even millennia of urban history folded into a single map. It's visually stunning, but actually quite flat and artificial. In the case of Teotihuacan, it gives an effect at once harmonious and misleading.

At the centre, anchoring the whole mirage, stand the great monuments – the two Pyramids and the Ciudadela (Citadel) containing the Temple of the Feathered Serpent. Extending for miles around are smaller but still impressively appointed residences that housed the city's population: some 2,000 multi-family apartments, finely built from stone masonry and organized on a tidy orthogonal grid, aligned to the ceremonial centre of the city. It is an almost perfectly functional image of civic prosperity and hierarchy. We are, it would seem, in the presence of something like More's *Utopia* or Campanella's *City of the Sun*. But there is a problem. The residences and pyramids do not strictly belong together, or at least not all of them. Their construction occupies different phases of time. Nor is the temple quite what it seems.

In fact, in historical terms it is all something of a grand illusion. To understand what's going on here we have to make some attempt, however tentative, to reconstruct a basic chronological sequence for the city's development.

HOW THE PEOPLE OF TEOTIHUACAN TURNED THEIR BACKS ON MONUMENT-BUILDING AND HUMAN SACRIFICE, AND INSTEAD EMBARKED ON A REMARKABLE PROJECT OF SOCIAL HOUSING

Teotihuacan's growth to urban dimensions began around the year 0. At that time, whole populations were on the move across the Basin of Mexico and Valley of Puebla, fleeing the effects of seismic activity on their southern frontiers, which included a Plinian eruption of the volcano Popocatépetl. From AD 50 to 150, the flow of people into Teotihuacan siphoned life from surrounding areas. Villages and towns were abandoned, and also whole cities, like Cuicuilco, with its early traditions of pyramid-building. Under several feet of ash lie the ruins of other abandoned settlements. At the Pueblan site of Tetimpa, just eight miles from Popocatépetl, archaeologists have unearthed houses that foreshadow – on a smaller scale – the civic architecture of Teotihuacan.[21]

Here the later chronicles do provide some useful, or at least thought-provoking accounts. Folk memories of a mass exodus survived right up to the time of the *conquista*. One tradition, preserved in the work of the Franciscan friar Bernardino de Sahagún, tells how Teotihuacan was founded by a coalition of elders, priests and wise men from other settlements. As the city grew it incorporated these smaller traditions, maize gods and village ancestors rubbing shoulders with urban deities of fire and rain.

What we can refer to as Teotihuacan's 'Old City' was organized on a parish system, with local shrines serving particular neighbourhoods. The layout of these district temples – three buildings around a central plaza – also follows the plan of earlier structures at Tetimpa, which

Teotihuacan: residential apartments surrounding major monuments in the central districts

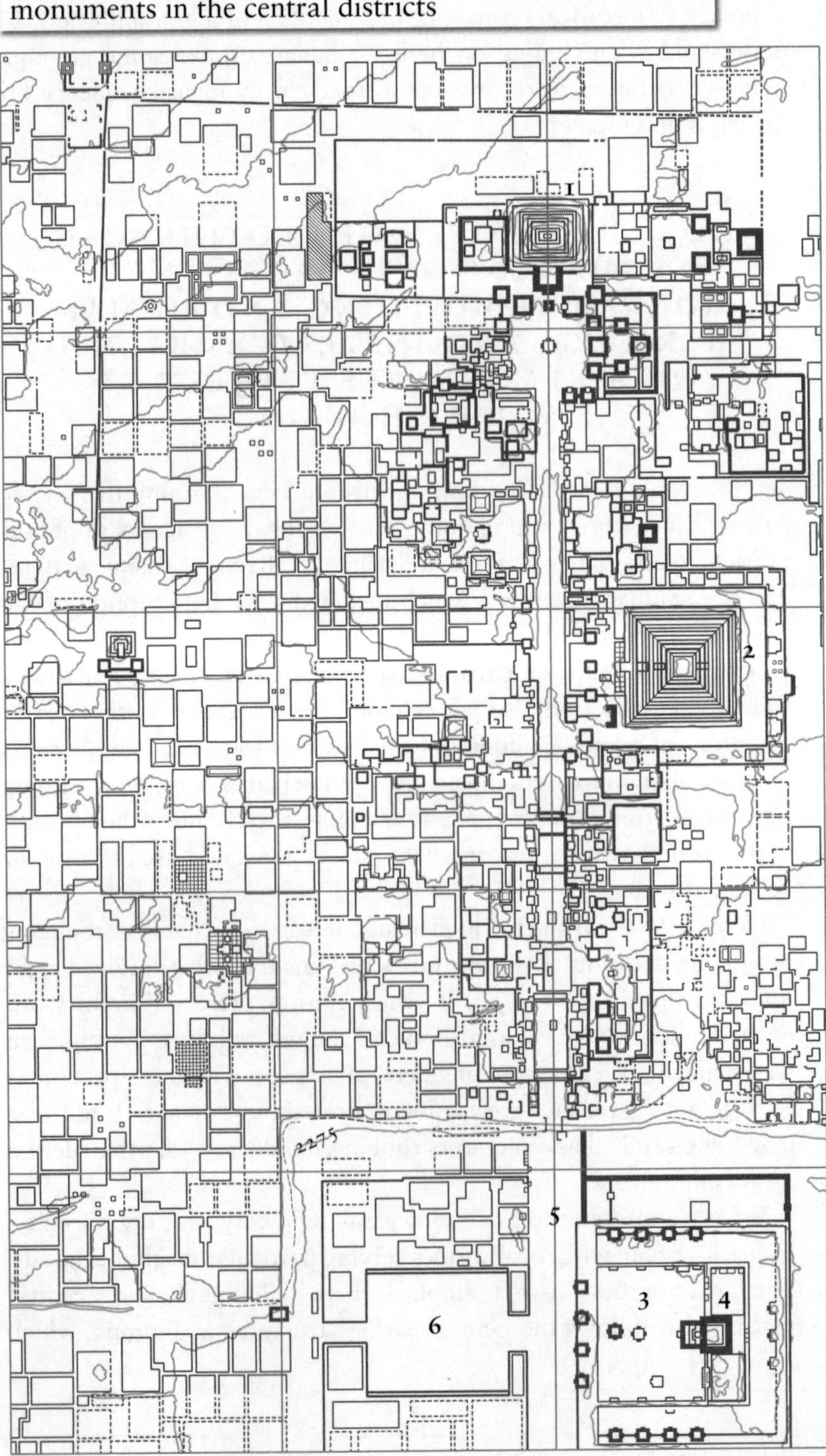

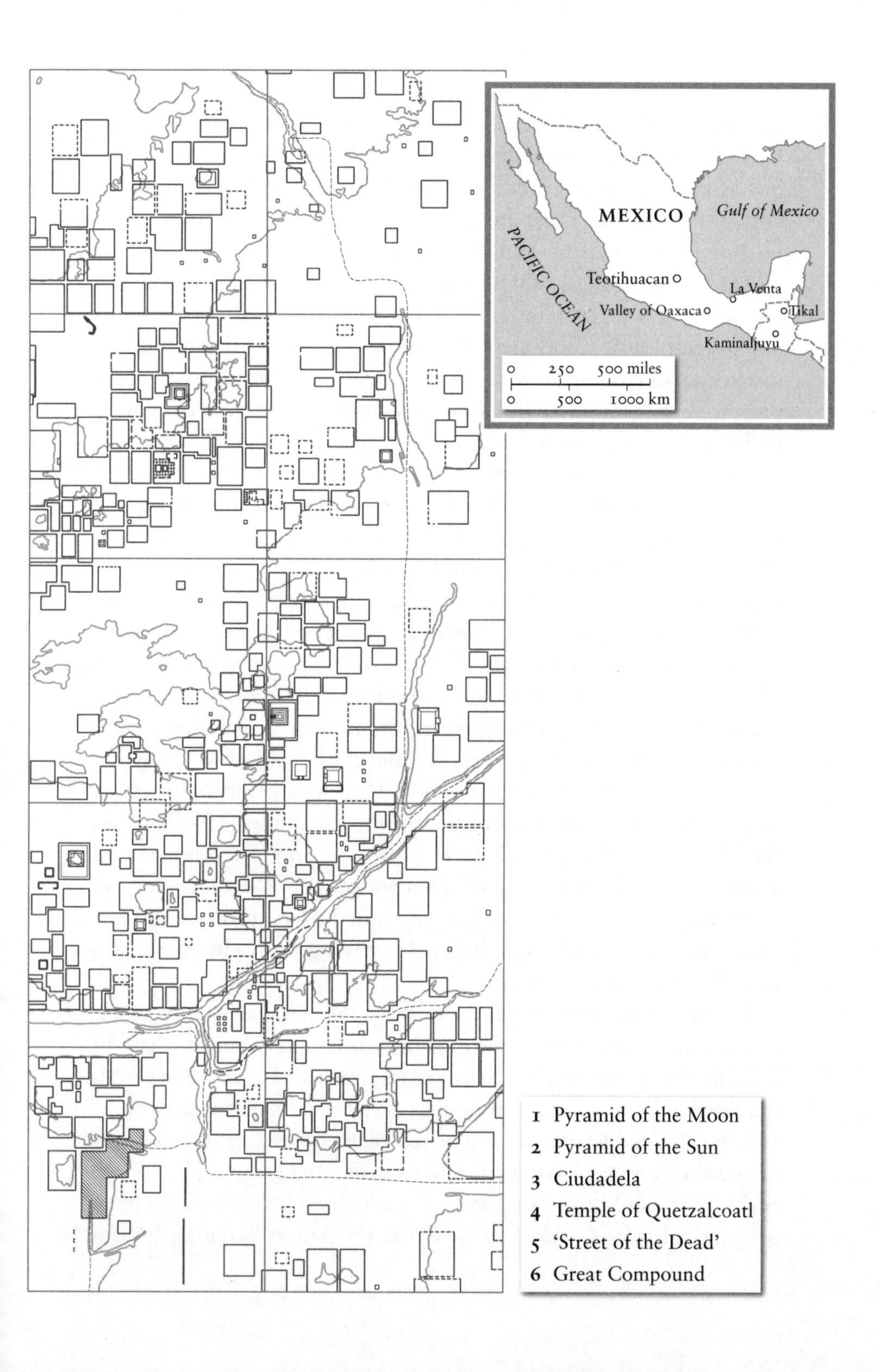
MEXICO
Gulf of Mexico
PACIFIC OCEAN
Teotihuacan
La Venta
Valley of Oaxaca
Tikal
Kaminaljuyu
0 250 500 miles
0 500 1000 km
1 Pyramid of the Moon
2 Pyramid of the Sun
3 Ciudadela
4 Temple of Quetzalcoatl
5 'Street of the Dead'
6 Great Compound

housed the cults of village ancestors.[22] In these early days, from AD 100 to 200, the residential quarters of Teotihuacan may well have looked like an enormous shanty town – but we don't really know,[23] just as we have no clear idea how the fledgling city divided access to arable land and other resources among its citizens. Maize was widely farmed, to be eaten by humans and domestic animals. People kept and ate turkeys, dogs, rabbits and hares. They also grew beans, and enjoyed access to whitetail deer and peccaries, as well as wild fruits and vegetables. Seafood arrived from the distant coast, presumably smoked or salted; but how far the various sectors of the urban economy were integrated at this time, and how exactly resources were pulled in from a wider hinterland, is altogether unclear.[24]

What we can say is that the Teotihuacanos' efforts to create a civic identity focused initially on the building of monuments: the raising of a sacred city in the midst of the wider urban sprawl.[25] This meant the creation of an entirely new landscape in the centre of Teotihuacan, requiring the work of some thousands of labourers. Pyramid-mountains and artificial rivers went up, providing a stage for the performance of calendrical rituals. In a colossal feat of civil engineering, the channels of the Rio San Juan and Rio San Lorenzo were diverted, tying them to the city's orthogonal grid and transforming their marshy banks into solid foundations (all this, recall, without the benefit of working animals or metal tools). This in turn laid the basis for a grand architectural programme which saw the erection of the Pyramids of the Sun and the Moon and the Temple of the Feathered Serpent. The temple faced a sunken plaza that captured the floodwaters of the San Juan to form a seasonal lake, its waters lapping at painted carvings of plumed serpents and shells on the temple façade, making them glisten as rains began to fall in late spring.[26]

All that effort of monumental construction required sacrifices, not just of labour and resources but of human life. Each major phase of building is associated with archaeological evidence of ritual killing. Adding together human remains from the two pyramids and the temple, the victims can be counted in the hundreds. Their bodies were placed in pits or trenches arranged symmetrically to define the ground plan of the edifice that would rise over them. At the corners of the Sun Pyramid, offerings of infants were found; under the Moon Pyramid, foreign captives, some decapitated or otherwise mutilated; and in the foundations of the Temple of the Feathered Serpent lay the corpses of male warriors, arms tied back at point of death, buried with the tools and trophies of their former trade. Among the bodies were found obsidian knives and spearheads, trinkets of shell and greenstone and collars made of human teeth and jawbones (some, as it turns out, cunningly faked in shell).[27]

You would think that at this point – around AD 200 – the fate of Teotihuacan was sealed: its destiny to join the ranks of 'classic' Mesoamerican civilizations with their strong traditions of warrior aristocracy and city-states governed by hereditary nobles. What we might then expect to see next, in the archaeological record, is a concentration of power around the city's focal monuments: the rise of luxurious palaces, inhabited by rulers who were the font of wealth and privilege, with attached quarters for elite kinsmen; and the development of monumental art to glorify their military conquests, the lucrative tribute it generated and their services to the gods. But the evidence tells a very different story, because the citizens of Teotihuacan chose a different path.

In fact, the entire trajectory of Teotihuacan's political development seems to have gone off on a remarkable tangent. Instead of building palaces and elite quarters, the citizens embarked on a remarkable project of urban renewal, supplying high-quality apartments for nearly all the city's population, regardless of wealth or status.[28] Without written sources, we can't really say why. Archaeologists are not yet able to distinguish the precise sequence of events with any confidence. But nobody doubts that something did happen, and what we will try to do now is sketch out what it was.

The big turnaround in Teotihuacan's fortunes seems to have begun around AD 300. At that time, or shortly after, the Temple of the Feathered Serpent was desecrated and its stores of offerings looted. Not only was it set on fire; many of the gargoyle-like heads of the Feathered Serpent on its façade were smashed or ground to a stump. A large-stepped platform was then constructed to its west, which made what was left of the temple invisible from its main avenue. If you visit the heavily reconstructed ruins of Teotihuacan today and wish to see what remains of its goggle-eyed gods and plumed snakeheads you will have to stand on top of this platform, which archaeologists call the *adosada*.[29]

At this point all new pyramid construction stopped permanently, and there is no further evidence of ritually sanctioned killing at the established Pyramids of the Sun and Moon, which remained in use as civic monuments until around AD 550 – albeit for other, less lethal purposes about which we know little.[30] Instead, what we see after AD 300 is an extraordinary flow of urban resources into the provision of excellent stone-built housing, not just for the wealthy or privileged but for the great majority of Teotihuacan's population. These impressive apartments, laid out in regular plots from one end of the city to the other, were probably not an innovation of this period. Their construction on a city-grid may have begun a century or so earlier, as did the razing of older and more ramshackle dwellings to make way for them.[31]

Archaeologists at first considered the masonry apartments to be palaces, and it is possible that is exactly how they began around AD 200, when the city seemed set on a course of political centralization. But after AD 300, when the Temple of the Feathered Serpent was desecrated, their construction continued apace, until most of the city's 100,000 or so residents were effectively living in 'palatial', or at least very comfortable, conditions.[32] So what were these apartments like, and what kind of homes did people make in them?

The evidence suggests we should picture small groups of nuclear families, living comfortable lives in single-storey buildings, each equipped with integral drainage facilities and finely plastered floors and walls. Each family seems to have had its own set of rooms within the larger apartment block, complete with private porticoes where light entered the otherwise windowless rooms. We can deduce that the average apartment compound would have housed in total around 100 people, who would have encountered each other routinely in a central courtyard, which also seems to have been the focus of domestic rituals, perhaps jointly observed. Most of these communal spaces were fitted with altars in the standard style of civic construction (known as *talud-tablero*), and the walls were often brightly painted with murals. Some courtyards had pyramid-shaped shrines, suggesting this architectural form had taken on new and less exclusive roles within the city.[33]

René Millon, the archaeologist responsible for producing the first detailed map of Teotihuacan's layout, felt that the apartment compound was actually invented as a form of social housing, 'designed for urban life in a city that was becoming increasingly crowded, perhaps approaching the chaotic'.[34] Each block was initially laid out to similar scale and dimensions, on plots of roughly 3,600 square metres, although some deviated from this ideal scheme. Strict uniformity was avoided in the arrangement of rooms and courtyards, so in the last resort each compound was unique. Even the more modest apartments show signs of a comfortable lifestyle, with access to imported goods and a staple diet of corn tortillas, eggs, turkey and rabbit meat, and the milk-hued drink known as *pulque* (an alcoholic beverage fermented from the spiky agave plant).[35]

In other words, few were deprived. More than that, many citizens enjoyed a standard of living that is rarely achieved across such a wide sector of urban society in any period of urban history, including our own. Teotihuacan had indeed changed its course away from monarchy and aristocracy to become instead a 'Tollan of the people'.

But how was this remarkable transformation achieved? Apart from spoilage of the Temple of the Feathered Serpent, there are few signs of violence. Land and resources appear to have been allocated to family groups who became neighbours. In this multi-ethnic city, each co-residential group of between sixty and 100 people would have enjoyed two kinds of communal life. One was based on kinship, with family ties extending far beyond the apartment block and often beyond the city – ties which could have troublesome implications, as we'll shortly see. The other was based more strictly on co-residence in apartments and neighbourhoods, often reinforced by shared craft specializations such as garment-making or obsidian-working.

Both forms of urban community, existing alongside one another, retained a human scale, a world away from our modern conception of the 'housing estate' in which nuclear families are sequestered by the thousands in multi-storey monoliths. So we are back to the question with which we started: what held this 'New Teotihuacan' together, if not a hereditary elite or some other type of governing class?

Without written evidence it may never be possible to reconstruct the details, but by now we can probably rule out any sort of top-down system in which elite cadres of royal administrators or priests drew up plans and sent out orders. A more likely possibility is that authority was distributed among local assemblies, perhaps answerable to a governing council. If any trace of these community associations survives it is in the district shrines known as 'three-temple complexes'. At least twenty such complexes were dispersed throughout the city, serving a total of 2,000 apartments, one for every 100 apartment blocks.[36]

This might imply the delegation of government to neighbourhood councils with constituencies similar in size to those of Mesopotamian city-wards, or the assembly houses of Ukrainian mega-sites we discussed in Chapter Eight, or for that matter the *barrios* of later Mesoamerican towns. It may seem hard to imagine a city this size running successfully in this way for centuries without strong leaders or an extensive bureaucracy; but as we'll see, first-hand accounts of later cities from the time of the Spanish conquest lend credence to the idea.

Another, more ebullient face of Teotihuacan's civic identity is revealed in its mural art. Despite efforts to see them as sombre religious iconography, these playful pictorial scenes – painted on the interior walls of apartment compounds from around AD 350 – often seem veritably psychedelic.[37] Streaming effigies emerge from clustered plant, human and animal bodies, framed by figures with elaborate costumes, sometimes grasping hallucinogenic seeds and mushrooms; and among the crowd scenes we find flower eaters with rainbows bursting from their heads.[38] Such scenes often depict human figures all at roughly the same size, with no individual raised up over another.[39]

Of course, these murals represent Teotihuacanos as they liked to imagine themselves; social realities are always more complex. Archaeological excavations in a part of the city known as Teopancazco, lying south of the city centre, show just how complex those realities could actually get. Traces of domestic life in Teopancazco dating to around AD 350 reveal the affluent life of its inhabitants, whose shell-ornamented cotton dress suggests they originally came from the Gulf

Coast and continued to trade with that region. From there they also brought with them certain customs, including unusually violent rituals, which are not so far documented elsewhere in the city. These seem to have involved the capture and decapitation of foreign enemies, whose heads were kept and buried in offering vessels, found within their private homes.[40]

Now here we have something going on that would obviously be very difficult to square with the idea of communal living on a large scale; and this is precisely our point. Below the surface of civil society at Teotihuacan there must have been all sorts of social tensions simmering away among groups of radically different ethnic and linguistic backgrounds who were constantly moving in and out, consolidating relationships with foreign trading partners, cultivating alter egos in remote places and sometimes bringing those forms of identity back with them. (We might allow ourselves to imagine what would happen should a Teotihuacano freebooter who'd managed to make himself King of Tikal ever have returned home.) By around AD 550, the social fabric of the city had begun to come apart at the seams. There is no compelling evidence of foreign invasion. Things seems to have disintegrated from within. Almost as suddenly as it had once coalesced some five centuries previously, the city's population dispersed again, leaving their Tollan behind them.[41]

The rise and decline of Teotihuacan set in motion a roughly cyclical pattern of demographic concentration and dispersal in central Mexico which repeated itself a number of times between AD 300 and 1200, down to the disintegration of Tula and the fall of the Toltec state.[42] Over this longer span of time, what was the legacy of Teotihuacan and its grand urban experiment? Should we view the whole episode as a passing deviation, a blip (albeit an extremely large blip) on the road that led from Olmec hierarchy to Toltec aristocracy and eventually Aztec imperialism? Or might the egalitarian aspects of Teotihuacan have a distinct legacy of their own? Few have really considered the latter possibility, but there are good reasons to ask, especially since early Spanish accounts of the Mexican highlands provide some extraordinarily suggestive material – including descriptions of indigenous cities which, to European eyes, could only be understood as republics, or even democracies.

ON THE CASE OF TLAXCALA, AN INDIGENOUS REPUBLIC THAT RESISTED THE AZTEC EMPIRE THEN CAME TO JOIN FORCES WITH SPANISH INVADERS, AND HOW ITS FATEFUL DECISION EMERGED FROM DEMOCRATIC DELIBERATIONS IN AN URBAN PARLIAMENT (AS OPPOSED TO THE DAZZLING EFFECTS OF EUROPEAN TECHNOLOGY ON 'INDIAN MINDS')

With this in mind, let's now consider a very different case of cultural contact, which takes us forwards in time to the beginnings of European expansion in the Americas. It concerns an indigenous city-state by the name of Tlaxcala, adjacent to what's now the Mexican state of Puebla, which played a pivotal role in the Spanish conquest of the Aztec Empire or Triple Alliance. Here is how Charles C. Mann, in his acclaimed *1491: New Revelations of the Americas before Columbus* (2005), describes what happened in 1519 when Hernán Cortés passed through:

> Marching inland from the sea, the Spanish at first fought repeatedly with Tlaxcala, a confederation of four small kingdoms that had maintained its independence despite repeated Alliance incursions. Thanks to their guns, horses, and steel blades, the foreigners won every battle, even with Tlaxcala's huge numerical advantage. But Cortés's forces shrank with every fight. He was on the verge of losing everything when the four Tlaxcala kings abruptly reversed course. Concluding from the results of their battles that they could wipe out the Europeans, though at great cost, the Indian leaders offered what seemed a win-win deal: they would stop attacking Cortés, sparing his life, the lives of the surviving Spaniards, and those of many Indians, if he in return would join with Tlaxcala in a united assault on the hated Triple Alliance.[43]

Now there is a basic problem with this account. There were no kings in Tlaxcala. Therefore, it could not in any sense be described as a confederation of kingdoms. So how did Mann come to think there were? As an award-winning science journalist, but not a specialist in the history of sixteenth-century Mesoamerica, he was at the mercy of secondary sources; and this, it turns out, is where much of the problem begins.

No doubt Mann must have assumed (as would any reasonable person) that if Tlaxcala were anything other than a kingdom – say a republic or a democracy, or even some form of oligarchy – then the secondary literature would have been full of lively debates about what this implies, not just for our understanding of the Spanish conquest as a key turning point in modern world history, but for the development of indigenous societies in Mesoamerica, or indeed for political theory in general. Oddly, he'd have been wrong to assume this.[44] Finding ourselves in a similar position, we decided to delve a little deeper. What we found, we must admit, was rather startling, even to us. Let's begin by comparing Mann's account to the one Cortés himself addressed to his king, the Holy Roman Emperor Charles V.

In his *Five Letters of Relation*, written between 1519 and 1526, Hernán Cortés recounts his entry to the mountain-ringed Valley of Puebla, on the southern tip of the Mexican *altiplano*. The valley at that time sheltered numerous native cities, of which the largest included pyramid-studded Cholula, and also the city of Tlaxcala. It was indeed in Tlaxcala that Cortés found local allies who fought alongside him, advancing first on Cholula and then going on to defeat the armies of Moctezuma the Younger and lay waste to the Aztec capital of Tenochtitlan, in the neighbouring Valley of Mexico. Cortés estimated the population of Tlaxcala and its rural dependencies at 150,000. 'There is a market in this city,' he reported back to Charles V, 'in which more than thirty thousand people are occupied in buying and selling,' and the province 'contains many wide-spreading fertile valleys all tilled and sown, no part of it being left wild, and measures some ninety leagues in circumference'. Also, the 'order of government so far observed among the people resembles very much the republics of Venice, Genoa, and Pisa for there is no supreme overlord.'[45]

Cortés was a minor aristocrat from a part of Spain where even municipal councils were still something of a novelty; one might argue he had little real knowledge of republics and therefore would hardly be the most reliable judge of such matters. Perhaps so; but by 1519 he had considerable experience in identifying Mesoamerican kings and either recruiting or neutralizing them, since this is largely what he had been doing since his arrival on the mainland. In Tlaxcala, he couldn't find any. Instead, after an initial clash with Tlaxcalteca warriors, he found himself engaged with representatives of a popular urban council whose every decision had to be collectively ratified. Here is where things become decidedly strange, in terms of how the history of these events has come down to us.

It is worth emphasizing again that we are dealing here with what is, by most estimations, one of the pivotal episodes of modern world history: the events leading directly up to the Spanish conquest of the Aztec Empire, and a blueprint for subsequent European conquests throughout the Americas. We assume that nobody – even the most ardent believer in the forces of technological progress, or 'guns, germs, and steel' – would go so far as to claim that fewer than 1,000 Spaniards could ever have conquered Tenochtitlan (a highly organized city, covering over five square miles, containing roughly a quarter of a million people) without the help of these indigenous allies, who included some 20,000 warriors from Tlaxcala. In which case, to understand what was really going on it becomes crucial to understand why the Tlaxcalteca decided to joined forces with Cortés, and how – with a population of tens of thousands, and no supreme overlord to govern them – they arrived at a decision to do so.

On the first matter, our sources are clear. The Tlaxcalteca were out to settle old scores. From their perspective, an alliance with Cortés might bring to a favourable end their struggles against the Aztec Triple Alliance, and the so-called 'Flowery Wars' between the Valleys of Puebla and Mexico.[46] As usual, most of our sources reflect the perspective of Aztec elites, who liked to portray Tlaxcala's long-standing resistance to their imperial yoke as something between a game and imperial largesse (they allowed the Tlaxcalteca to remain independent, the Aztecs later insisted to their Spanish conquerors, because, after all, the empire's soldiers needed somewhere to train; their priests needed a stockyard of human victims for sacrifice to the gods, and so forth). But this was braggadocio. In fact, Tlaxcala and its Otomí guerrilla units had been holding the Aztecs successfully at bay for generations. Their resistance was not just military. Tlaxcalteca cultivated a civic ethos that worked against the emergence of ambitious leaders, and hence potential quislings – a counter-example to Aztec principles of governance.

Here we come to the crux of the problem.

Politically, the Aztec capital of Tenochtitlan and the city-state of Tlaxcala embodied opposite ideals (no less than, say, ancient Sparta and Athens). Still, little of this history is known, because the story we've become used to telling about the conquest of the Americas is an entirely different one. The fall of Tenochtitlan in 1521 is often used to illustrate what some feel to be deeper, underlying currents of change in human societies: the forces that give history its overall shape and direction. Starting with Alfred Crosby and Jared Diamond,[47] writers in this vein have repeatedly pointed out that the conquistadors had something akin to manifest destiny on their side. Not the divinely ordained sort of destiny they envisaged for themselves, but rather the unstoppable force of an invisible army of Neolithic Old World microbes, marching alongside the Spaniards, carrying waves of smallpox to decimate indigenous populations, and a Bronze Age legacy of metal weapons, guns and horses to shock and awe the helpless natives.

We like to tell ourselves that Europeans introduced the Americas not just to these agents of destruction but also to modern industrial democracy, ingredients for which were nowhere to be found there, not even in embryo. All this supposedly came as a single cultural package: advanced metallurgy, animal-powered vehicles, alphabetic writing systems and a certain penchant for freethinking that is seen as necessary for technological progress. 'Natives', in contrast, are assumed to have existed in some sort of alternative, quasi-mystical universe. They could not, by definition, be arguing about political constitutions or engaging in processes of sober deliberation over decisions that changed the course of world history; and if European observers report them doing so, they must either be mistaken, or were simply projecting on to 'Indians' their own ideas about democratic governance, even when those ideas were hardly practised in Europe itself.

As we've also seen, this way of reading history would have been quite alien to Enlightenment philosophers, who were more inclined to think their ideals of freedom and equality owed much to the peoples of the New World and were by no means certain if those ideals were at all compatible with industrial advance. We are dealing, again, with powerful modern myths. Such myths don't merely inform what people say: to an even greater extent, they ensure certain things go unnoticed. Some of the key early sources on Tlaxcala have never even appeared in translation, and new data emerging in recent years has not really been noticed outside specialist circles. Let's see if we can't set the record straight.

How, exactly, did the Tlaxcalteca arrive at a decision to ally with Cortés on the field of battle, thereby ensuring the Spaniards' victory over the Aztec Empire? It is clear the matter was fraught and deeply divisive (as it was in other Pueblan cities as well: in Cholula, for example, the same dilemma occasioned a rupture between the leaders of six *calpolli* – urban wards – three of whom took the others hostage, whereupon the latter absconded to Tlaxcala).[48] In Tlaxcala itself, though, the argument took a very different form to what happened in Cholula.

Some of the evidence is to be found in Bernal Díaz's famous *Historia verdadera de la conquista de la Nueva España* (1568), which contains lengthy passages on the Spaniards' interactions with warriors and emissaries from Tlaxcala. Another much-used source is the illustrated codex known as *Historia de Tlaxcala* (1585), by the mestizo historian Diego Muñoz Camargo; and there are also important writings by the Francisan friar Toribio of Benavente. But the most detailed source – in our minds the key one – is a book that is hardly ever cited; in fact, it is hardly ever read, at least by historians (though specialists in Renaissance humanism sometimes comment on its literary style). We are referring to the unfinished *Crónica de la Nueva España*, composed between 1558 and 1563 by Francisco Cervantes de Salazar, one of the first rectors of the University of Mexico.[49]

Cervantes de Salazar was born around 1515 in the Spanish city of Toledo and studied at the prestigious University of Salamanca, where his scholarly reputation was second to none. After a time in Flanders he became Latin secretary to the Archbishop of Seville; this gained him entry to the court of Charles V, where he heard Hernán Cortés relating his experiences of the New World. This young and gifted scholar soon became a devotee of the conquistador, and within a few years of Cortés's death in 1547 Cervantes de Salazar set sail for Mexico. On arrival, he taught Latin on premises owned by Cortés's son and heir, but soon became a central figure in the newly established university while also taking holy orders; he would attempt to juggle ecclesiastical and scholarly duties for the remainder of his life, with mixed success.

In 1558 the Municipality of Mexico, composed mainly of first-generation conquistadors or their descendants, was sufficiently impressed with Cervantes de Salazar's scholarly abilities to grant him his greatest wish: an annual stipend of 200 gold pesos to support his composition of a general history of New Spain, focusing on the themes of discovery and conquest. This was quite the endorsement and, two years later, Cervantes de Salazar (already some way into his manuscript) won a further grant, which was specifically intended to support a period of fieldwork. He must have visited Tlaxcala and its environs during that time in order to obtain valuable historical evidence directly from local caciques who lived through the *conquista*, and from their immediate descendants.[50]

The municipality appears to have kept its appointed chronicler on a tight leash, demanding three-monthly updates on his manuscript. His last submission came in 1563, by which time, despite his best efforts, he was embroiled in a bitter ecclesiastical dispute that put him on the wrong side of the General Inquisitor, the powerful Pedro Moya de Contreras. In those acrimonious years, Cervantes de Salazar saw Martín Cortés and many of his other close associates variously imprisoned, tortured or exiled as rebels against the Spanish Crown. Cervantes de Salazar made sufficient compromises to escape such a fate; but his reputation suffered, and to this day he is often regarded as a minor academic source by comparison, say, with Bernardino de Sahagún. Ultimately, both scholars' work would meet a similar fate, delivered to the imperial councils of the Indies and the Inquisition in Spain for obligatory censorship of matters relating to 'idolatrous practices' (though not, it seems, on matters of indigenous politics), without allowing any original or copy to remain in circulation.[51]

The result was that, for a period of centuries, Cervantes de Salazar's *Crónica* was effectively hiding in plain sight.[52] It is largely to the remarkable efforts of Zelia Maria Magdalena Nuttall (1857–1933) – pioneering archaeologist, anthropologist and finder of lost codices – that we owe not just the rediscovery of Cervantes de Salazar's unfinished *Crónica de la Nueva España*, which she identified in the Biblioteca Nacional in Madrid in 1911, but also most of the surviving details of his life and the circumstances of its composition, which she extracted from the archives of the town council in the city of Mexico, finding (to her astonishment) that less careful historians who went before her had discovered nothing worthy of note there. It was only in 1914 that the *Crónica* saw publication. To this day there is still no critical introduction or commentary to guide readers through its sixteenth-century prose, or point them towards its significance as a record of political affairs in an indigenous Mesoamerican city.[53]

Critics have emphasized that Cervantes de Salazar was writing a few decades after the facts he described, basing his chronicle on earlier accounts – but this is equally true of other key sources regarding the Spanish conquest. They also note he wasn't a particularly competent ethnographer in the mould of, say, Sahagún, being more steeped in the works of Horace and Livy than the indigenous traditions of Mexico. All this may be true, just as it is true that the literary tradition prevailing at the time tended to invoke Greek and especially Roman examples at the drop of a hat. Still, the *Crónica* is clearly not some kind of projection of Salazar's classical training. It contains rich descriptions of indigenous figures and institutions from the time of the Spanish invasion which bear no resemblance to any classical sources and which in many cases are corroborated by first-hand accounts. What are not, apparently, in those other accounts are the details that Cervantes de Salazar provides.

Of special interest to us are those extended sections of the *Crónica* that deal directly with the governing council of Tlaxcala, and its deliberations over whether to ally with the Spanish invaders. They include lengthy accounts of speeches and diplomatic gifts going back and forth between representatives of the Spaniards and their Tlaxcalteca counterparts, whose oratory in council occasioned much admiration. According to Cervantes de Salazar, those who spoke for Tlaxcala included elder statesmen – such as Xicotencatl the Elder, father to the military general also named Xicotencatl who is still lionized in the state of Tlaxcala to this day[54] – but also masters of commerce, religious experts and the top legal authorities of the time. What Salazar describes in these remarkable passages is evidently not the workings of a royal court but of a mature urban parliament, which sought consensus for its decisions through reasoned argument and lengthy deliberations – carrying on, when necessary, for weeks at a time.

The key passages are in Book Three. Cortés is still encamped outside the city with his newfound Totonac allies. Ambassadors move back and forth between the Spaniards and the *Ayuntamiento* (city council) of Tlaxcala, where deliberations commence. After many welcomes and much kissing of hands, a lord named Maxixcatzin – well known for his 'great prudence and affable conversation' – gets the ball rolling with an eloquent appeal for the Tlaxcalteca to follow his lead (indeed, to follow what the gods and ancestors ordained), and ally with Cortés to rise up against their common Aztec oppressors. His reasoning is widely accepted in the council, until that is, Xicotencatl the Elder – by then over 100 years old and almost blind – intervenes.

A chapter follows, detailing 'the brave speech that Xicotencatl made, contradicting Maxixcatzin'. Nothing, he reminds the council, is harder to resist than an 'enemy within', which is what the newcomers will likely become if welcomed into town. Why, asks Xicotencatl,

> ... does Maxixcatzin deem these people gods, who seem more like ravenous monsters thrown up by the intemperate sea to blight us, gorging themselves on gold, silver, stones, and pearls; sleeping in their own clothes; and generally acting in the manner of those who would one day make cruel masters ... There are barely enough chickens, rabbits, or corn-fields in the entire land to feed their bottomless appetites, or those of their ravenous 'deer' [the Spanish horses]. Why would we – who live without servitude, and never acknowledged a king – spill our blood, only to make ourselves into slaves?[55]

Members of the council, we learn, were swayed by Xicotencatl's words: 'a murmur began among them, speaking with each other, the voices were rising, each one declaring what he felt.' The council was divided, and without consensus. What followed would be familiar to anyone who has participated in a process of consensus decision-making: when matters seem to come to loggerheads, rather than putting it to a vote someone proposes a creative synthesis. Temilotecutl – one of the city's four 'senior justices' – stepped in with a cunning plan. To satisfy both sides of the debate, Cortés would be invited into the city, but as soon as he entered Tlaxcalteca territory the city's leading general, Xicotencatl the Younger, would ambush him, together with a contingent of Otomí warriors. If the ambush succeeded, they would be heroes. If it failed, they would blame it on the uncouth and impulsive Otomí, make their excuses, and ally themselves with the invaders.

We need not rehearse here the events leading to an alliance between Tlaxcala and Cortés;[56] we have said enough to give the reader a flavour of our sources concerning the democracy of Tlaxcala, and the facility of its politicians in reasoned debate. Such accounts have not fared well in the hands of modern historians. Few would go so far as to suggest that what de Salazar described never really happened, or was simply his own imagination of a scene from some ancient Greek agora or Roman senate, placed into the mouths of 'Indians'. Yet on those rare occasions when the *Crónica* is considered by scholars today, it is mostly as a contribution to the literary genre of early Catholic humanism rather than as a source of historical information about indigenous forms of government – in much the same way that commentators on the writings of Lahontan never really concern themselves with what Kandiaronk might actually have argued, but dwell on the possibility that some passages might be inspired by Greek satirists like Lucian.[57]

There is a subtle snobbery at play here. It's not so much that anyone denies outright that accounts of deliberative politics reflect historical reality; it's just that no one seems to find this fact particularly interesting. What seems interesting to historians is invariably the relation of these accounts to European textual traditions, or European expectations. Much the same occurs with the treatment of later texts from Tlaxcala: extant, detailed written records of the proceedings at its municipal council in the decades following the Spanish conquest, the *Tlaxcalan Actas*, which affirm at length both the oratorical skills of indigenous politicians and their facility with principles of consensus decision-making and reasoned debate.[58]

You might think all this would be of interest to historians. Instead, what really seems to strike them as worthy of debate is the degree to which democratic mores displayed in the texts might be some sort of near-miraculous adaptation by 'astute Indians' to the political expectations of their European masters: effectively some kind of elaborate play-acting.[59] Why such historians imagine that a collection of sixteenth-century Spanish friars, petty aristocrats and soldiers were likely to know anything about democratic procedure (much less, be impressed by it) is unclear, because educated opinion in Europe was almost uniformly anti-democratic at the time. If anyone was learning something new from the encounter, it was surely the Spaniards.

In the current intellectual climate, to suggest the Tlaxcalteca were anything but cynics or victims is considered just a tiny bit dangerous: one is opening oneself up to charges of naive romanticism.[60] In fact, these days more or less any attempt to suggest that Europeans learned anything at all of moral or social value from Native American people is likely to be met with mild derision and accusations of indulging in 'noble savage' tropes, or occasionally almost hysterical condemnation.[61]

But a strong case can be made that the deliberations recorded in Spanish sources are exactly what they seem to be – a glimpse into the mechanics of collective indigenous government – and if these deliberations bear any superficial resemblance to debates recorded in Thucydides or Xenophon, this is because, well, there are really only so many ways to conduct a political debate. At least one Spanish source provides explicit confirmation in this regard. Here we turn to Friar Toribio of Benavente, called by locals Motolinía (the 'afflicted one') for his ragged appearance, a sobriquet he seems to have happily adopted. It is to Motolinía and his Tlaxcalteca informants – who included Antonio Xicotencatl, most likely a grandson of Xicotencatl the Elder – that we owe the *Historia de los Indios de la Nueva España* (1541).[62]

Motolinía confirms Cortés's original observation: that Tlaxcala was indeed an indigenous republic governed not by a king, nor even by rotating office holders (as at Cholula), but by a council of elected officials (*teuctli*) answerable to the citizenry as a whole. Exactly how many sat on the high council of Tlaxcala is not clear: Spanish sources suggest any number from fifty to 200. Perhaps it depended on the matter at hand. Neither, unfortunately, does he tell us anything in detail about how these individuals were selected for office, or who was eligible (other Pueblan cities, including royal ones, rotated official duties among representatives of *calpolli*). On the topic of Tlaxcalteca modes of political training and instruction, however, Motolinía's account comes alive.

Those who aspired to a role on the council of Tlaxcala, far from being expected to demonstrate personal charisma or the ability to outdo rivals, did so in a spirit of self-deprecation – even shame. They were required to subordinate themselves to the people of the city. To ensure that this subordination was no mere show, each was subject to trials, starting with mandatory exposure to public abuse, regarded as the proper reward of ambition, and then – with one's ego in tatters – a long period of seclusion, in which the aspiring politician suffered ordeals of fasting, sleep deprivation, bloodletting and a strict regime of moral instruction. The initiation ended with a 'coming out' of the newly constituted public servant, amid feasting and celebration.[63]

Clearly, taking up office in this indigenous democracy required personality traits very different to those we take for granted in modern electoral politics. On this latter point, it is worth recalling that ancient Greek writers were well aware of the tendency for elections to throw up charismatic leaders with tyrannical pretensions. This is why they considered elections an aristocratic mode of political appointment, quite at odds with democratic principles; and why for much of European history the truly democratic way of filling offices was assumed to be by lottery.

Cortés may have praised Tlaxcala as an agrarian and commercial arcadia but, as Motolinía explains, when its citizens thought about their own political values, they actually saw those values as coming from the desert. Like other Nahuatl speakers, including the Aztecs, Tlaxcalteca liked to claim they were descended from Chichimec. These were considered the original hunter-gatherers who lived ascetic lives in deserts and forests, dwelling in primitive huts, ignorant of village or city life, rejecting corn and cooked food, bereft of clothing or organized religion, and living on wild things alone.[64] The ordeals endured by aspiring councillors in Tlaxcala were reminders of the need to cultivate Chichimec qualities (ultimately to be balanced by the Toltec virtues of an urbane warrior; and just where the correct balance lay was much debated among the Tlaxcalteca).

If all this sounds a little familiar, we must ask ourselves why. The Spanish friars will no doubt have heard echoes in these tales of Old World tropes for republican virtue – that same atavistic streak running from the biblical prophets through to Ibn Khaldun, not to mention their own ethic of world renunciation. The correspondences are so close that one begins to wonder if, in their auto-ethnography, the Tlaxcalteca in this case actually did present themselves to Spaniards in terms they knew would be instantly recognized and understood. Certainly, we know that the citizens of Tlaxcala staged some remarkable theatrical spectacles for the benefit of their conquerors, including a 1539 pageant of the Crusader *Conquest of Jerusalem*, in which the climax was a mass baptism of (actual) pagans, dressed up as Moors.[65]

Spanish observers may even have learned from Tlaxcalteca or Aztec sources what it means to have once been a 'noble savage'. Nor can we rule out the possibility that indigenous Mexican ideas on the subject entered wider streams of European political thought that gathered force only in the days of Rousseau, whose State of Nature maps with alarming fidelity on to Motolinía's account of the Chichimec, right down to the 'primitive hut dwellings' in which they were supposed to have lived. Perhaps some of the seeds of our own evolutionary story about how it all began with simple, egalitarian hunter-gatherers were sown right there, in the imaginations of city-dwelling Amerindians.

But we digress.

Amid all this mutual positioning, what can we really conclude about the political constitution of Tlaxcala at the time of the Spanish conquest? Was it really a functioning urban democracy and, if so, how many other such democracies might have existed in the pre-Columbian Americas? Or are we confronting a mirage, a strategic conjuring of the 'ideal commonwealth', supplied to a receptive audience of millenarian friars? Were elements of history and mimesis both at work?

If all we had to go on were written sources, there would always be room for doubt; but archaeologists confirm that by the fourteenth century AD the city of Tlaxcala was, in fact, already organized on an entirely different basis to Tenochtitlan. There is no sign of a palace or central temple, and no major ball-court (an important setting, recall, for royal ritual in other Mesoamerican cities). Instead, archaeological survey reveals a cityscape given over almost entirely to the well-appointed residences of its citizens, constructed to uniformly high standards around more than twenty district plazas, all raised up on grand earthen terraces. The largest municipal assemblies were housed in a civic complex called Tizatlan, but this was located outside the city itself, with spaces for public gatherings entered via broad gateways.[66]

Modern archaeological investigations thus confirm the existence of an indigenous republic at Tlaxcala long before Cortés set foot on Mexican soil, while later written sources leave us in little doubt as to its democratic credentials. The contrasts with other known Mesoamerican cities of the time are quite striking – though it should also be said that fifth-century Athens was something of an outlier, surrounded by petty kingdoms and oligarchies. Nor should these contrasts be overdrawn. What we have learned in this chapter is that the political traditions of Tlaxcala are not an anomaly, but lie in one broad stream of urban development which can be traced back, in outline, to the experiments in social welfare undertaken 1,000 years earlier at Teotihuacan. Despite Aztec claims to a special relationship with that abandoned city, Tlaxcala was at least as much a part of its legacy as the Aztec capital of Tenochtitlan – and in most really significant ways, more so.

After all, it was the Aztec rulers of Tenochtitlan who finally broke with tradition, creating a predatory empire that was in some ways closer to the dominant European political models of the time, or what has since come to be known as 'the state'. In the next chapter, we intend to turn back and consider this term. What precisely is a state? Does it really mark an entirely new phase of human history? Is the term even useful any more?

10

Why the State Has No Origin

The humble beginnings of sovereignty, bureaucracy and politics

The quest for the 'origins of the state' is almost as long-standing, and hotly contested, as the pursuit of the 'origins of social inequality' – and in many ways, it is just as much of a fool's errand. It is generally accepted that, today, pretty much everyone in the world lives under the authority of a state; likewise, a broad feeling exists that past polities such as Pharaonic Egypt, Shang China, the Inca Empire or the kingdom of Benin qualify as states, or at least as 'early states'. However, with no consensus among social theorists about what a state actually is, the problem is how to come up with a definition that includes all these cases but isn't so broad as to be absolutely meaningless. This has proved surprisingly hard to do.

Our term 'the state' only came into common usage in the late sixteenth century, when it was coined by a French lawyer named Jean Bodin, who also wrote, among many other things, an influential treatise on witchcraft, werewolves and the history of sorcerers. (He is further remembered today for his profound hatred of women.) But perhaps the first to attempt a systematic definition was a German philosopher named Rudolf von Ihering, who, in the late nineteenth century, proposed that a state should be defined as any institution that claims a monopoly on the legitimate use of coercive force within a given territory (this definition has since come to be identified with the sociologist Max Weber). On this definition, a government is a 'state' if it lays claim to a certain stretch of land and insists that, within its borders, it is the only institution whose agents can kill people, beat them up, cut off parts of their body or lock them in cages; or, as von Ihering emphasized, that can decide who else has the right to do so on its behalf.

Von Ihering's definition worked fairly well for modern states. However, it soon became clear that for most of human history, rulers either didn't make such grandiose claims – or, if they did, their claims to a monopoly on coercive force held about the same status as their claims to control the tides or the weather. To retain von Ihering and Weber's definition one would either have to conclude that, say, Hammurabi's Babylon, Socrates' Athens or England under William the Conqueror weren't states at all – or come up with a more flexible or nuanced definition. Marxists offered one: they suggested that states make their first appearance in history to protect the power of an emerging ruling class. As soon as one has a group of people living routinely off the labour of another, the argument ran, they will necessarily create an apparatus of rule, officially to protect their property rights, in reality to preserve their advantage (a line of thinking very much in the tradition of Rousseau). This definition brought Babylon, Athens and medieval England back into the fold, but also introduced new conceptual problems, such as how to define exploitation. And it was unpalatable to liberals, ruling out any possibility that the state could ever become a benevolent institution.

For much of the twentieth century, social scientists preferred to define a state in more purely functional terms. As society became more complex, they argued, it was increasingly necessary for people to create top-down structures of command in order to co-ordinate everything. This same logic is still followed in essence by most contemporary theorists of social evolution. Evidence of 'social complexity' is automatically treated as evidence for the existence of some sort of governing apparatus. If one can speak, say, of a settlement hierarchy with four levels (e.g. cities, towns, villages, hamlets), and if at least some of those settlements also contained full-time craft specialists (potters, blacksmiths, monks and nuns, professional soldiers or musicians), then whatever apparatus administered it must *ipso facto* be a state. And even if that apparatus did not claim a monopoly of force, or support a class of elites living off the toil of benighted labourers, this was inevitably going to happen sooner or later. This definition, too, has its advantages, especially when speculating about very ancient societies, whose nature and organization has to be teased out from fragmentary remains; but its logic is entirely circular. Basically, all it says is that, since states are complicated, any complicated social arrangement must therefore be a state.

Actually, almost all these 'classic' theoretical formulations of the last century started off from exactly this assumption: that any large and complex society necessarily required a state. The real bone of contention was, why? Was it for good practical reasons? Or was it because any such society would necessarily produce a material surplus, and if there was a material surplus – like, for instance, all that smoked fish on the Pacific Northwest Coast – then there would also, necessarily, be people who managed to grab hold of a disproportionate share?

As we've already seen in Chapter Eight, these assumptions don't hold up particularly well for the earliest cities. Early Uruk, for example, does not appear to have been a 'state' in any meaningful sense of the word; what's more, when top-down rule does emerge in the region of ancient Mesopotamia, it's not in the 'complex' metropolises of the lowland river valleys, but among the small, 'heroic' societies of the surrounding foothills, which were averse to the very principle of administration and, as a result, don't seem to qualify as 'states' either. If there is a good ethnographic parallel for these latter groups it might be the societies of the Northwest Coast, since there too political leadership lay in the hands of a boastful and vainglorious warrior aristocracy, competing in extravagant contests over titles, treasures, the allegiance of commoners and the ownership of slaves. Recall, here, that Haida, Tlingit and the rest not only lacked anything that could be called a state apparatus; they lacked any kind of formal governmental institutions.[1]

One might then argue that 'states' first emerged when the two forms of governance (bureaucratic and heroic) merged together. A case could be made. But equally we might ask if this is really such a significant issue in the first place? If it is possible to have monarchs, aristocracies, slavery and extreme forms of patriarchal domination, even without a state (as it evidently was); and if it's equally possible to maintain complex irrigation systems, or develop science and abstract philosophy without a state (as it also appears to be), then what do we actually learn about human history by establishing that one political entity is what we would like to describe as a 'state' and another isn't? Are there not more interesting and important questions we could be asking?

In this chapter we are going to explore the possibility that there are. What would history look like if – instead of assuming that there must be some deep internal resemblance between the governments of, say, ancient Egypt and modern Britain, and our task is therefore to figure out precisely what it is – we were to look at the whole problem with new eyes. There is no doubt that, in most of the areas that saw the rise of cities, powerful kingdoms and empires also eventually emerged. What did they have in common? Did they, in fact, have anything in common? What does their appearance really tell us about the history of human freedom and equality, or its loss? In what way, if any, do they mark a fundamental break with what came before?

IN WHICH WE LAY OUT A THEORY CONCERNING THE THREE ELEMENTARY FORMS OF DOMINATION, AND BEGIN TO EXPLORE ITS IMPLICATIONS FOR HUMAN HISTORY

The best way to go about this task, we suggest, is by returning to first principles. We have already talked about fundamental, even primary, forms of freedom: the freedom to move; the freedom to disobey orders; the freedom to reorganize social relations. Can we speak similarly about elementary forms of domination?

Recall how Rousseau, in his famous thought experiment, felt that everything went back to private property, and especially property in land: in that terrible moment when a man first threw up a barrier and said, 'This territory is mine, and mine alone', all subsequent forms of domination – and therefore, all subsequent catastrophes – became inevitable. As we've seen, this obsession with property rights as the basis of society, and as a foundation of social power, is a peculiarly Western phenomenon – indeed, if 'the West' has any real meaning, it would probably refer to that legal and intellectual tradition which conceives society in those terms. So, to begin a thought experiment of a slightly different kind, it might be good to start right here. What are we really saying when we say that the power of a feudal aristocracy, or a landed gentry, or absentee landlords is 'based on land'?

Often we use such language as a way of cutting through airy abstractions or high-minded pretensions to address simple material realities. For example, the two dominant political parties in nineteenth-century England, the Whigs and the Tories, liked to represent themselves as arguing about ideas: a certain conception of free-market liberalism versus a certain notion of tradition. An historical materialist might object that, in fact, Whigs represented the interests of the commercial classes, and the Tories those of the landowners. They are of course right. It would be foolhardy to deny it. What we might question, however, is the premise that 'landed' (or any other form of) property is itself particularly material. Yes: soil, stones, grass, hedges, farm buildings and granaries are all material things; but when one speaks of 'landed property' what one is really talking about is an individual's claim to exclusive access and control over all the soil, stones, grass, hedges, etc. within a specific territory. In practice, this means a legal right to keep anyone else off it. Land is only really 'yours', in this sense, if no one would think to challenge your claim over it, or if you have the capacity to summon at will people with weapons to threaten or attack anyone who disagrees, or just enters without permission and refuses to leave. Even if you shoot the trespassers yourself, you still need others to agree you were within your rights to do so. In other words, 'landed property' is not actual soil, rocks or grass. It is a legal understanding, maintained by a subtle mix of morality and the threat of violence. In fact, land ownership illustrates perfectly the logic of what Rudolf von Ihering called the state's monopoly of violence within a territory – just within a much smaller territory than a nation state.

All this might sound a little abstract, but it is a simple description of what happens in reality, as any reader who has ever tried to squat a piece of land, occupy a building or for that matter overthrow a government will be keenly aware. Ultimately, everyone knows it all comes down to whether someone will eventually be given orders to remove you by force, and if it does, then everything comes down to whether that someone is actually willing to follow orders. Revolutions are rarely won in open combat. When revolutionaries win, it's usually because the bulk of those sent to crush them refuse to shoot, or just go home.

So does that mean property, like political power, ultimately derives (as Chairman Mao so delicately put it) 'from the barrel of a gun' – or, at best, from the ability to command the loyalties of those trained to use them?

No. Or not exactly.

To illustrate why not, and continue our thought experiment, let's take a different sort of property. Consider a diamond necklace. If Kim Kardashian walks down the street in Paris wearing a diamond necklace worth millions of dollars, she is not only showing off her wealth, she is also flaunting her power over violence, since everyone assumes she would not be able to do so without the existence, visible or not, of an armed personal security detail, trained to deal with potential thieves. Property rights of all sorts are ultimately backed up by what legal theorists like von Ihering euphemistically called 'force'. But let us imagine, for a moment, what would happen if everyone on earth were suddenly to become physically invulnerable. Say they all drank a potion which made it impossible for anyone to harm anyone else. Could Kim Kardashian still maintain exclusive rights over her jewellery?

Well, perhaps not if she showed it off too regularly, since someone would presumably snatch it; but she certainly could if she normally kept it hidden in a safe, the combination of which she alone knew and only revealed to trusted audiences at events which were not announced in advance. So there is a second way of ensuring that one has access to rights others do not have: the control of information. Only Kim and her closest confidants know where the diamonds are normally kept, or when she is likely to appear wearing them. This obviously applies to all forms of property that are ultimately backed up by the 'threat of force' – landed property, wares in stores, and so forth. If humans were incapable of hurting each other, no one would be able to declare something absolutely sacred to themselves or to defend it against 'all the world'. They could only exclude those who agreed to be excluded.

Still, let us take the experiment a step further and imagine everyone on earth drank another potion which rendered them all incapable of keeping a secret, but still unable to harm one another physically as well. Access to information, as well as force, has now been equalized. Can Kim still keep her diamonds? Possibly. But only if she manages to convince absolutely everyone that, being Kim Kardashian, she is such a unique and extraordinary human being that she actually deserves to have things no one else can.

We would like to suggest that these three principles – call them control of violence, control of information, and individual charisma – are also the three possible bases of social power.[2] The threat of violence tends to be the most dependable, which is why it has become the basis for uniform systems of law everywhere; charisma tends to be the most ephemeral. Usually, all three coexist to some degree. Even in societies where interpersonal violence is rare, one may well find hierarchies based on knowledge. It doesn't even particularly matter what that knowledge is about: maybe some sort of technical know-how (say, of smelting copper, or using herbal medicines); or maybe something we consider total mumbo jumbo (the names of the twenty-seven hells and thirty-nine heavens, and what creatures one would be likely to meet if one travelled there).

Today, it is quite commonplace – for instance, in parts of Africa and Papua New Guinea – to find initiation ceremonies that are so complex as to require bureaucratic management, where initiates are gradually introduced to higher and higher levels of arcane knowledge, in societies where there are otherwise no formal ranks of any sort. Even where such hierarchies of knowledge do not exist, there will obviously always be individual differences. Some people will be considered more charming, funny, intelligent or physically attractive than others. This will always make some sort of difference, even within groups that develop elaborate safeguards to ensure that it doesn't (as, for instance, with the ritual mockery of successful hunters among 'egalitarian' foragers like the Hadza).

As we've noted, an egalitarian ethos can take one of two directions: it can either deny such individual quirks entirely, and insist that people are (or at least should be) treated as if they were exactly the same; or it can celebrate their quirks in such a way as to imply that everyone is so profoundly different that any overall ranking would be inconceivable. (After all, how do you measure the best fisherman against the most dignified elder, against the person who tells the funniest jokes, and so on?). In such cases, it might happen that certain 'extreme individuals' – if we may call them that – do gain an outstanding, even leadership role. Here we might think of Nuer prophets, or certain Amazonian shamans, Malagasy *mpomasy* or astrologer-magicians, or for that matter the 'rich' burials of the Upper Palaeolithic, which so often focus on individuals with strikingly anomalous physical (and probably other) attributes. As those examples imply, however, such characters are so highly unusual that it would be difficult to turn their authority into any sort of ongoing power.

What really concerns us about these three principles is that each has become the basis for institutions now seen as foundational to the modern state. In the case of control over violence, this is obvious. Modern states are 'sovereign': they hold the power once held by kings, which in practice translates into von Ihering's monopoly on the legitimate use of coercive force within their territory. In theory, a true sovereign exercised a power that was above and beyond the law. Ancient kings were rarely able to enforce this power systematically (often, as we've seen, their supposedly absolute power really just meant they were the only people who could mete out arbitrary violence within about 100 yards of where they were standing or sitting at any given time). In modern states, the very same kind of power is multiplied a thousand times because it is combined with the second principle: bureaucracy. As Weber, the great sociologist of bureaucracy, observed long ago, administrative organizations are always based not just on control of information, but also on 'official secrets' of one sort or another. This is why the secret agent has become the mythic symbol of the modern state. James Bond, with his licence to kill, combines charisma, secrecy and the power to use unaccountable violence, underpinned by a great bureaucratic machine.

The combination of sovereignty with sophisticated administrative techniques for storing and tabulating information introduces all sorts of threats to individual freedom – it makes possible surveillance states and totalitarian regimes – but this danger, we are always assured, is offset by a third principle: democracy. Modern states are democratic, or at least it's generally felt they really should be. Yet democracy, in modern states, is conceived very differently to, say, the workings of an assembly in an ancient city, which collectively deliberated on common problems. Rather, democracy as we have come to know it is effectively a game of winners and losers played out among larger-than-life individuals, with the rest of us reduced largely to onlookers.

If we are seeking an ancient precedent to *this* aspect of modern democracy, we shouldn't turn to the assemblies of Athens, Syracuse or Corinth, but instead – paradoxically – to aristocratic contests of 'heroic ages', such as those described in the *Iliad* with its endless *agons*: races, duels, games, gifts and sacrifices. As we noted in Chapter Nine, the political philosophers of later Greek cities did not actually consider elections a democratic way of selecting candidates for public office at all. The democratic method was sortition, or lottery, much like modern jury duty. Elections were assumed to belong to the aristocratic mode (aristocracy meaning 'rule of the best'), allowing commoners – much like the retainers in an old-fashioned, heroic aristocracy – to decide who among the well born should be considered best of all; and well born, in this context, simply meant all those who could afford to spend much of their time playing at politics.[3]

Just as access to violence, information and charisma defines the very possibilities of social domination, so the modern state is defined as a combination of sovereignty, bureaucracy and a competitive political field.[4] It seems only natural, then, that we should examine history in this light too; but as soon as we try to do so, we realize there is no actual reason why these three principles should go together, let alone reinforce each other in the precise fashion we have come to expect from governments today. For one thing, the three elementary forms of domination have entirely separate historical origins. We've already seen this in ancient Mesopotamia, where initially the bureaucratic-commercial societies of the river valleys existed in tension with the heroic polities of the hills and their endless petty princelings, vying for the loyalty of retainers through spectacular contests of one sort or another; while the hill people, in turn, rejected the very principle of administration.

Nor is there any compelling evidence that ancient Mesopotamian cities, even when ruled by royal dynasties, achieved any measure of real territorial sovereignty, so we are still a long way here from anything like an embryonic version of the modern state.[5] In other words, they simply weren't states in von Ihering's sense of the term; and even if they had been, it makes little sense to define a state simply in terms of sovereignty. Recall the example of the Natchez of Louisiana, whose Great Sun wielded absolute power within his own (rather small) Great Village, where he could order summary executions and appropriate goods pretty much as he had a mind to, but whose subjects largely ignored him when he wasn't around. The divine kingship of the Shilluk, a Nilotic people of East Africa, worked on similar lines: there were very few limits on what the king could do to those in his physical presence, but there was also nothing remotely resembling an administrative apparatus to translate his sovereign power into something more stable or extensive: no tax system, no system to enforce royal orders, or even report on whether or not they had been obeyed.

As we can now begin to see, modern states are, in fact, an amalgam of elements that happen to have come together at a certain point in human history – and, arguably, are now in the process of coming apart again (consider, for instance, how we currently have planetary bureaucracies, such as the WTO or IMF, with no corresponding principle of global sovereignty). When historians, philosophers or political scientists argue about the origin of the state in ancient Peru or China, what they are really doing is projecting that rather unusual constellation of elements backwards: typically, by trying to find a moment when something like sovereign power came together with something like an administrative system (the competitive political field is usually considered somewhat optional). What interests them is precisely how and why these elements came together in the first place.

For instance, a standard story of human political evolution told by earlier generations of scholars was that states arose from the need to manage complex irrigation systems, or perhaps just large concentrations of people and information. This gave rise to top-down power, which in turn came to be tempered, eventually, by democratic institutions. That would imply a sequence of development somewhat like this:

Administration ⟶ *Sovereignty* ⟶ *(eventually) Charismatic Politics*

As we showed in Chapter Eight, contemporary evidence from ancient Eurasia now points to a different pattern, where urban administrative systems inspire a cultural counter-reaction (a further example of schismogenesis), in the form of squabbling highland princedoms ('barbarians', from the perspective of the city dwellers),[6] which eventually leads to some of those princes establishing themselves in cities and systematizing their power:

Administration ⟶ *Charismatic Politics* ⟶ *Sovereignty*
(by schismogenesis)

This may well have happened in some cases – Mesopotamia, for example – but it seems unlikely to be the only way in which such developments might culminate in something that (to us at least) resembles a state. In other places and times – often in moments of crisis – the process may begin with the elevation to pre-eminent roles of charismatic individuals who inspire their followers to make a radical break with the past. Eventually, such figureheads assume a kind of absolute, cosmic authority, which is finally translated into a system of bureaucratic roles and offices.[7] The path then might look more like this:

Charismatic vision ⟶ *Sovereignty* ⟶ *Administration*

What we are challenging here is not any particular formulation, but the underlying teleology. All these accounts seem to assume that there is only one possible end point to this process: that these various types of domination were somehow bound to come together, sooner or later, in something like the particular form taken by modern nation states in America and France at the end of the eighteenth century, a form which was gradually imposed on the rest of the world after both world wars.

What if this wasn't true?

What we are going to do here is to see what happens if we approach the history of some of the world's first kingdoms and empires without any such preconceptions. Along with the origins of the state, we will also be putting aside such similarly vague and teleological notions as the 'birth of civilization' or the 'rise of social complexity' in order to take a closer look at what actually happened. How did large-scale forms of domination first emerge, and what did they actually look like? What, if anything, do they have to do with arrangements that endure to this day?

Let's start by examining those few cases in the pre-Columbian Americas which even the greatest sticklers for definition tend to agree were 'states' of some kind.

ON AZTECS, INCA AND MAYA (AND THEN ALSO SPANIARDS)

The general consensus is that there were only two unambiguous 'states' in the Americas at the time of the Spanish conquest: the Aztecs and the Inca. Of course, that is not how the Spanish would have referred to them. Hernán Cortés, in his letters and communications, wrote of cities, kingdoms and occasionally republics. He hesitated to refer to the Aztec ruler, Moctezuma, as an 'emperor' – presumably so as not to risk ruffling the feathers of his own lord, the 'most Catholic emperor Charles V'. But it would never have occurred to him to ponder whether any of these kingdoms or cities qualified as 'states', since the concept barely existed at the time. Nonetheless, this is the question which has preoccupied modern scholars, so let us consider each of these polities in turn.

We will begin with an anecdote, recorded in a Spanish source not long after the *conquista*, about the raising of children in the Aztec capital of Tenochtitlan, shortly before it fell to Spanish forces: 'at birth boys were given a shield with four arrows. The midwife prayed that they might be courageous warriors. They were presented four times to the sun and told of the uncertainties of life and the need to go to war. Girls, on the other hand, were given spindles and shuttles as a symbol of their future dedication to homely tasks.'[8] It is hard to say how widespread this practice was, but it points to something fundamental in Aztec society. Women still occupied important positions in Tenochtitlan as merchants, doctors and priestesses; but they were excluded from an ascendant class of aristocrats whose power was based on warfare, predation and tribute. How far back this erosion of female political power went among the Aztecs is unclear (certain lines of evidence, such as the obligation for high-ranking advisors at court to take on the cultic role of Cihuacóatl – or 'Snake Woman' – suggest not far at all). What we do know is that masculinity, often expressed through sexual violence, became part of the dynamics of imperial expansion.[9] Indeed, the rape and enslavement of conquered women were among the primary grievances reported to Cortés and his men by Aztec subjects in Veracruz,[10] who by 1519 were willing to take their chances with a band of unknown Spanish freebooters.

Male nobility among the Aztecs or Mexica seem to have viewed life as an eternal contest, or even conquest – a cultural tendency which they traced back to their origins as an itinerant community of warriors and colonizers. Theirs seems to have been a 'capturing society' not unlike some of the other, more recent Amerindian societies we've explored, but on an infinitely greater scale. Enemies taken in war were kept, nurtured to ensure their vitality – sometimes in luxurious circumstances – but then finally killed by ritual specialists to repay a primordial debt of life to the gods, and presumably for any number of other reasons too. At Tenochtitlan's Templo Mayor the result was a veritable industry of pious bloodletting, which some Spanish observers took as clear proof that the Aztec ruling classes were in league with Satan.[11]

This is how the Aztecs attempted to impress their neighbours, and it is still how they impress themselves on the human imagination today: the image of thousands of prisoners, waiting in line to have their hearts torn out by masked god-impersonators, is, admittedly, difficult to get out of one's head. In other respects, however, the sixteenth-century Aztecs seemed to the Spaniards to present a rather familiar picture of human government; certainly, more familiar than anything they encountered in the Caribbean or in the swamps and savannahs of Yucatán. Monarchy, ranks of officialdom, military cadres and organized religion (however 'demonic') were all highly developed. Urban planning in the Valley of Mexico, as some Spaniards remarked, seemed superior to what was found in their Castilian cities back home. Sumptuary laws, no less elaborate than in Spain, kept a respectable distance between governing and governed, dictating everything from fashion to sexual mores. Tribute and taxation were overseen by *calpixque* who, appointed from among the ranks of commoners, were unable to turn their knowledge of administration into political power (a preserve of noblemen and warriors). In the conquered territories local nobles were kept in place, obedience being assured by a patronage system that tied them to sponsors at the Aztec court. Here too the Spaniards found resonance with their practice of *aeque principali*, which granted autonomy to newly acquired territories so long as their local headmen supplied annual tithes to the Crown.[12]

Like the Spanish Habsburgs, who became their overlords, the Aztec warrior aristocracy had risen from relatively humble origins to create one of the world's largest empires. Even their Triple Alliance paled, however, when compared with what the conquistadors found in the Peruvian Andes.

In Spain, as in much of Eurasia, mountains offered refuge from the coercive power of kings and emperors; rebels, bandits and heretics hid out in the highlands. But in Inca Peru, everything seemed to work the other way round. Mountains formed the backbone of imperial power. This upside-down (to European eyes) political world, conceived atop the Andean Cordillera, was the super-kingdom of Tawantinsuyu, meaning 'quarters closely bound'.[13]

More precisely, Tawantinsuyu refers to the four *suyus* or major administrative units of the Sapa Inca's domain. From their capital at Cuzco, where it was said even grass was made of gold, Inca of royal blood extracted periodic *mit'a* – a rotating labour tribute, or corvée – from some millions of subjects distributed across the western littoral of South America, from Quito to Santiago.[14] Exercising a degree of sovereignty over eighty contiguous provinces and countless ethnic groups, by the end of the fifteenth century the Inca had achieved something like the 'universal monarchy' (*monarchia universalis*) that the Habsburgs, rulers of numerous scattered territories, could only conjure in their dreams. Nevertheless, if Tawantinsuyu is to be considered a state, it was still very much a state in formation.

Just as the popular image of the Aztecs turns on mass carnage, popular images of the Inca tend to portray them as master administrators: as we've seen, Enlightenment thinkers like Madame de Graffigny and her readers formed their first impression of what a welfare state, or even state socialism, might be like by contemplating accounts of the empire in the Andes. In reality, Inca efficiency was decidedly uneven. The empire, after all, was over 2,500 miles long. In villages at any appreciable distance from Cuzco, Chan Chan or other centres of royal power, the imperial apparatus made, at best, a sporadic appearance and many villages remained largely self-governing. Chroniclers and officials like Juan Polo de Ondegardo y Zárate were intrigued to discover that while typical Andean villages did indeed have a complex administrative apparatus, that apparatus appeared to be entirely home-grown, based on collective associations called *ayllu*. In order to accommodate imperial demands for tribute or corvée labour, local communities had merely tweaked these collectives slightly.[15]

The imperial centre of the Inca Empire forms a stark contrast with that of the Aztec. Moctezuma, despite his grandeur (his palace contained everything from an aviary to quarters for troupes of comic dwarfs), was officially just the *tlatoani* or 'first speaker' in a council of aristocrats, and his empire officially a Triple Alliance of three cities. For all the bloodthirsty spectacle, the Aztec Empire was really a confederation of noble families. Indeed, the spectacle itself seems to have been at least partly rooted in the same spirit of aristocratic one-upmanship that spurred Aztec nobles to compete in public ball games, or for that matter philosophical debate. The Inca, in contrast, insisted their sovereign was himself the incarnate Sun. All authority derived from a single point of radiance – the person of *Sapa Inca* (Unique Inca) himself – cascading downwards through ranks of royal siblings. The Inca court was an incubator, a hothouse for sovereignty. Compressed within its walls were not only the household of the living king and his sister, who was also his *Coya* (queen), but also the administrative heads, chief priests and imperial guard of the kingdom, most of them blood relatives of the king.

Being a god, the *Sapa Inca* never really died. The bodies of former kings were preserved, wrapped and mummified, much like the pharaohs of ancient Egypt; like the pharaohs, too, they held court from beyond the grave, receiving regular offerings of food and clothing from their former rural estates – though unlike the mummified bodies of Egyptian pharaohs, which at least remained confined to their tombs, their Peruvian equivalents were wheeled out to attend public events and sponsored festivals.[16] (One reason why each new ruler was obliged to expand the empire was precisely this: they only inherited the old ruler's army. His court, lands and retainers remained in the dead Inca's hands.) This extraordinary concentration of power around the Inca's own body had a flip side: royal authority was extremely difficult to delegate.

The most important officials were 'honorary Inca' who, while not directly related to the sovereign, were allowed to wear the same ear ornaments and were otherwise seen as an extension of his personage. Statue doubles or other substitutes might also be employed – there was an elaborate ritual protocol surrounding these – but to do anything important, the Sapa Inca's personal presence was required, meaning the court was continually on the move, with the royal person being regularly carried through the 'four quarters' in a litter lined with silver and feathers. This, as much as the need to carry armies and supplies, required enormous investment in road systems, converting one of the world's most complex and rugged terrains into a continuous network of well-maintained highways and stepped paths, punctuated by shrines (*huacas*) and way stations, stocked and staffed from the royal coffers.[17] It was on one such annual tour, far from the walls of Cuzco, that the last Sapa Inca, Atahualpa, was abducted by Pizarro's men and subsequently killed.

As with the Aztecs, consolidation of the Inca's empire seems to have involved a great deal of sexual violence, and resulting changes in gender roles. In this case, what began as a customary system of marriage became a template for class domination. Traditionally, in those parts of the Andes where people were divided by social rank, women were expected to marry into families of higher status than their own. In doing so the bride's lineage was said to be 'conquered' by the groom's. What began as a kind of ritual figure of speech seems to have been turned into something more literal and systematic. In each newly conquered territory, the Inca immediately built a temple and forced a quota of local virgins to become 'Brides of the Sun': women cut off from their families, kept either as permanent virgins or dedicated to the Sapa Inca, for him to exploit and dispose of as he pleased. In consequence, the king's subjects could be referred to collectively as 'conquered women',[18] and local nobles jockeyed for position by trying to place their daughters in prominent roles at court.

What, then, of the famous Inca administrative system? It did, certainly, exist. Records were kept largely in the form of knotted strings called *khipu* (or *quipu*), described in Pedro Cieza de León's *Crónica del Perú* (1553):

> In each provincial center they had accountants who were called 'knot-keepers/orderers' [*khipukamayuqs*], and by means of these knots they kept the record and account of what had been given in tribute by those [people] in that district, from the silver, gold, clothing, herd animals, to the wool and other things down to the smallest items, and by the same knots they commissioned a record of what was given over one year, or ten or twenty years and they kept the accounts so well that they did not lose a pair of sandals.[19]

Spanish chroniclers provided few details, however, and after the use of *khipu* was officially banned in 1583, local specialists had little incentive to commit their lore to writing. We don't know exactly how it worked, although new sources of information are still emerging from remote Andean communities, where it turns out Inca-style *khipus* and their associated forms of knowledge were kept in use until much more recent times.[20] Scholars argue about whether *khipu* should be considered a form of writing. What sources we do have mainly describe the numerical system, noting the hierarchical arrangement of colour-coded knots into decimal units, from 1 to 10,000; but it seems the most elaborate string bundles encoded records of topography and genealogy, and most likely also narratives and songs.[21]

In many ways, these two great polities – Aztec and Inca – were ideal targets for conquest. Both were organized around easily identifiable capitals, inhabited by easily identifiable kings who could be captured or killed, and surrounded by peoples who were either long accustomed to obeying orders or, if they had any inclination to shrug off power from the centre, were likely to do so precisely by joining forces with would-be conquistadors. If an empire is based largely on military force, it is relatively easy for a superior force of the same kind to seize control of its territory, since if one takes control of that centre – as Cortés did by laying siege to Tenochtitlan in 1521, or Pizarro by seizing Atahualpa at Cajamarca in 1532 – everything else falls readily into place. There might be stubborn resistance (the siege of Tenochtitlan took over a year of gruelling house-to-house fighting) but, once it was over, the conquerors could take over many of the mechanisms of rule that already existed and start conveying orders to subjects schooled in obedience.

Where there are no such powerful kingdoms – either because they had never existed, as in much of North America or Amazonia, or because a population had consciously rejected central government – things could get decidedly trickier.

A good example of such decentralization is the territory inhabited by speakers of the various Maya languages: the Yucatán Peninsula and the highlands of Guatemala and Chiapas to its south.[22] At the time of the initial Spanish incursion, the region was divided into what seemed to the settlers an endless succession of tiny principalities, townships, villages and seasonal hamlets. Conquest was a long and laborious business, and no sooner was it completed (or at least, no sooner had the Spanish decided it was completed),[23] than the new authorities faced an apparently endless series of popular revolts.

As early as 1546, a coalition of Maya rebels rose up against Spanish settlers and, despite brutal reprisals, resistance never really died down. Prophetic movements brought a second major wave of insurrections in the eighteenth century; and in 1848, a mass rising almost drove the settlers' descendants out of Yucatán entirely, until the siege of their capital, at Mérida, was interrupted by the planting season. The resulting 'Caste War', as it was called, continued for generations. There were still rebels holding out in parts of Quintana Roo at the time of the Mexican Revolution in the second decade of the twentieth century; indeed, you could argue that the same rebellion continues, in another form, with the Zapatista movement that controls large parts of Chiapas today. As the Zapatistas also show, it was in these territories, where no major state or empire had existed for centuries, that women came most prominently to the fore in anti-colonial struggles, both as organizers of armed resistance and as defenders of indigenous tradition.

Now, this anti-authoritarian streak might come as something of a surprise to those who know the Maya as one of a triumvirate of New World civilizations – Aztecs, Maya, Inca – familiar from books on art history. Much of the art from what's called the Classic Maya period, roughly AD 150–900, is exquisitely beautiful. Most derives from cities that once existed in what are now the tangled rainforests of Petén. On first appraisal, the Maya in this period seem to have been organized into kingdoms much like those of the Andes or central Mexico, only smaller; but then our picture, until quite recently, was dominated by sculpted monuments and glyphic inscriptions commissioned by the ruling elites themselves.[24] These focus, predictably enough, on the deeds of great rulers (holders of the title *ajaw*), especially their conquests, as alliances of independent city-states vied for hegemony over the lowlands under the leadership of two rival dynasties – those of Tikal and the 'snake kings' of Calakmul.[25]

These monuments tell us a great deal about the rituals such rulers conducted to commune with their divinized ancestors[26] – but precious little about what ordinary life was like for their subjects, let alone how those subjects felt about their rulers' claims to cosmic power. If there were prophetic movements or periodic insurrections during the Classic Maya period, as there were in the colonial period, we would currently have few ways to know about them; although archaeological research may yet change this picture. What we do know is that, in the final centuries of the Classic period, women attain a new visibility in sculpture and inscription, appearing not just as consorts, princesses and queen mothers but also as powerful rulers and spirit mediums in their own right. We also know that at some point in the ninth century the Classic Maya political system came apart, and most of the great cities were abandoned.

Archaeologists argue about what happened. Some theories assume that popular resistance – some combination of defection, mass movements or outright rebellion – must have played a part, even if most are understandably reluctant to draw too firm a line between cause and consequence.[27] It is significant that one of the few urban societies which endured, even grew, was located in the northern lowlands around the city of Chichén Itza. Here, kingship seems to have dramatically changed its character, becoming a more purely ceremonial or even theatrical affair – so hedged about by ritual that any serious political intervention was no longer possible – while day-to-day governance apparently passed largely into the hands of a coalition that formed among collectives of prominent warriors and priests.[28] Indeed, some of what were once assumed to be royal palaces in this 'Post-Classic' period are now being reinterpreted as assembly halls (*popolna*) for local representatives.[29]

By the time the Spaniards arrived, six centuries after the collapse of cities in Petén, Mayan societies were thoroughly decentralized, parsed into a bewildering variety of townships and principalities, many without kings.[30] The books of *Chilam Balam*, prophetic annals written down in the late sixteenth century, dwell endlessly on the disasters and miseries that befall oppressive rulers. In other words, there's every reason to believe that the spirit of rebellion which has marked this particular region can be traced back to at least the time of Charlemagne (the eighth century AD); and that across the centuries, overbearing Maya rulers were quite regularly and repeatedly disposed of.

Undoubtedly, the Classic Maya artistic tradition is magnificent, one of the greatest the world has ever seen. By comparison, artistic products from the 'Post-Classic' – as the period from roughly AD 900 to 1520 is known – often seem clumsy and less worthy of appreciation. On the other hand, how many of us would really prefer to live under the arbitrary power of a petty warlord who, for all his patronage of fine arts, counts tearing the hearts out of living human bodies among his most significant accomplishments? Of course, history is not usually thought about in such terms, and it is worth asking why. Part of the reason is simply the designation 'Post-Classic', which suggests little more than an afterthought. It may seem a trivial issue – but it matters, because such habits of thought are one reason why periods of relative freedom and equality tend to get sidelined in the larger sweep of history.

This is important: let's look at it further, before we return to our three forms of domination.

IN WHICH WE OFFER A DIGRESSION ON 'THE SHAPE OF TIME',[31] AND SPECIFICALLY HOW METAPHORS OF GROWTH AND DECAY INTRODUCE UNNOTICED POLITICAL BIASES INTO OUR VIEW OF HISTORY

History and archaeology abound with terms like 'post' and 'proto', 'intermediate' or even 'terminal'. To some degree, these are products of early-twentieth-century cultural theory. Alfred Kroeber, a pre-eminent anthropologist of his day, spent decades on a research project aimed at determining if identifiable laws lie behind the rhythms and patterns of cultural growth and decay: whether systematic relations could be established between artistic fashions, economic booms and busts, periods of intellectual creativity and conservatism, and the expansion and collapse of empires. It was an intriguing question but, after many years, his ultimate conclusion was: no, there were no such laws. In his *Configurations of Cultural Growth* (1944) Kroeber examined the relation of the arts, philosophy, science and population across human history and found no evidence for any consistent pattern; nor has any such pattern been successfully discerned in those few more recent studies which continue to plough the same furrow.[32]

Despite this, when we write about the past today we almost invariably organize our thinking as if such patterns really did exist. Civilizations are typically represented either as flower-like – growing, blooming and then shrivelling up – or else as like some grand building, painstakingly constructed but prone to sudden 'collapse'. The latter term tends to be used indiscriminately for situations like the Classic Maya collapse, which did indeed involve a rapid abandonment of some hundreds of settlements and the disappearance of millions of people; but equally it's used for the 'collapse' of the Egyptian Old Kingdom, where the only thing that really seems to have declined precipitously is the power of Egypt's elites ruling from the northern city of Memphis.

Even in the Maya case, to describe the entire period between AD 900 and 1520 as 'Post-Classic' is to suggest that the only really significant thing about it is the degree to which it can be seen as the waning of a Golden Age. In a similar way, terms like 'Proto-palatial Crete', 'Predynastic Egypt' or 'Formative Peru' convey a sense of impatience, as if Minoans, Egyptians or Andean peoples spent centuries doing little but laying the groundwork for such a Golden Age – and, it is implied, for strong, stable government – to come about.[33] We've already seen how this played out in Uruk, where at least seven centuries of collective self-rule (also termed 'Predynastic' in earlier scholarship) comes to be written off as a mere prelude to the 'real' history of Mesopotamia – which is then presented as a history of conquerors, dynasts, lawgivers and kings.

Some periods are dismissed as prefaces, others as postfaces. Still others become 'intermediary'. The ancient Andes and Mesoamerica are cases in point, but probably the most familiar – and the most striking – example is again that of Egypt. Museumgoers will no doubt be familiar with the division of ancient Egyptian history into Old, Middle and New Kingdoms. Each is separated by an 'intermediate' period, often described as epochs of 'chaos and cultural degeneration'. In fact, these were simply periods when there was no single ruler of Egypt. Authority devolved to local factions or, as we will shortly see, changed its nature altogether. Taken together, these intermediate periods span about a third of Egypt's ancient history, down to the accession of a series of foreign or vassal kings (known simply as the Late period), and they saw some very significant political developments of their own.

To take just one example, at Thebes between 754 and 525 BC – spanning the Third Intermediate and Late periods – a series of five unmarried, childless princesses (of Libyan and Nubian descent) were elevated to the position of 'god's wife of Amun', a title and role which acquired not just supreme religious, but also great economic and political weight at this time. In official representations, these women are given 'throne names' framed by cartouches, just like kings, and appear leading royal festivals and making offerings to the gods.[34] They also owned some of the richest estates in Egypt, including extensive lands and a large staff of priests and scribes. To have a situation in which women not only command power on such a scale, but in which this power is linked to an office reserved explicitly for single women, is historically unusual. Yet this political innovation is little discussed, partly because it is already framed within an 'intermediate' or 'late' period that signals its transitory (or even decadent) nature.[35]

One might assume the division into Old, Middle and New Kingdoms is itself very ancient, perhaps going back thousands of years to Greek sources like the third-century BC *Aegyptiaca*, composed by Egyptian chronicler Manetho, or even to the hieroglyphic records themselves. Not so. In fact, the tripartite division only began to be proposed by modern Egyptologists in the late nineteenth century, and the terms they introduced (initially 'Reich' or 'empire', later 'kingdom') were explicitly modelled on European nation states. German, particularly Prussian, scholars played a leading role here. Their tendency to perceive ancient Egypt's past as a series of cyclical alternations between unity and disintegration clearly echoes the political concerns of Bismarck's Germany, where an authoritarian government was trying to assemble a unified nation state from an endless variety of tiny statelets. After the First World War, as Europe's own regime of old monarchies was coming apart, prominent Egyptologists such as Adolf Erman granted the 'Intermediate' periods their own place in history, drawing comparisons between the end of the Old Kingdom and the Bolshevik Revolution of their own time.[36]

With hindsight, it's easy to see just how much these chronological schemes reflect their authors' political concerns. Or even, perhaps, a tendency – when casting their minds back in time – to imagine themselves either as part of the ruling elite, or as having roles somewhat analogous to ones they had in their own societies: the Egyptian or Maya equivalents of museum curators, professors and middle-range functionaries. But why, then, have these schemes become effectively canonical?

Consider the Middle Kingdom (2055–1650 BC), represented in standard histories as a time when Egypt moved from the supposed chaos of the First Intermediate period into a renewed phase of strong and stable government, bringing with it an artistic and literary renaissance.[37] Even if we set aside the question of just how chaotic the 'intermediate period' really was (we'll get to that soon), the Middle Kingdom could equally well be represented as a period of violent disputes over royal succession, crippling taxation, state-sponsored suppression of ethnic minorities, and the growth of forced labour to support royal mining expeditions and construction projects – not to mention the brutal plundering of Egypt's southern neighbours for slaves and gold. However much future Egyptologists would come to appreciate them, the elegance of Middle Kingdom literature like *The Story of Sinuhe* and the proliferation of Osiris cults likely offered little solace to the thousands of military conscripts, forced labourers and persecuted minorities of the time, many of whose grandparents were living quite peaceful lives in the preceding 'dark ages'.

What is true of time, incidentally, is also true of space. For the last 5,000 years of human history – i.e. roughly the span of time we will be moving around in, over the course of this chapter – our conventional vision of world history is a chequerboard of cities, empires and kingdoms; but in fact, for most of this period these were exceptional islands of political hierarchy, surrounded by much larger territories whose inhabitants, if visible at all to historians' eyes, are variously described as 'tribal confederacies', 'amphictyonies' or (if you're an anthropologist) 'segmentary societies' – that is, people who systematically avoided fixed, overarching systems of authority. We know a bit about how such societies worked in parts of Africa, North America, Central or Southeast Asia and other regions where such loose and flexible political associations existed into recent times, but we know frustratingly little of how they operated in periods when these were by far the world's most common forms of government.

A truly radical account, perhaps, would retell human history from the perspective of the times and places in between. In that sense, this chapter is not truly radical: for the most part, we are telling the same old story; but we are at least trying to see what happens when we drop the teleological habit of thought, which makes us scour the ancient world for embryonic versions of our modern nation states. We are considering, instead, the possibility that – when looking at those times and places usually taken to mark 'the birth of the state' – we may in fact be seeing how very different kinds of power crystallize, each with its own peculiar mix of violence, knowledge and charisma: our three elementary forms of domination.

One way to test the value of a new approach is to see if it helps us explain what had previously seemed anomalous cases: that is, ancient polities which undeniably mobilized and organized enormous numbers of people, but which don't seem to fit any of the usual definitions of a state. Certainly, there are plenty of these. Let's start with the Olmec, generally seen as the first great Mesoamerican civilization.

ON POLITICS AS SPORT: THE OLMEC CASE

How precisely to describe the Olmec has proved a difficult problem for archaeologists to grapple with. Early-twentieth-century scholars referred to them as an artistic or cultural 'horizon', largely because it wasn't clear how else to describe a style – easily identifiable by certain common types of pottery, anthropomorphic figurines and stone sculpture – that seemed to pop up between 1500 and 1000 BC across an enormous area, straddling the Isthmus of Tehuantepec and including Guatemala, Honduras and much of southern Mexico, but whose meaning was otherwise uncertain. Whatever the Olmec were, they seemed to represent the 'mother culture', as it came to be known, of all later Mesoamerican civilizations, having invented the region's characteristic calendar systems, glyphic writing and even ball games.[38]

At the same time, there was no reason to assume the Olmec were a unified ethnic or even political group. There was much speculation about wandering missionaries, trading empires, elite fashion styles and much else besides. Eventually, archaeologists came to understand that there was, in fact, an Olmec heartland in the marshlands of Veracruz, where the swamp cities of San Lorenzo and La Venta arose along the fringes of Mexico's Gulf Coast. The internal structure of these Olmec cities is still poorly understood. Most seem to have been centred on ceremonial precincts – of uncertain layout, but including large earthen pyramid mounds – surrounded by extensive suburbs. These monumental epicentres stand in relative isolation, amid an otherwise fragmented and relatively unstructured landscape of small maize-farming settlements and seasonal forager camps.[39]

What can we really say, then, about the structure of Olmec society? We know it was in no sense egalitarian; there were clearly marked elites. The pyramids and other monuments suggest that, at least at certain times of year, these elites had extraordinary resources of skill and labour at their disposal. In every other respect, though, ties between centre and hinterland appear to have been surprisingly superficial. The collapse of the first great Olmec city at San Lorenzo, for instance, seems to have had very little impact on the wider regional economy.[40]

Any further assessment of Olmec political structure has to reckon with what many consider its signature achievement: a series of absolutely colossal sculpted heads. These remarkable objects are free-standing, carved from tons of basalt, and of a quality comparable with the finest ancient Egyptian stonework. Each must have taken untold hours of grinding to produce. These sculptures appear to be representations of Olmec leaders, but, intriguingly, they are depicted wearing the leather helmets of ball players. All the known examples are sufficiently similar that each seems to reflect some kind of standard ideal of male beauty; but, at the same time, each is also different enough to be seen as a unique portrait of a particular, individual champion.[41]

No doubt there were also actual ball-courts – though these have proved surprisingly elusive in the archaeological record – and while we obviously don't know what kind of game was played, if they were anything like later Maya and Aztec ball games it likely took place in a long and narrow court, with two teams from high-ranking families competing for fame and honour by striking a heavy rubber ball with the hips and buttocks. It seems both reasonable and logical to conclude that there was a fairly direct relationship between competitive games and the rise of an Olmec aristocracy.[42] Without written evidence it's hard to say much more, but looking a bit closer at later Mesoamerican ball games might at least give us a sense of how this worked in practice.

Stone ball-courts were common features of Classic Maya cities, alongside royal residences and pyramid-temples. Some were purely ceremonial; others were actually used for sport. The chief Maya gods were themselves ball players. In the K'iche Maya epic *Popol Vuh* a ball game provides the setting in which mortal heroes and underworld gods collide, leading to the birth of the Hero Twins Hunahpu and Xbalanque, who go on to beat the gods at their own deadly game and ascend to take their own place among the stars.

The fact that the greatest known Maya epic centres on a ball game gives us a sense of how central the sport was to Maya notions of charisma and authority. So too, in a more visceral way, does an inscribed staircase built at Yaxchilán to mark the accession (in AD 752) of what was probably its most famous king, known as Bird Jaguar the Great. On the central block he appears as a ball player. Flanked by two dwarf attendants, the king prepares to strike a huge rubber ball containing the body of a human captive – bound, broken and bundled – as it tumbles down a flight of stairs. Capturing high-ranking enemies to be held for ransom or, failing payment, to be killed at ball games was a major objective of Maya warfare. This particular unfortunate figure may be a certain Jewelled Skull, a noble from a rival city, whose humiliation was so important to Bird Jaguar that he also made it the central feature of a carved lintel on a nearby temple.[43]

In some parts of the Americas, competitive sports served as a substitute for war. Among the Classic Maya, one was really an extension of the other. Battles and games formed part of an annual cycle of royal competitions, played for life and death. Both are recorded on Maya monuments as key events in the lives of rulers. Most likely, these elite games were also mass spectacles, cultivating a particular sort of urban public – the sort that relishes gladiatorial contests, and thereby comes to understand politics in terms of opposition. Centuries later, Spanish conquistadors described Aztec versions of the ball game played at Tenochtitlan, where players confronted each other amid racks of human skulls. They reported how reckless commoners, carried away in the competitive fervour of the tournament, would sometimes lose all they had or even gamble themselves into slavery.[44] The stakes were so high that, should a player actually send a ball through one of the stone hoops adorning the side of the court (these were made so small as to render it nearly impossible; normally the game was won in other ways), the contest ended immediately, and the player who performed the miracle received all the goods wagered, as well as any others he might care to pillage from the onlookers.[45]

It is easy to see why the Olmec, with their intense fusion of political competition and organized spectacle, are nowadays seen as cultural progenitors of later Mesoamerican kingdoms and empires; but there is little evidence that the Olmec themselves ever created an infrastructure for dominating a large population. So far as anyone knows, their rulers did not command a stable military or administrative apparatus that might have allowed them to extend their power throughout a wider hinterland. Instead, they presided over a remarkable spread of cultural influence radiating from ceremonial centres, which may only have been densely occupied on specific occasions (such as ritual ball games) scheduled in concert with the demands of the agricultural calendar, and largely empty at other times of year.

In other words, if these were 'states' in any sense at all, then they are probably best defined as seasonal versions of what Clifford Geertz once called 'theatre states', where organized power was realized only periodically, in grand but fleeting spectacles. Anything we might consider 'statecraft', from diplomacy to the stockpiling of resources, existed in order to facilitate the rituals, rather than the other way round.[46]

CHAVÍN DE HUÁNTAR – AN 'EMPIRE' BUILT ON IMAGES?

In South America we find a somewhat analogous situation. Before the Inca, a whole series of other societies are identified tentatively by scholars as 'states' or 'empires'. All these societies occur within the area later controlled by the Inca: the Peruvian Andes and adjacent coastal drainages. None used writing, at least in any form we can recognize. Still, from AD 600 onwards many did employ knotted strings for record-keeping, and probably other forms of notation too.

Monumental centres of some kind were already appearing in the Rio Supe region in the third millennium BC.[47] Later, between 1000 and 200 BC, a single centre – at Chavín de Huántar, in the northern highlands of Peru – extended its influence over a much larger area.[48] This 'Chavín horizon' gave way to three distinct regional cultures. In the central highlands arose a militarized polity known as Wari. In parallel, on the shores of Lake Titicaca, a metropolis called Tiwanaku – at 420 hectares, roughly twice the size of Uruk or Mohenjo-daro – took form, using an ingenious system of raised fields to grow its crops on the freezing heights of the Bolivian *altiplano*.[49] On the north coast of Peru, a third culture, known as Moche, displays striking funerary evidence of female leadership: lavish tombs of warrior-priestesses and queens, drenched in gold and flanked by human sacrifices.[50]

The first Europeans to study these civilizations, in the late nineteenth and early twentieth centuries, assumed that any city or set of cities with monumental art and architecture, exerting its 'influence' over a surrounding region, must be the capitals of states or empires (they also assumed – just as wrongly, it turns out – that all the rulers were male). As with the Olmec, a surprisingly large proportion of that influence seems to have come in the form of images – distributed, in the Andean case, on small ceramic vessels, objects of personal adornment and textiles – rather than in the spread of administrative, military or commercial institutions and their associated technologies.

Consider Chavín de Huántar itself, located high in the Mosna valley of the Peruvian Andes. Archaeologists once believed it to have been the core of a pre-Inca empire in the first millennium BC: a state controlling a hinterland that stretched to the Amazonian rainforest to the east and the Pacific Coast to the west, and included all the intervening highlands and coastal drainages in between. Such power seemed commensurate with the scale and sophistication of Chavín's cut-stone architecture, its unrivalled abundance of monumental sculpture, and the appearance of Chavín motifs on pottery, jewellery and textiles across the wider region. But was Chavín really some kind of 'Rome of the Andes'?

In fact, little evidence has emerged since to suggest this. In order to get a sense of what might really have been going on at Chavín we must look more closely at the sort of images we're talking about, and what they tell us about the wider importance of vision and knowledge in Chavín notions of power.

The art of Chavín is not made up of pictures, still less pictorial narratives – at least, not in any intuitively recognizable sense. Neither does it appear to be a pictographic writing system. This is one reason why we can be fairly certain we are not dealing with an actual empire. Real empires tend to favour styles of figural art that are both very large but also very simple, so their meaning can be easily understood by anyone they wish to impress. If an Achaemenid Persian emperor carved his likeness into the side of a mountain, he did it in such a way that anyone, even an ambassador from lands as yet unknown to him (or an antiquarian of some remote future age), would be able to recognize that it is indeed the image of a very great king.

Chavín images, by contrast, are not for the uninitiated. Crested eagles curl in on themselves, vanishing into a maze of ornament; human faces grow snake-like fangs, or contort into a feline grimace. No doubt other figures escape our attention altogether. Only after some study do even the most elementary forms reveal themselves to the untrained eye. With due attention, we can eventually begin to tease out recurrent images of tropical forest animals – jaguars, snakes, caimans – but just as the eye attunes to them they slip back from our field of vision, winding in and out of each other's bodies or merging into complex patterns.[51]

Some of these images are described by scholars as 'monsters', but they have nothing in common with the simple composite figures of ancient Greek vases or Mesopotamian sculpture – centaurs, griffins and the like – or their Moche equivalents. We are in another kind of visual universe altogether. It is the realm of the shape-shifter, where no body is ever quite stable or complete, and diligent mental training is required to tease out structure from what at first seems to be visual mayhem. One reason why we can say any of this with a degree of confidence is because the arts of Chavín appear to be an early (and monumental) manifestation of a much wider Amerindian tradition, in which images are not meant to illustrate or represent, but instead serve as visual cues for extraordinary feats of memory.

Up until recent times, a great many indigenous societies were still using systems of broadly similar kinds to transmit esoteric knowledge of ritual formulae, genealogies or records of shamanic journeys to the world of chthonic spirits and animal familiars.[52] In Eurasia, similar techniques were developed in the ancient 'arts of memory', where those trying to memorize stories, speeches, lists or similar material would each have a familiar 'memory palace'. This consisted of a mental pathway or room in which a series of striking images could be arranged, each a cue to a particular episode, incident or name. One can only imagine what might happen if someone were to draw or carve one such set of visual cues, and a later archaeologist or art historian were to discover it, with no idea of the context, let alone what the story being memorized was actually about.

In the case of Chavín, we actually can be on fairly safe ground in assuming that these images were records of shamanic journeys; not just because of the peculiar nature of the images themselves, but also because of a wealth of circumstantial evidence relating to altered states of consciousness. At Chavín itself, snuff spoons, small ornate mortars and bone pipes have been found; and among its carved images are sculpted male figures with fangs and snake headdresses holding aloft the stalk of the San Pedro cactus. This plant is the basis of *Huachuma*, a mescaline-based infusion still made in the region today which induces psychoactive visions. Other carved figures, all of them apparently male, are surrounded by images of vilca leaves (*Anadenanthera sp.*), which contain a powerful hallucinogen. Released when the leaves are ground up and snorted, it induces a gush of mucus from the nose, as faithfully depicted on sculpted heads that line the walls of Chavín's major temples.[53]

In fact, nothing in Chavín's monumental landscape really seems concerned with secular government at all. There are no obvious military fortifications or administrative quarters. Almost everything, on the other hand, seems to have something to do with ritual performance and the revelation or concealment of esoteric knowledge.[54] Intriguingly, this is exactly what indigenous informants were still telling Spanish soldiers and chroniclers who arrived at the site in the seventeenth century. For as long as anyone could remember, they said, Chavín had been a place of pilgrimage but also one of supernatural danger, on which the heads of important families converged from different parts of the country to seek visions and oracles: the 'speech of the stones'. Despite initial scepticism, archaeologists have gradually come round to accepting that they were right.[55]

It's not just the evidence for ritual and altered states of mind, but also the extraordinary architecture of the place. The temples at Chavín contain stone labyrinths and hanging staircases which seem designed not for communal acts of worship but for individual trials, initiations and vision quests. They imply tortuous journeys ending at narrow corridors, large enough for only a single person, beyond which lies a tiny sanctum containing a monolith, carved with dense tangles of images. The most famous such monument, a stela called 'El Lanzón' ('the lance'), is a shaft of granite over thirteen feet tall, around which the Old Temple of Chavín was constructed. A well-lit replica of the stela, often assumed to represent a god who is also the *axis mundi*, or a central pillar connecting the polar ends of a shamanic universe, has pride of place in Peru's Museo de la Nación; but the 3,000-year-old original still resides at the heart of a darkened maze, illuminated by thin slats, where no single viewer could ever grasp the totality of its form or meaning.[56]

If Chavín – a remote precursor to the Inca – was an 'empire', it was one built on images linked to esoteric knowledge. Olmec was, on the other hand, an 'empire' built on spectacle, competition and the personal attributes of political leaders. Clearly, our use of the term 'empire' here is about as loose as it could possibly be. Neither was remotely similar to, say, the Roman or Han, or indeed the Inca and Aztec Empires. Nor do they fulfil any of the important criteria for 'statehood' – at least not on most standard sociological definitions (monopoly of violence, levels of administrative hierarchy, and so forth). The usual recourse is to describe such regimes instead as 'complex chiefdoms', but this too seems hopelessly inadequate – a shorthand way of saying, 'looks somewhat like a state, but it isn't one'. This tells us precisely nothing.

What makes more sense, we suggest, is to look at these otherwise puzzling cases through the lens of our three elementary principles of domination – control of violence (or sovereignty), control of knowledge, and charismatic politics – outlined at the start of the chapter. In doing so, we can see how each stresses a particular form of domination to an exceptional degree and develops it on an unusually large scale. Let's give it a go.

First, in the case of Chavín, power over a large and dispersed population was clearly about retaining control over certain kinds of knowledge: something perhaps not that far removed from the idea of 'state secrets' found in later bureaucratic regimes, although the content was obviously very different, and there was little in the way of military force to back it up. In the Olmec tradition, power involved certain formalized ways of competing for personal recognition, in an atmosphere of play laced with risk: a prime example of a large-scale, competitive political field, but again in the absence of territorial sovereignty or an administrative apparatus. No doubt there was a certain degree of personal charisma and jockeying at Chavín; no doubt among the Olmec, too, some obtained influence by their command of arcane knowledge; but neither case gives us reason to think anyone was asserting a strong principle of sovereignty.

We'll refer to these as 'first-order regimes' because they seem to be organized around one of the three elementary forms of domination (knowledge-control, for Chavín; charismatic politics for Olmec), to the relative neglect of the other two. The obvious next question, then, is whether examples of the third possible variant can also be found: i.e. cases of societies which develop a principle of sovereignty (that is, grant an individual or small group a monopoly on the right to use violence with impunity), and take it to extreme lengths, without either an apparatus for controlling knowledge or any sort of competitive political field. In fact, there are quite a lot of examples. Admittedly, the existence of such a society would probably be more difficult to establish from archaeological evidence alone, but to illustrate this third variant we can turn, fortunately, to more recent Amerindian societies where written documentation is available.

As always, we must be careful with such sources, since they are written by European observers who not only brought their own biases but tended to describe societies already enmeshed in the chaotic destruction that Europeans themselves almost invariably brought in their wake. Still, French accounts of the Natchez of southern Louisiana in the eighteenth century seem to describe exactly the sort of arrangement we are interested in. By general consent, the Natchez (who called themselves *Théoloël*, or 'People of the Sun') represent the only undisputed case of divine kingship north of the Rio Grande. Their ruler enjoyed an absolute power of command that would have satisfied a Sapa Inca or Egyptian pharaoh; but they had a minimal bureaucracy, and nothing like a competitive political field. As far as we know it has never occurred to anyone to refer to this arrangement as a 'state'.

ON SOVEREIGNTY WITHOUT 'THE STATE'

Let us turn to the work of a French Jesuit, Father Maturin Le Petit, who gave an account of the Natchez in the early eighteenth century. Le Petit found the Natchez to be nothing like the people Jesuits had encountered in what is now Canada. He was especially struck by their religious practices. These revolved around a settlement all the French sources refer to as the Great Village, which centred on two great earthen platforms separated by a plaza. On one platform was a temple; on the other a kind of palace, the house of a ruler called the Great Sun, large enough to contain up to 4,000 people, roughly the size of the entire Natchez population at the time.

The temple, in which an eternal fire burned, was dedicated to the founder of the royal dynasty. The current ruler, together with his brother (called 'the Tattooed Serpent') and eldest sister ('the White Woman'), were for their own parts treated with something that seemed very much like worship. Anyone who came into their presence was expected to bow and wail, and to retreat backwards. No one, not even the king's wives, was allowed to share a meal with him; only the most privileged could even see him eat. What this meant in practice was that members of the royal family lived out their lives largely within the confines of the Great Village itself, rarely venturing beyond.[57] The king himself emerged mainly during major rituals or times of war.

Le Petit and other French observers – who at the time lived under the suzerainty of Louis XIV, who of course also fancied himself a 'Sun King' – were quite fascinated by the parallels: as a result, they described the goings-on in the Great Village in some detail. The Natchez Great Sun might not have had the grandeur of Louis XIV, but what he lacked in that regard he appeared to make up for in terms of sheer personal power. French observers were particularly struck by the arbitrary executions of Natchez subjects, the property confiscations and the way in which, at royal funerals, court retainers would – often, apparently, quite willingly – offer themselves up to be strangled to accompany the Great Sun and his closest family members in death. Those sacrificed on such occasions consisted largely of people who were, up to that point, immediately responsible for the king's care and his physical needs – including his wives, who were invariably commoners (the Natchez were matrilineal, so it was the White Woman's children that succeeded to the throne). Many, according to French accounts, went to their deaths voluntarily, even joyfully. One wife remarked how she dreamed of finally being able to share a meal with her husband, in another world.

One paradoxical outcome of these arrangements was that, for most of the year, the Great Village was largely depopulated. As noted by another observer, Father Pierre de Charlevoix, 'The great Village of the Natchez is at present reduced to a very few Cabins. The Reason which I heard for this is that the Savages, from whom the Great Chief has the Right to take all they have, get as far away from him as they can; and therefore, many Villages of this Nation have been formed at some Distance from this.'[58]

Away from the Great Village, ordinary Natchez appear to have led very different lives, often showing blissful disregard for the wishes of their ostensible rulers. They conducted their own independent commercial and military ventures, and sometimes flatly refused royal commands conveyed by the Great Sun's emissaries or relatives. Archaeological surveys of the Natchez Bluffs region bear this out, showing that the eighteenth-century 'kingdom' in fact comprised semi-autonomous districts, including many settlements that were both larger and wealthier in trade goods than the Great Village itself.[59]

How exactly are we to understand this situation? It might seem paradoxical – but historically such arrangements are not particularly unusual. The Great Sun was a sovereign in the classical sense of the term, which is to say he embodied a principle that was seen as higher than law. Therefore no law applied to him. This is a very common bit of cosmological reasoning that we find, in some form or another, almost anywhere from Bologna to Mbanza Congo. Just as gods (or God) are not seen as bound by morality – since only a principle existing beyond good and evil could have created good and evil to begin with – so 'divine kings' cannot be judged in human terms; behaving in arbitrarily violent ways to anyone around them is itself proof of their transcendent status. Yet at the same time, they are expected to be creators and enforcers of systems of justice. Such with the Natchez too. The Great Sun was said to be descended from a child of the Sun who came to earth bearing a universal code of laws, among the most prominent of which were proscriptions against theft and murder. Yet the Great Sun himself ostentatiously violated those laws on a regular basis, as if to prove his identification with a principle prior to law and, therefore, able to create it.

The problem with this sort of power (at least, from the sovereign's vantage point) is that it tends to be intensely personal. It is almost impossible to delegate. The king's sovereignty extends about as far as the king himself can walk, reach, see or be carried. Within that circle it is absolute. Outside it, it attenuates rapidly. As a result, in the absence of an administrative system (and the Natchez king had only a handful of assistants), claims to labour, tribute or obedience could, if considered odious, be simply ignored. Even the 'absolutist' monarchs of the Renaissance, like Henry VIII or Louis XIV, had a great deal of trouble delegating their authority – that is, convincing their subjects to treat royal representatives as deserving anything like the same deference and obedience due to the king himself. Even if one does develop an administrative apparatus (as they of course did), there is the additional problem of how to get the administrators actually to do what they're told – and, by the same token, how to get anyone to tell you if they aren't. As late as the 1780s, as Max Weber liked to point out, Frederick the Great of Prussia found that his repeated efforts to free the country's serfs came to nothing because bureaucrats would simply ignore the decrees or, if challenged by his legates, insisted the words of the decree should be interpreted as saying the exact opposite of what was obviously intended.[60]

In this sense, French observers were not entirely off the mark: the Natchez court really could be considered a sort of hyper-concentrated version of Versailles. On the one hand, the Great Sun's power in his immediate presence was even more absolute (Louis could not actually snap his fingers and order someone executed on the spot); while on the other, his ability to extend that power was even more restricted (Louis did, after all, have an administration at his disposal, though a fairly limited one compared to modern nation states). Natchez sovereignty was, effectively, bottled up. There was even a suggestion that this power, and particularly its benevolent aspect, was in some way dependent on being bottled up. According to one account, the main ritual role of the king was to seek blessings for his people – health, fertility, prosperity – from the original lawgiver, a being who in his lifetime was so terrifying and destructive that he eventually agreed to be turned into a stone statue and hidden in a temple where no one would see him.[61] In a similar way, the king was sacred, and could be a conduit for such blessings, precisely insofar as he could be contained.

The Natchez case illustrates, with unusual clarity, a more general principle whereby the containment of kings becomes one of the keys to their ritual power. Sovereignty always represents itself as a symbolic break with the moral order; this is why kings so often commit some kind of outrage to establish themselves, massacring their brothers, marrying their sisters, desecrating the bones of their ancestors or, in some documented cases, literally standing outside their palace and gunning down random passers-by.[62] Yet that very act establishes the king as potential lawmaker and high tribunal, in much the same way that 'High Gods' are so often represented as both throwing random bolts of lightning, and standing in judgment over the moral acts of human beings.

People have an unfortunate tendency to see the successful prosecution of arbitrary violence as in some sense divine, or at least to identify it with some kind of transcendental power. We might not fall on our knees before any thug or bully who manages to wreak havoc with impunity (at least, if he isn't actually in the room), but insofar as such a figure does manage to establish themselves as genuinely standing above the law – in other words, as sacred or set apart – another apparently universal principle kicks in: in order to keep him apart from the muck and mire of ordinary human life, that same figure becomes surrounded with restrictions. Violent men generally insist on tokens of respect, but tokens of respect taken to the cosmological level – 'not to touch the earth', 'not to see the sun' – tend to become severe limits on one's freedom to act, violently or indeed in most other ways.[63]

For most of history, this was the internal dynamic of sovereignty. Rulers would try to establish the arbitrary nature of their power; their subjects, insofar as they were not simply avoiding the kings entirely, would try to surround the godlike personages of those rulers with an endless maze of ritual restrictions, so elaborate that the rulers ended up, effectively, imprisoned in their palaces – or even, as in some of the cases of 'divine kingship' first made famous by Sir James Frazer's *The Golden Bough*, facing ritual death themselves.

So far, then, we have seen how each of the three principles we began with – violence, knowledge and charisma – could, in first-order regimes, become the basis for political structures which, in some ways, resemble what we think of as a state, but in others clearly don't. None could in any sense be described as 'egalitarian' societies – they were all organized around a very clearly demarcated elite – but at the same time, it's not at all clear how far the existence of such elites restricted the basic freedoms we described in earlier chapters. There is little reason to believe, for instance, that such regimes did much to impair freedom of movement: Natchez subjects seemed to have faced little opposition if they chose simply to move away from the proximity of the Great Sun, which they generally did. Neither do we find any clear sense of the giving or taking of orders, except in the sovereign's immediate (and decidedly limited) ambit.

Another instructive case of sovereignty without the state is found in the recent history of South Sudan, among the Shilluk, a Nilotic people living alongside the Nuer. To recap, the early-twentieth-century Nuer were a pastoral society, of the sort often referred to in the anthropological literature as 'egalitarian' (though not, in fact, entirely so), because of their extreme distaste for any situation that might even suggest the giving and taking of orders. The Shilluk speak a western Nilotic language closely related to Nuer, and most believe that at some point in the past they were one people. While the Nuer occupied lands best fit for cattle-grazing, the Shilluk found themselves living along a fertile stretch of the White Nile, which allowed them to grow the local grain known as durra, and support dense populations. However, the Shilluk – unlike the Nuer – had a king. Known as the *reth*, this Shilluk monarch could also be seen as embodying sovereignty in the raw, in much the same way as the Natchez Great Sun.

Both the Great Sun and the Shilluk *reth* could act with total impunity, but only towards those in their immediate presence. Each normally resided in an isolated capital, where he conducted regular rituals to guarantee fertility and well-being. According to one Italian missionary, writing in the early twentieth century:

> The Reth lives isolated, as a rule, with some of his wives in the small but famous hill-village of Pacooda, known as Fashoda ... His person is sacred and can be approached only with difficulty by ordinary people, and only with elaborate etiquette by the higher class. His appearance among the people, as for a journey, is rare and awe-inspiring, so that most people used to go into hiding or keep out of his path; girls especially do so.[64]

The latter presumably for fear of being snatched up and carried off to the royal harem. Yet to be a royal wife was not without advantages, as the college of royal wives was effectively what substituted for an administration, maintaining connections between Fashoda and their natal villages; and it was powerful enough, if the wives came to consensus, to order the king's execution. Then again, the *reth* also had his henchmen: often these were orphans, criminals, runaways and other unattached persons who would gravitate to him. If the king attempted to mediate a local dispute and one party refused to comply, he would occasionally throw in his lot with the other side, raid the offending village and carry off what cattle and other things of value his men could get their hands on. The royal treasury thus consisted almost entirely of wealth that had been stolen, plundered in raids on foreigners or on the king's own subjects.

All this might seem a pretty poor model for a free society – but in fact, in everyday affairs ordinary Shilluk appear to have maintained the same fiercely independent attitude as Nuer, and to have been just as averse to taking orders. Even the members of the 'higher class' (basically, descendants of earlier kings) could expect only a few gestures of deference, certainly not obedience. An old Shilluk legend sums it up nicely:

> There was once a cruel king, who killed many of his subjects, he even killed women. His subjects were terrified of him. One day, to demonstrate that his subjects were so afraid they would do anything he asked, he assembled the Shilluk chiefs and ordered them to wall him up inside a house with a young girl. Then he ordered them to let him out again. They didn't. So he died.[65]

If such oral traditions are anything to go on, Shilluk appear to have made a conscious choice that the sporadic appearance of an arbitrary and sometimes violent sovereign was preferable to any gentler but more systematic method of rule. Whenever a *reth* attempted to set up an administrative apparatus, even if only to collect tribute from defeated peoples, his actions were met with overwhelming waves of popular protest that either forced him to abandon the project or ousted him entirely.[66]

Unlike the Shilluk *reth*, Chavín and Olmec elites were able to mobilize enormous amounts of labour, but it's not at all clear if they did so through chains of command. As we've seen in ancient Mesopotamia, corvée or periodic labour service could also be a festive, public-spirited, even levelling occasion. (And as we shall see in the case of ancient Egypt, the most authoritarian regimes still often ensured it continued to have something of the same spirit.) Lastly, then, we should consider the impact of such first-order regimes on our third basic form of freedom: the freedom to shift and renegotiate social relations, either seasonally or permanently. This is, of course, the hardest to assess. Certainly, most of these new forms of power had a decidedly seasonal element. During certain times of year, as with the makers of Stonehenge, the entire social apparatus of authority would dissolve away and effectively cease to exist. What seems most difficult to comprehend is how these strikingly new institutional arrangements, and the physical infrastructure that sustained them, came into being in the first place.

Who came up with the design for the labyrinthine temple of Chavín de Huántar, or the royal compounds of La Venta? Insofar as they were collectively conceived – as they may well have been[67] – such grand fabrications may themselves be considered extraordinary exercises in human freedom. None of these first-order regimes could be considered examples of state formation – few now would even claim they were. So let's turn instead to one of the only cases that pretty much everyone agrees *can* be considered a state, and which has served, in many ways, as a paradigm for all subsequent states: ancient Egypt.

HOW CARING LABOUR, RITUAL KILLING AND 'TINY BUBBLES' ALL CAME TOGETHER IN THE ORIGINS OF ANCIENT EGYPT

If we had no written accounts to go by, but only the archaeological remains of the Natchez, would we have any way of knowing that a figure like the Great Sun even existed in Natchez society? Conceivably not. We would know that there were some fairly large mounds in the Great Village, built up in various stages, and no doubt post-holes would provide evidence for some large wooden structures built on them. Inside those structures, a number of hearths, refuse pits and scattered artefacts would undoubtedly point to some of the activities that went on there.[68] Perhaps the only compelling evidence of kingship, though, would come in the form of burials of richly decorated bodies surrounded by sacrificed retainers – if, that is, archaeologists happened to locate them.[69]

For some readers, the idea of a dead monarch sent off to the afterlife amid the corpses of his retainers might evoke images of early pharaohs. Some of Egypt's earliest known kings, those of the First Dynasty around 3000 BC (who, in fact, were not yet referred to as 'pharaoh'), were indeed buried in this way.[70] But Egypt is not alone in this respect. Burials of kings surrounded by dozens, hundreds, on some occasions even thousands of human victims killed specially for the occasion can be found in almost every part of the world where monarchies did eventually establish themselves, from the early dynastic city-state of Ur in Mesopotamia to the Kerma polity in Nubia to Shang China. There are also credible literary descriptions from Korea, Tibet, Japan and the Russian steppes. Something similar seems to have occurred as well in the Moche and Wari societies of South America, and the Mississippian city of Cahokia.[71]

We might do well to think a bit more about these mass killings, because most archaeologists now treat them as one of the more reliable indications that a process of 'state formation' was indeed under way. They follow a surprisingly consistent pattern. Almost invariably, they mark the first few generations of the founding of a new empire or kingdom, often being imitated by rivals in other elite households; then the practice gradually fades away (though sometimes surviving in very attenuated versions, as in *sati* or widow-suicide among largely *kshatriya* – warrior-caste – families in much of South Asia). In the initial moment, the practice of ritual killing around a royal burial tends to be spectacular: almost as if the death of a ruler meant a brief moment when sovereignty broke free of its ritual fetters, triggering a kind of political supernova that annihilates everything in its path, including some of the highest and mightiest individuals in the kingdom.

Often, in that moment, close members of the royal family, high-ranking military officers and government officials are counted among the victims. Of course, if looking at a burial without written records, it's often hard to tell when we're dealing with the bodies of royal wives, viziers or court musicians, as opposed to those of war captives, slaves or commoners seized randomly on the road (as we know was sometimes done in Buganda or Benin) – or even entire military units (as was sometimes the case in China). Perhaps, indeed, the individuals named as kings and queens in the famous Royal Tombs of Ur were not really that at all, but just hapless victims, substitute figures or maybe high-ranking priests and priestesses dressed up as royalty.[72]

Even if some cases were just a particularly bloody form of costume drama, others clearly weren't, so the question remains: why did early kingdoms ever do this sort of thing at all? And why did they stop doing it once their power became more established?

At the Shang capital of Anyang, on the central Chinese Plain, rulers tended to make their way into the afterlife accompanied by a few important retainers, who went voluntarily – if not always happily – to their deaths and were interred with due honours. These were only a small proportion of the bodies that went with them. It was also a royal prerogative to have one's tomb surrounded by the bodies of sacrificial victims.[73] Often these appear to be war captives taken from rival lineages and – unlike the retainers – their bodies were systematically mutilated, usually in mocking rearrangements of the victims' heads. For the Shang, this seems to have been a way of denying their victims the possibility of becoming dynastic ancestors, thereby rendering the living members of their lineage unable to take part in the care and feeding of their own dead kin, ordinarily one of the fundamental duties of family life. Cast adrift, and socially scarred, the survivors were more likely to fall under the sway of the Shang court. The ruler became a greater ancestor, in effect, by preventing others from becoming ancestors at all.[74]

It's interesting to bear this in mind when we turn to Egypt, because on the surface what we observe in the earliest dynasties seems the exact opposite. The first Egyptian kings, and at least one queen, are indeed buried surrounded by sacrificial victims, but those victims seem to have been drawn almost exclusively from their own inner circles. Our evidence for this derives from a series of 5,000-year-old burial chambers, looted in antiquity but still visible near the site of the ancient city of Abydos in the low desert of southern Egypt. These were the tombs of Egypt's First Dynasty.[75] Around each royal tomb lie long rows of subsidiary burials, numbering in the hundreds, forming a kind of perimeter. Such 'retainer burials' – including royal attendants and courtiers, killed in the prime of life – were placed in smaller brick compartments of their own, each marked with a gravestone inscribed with the individual's official titles.[76] There do not appear to be any dead captives or enemies among the buried. On the death of a king, then, his successor appears to have presided instead over the death of his predecessor's courtly entourage, or at least a sizeable portion of it.

So why all this ritual killing at the birth of the Egyptian state? What was the actual purpose of subsidiary burials? Was it to protect the dead king from the living, or the living from the dead king? Why did those sacrificed include so many who had evidently spent their lives caring for the king: most likely including wives, guards, officials, cooks, grooms, entertainers, palace dwarfs and other servants, grouped by rank around the royal tomb, according to their roles or occupation? There is a terrible paradox here. On the one hand, we have a ritual that appears to be the ultimate expression of love and devotion, as those who on a day-to-day basis made the king into something king-like – fed him, clothed him, trimmed his hair, cared for him in sickness and kept him company when he was lonely – went willingly to their deaths, to ensure he would continue to be king in the afterlife. At the same time, these burials are the ultimate demonstration that for a ruler, even his most intimate subjects could be treated as personal possessions, casually disposed of like so many blankets, gaming boards or jugs of spelt. Many have speculated about what it all means. Likely as not, 5,000 years ago, many of those laying out the bodies wondered too.

Written records from the time don't give us much sense of the official motives, but one thing that's quite striking in the evidence we do have – largely, a list of names and titles – is the very mixed composition of these royal cemeteries. They seem to include both blood relatives of the early kings and queens, notably some female members of the royal family, and a good number of other individuals who were taken in as members of the royal household owing to their unusual skills or striking personal qualities, and who thus came to be seen as members of the king's extended family. The violence and shedding of blood that attended these mass funerary rituals must have gone some way to effacing those differences, melding them into a single unit, turning servants into relatives and relatives into servants. In later times the king's close kin represented themselves in exactly this way, by placing in their tombs some humble replicas of themselves engaged in acts of menial labour, such as grinding grain or cooking meals.[77]

When sovereignty first expands to become the general organizing principle of a society, it is by turning violence into kinship. The early, spectacular phase of mass killing in both China and Egypt, whatever else it may be doing, appears to be intended to lay the foundations of what Max Weber referred to as a 'patrimonial system': that is, one in which all the kings' subjects are imagined as members of the royal household, at least to the degree that they are all working to care for the king. Turning erstwhile strangers into part of the royal household, or denying them their own ancestors, are thereby ultimately two sides of the same coin.[78] Or to put things another way, a ritual designed to produce kinship becomes a method of producing kingship.

These extreme forms of ritual killing around royal burials ended fairly abruptly in the course of Egypt's Second Dynasty. However, the patrimonial polity continued to expand – not so much in the sense of expanding Egypt's external borders, which were established early on through outward violence directed at neighbours in Nubia and elsewhere,[79] but more in terms of reshaping the lives of its internal subjects. Within a few generations we find the valley and delta of the

Nile divided into royal estates, each dedicated to provisioning the mortuary cults of different former rulers; and, not long after that, the foundation of entire 'workers' towns' devoted to the construction of the pyramids on the Giza Plateau, drawing corvée labour from up and down the country.[80]

At this point, with the construction of the great pyramids at Giza, surely no one could deny that we are in the presence of some sort of state; but the pyramids, of course, were also tombs. In the case of Egypt, it seems, 'state formation' began with some kind of Natchez or Shilluk-like principle of individual sovereignty, bursting out of its ritual cages precisely through the vehicle of the sovereign's demise in such a way that royal death ultimately became the basis for reorganizing much of human life along the length of the Nile. To understand how this could happen, we need to look at what Egypt was like well before the First Dynasty tombs at Abydos.

Before we consider what happened in the centuries directly preceding Egypt's First Dynasty – the so-called Predynastic and Proto-dynastic periods, from around 4000 to 3100 BC – it is worth casting our minds back to an even earlier phase of prehistory in the same region.

Let's recall that the African Neolithic, including that of the Nile valley – Egyptian and Sudanese – took a different form to that of the Middle East. In the fifth millennium BC, there was less of an emphasis on cereal agriculture and more on cattle, along with the wide variety of wild and cultivated food sources typical of the period. Perhaps the best modern comparison we have – though it's very far from exact – is with Nilotic peoples like the Nuer, Dinka, Shilluk or Anuak, who grow crops but think of themselves as pastoralists, shifting back and forth each season between camps improvised for the occasion. If we might hazard a very broad generalization, where in the Middle Eastern Neolithic (the Fertile Crescent) the cultural focus – in the sense of decorative arts, care and attention – was on houses, in Africa it was on bodies: from very early on we have burials with beautifully worked objects of personal grooming and highly elaborate sets of body ornamentation.[81]

It's no coincidence that many centuries later, when the Egyptian First Dynasty took form, among the very first objects with royal inscriptions we find the 'ivory comb of King Djet' and the famous 'palette of King Narmer' (stone palettes being used, both by men and women, for grinding and mixing make-up). These are basically spectacular versions of the sort of objects Neolithic Nile dwellers used to beautify themselves millennia earlier and, not coincidentally, to offer as gifts to the ancestral dead; and in Neolithic and Predynastic times, such objects were widely available to women, men and children. In fact from those very early times, in Nilotic society the human body itself became a sort of monument. Experiments with techniques of mummification took place long before the First Dynasty; as early as the Neolithic period, Egyptians were already mixing aromatics and preservative oils to produce bodies that could last forever and whose places of burial were the fixed points of reference in an ever-shifting social landscape.[82]

How, then, do we get from such a remarkably fluid state of affairs to the spectacular appearance of the First Dynasty almost 2,000 years later? Territorial kingdoms don't come out of nowhere.[83] Until quite recently, we had little more than fragmentary hints of what must have been happening during what are technically referred to as the Predynastic and Proto-dynastic periods – that is, roughly the fourth millennium, before King Narmer appears around 3100 BC. In such cases, it is tempting to revert to analogies with more recent situations. As we've seen, modern Nilotic peoples, and particularly the Shilluk, show how relatively mobile societies that place great value on individual freedom might, nonetheless, prefer an arbitrary despot – who could eventually be got rid of – to any more systematic or pervasive form of rule. This is especially true if, like so many peoples whose ancestors organized their lives around livestock, they tend toward patriarchal forms of organization.[84] One could imagine the prehistoric Nile valley as dominated by a collection of Shilluk-like *reths*, each with their own settlement which was, essentially, an extended patriarchal family; bickering and feuding with one another, but otherwise, as yet, making fairly little difference to the lives of those over whom they ostensibly ruled.

Still, there is no substitute for actual archaeological evidence – and in recent years it has been building up apace. New discoveries show that, by no later than 3500 BC – and so still some five centuries before the First Dynasty – we do indeed find burials of petty monarchs at various locations throughout the valley of the Nile, and also down into Nubia. We don't know any of their names, since writing had barely developed yet. Most of these kingdoms appear to have been extremely small. The largest we know of centred on Naqada and Abydos, near the great bend of the Nile in Upper Egypt; on Hierakonpolis further to the south; and on the site of Qustul in Lower Nubia – but even those do not seem to have controlled extensive territories.[85]

What preceded the First Dynasty, then, was not so much a lack of sovereign power as a superfluity of it: a surfeit of tiny kingdoms and miniature courts, always with a core of blood relatives and a motley collection of henchmen, wives, servants and assorted hangers-on. Some of these courts appear to have been quite magnificent in their own way, leaving behind large tombs and the bodies of sacrificed retainers. The most spectacular, at Hierakonpolis, includes not only a male dwarf (they seem to have become a fixture of courtly society very early on), but a significant number of teenage girls, and what seem to be the remains of a private zoo: a menagerie of exotic animals including two baboons and an African elephant.[86] These kings give every sign of making grandiose, absolute, cosmological claims; but little sign of maintaining administrative or military control over their respective territories.

How do we get from here to the massive agrarian bureaucracy of later, dynastic times in Egypt? Part of the answer lies in a parallel process of change that archaeology also allows us to untangle, around the middle of the fourth millennium BC – we might imagine it as a kind of extended argument or debate about the responsibilities of the living to the dead. Do dead kings, like live ones, still need us to take care of them? Is this care different from the care accorded ordinary ancestors? Do ancestors get hungry? And if so, what exactly do they eat? For whatever reasons, the answer that gained traction across the Nile valley around 3500 BC was that ancestors do indeed get hungry, and what they required was something which, at that time, can only have been considered a rather exotic and perhaps luxurious form of food: leavened bread and fermented wheat beer, the pot-containers for which now start to become standard fixtures of well-appointed grave assemblages. It is no coincidence that arable wheat-farming – though long familiar in the valley and delta of the Nile – was only refined and intensified around this time, at least partly in response to the new demands of the dead.[87]

The two processes – agronomic and ceremonial – were mutually reinforcing, and the social effects epochal. In effect, they led to the creation of what might be considered the world's first peasantry. As in so many parts of the world initially favoured by Neolithic populations, the periodic flooding of the Nile had at first made permanent division of lands difficult; quite likely, it was not ecological circumstances but the social requirement to provide bread and beer on ceremonial occasions that allowed such divisions to become entrenched. This was not just a matter of access to sufficient quantities of arable land, but also the means to maintain ploughs and oxen – another introduction of the late fourth millennium BC. Families who found themselves unable to command such resources had to obtain beer and loaves elsewhere, creating networks of obligation and debt. Hence important class distinctions and dependencies did, in fact, begin to emerge,[88] as a sizeable sector of Egypt's population found itself deprived of the means to care independently for ancestors.

If any of this seems fanciful, we need only compare what happened with the extension of Inca sovereignty in Peru. Here, too, we find a contrast between the traditional, varied and flexible regime of everyday foodstuffs – in this case centring on cuisine made from freeze-dried potatoes (*chuño*) – and the introduction of a completely different sort of food, in this case, maize beer (*chicha*), which was considered fit for the gods and also gradually became, as it were, the food of empire.[89] By the time of the Spanish conquest, maize was a ritual necessity for rich and poor alike. Gods and royal mummies dined on it; armies marched on it; and those too poor to grow it – or who lived too high up on the *altiplano* – had to find other ways of obtaining it, often ending up in debt to the royal estate as a result.[90]

In the case of Peru, we have the Spanish chroniclers to help us understand how an intoxicant could gradually become the lifeblood of an empire; in Egypt, 5,000 years ago, we can really only guess at the details. It is a remarkable tribute to the discipline of archaeology that we know as much as we do, and we are starting to put the pieces together. For instance, it is around 3500 BC that we begin to find remains of facilities used for both baking and brewing – first alongside cemeteries, and within a few centuries attached to palaces and grand tombs.[91] A later depiction, from the tomb of an official called Ty, shows how they could have operated, with pot-baked bread and beer produced by a single process. The gradual extension of royal authority, and also administrative reach, throughout Egypt began around the time of the First Dynasty or a little before, with the creation of estates ostensibly dedicated to organizing the provision, not so much of living kings but of dead ones, and eventually dead royal officials too. By the time of the Great Pyramids (*c.*2500 BC), bread and beer were being manufactured on an industrial scale to supply armies of workers during their seasonal service on royal construction projects, when they too got to be 'relatives' or at least care-givers of the king, and as such were at least temporarily well provisioned and well cared for.

The workers' town at Giza produced some thousands of ceramic moulds. These were used to make the huge communal loaves known as *bedja* bread, eaten in large groups with copious amounts of meat supplied by royal livestock pens and washed down with spiced beer.[92] The latter was of special importance for the solidarity of seasonal work crews in Old Kingdom Egypt. The facts emerge with disarming simplicity, from graffiti on the reverse sides of building blocks used in the construction of royal pyramids. 'Friends of [the king] Menkaure' reads one such, 'Drunkards of Menkaure' another. These seasonal work units (or *phyles*, as Egyptologists call them) seem to have been made up only of men who passed through special age-grade rituals, and who modelled themselves on the organization of a boat's crew.[93] Whether such ritual brotherhoods ever took to the water together isn't clear, but there are notable parallels between the team skills used in maritime engineering and those used in manipulating multi-ton blocks of limestone and granite for royal pyramid-temples or other such monuments.[94]

There may be interesting parallels to explore here with what happened in the Industrial Revolution, when techniques of discipline, transforming crews of people into clock-like machines, were first pioneered on sailing ships and only later transferred to the factory floor. Were ancient Egyptian boat crews the model for what have been called the world's first production-line techniques, creating vast monuments, far more impressive than anything the world had yet seen, by dividing tasks into an endless variety of simple, mechanical components: cutting, dragging, hoisting, polishing? This is how the pyramids were actually built: by rendering subjects into great social machines, afterwards celebrated by mass conviviality.[95]

We have just described, in broad outline, what's widely treated as the world's first known example of 'state formation'. It would be easy to go on from here to generalize. Perhaps this is what a state actually is: a combination of exceptional violence and the creation of a complex social machine, all ostensibly devoted to acts of care and devotion.

There is obviously a paradox here. Caring labour is in a way the very opposite of mechanical labour: it is about recognizing and understanding the unique qualities, needs and peculiarities of the cared-for – whether child, adult, animal or plant – in order to provide what they require to flourish.[96] Caring labour is distinguished by its particularity. If those institutions we today refer to as 'states' really do have any common features, one must certainly be a tendency to displace this caring impulse on to abstractions; today this is usually 'the nation', however broadly or narrowly defined. Perhaps this is why it's so easy for us to see ancient Egypt as a prototype for the modern state: here too, popular devotion was diverted on to grand abstractions, in this case the ruler and the elite dead. This process is what made it possible for the whole arrangement to be imagined, simultaneously, as a family and as a machine, in which everyone (except of course the king) was ultimately interchangeable. From the seasonal work of tomb-building to the daily servicing of the ruler's body (recall again how the first royal inscriptions are found on combs and make-up palettes), most of human activity was directed upwards, either towards tending rulers (living and dead) or assisting them with their own task of feeding and caring for the gods.[97] All this activity was seen as generating a downward flow of divine blessings and protection, which occasionally took material form in the great feasts of the workers' towns.

The problems come when we try to take this paradigm and apply it almost anywhere else. True, as we've noted, there are some interesting parallels between Egypt and Peru (all the more remarkable, considering their strikingly different topographies – the flat and easily navigable Nile as against the 'vertical archipelagos' of the Andes). These parallels appear in uncanny details, like the mummification of dead rulers and the way in which such mummified rulers continue to maintain their own rural estates; the way living kings are treated as gods who have to make periodic tours of their domains. Both societies too shared a certain antipathy to urban life. Their capitals were really ceremonial centres, stages for royal display, with relatively few permanent residents, and their ruling elites preferred to imagine their subjects as living in a realm of bucolic estates and hunting grounds.[98] But all this only serves to underline the degree to which other cases referred to in the literature as 'early states' were entirely different.

IN WHICH WE REFLECT ON THE DIFFERENCES BETWEEN WHAT ARE USUALLY CALLED 'EARLY STATES', FROM CHINA TO MESOAMERICA

The kingdom of Egypt and the Inca Empire demonstrate what can happen when the principle of sovereignty arms itself with a bureaucracy and manages to extend itself uniformly across a territory. As a result, they are very often invoked as primordial examples of state formation, even though they are dramatically separated in time and space. Almost none of the other canonical 'early states' appear to have taken this approach.

Early Dynastic Mesopotamia, for instance, was made up of dozens of city-states of varying sizes, each governed by its own charismatic warrior-king – whose special, individual qualities were said to be recognized by the gods, and physically marked in the outstanding virility and allure of his body – all vying constantly for dominance. Only occasionally would one ruler gain enough of an upper hand to create something that might be described as the beginnings of a unified kingdom or empire. It's not clear whether any of these early Mesopotamian rulers actually claimed 'sovereignty' – at least in the absolute sense of standing outside the moral order and thus being able to act with impunity, or to create entirely new social forms of their own volition. The cities they ostensibly ruled over had been around for centuries: commercial hubs with strong traditions of self-governance, each with its own city gods who presided over local systems of temple administration. Kings, in this case, almost never claimed to be gods themselves, but rather the gods' vicegerents, and sometimes heroic defenders on earth: in short, delegates of sovereign power that resided properly in heaven.[99] The result was a dynamic tension between two principles which, as we've seen, originally arose in opposition to one another: the administrative order of the river valleys and the heroic, individualistic politics of the surrounding highlands. Sovereignty, in the last resort, belonged to the gods alone.[100]

The Maya lowlands were different again. To be a Classic Maya ruler (*ajaw*) was to be a hunter and god-impersonator of the first rank, a warrior whose body, on entering battle or during dance rituals, became host to the spirit of an ancestral hero, deity or dreamlike monsters. *Ajaws* were, effectively, like tiny squabbling gods. If anything was projected into the cosmos, in the Classic Maya case, it was precisely the principle of bureaucracy. Most Mayanists would agree that Classic-period rulers lacked a sophisticated administrative apparatus, but they imagined the cosmos as itself a kind of administrative hierarchy, governed by predictable laws:[101] an intricate set of celestial or subterranean wheels within wheels, such that it was possible to establish the exact birth and death dates of major deities thousands of years in the past (the deity Muwaan Mat, for instance, was born on 7 December 3121 BC, seven years before the creation of the current universe), even if it would never occur to them to register the numbers, wealth, let alone birthdates of their own subjects.[102]

So do these 'early states' have any common features at all? Obviously, some basic generalizations can be made. All deployed spectacular violence at the pinnacle of the system; all ultimately depended on and to some degree mimicked the patriarchal organization of households. In every case, the apparatus of government stood on top of some kind of division of society into classes. But as we've seen in earlier chapters, these elements could just as well exist without or prior to the creation of central government – and even when such government was established, they could take very different forms. In Mesopotamian cities, for instance, social class was often based on land tenure and mercantile wealth. Temples doubled as city banks and factories. Their gods might only leave the temple grounds on festive occasions, but priests moved in broader circles, making interest-bearing loans to traders, watching over armies of female weavers and jealously guarding their fields and flocks. There were powerful societies of merchants. We know much less about such matters in the Maya lowlands, but what we do know suggests that power was based less on the control of land or commerce than on the ability to control flows of people and loyalty directly, through intermarriage and the intensely personal bonds that obtained between lords and lesser nobles. Hence the focus, in Classic Maya politics, on capturing high-status rivals in warfare as a form of 'human capital' (something which hardly features in Mesopotamian sources).[103]

Looking at China only seems to complicate things even further. In the time of the late Shang, from 1200 to 1000 BC, Chinese society did share certain features with the other canonical 'early states' but, considered as an integrated whole, it's entirely unique. Like Inca Cuzco, the Shang capital at Anyang was designed as a 'pivot of the four quarters' – a cosmological anchor for the entire kingdom, laid out as a grand stage for royal ritual. Like both Cuzco and the Egyptian capital of Memphis (and later Thebes), the city was suspended between the worlds of the living and the dead, serving as home to the royal cemeteries and their attached mortuary temples, as well as a living administration. Its industrial quarters produced enormous quantities of bronze vessels and jades, the tools used in communing with ancestors.[104] But in most important ways, we find little similarity between the Shang and either Old Kingdom Egypt or Inca Peru. For one thing, Shang rulers did not claim sovereignty over an extended area. They couldn't travel safely, let alone issue commands, outside a narrow band of territories clustered on the middle and lower reaches of the Yellow River, not far from the royal court.[105] Even there one is left with a sense that they didn't really claim sovereignty in the same sense as Egyptian, Peruvian or even Mayan rulers. The clearest evidence is the exceptional importance of divination in the early Chinese state, which stands in striking contrast to pretty much all the other examples we've been looking at.[106]

Effectively, any royal decision – whether war, alliance, the founding of new cities, or even such apparently trivial matters as extending royal hunting grounds – could only proceed if approved by the ultimate authorities, who were the gods and ancestral spirits; and there was no absolute assurance that such approval would be forthcoming in any given case. Shang diviners appealed to gods through the medium of burnt offerings. The process was as follows: when hosting gods or ancestors at a ritual meal, kings or their diviners put turtle shells and ox scapulae on the fire, then 'read' the cracks that broke out on their surfaces as a kind of oracular writing. The proceedings were quite bureaucratic. Once an answer had been obtained, the diviner or an appointed scribe would then authorize the reading by etching an inscription on to bone or shell, and the resulting oracle would be stored for later consultation.[107] These oracle texts are the first written inscriptions in China we actually know about, and while it is very possible that writing was used for everyday purposes on perishable media that don't survive, there remains as yet no clear evidence for the other forms of administrative activity or archives that became so typical of later Chinese dynasties, nor much in the way of an elaborate bureaucratic apparatus at all.[108]

Like the Maya, Shang rulers routinely waged war to acquire stocks of living human victims for sacrifices. Rival courts to the Shang had their own ancestors, sacrifices and diviners, and while they appear to have recognized the Shang as paramount – especially in ritual contexts – there seemed to be no contradiction between this and actually going to war with them, if they felt there was sufficient cause. Such rivalries help explain the lavishness of Shang funerals and mutilation of captive bodies; their rulers were still in a sense playing the agonistic games typical of a 'heroic society', competing to outshine and humiliate their rivals. Such a situation is inherently unstable and eventually one rival dynasty, the Western Zhou, did manage definitively to defeat the Shang, and claimed for itself the Mandate of Heaven.[109]

*

At this point it should be clear that what we are really talking about, in all these cases, is not the 'birth of the state' in the sense of the emergence, in embryonic form, of a new and unprecedented institution that would grow and evolve into modern forms of government. We are speaking instead of broad regional systems; it just happens, in the case of Egypt and the Andes, that an entire regional system became united (at least some of the time) under a single government. This was actually a fairly unusual arrangement. More common were arrangements such as those in Shang China, where unification was largely theoretical; or Mesopotamia, where regional hegemony rarely lasted for longer than a generation or two; or the Maya, where there was a protracted struggle between two main power blocs, neither of which could ever quite overcome the other.[110]

In terms of the specific theory we've been developing here, where the three elementary forms of domination – control of violence, control of knowledge, and charismatic power – can each crystallize into its own institutional form (sovereignty, administration and heroic politics), almost all these 'early states' could be more accurately described as 'second-order' regimes of domination. First-order regimes like the Olmec, Chavín or Natchez each developed only one part of the triad. But in the typically far more violent arrangements of second-order regimes, two of the three principles of domination were brought together in some spectacular, unprecedented way. Which two it was seems to have varied from case to case. Egypt's early rulers combined sovereignty and administration; Mesopotamian kings mixed administration and heroic politics; Classic Maya *ajaws* fused heroic politics with sovereignty.

We should emphasize that it's not as if any of these principles, in their elementary forms, were entirely absent in any one case: in fact, what seems to have happened is that two of them crystallized into institutional forms – fusing in such a way as to reinforce one another as the basis of government – while the third form of domination was largely pushed out of the realm of human affairs altogether and displaced on to the non-human cosmos (as with divine sovereignty in Early Dynastic Mesopotamia, or the cosmic bureaucracy of the Classic Maya). Keeping all this in mind, let's return briefly to Egypt to clarify some remaining points.

IN WHICH WE RECONSIDER THE EGYPTIAN CASE IN LIGHT OF OUR THREE ELEMENTARY PRINCIPLES OF DOMINATION, AND ALSO REVISIT THE PROBLEM OF 'DARK AGES'

The architects of Egypt's Old Kingdom clearly saw the world they were creating as something like a cultured pearl, reared in precious isolation. Their vision is vividly documented in relief carvings of stone, lining the walls of royal temples, which served the mortuary cults of kings such as Djoser, Menkaure, Sneferu and Sahure. Here Egypt, the 'Two Lands', is always represented as both a celestial theatre-state, in which king and gods share equal billing, and an earthly domain: a world of rural estates and hunting grounds, mapped out in a cartography of compliance, each parcel of land personified as a lady-in-waiting who brings her bounty to the feet of the king. The governing principle of this vision of Egypt is the monarch's absolute sovereignty over everything, symbolized in his gigantic funerary monuments, his defiant assertion that there was nothing he could not conquer, even death.

Egyptian kingship was, however, Janus-faced. Its inner visage was that of supreme patriarch, standing guard over a vastly extended family – a Great House (the literal meaning of 'pharaoh'). Its outer face is shown in depictions of the king as a war leader or hunt leader asserting control over the country's wild frontiers; all were fair game when the king turned his violence upon them.[111] This is very different, however, to heroic violence. In a way, it's the opposite. In a heroic order, the warrior's honour is based on the fact that he *might* lose; his reputation means so much to him that he is willing to stake his life, dignity and freedom to defend it. Egyptian rulers, in these early periods, never represent themselves as heroic figures in this sense. They could not, conceivably, lose. As a result, wars are not represented as 'political' contests, which imply a match between potential equals. Instead, combat and the chase alike were assertions of ownership, endless rehearsals of the same sovereignty the king exercised over his people and which ultimately derived from his kinship with the gods.

As we've already had occasion to observe, any form of sovereignty at once so absolute and so personal as a pharaoh's will necessarily pose severe problems of delegation. Here, too, all state officials had to be in some sense appendages of the king's own person. Major landowners, military commanders, priests, administrators and other senior government officials also held titles like 'Keeper of the King's Secrets', 'Beloved Acquaintance of the King', 'Director of Music to the Pharaoh', 'Overseer of the Palace Manicurists' or even 'of the King's Breakfast'. We are not suggesting that power games were absent here; no doubt there's never been a royal court without jockeying for position, tricks and double-dealing and political intrigue. The point is that these were not public contests, and no sanctioned space existed for open competition. Everything remained confined to life at court. This is abundantly clear in the 'tomb biographies' of Old Kingdom officials, which describe their life achievements almost exclusively in terms of their relationship to and their care for the king, rather than any personal qualities or attainments.[112]

What we have in this case, then, seems to be a hypertrophy of the principles of sovereignty and administration and an almost complete absence of competitive politics. Dramatic public contests of any sort, political or otherwise, were well-nigh non-existent. There is nothing in the official sources of the Egyptian Old Kingdom (nor much in later periods of ancient Egyptian history) that is remotely reminiscent of, say, Roman chariot-racing or Olmec or Zapotec ball games. In the royal jubilee or *sed* festival, when Egyptian kings ran a circuit to celebrate the unification of the Two Lands of Upper and Lower Egypt, it took the form of a solo performance, the outcome of which was never in doubt. Insofar as competitive politics appears in later Egyptian literature (which it occasionally does), it takes place precisely *between* the gods, as in works like the *Contendings of Horus and Seth*. Dead kings, perhaps, compete with one another; but by the time sovereignty comes down to the domain of mortals, matters have already been settled.

Just to be utterly clear about what we are saying here, when we speak of an absence of charismatic politics we are talking about the absence of a 'star system' or 'hall of fame', with institutionalized rivalries between knights, warlords, politicians and so on. We are most certainly not speaking about an absence of individual personalities. It's just that in a pure monarchy there is only one person, or at best a handful of individuals, who really matter. Indeed, if we are trying to understand the appeal of monarchy as a form of government – and it cannot be denied that for much of recorded human history it was a very popular one – then likely it has something to do with its ability to mobilize sentiments of a caring nature and abject terror at the same time. The king is both the ultimate individual, his quirks and fancies always to be indulged like a spoilt baby, and at the same time the ultimate abstraction, since his powers over mass violence, and often (as in Egypt) mass production, can render everyone the same.

It is also worth observing that monarchy is probably the only prominent system of government we know of in which children are crucial players, since everything depends on the monarch's ability to continue the dynastic line. The dead can be worshipped under any regime – even the United States, which frames itself as a beacon of democracy, creates temples to its Founding Fathers and carves portraits of dead presidents into the sides of mountains – but infants, pure objects of love and nurture, are only politically important in kingdoms and empires.

If the ancient Egyptian regime is often held out as the first true state and a paradigm for all future ones, it is largely because it was capable of synthesizing absolute sovereignty – the monarch's ability to stand apart from human society and engage in arbitrary violence with impunity – with an administrative apparatus which, at certain moments at least, could reduce almost everyone to cogs in a single great machine. Only heroic, competitive politics was lacking, pushed off into the worlds of gods and the dead. But there was, of course, a great exception to this which comes precisely in those periods when central authority broke down, the supposed 'dark ages', beginning with the First Intermediate period (*c.* 2181–2055 BC).

Already towards the end of the Old Kingdom, 'nomarchs' or local governors had made themselves into de facto dynasties.[113] When the central government split between rival centres at Herakleopolis and Thebes, such local leaders began to take over most functions of government. Often referred to as 'warlords', these nomarchs were in fact nothing like the petty kings of the Predynastic period. At least in their own monuments, they represent themselves as something closer to popular heroes, even saints. Neither was this always just idle boasting; some were indeed revered as saints for centuries to come. No doubt charismatic local leaders had always existed in Egypt; but with the breakdown of the patrimonial state, such figures could begin to make open claims of authority based on their personal achievements and attributes (bravery, generosity, oratorical and strategic skills) and – crucially – redefine social authority itself as based on qualities of public service and piety to the gods of their local town, and the popular support those qualities inspired.

In other words, whenever state sovereignty broke down, heroic politics returned – with charismatic figures just as vainglorious and competitive, perhaps, as those we know from ancient epics, but far less bloodthirsty. The change is clearly visible in autobiographical inscriptions, like those in the rock-cut tomb of the nomarch Ankhtifi at El-Mo'alla, south of Thebes. Here's how he narrates his role in war: 'I was one who found the solution when it was lacking, thanks to my vigorous plans; one with commanding words and untroubled mind on the day when the nomes [administered territories] allied together (to wage war). I am the hero without equal; one who spoke freely while people were silent on the day when fear was spread and Upper Egypt did not dare to speak.' Even more striking, here's how he celebrates his social achievements:

> I gave bread to the hungry and clothing to the naked; I anointed those who had no cosmetic oil; I gave sandals to the barefooted; I gave a wife to him who had no wife. I took care of the towns of Hefat [El-Mo'alla] and Hor-mer in every [crisis, when] the sky was clouded and the earth [was parched? And people died] of hunger on this sandbank of Apophis. The south came with its people and the north with its children; they brought finest oil in exchange for barley which was given to them . . . All of Upper Egypt was dying of hunger and people were eating their children, but I did not allow anybody to die of hunger in this nome . . . never did I allow anybody in need to go from this nome to another one. I am the hero without equal.[114]

It's only at this point, in the First Intermediate period, that we see a hereditary aristocracy coming into its own in Egypt, as local magnates like Ankhtifi began transferring their powers to their offspring and extended families. Aristocracy and personal politics had no such recognized place in the Old Kingdom, precisely because they came into conflict with the principle of sovereignty. In summary, the transition from Old Kingdom to First Intermediate period was not so much a shift from 'order' to 'chaos' – as Egyptological orthodoxy once had it – as a swing from 'sovereignty' to 'charismatic politics' as different ways of framing the exercise of power. With that came a shift in emphasis, from the people's care of god-like rulers to the care of the people as a legitimate path to authority. In ancient Egypt, as so often in history, significant political accomplishments occur in precisely those periods (the so-called 'dark ages') that get dismissed or overlooked because no one was building grandiose monuments in stone.

IN WHICH WE GO IN SEARCH OF THE REAL ORIGINS OF BUREAUCRACY, AND FIND THEM ON WHAT APPEARS TO BE A SURPRISINGLY SMALL SCALE

At this point it should be easy enough to understand why ancient Egypt is so regularly held out as the paradigmatic example of state formation. It's not just that it is chronologically the earliest of what we've called second-order regimes of domination; aside from the much later Inca Empire, it's also just about the only case where the two principles that came together were sovereignty and administration. In other words, it's the only case from a suitably distant phase of history that perfectly fits the model of what *should* have happened. All such assumptions really go back to a certain kind of social theory – or, maybe better put, a theory of organization – that we described at the start of Chapter Eight. Small, intimate groups (the argument goes) might be able to adopt informal, egalitarian means of problem-solving, but as soon as large numbers of people are assembled together in a city or a kingdom everything changes.

It's simply assumed, in this kind of theory, that once societies scale up they will need, as Robin Dunbar puts it, 'chiefs to direct, and a police force to ensure that social rules are adhered to'; or as Jared Diamond says, 'large populations can't function without leaders who make the decisions, executives who carry out the decisions, and bureaucrats who administer the decisions and laws.'[115] In other words, if you want to live in a large-scale society you need a sovereign and an administration. It is more or less taken for granted that some kind of monopoly of coercive force (again, the ability to threaten everyone with weapons) is ultimately required in order to do this. Writing systems, in turn, are almost invariably assumed to have developed in the service of impersonal bureaucratic states, which were the result of the whole process.

Now, as we've already seen, none of this is really true, and predictions based on these assumptions almost invariably turn out to be wrong. We saw one dramatic example in Chapter Eight. It was once widely assumed that if bureaucratic states tend to arise in areas with complex irrigation systems, it must have been because of the need for administrators to co-ordinate the maintenance of canals and regulate the water supply. In fact, it turns out that farmers are perfectly capable of co-ordinating very complicated irrigation systems all by themselves, and there's little evidence, in most cases, that early bureaucrats had anything to do with such matters. Urban populations seem to have a remarkable capacity for self-governance in ways which, while usually not quite 'egalitarian', were likely a good deal more participatory than almost any urban government today. Meanwhile most ancient emperors, as it turns out, saw little reason to interfere, as they simply didn't care very much about how their subjects cleaned the streets or maintained their drainage ditches.

We've also observed that when early regimes do base their domination on exclusive access to forms of knowledge, these are often not the kinds of knowledge we ourselves would consider particularly practical (the shamanic, psychotropic revelations that seem to have inspired the builders of Chavín de Huántar would be one such example). In fact, the first forms of functional administration, in the sense of keeping archives of lists, ledgers, accounting procedures, overseers, audits and files, seem to emerge in precisely these kinds of ritual contexts: in Mesopotamian temples, Egyptian ancestor cults, Chinese oracle readings and so forth.[116] So one thing we can now say with a fair degree of certainty is that bureaucracy did *not* begin simply as a practical solution to problems of information management, when human societies advanced beyond a particular threshold of scale and complexity.

This, however, raises the interesting question of where and when such technologies did first arise, and for what reason. Here there's some surprising new evidence too. Our emerging archaeological understanding suggests that the first systems of specialized administrative control actually emerged in very small communities. The earliest clear evidence of this appears in a series of tiny prehistoric settlements in the Middle East, dating over 1,000 years after the Neolithic site of Çatalhöyük was founded (at around 7400 BC), but still more than 2,000 years before the appearance of anything even vaguely resembling a city.

The best example of such a site is Tell Sabi Abyad, investigated by a team of Dutch archaeologists working in Syria's Balikh valley in the province of Raqqa. Around 8,000 years ago (*c.*6200 BC), in what was prehistoric Mesopotamia, a one-hectare village was destroyed there by fire, baking its mud walls and many of their clay contents, thus preserving them. While obviously a very bad bit of luck for the inhabitants, it was a stroke of brilliant luck for future researchers, since it has left us a unique insight into the organization of a Late Neolithic community, comprising perhaps around 150 individuals.[117] What the excavators discovered is that not only did the inhabitants of this village erect central storage facilities, including granaries and warehouses; they also employed administrative devices of some complexity to keep track of what was in them. These devices included economic archives, which were miniature precursors to the temple archives at Uruk and other later Mesopotamian cities.

These were not written archives: writing, as such, would not appear for another 3,000 years. What did exist were geometric tokens made of clay, of a sort that appear to have been used in many similar Neolithic villages, most likely to keep track of the allocation of particular resources.[118] At Tell Sabi Abyad, miniature seals bearing engraved designs were used alongside them to stamp and mark the clay stoppers of household vessels with identifying signs.[119] Perhaps most remarkably, the stoppers themselves, once removed from the vessels, were kept and archived in a special building – an office or bureau of sorts – near the centre of the village for later reference.[120] Ever since these discoveries were reported in the 1990s, archaeologists have been debating in whose interests and for what purpose such 'village bureaucracies' functioned.

In trying to answer this question, it's important to note that the central bureau and depot of Tell Sabi Abyad is not associated with any kind of unusually large residence, rich burials or other signs of personal status. If anything, what's striking about the remains of this community is their uniformity: the surrounding dwellings, for instance, are all roughly equal in size, quality and surviving contents. The contents themselves suggest small family units which maintained a complex division of labour, often including tasks that would have required the co-operation of multiple households. Flocks had to be pastured, a variety of cereal crops sown, harvested and threshed, as well as flax for weaving, which was practised alongside other household crafts such as potting, bead-making, stone-carving and simple forms of metalworking. And of course there were children to raise, old people to care for, houses to build and maintain, marriages and funerals to co-ordinate, and so on.

Careful scheduling and mutual aid would have been vital for the successful completion of an annual round of productive activities, while evidence of obsidian, metals and exotic pigments indicates that villagers also interacted regularly with outsiders, no doubt through intermarriage as well as travel and trade.[121] As we've already observed in the case of traditional Basque villages, these sorts of activity could well involve quite complicated mathematical calculations. Still, this in itself doesn't explain why there was a need to fall back on precise systems of measurement and archiving. After all, there are untold thousands of agricultural communities across human history who juggled similarly complex combinations of tasks and responsibilities without having to create new techniques of record-keeping.

Whatever the reason, the effect of introducing such techniques seems to have been profound for villages in prehistoric Mesopotamia and the surrounding hill country. Recall that 2,000 years separate Tell Sabi Abyad from the earliest cities, and during that long span of time village life in the Middle East underwent a series of remarkable changes. In some ways, people living in small-scale communities began to act as if they were already living in mass societies of a certain kind, even though nobody had ever seen a city. It sounds counter-intuitive – but it is what we see in the intervening centuries in the evidence of villages scattered across a large region, from southwestern Iran through much of Iraq and all the way over to the Turkish highlands. In many ways this phenomenon was another version of the kind of 'culture areas' or hospitality zones that we discussed in earlier chapters, but there was a different element: affinities between distant households and families seem to have been increasingly based on a principle of cultural uniformity. In a sense, then, this was the first era of the 'global village'.[122]

What it all looks like, in the archaeological record, is impossible to miss. We write from first-hand experience here, since one of us has conducted archaeological investigations of prehistoric villages in Iraqi Kurdistan, dating before and after the great transformation took place. What you find, in the fifth millennium BC, is the gradual disappearance from village life of most outward signs of difference or individuality, as administrative tools and other new media technologies spread across a large swathe of the Middle East. Households were now built to increasingly standard tripartite plans, and pottery, which had once been a way of expressing individual skill and creativity, now seems to have been made deliberately drab, uniform and in some cases almost standardized. Craft production in general became more mechanical, and female labour was subject to new forms of spatial control and segregation.[123]

In fact this entire period, lasting around 1,000 years (archaeologists call it the 'Ubaid, after the site of Tell al-'Ubaid in southern Iraq), was one of innovation in metallurgy, horticulture, textiles, diet and long-distance trade; but from a social vantage point, everything seems to have been done to prevent such innovations becoming markers of rank or individual distinction – in other words, to prevent the emergence of obvious differences in status, both within and between villages. Intriguingly, it is possible that we are witnessing the birth of an overt ideology of equality in the centuries prior to the emergence of the world's first cities, and that administrative tools were first designed not as a means of extracting and accumulating wealth but precisely to prevent such things from happening.[124] To get a sense of how such small-scale bureaucracies might have worked in practice we can briefly consider again the *ayllu*, those Andean village associations which, as we mentioned earlier, had their own home-grown administration.

Ayllu too were based on a strong principle of equality; their members literally wore uniforms, with each valley having its own traditional design of cloth. One of the *ayllu*'s main functions was to redistribute agricultural land as families grew larger or smaller, to ensure none grew richer than any other – indeed, to be a 'rich' household meant, in practice, to have a large number of unmarried children, hence much land, since there was no other basis for comparing wealth.[125] *Ayllu* also helped families avoid seasonal labour crunches and kept track of the number of able-bodied young men and women in each household, so as to ensure not only that none were short-handed at critical moments, but also that the aged or infirm, widows, orphans or disabled were taken care of.

Between households, responsibilities came down to a principle of reciprocity: records were kept and at the end of each year all outstanding credits and debts were to be cancelled out. This is where the 'village bureaucracy' comes in. To do that meant units of work had to be measured in a way which allowed clear resolution to the inevitable arguments that crop up in such situations – about who did what for whom, and who owed what to whom.[126] Each *ayllu* appears to have had its own *khipu* strings, which were constantly knotted and re-knotted to keep track as debts were registered or cancelled out. It's possible that *khipu* were invented for such purposes. In other words, although the actual administrative tools used were different, the reason for their existence was quite similar to what we envisage for the village accounting systems in prehistoric Mesopotamia, and rooted in a similarly explicit ideal of equality.[127]

Of course, the danger of such accounting procedures is that they can be turned to other purposes: the precise system of equivalence that underlies them has the potential to give almost any social arrangement, even those founded on arbitrary violence (e.g. 'conquest'), an air of even-handedness and equity. That is why sovereignty and administration make such a potentially lethal combination, taking the equalizing effects of the latter and transforming them into tools of social domination, even tyranny.

Under the Inca, let's recall, all *ayllus* were reduced to the status of 'conquered women' and *khipu* strings were employed to keep track of labour debts owed to the central Inca administration. Unlike the local string records, these were fixed and non-negotiable; the knots were never unravelled and retied. Here it is necessary to overcome a few myths about the Inca, who are often portrayed as the mildest of empires – even a kind of benevolent proto-socialist state. In fact, it was the pre-existing *ayllu* system that continued to provide social security under Inca rule. By contrast, the overarching administrative structure put in place by the Inca court was largely extractive and exploitative in nature (even if local officers of the court preferred to misrepresent it as an extension of *ayllu* principles): for purposes of central monitoring and recording, households were grouped into units of 10, 50, 100, 500, 1,000, 5,000 and so on, each responsible for labour obligations over and above those they already owed to their community, in a way that could only play havoc with existing allegiances, geography and communal organization.[128] Corvée duties were assigned uniformly according to a rigid scale of measurement; work tasks might simply be invented if there was nothing that needed doing; scofflaws faced severe punishment.[129]

The results were predictable, and we can see them clearly reflected in the first-hand accounts supplied by Spanish chroniclers of the time, who took an obvious interest in Inca strategies of conquest and domination and their local workings. Community leaders became de facto state agents, and either took advantage of legalisms to get rich or tried to shield their wards and themselves if they got in to trouble. Those who were unable to meet labour debts or who tried unsuccessfully to flee or rebel, were reduced to the status of servants, retainers and concubines for Inca courts and officials.[130] This new class of hereditary peons was growing rapidly at the time of Spanish conquest.

None of which is to say the Inca reputation as adept administrators is unfounded. They apparently were capable of keeping exact track of births and deaths, adjusting household numbers at yearly festivals and so on. Why, then, impose such an oddly clumsy and monolithic system on to an existing one (the *ayllu*) which was clearly more nuanced? It's hard to escape the impression that in all such situations, the apparent heavy-handedness, the insistence on following the rules even when they make no sense, is really half the point. Perhaps this is simply how sovereignty manifests itself, in bureaucratic form. By ignoring the unique history of every household, each individual, by reducing everything to numbers one provides a language of equity – but simultaneously ensures that there will always be some who fail to meet their quotas, and therefore that there will always be a supply of peons, pawns or slaves.

In the Middle East, very similar things appear to have happened in later periods of history. Most famously, perhaps, the books of the Prophets in the Hebrew Bible preserve memories of powerful protests that ensued as demands for tribute drove farmers into penury, forced them to pawn their flocks and vineyards, and ultimately surrender their children into debt peonage. Or wealthy merchants and administrators took advantage of crop failures, floods, natural disasters or neighbours' simple bad luck to offer interest-bearing loans that led to the same results. Similar complaints are recorded in China and India as well. The first establishment of bureaucratic empires is almost always accompanied by some kind of system of equivalence run amok. This is not the place to outline a history of money and debt[131] – only to note that it's no coincidence that societies like those of Uruk-period Mesopotamia were, simultaneously, commercial and bureaucratic. Both money and administration are based on similar principles of *impersonal* equivalence. What we wish to emphasize at this point is how frequently the most violent inequalities seem to arise, in the first instance, from such fictions of legal equality. All citizens of a city, or all worshippers of its god, or all subjects of its king were considered ultimately the same – at least in that one specific way. The same laws, the same rights, the same responsibilities applied to all of them, whether as individuals or, in later and more patriarchal times, as families under the aegis of some *paterfamilias*.

What's important here is the fact that this equality could be viewed as making people (as well as things) interchangeable, which in turn allowed rulers, or their henchmen, to make impersonal demands that took no consideration of their subjects' unique situations. This is of course what gives the word 'bureaucracy' such distasteful associations almost everywhere today. The very term evokes mechanical stupidity. But there's no reason to believe that impersonal systems were originally, or are necessarily, stupid. If the calculations of a Bolivian *ayllu* or Basque council – or presumably a Neolithic village administration like that of Tell Sabi Abyad, and its urban successors in Mesopotamia – produced an obviously impossible or unreasonable result, matters could always be adjusted. As anyone knows who has spent time in a rural community, or serving on a municipal or parish council of a large city, resolving such inequities might require many hours, possibly days of tedious discussion, but almost always a solution will be arrived at that no one finds entirely unfair. It's the addition of sovereign power, and the resulting ability of the local enforcer to say, 'Rules are rules; I don't want to hear about it' that allows bureaucratic mechanisms to become genuinely monstrous.

Over the course of this book we have had occasion to refer to the three primordial freedoms, those which for most of human history were simply assumed: the freedom to move, the freedom to disobey and the freedom to create or transform social relationships. We also noted how the English word 'free' ultimately derives from a Germanic term meaning 'friend' – since, unlike free people, slaves cannot have friends because they cannot make commitments or promises. The freedom to make promises is about the most basic and minimal element of our third freedom, much as physically running away from a difficult situation is the most basic element of the first. In fact, the earliest word for 'freedom' recorded in any human language is the Sumerian term *ama(r)-gi*, which literally means 'return to mother' – because Sumerian kings would periodically issue decrees of debt freedom, cancelling all non-commercial debts and in some cases allowing those held as debt peons in their creditors' households to return home to their kin.[132]

One might ask, how could that most basic element of all human freedoms, the freedom to make promises and commitments and thus build relationships, be turned into its very opposite: into peonage, serfdom or permanent slavery? It happens, we'd suggest, precisely when promises become impersonal, transferable – in a nutshell, bureaucratized. It is one of history's great ironies that Madame de Graffigny's notion of the Inca state as a model of a benevolent, bureaucratic order actually derives from a misreading of the sources, if a very common one: mistaking the social benefits of local, self-organized administrative units (*ayllu*) for an imperial, Inca structure of command, which in reality served almost exclusively to provision the army, priesthood and administrative classes.[133] Mesopotamian and later Chinese kings also tended to represent themselves, like the Egyptian nomarchs, as protectors of the weak, feeders of the hungry, solace of widows and orphans.

As money is to promises, we might say, state bureaucracy is to the principle of care: in each case we find one of the most fundamental building blocks of social life corrupted by a confluence of maths and violence.

IN WHICH, ARMED WITH NEW KNOWLEDGE, WE RETHINK SOME BASIC PREMISES OF SOCIAL EVOLUTION

Social scientists and political philosophers have been debating the 'origins of the state' for well over a century. These debates are never resolved and are unlikely ever to be. At this point, at least we can understand why. Much like the search for the 'origins of inequality', seeking the origins of the state is little more than chasing a phantasm. As we noted at the beginning of the chapter, it never occurred to Spanish conquistadors to ask whether or not they were dealing with 'states' since the concept didn't really exist at the time. The language they used, of kingdoms, empires and republics, serves just as well, and in many ways rather better.

Historians, of course, still speak of kingdoms, empires and republics. If social scientists have come to prefer the language of 'states' and 'state formation' it's largely because this is taken to be more scientific – despite the lack of consistent definition. It's not clear why. Part of the reason might be that the notions of 'the state' and of modern science both emerged around the same time and were to a certain degree entangled with one another. Whatever the cause, because the existing literature is so relentlessly focused on a single narrative of increasing complexity, hierarchy and state formation, it becomes very difficult to use the term 'state' for any other purpose.

The fact that our planet is, at the present time, almost entirely covered by states obviously makes it easy to write as if such an outcome was inevitable. Yet our present situation regularly leads people to make 'scientific' assumptions about how we got here that have almost nothing to do with the actual data. Certain salient features of current arrangements are just projected backwards, presumed to exist once society has attained a certain degree of complexity – unless definitive evidence of their absence can be produced.

For example, it is often simply assumed that states begin when certain key functions of government – military, administrative and judicial – pass into the hands of full-time specialists. This makes sense if you accept the narrative that an agricultural surplus 'freed up' a significant portion of the population from the onerous responsibility of securing adequate amounts of food: a story that suggests the beginning of a process that would lead to our current global division of labour. Early states might have used this surplus largely to support full-time bureaucrats, priests, soldiers and the like, but – we are always reminded – its existence also allowed for full-time sculptors, poets and astronomers.

It's a compelling story. It is also quite true when applied to our present-day situation (at least, only a small percentage of us are now involved in the production and distribution of foodstuffs). However, almost none of the regimes we've been considering in this chapter were actually staffed by full-time specialists. Most obviously, none seem to have had a standing army. Warfare was largely a business for the agricultural off-season. Priests and judges rarely worked full-time either; in fact, most government institutions in Old Kingdom Egypt, Shang China, Early Dynastic Mesopotamia or for that matter classical Athens were staffed by a rotating workforce whose members had other lives as managers of rural estates, traders, builders or any number of different occupations.[134]

One could go further. It's not clear to what degree many of these 'early states' were themselves largely seasonal phenomena (recall that, at least as far back as the Ice Age, seasonal gatherings could be stages for the *performance* of something that looks to us a bit like kingship; rulers held court only during certain periods of the year; and some clans or warrior societies were given state-like police powers only during the winter months).[135] Like warfare, the business of government tended to concentrate strongly upon certain times of year: there were months full of building projects, pageants, festivals, census-taking, oaths of allegiance, trials and spectacular executions; and other times when a king's subjects (and sometimes even the king himself) scattered to attend to the more urgent needs of planting, harvesting and pasturage. This doesn't mean these kingdoms weren't real: they were capable of mobilizing, or for that matter killing and maiming, thousands of human beings. It just means that their reality was, in effect, sporadic. They appeared and then dissolved away.

Could it be that, in the same way that play farming – our term for those loose and flexible methods of cultivation which leave people free to pursue any number of other seasonal activities – turned into more serious agriculture, play kingdoms began to take on more substance as well? The evidence from Egypt might be interpreted along these lines. But it's also possible that both these processes, when they did happen, were ultimately driven by something else, such as the emergence of patriarchal relations and the decline of women's power within the household. Surely these are the kinds of questions we *should* be asking. Ethnography also teaches us that kings are rarely content with the idea of being a sporadic presence in most of their subjects' lives. Even rulers of kingdoms that nobody would describe as a state, like the Shilluk *reth* or rulers of minor principalities in Java or Madagascar, would try to insert themselves into the rhythms of ordinary social life by insisting that no one can swear an oath, or marry, or even greet one another without invoking their name. In this manner, the king would become the necessary means by which his subjects established relations with each other, in much the same way as later heads of state would insist on putting their faces on money.

In 1852 the Wesleyan minister and missionary Richard B. Lyth described how in the Fijian kingdom of Cakaudrove there was a daily rule of absolute silence at sunrise. Then the king's herald would proclaim that he was about to chew his kava root, whereon all his subjects shouted, 'Chew it!' This was followed by a thunderous roar when the ritual was completed. The ruler was the Sun, who gave both life and order to his people. He recreated the universe each day. In fact, most scholars nowadays insist this king wasn't even a king, but merely the head of a 'confederacy of chiefdoms' who ruled over perhaps a few thousand people. Such cosmic claims are regularly made in royal ritual almost everywhere in the world, and their grandeur seems to bear almost no relation to a ruler's actual power (as in their ability to make anyone do anything they don't want to do). If 'the state' means anything, it refers to precisely the totalitarian impulse that lies behind all such claims, the desire effectively to make the ritual last forever.[136]

Monuments like the Egyptian pyramids seem to have served a similar purpose. They were attempts to make a certain kind of power seem eternal – the kind that only really manifested itself in those particular months when pyramid construction was under way. Inscriptions or objects designed to project an image of cosmic power – palaces, mausoleums, lavish stelae with godlike figures announcing laws or boasting of their conquests – are precisely the ones most likely to endure, thereby forming the core of the world's major heritage sites and museum collections today. Such is their power that even now we risk falling under their spell. We don't really know how seriously to take them. After all, the Fijian subjects of the King of Cakaudrove must at least have been willing to play along with the daily sunrise ritual, since he lacked much in the way of means to compel them. Yet rulers such as Sargon the Great of Akkad or the First Emperor of China had many such means at their disposal, and as a result we can say even less about what their subjects really made of their more grandiose claims.[137]

To understand the realities of power, whether in modern or ancient societies, is to acknowledge this gap between what elites claim they can do and what they are actually able to do. As the sociologist Philip Abrams pointed out long ago, failure to make this distinction has led social scientists up countless blind alleys, because the state is 'not the reality which stands behind the mask of political practice. It is itself the mask which prevents our seeing political practice as it is.' To understand the latter, he argued, we must attend to 'the senses in which the state does not exist rather than to those in which it does'.[138] We can now see that these points apply just as forcefully to ancient political regimes as they do to modern ones – if not more so.

An origin for 'the state' has long been sought in such diverse places as ancient Egypt, Inca Peru and Shang China, but what we now regard as states turn out not to be a constant of history at all; not the result of a long evolutionary process that began in the Bronze Age, but rather a confluence of three political forms – sovereignty, administration and charismatic competition – that have different origins. Modern states are simply one way in which the three principles of domination happened to come together, but this time with a notion that the power of kings is held by an entity called 'the people' (or 'the nation'), that bureaucracies exist for the benefit of said 'people', and in which a variation on old, aristocratic contests and prizes has come to be relabelled as 'democracy', most often in the form of national elections. There was nothing inevitable about it. If proof of that were required, we need only observe how much this particular arrangement is currently coming apart. As we noted, there are now planetary bureaucracies (public and private, ranging from the IMF and WTO to J. P. Morgan Chase and various credit-rating agencies) without anything that resembles a corresponding principle of global sovereignty or global field of competitive politics; and everything from cryptocurrencies to private security agencies, undermining the sovereignty of states.

If anything is clear by now it's this. Where we once assumed 'civilization' and 'state' to be conjoined entities that came down to us as a historical package (take it or leave it, forever), what history now demonstrates is that these terms actually refer to complex amalgams of elements which have entirely different origins and which are currently in the process of drifting apart. Seen this way, to rethink the basic premises of social evolution is to rethink the very idea of politics itself.

CODA: ON CIVILIZATION, EMPTY WALLS AND HISTORIES STILL TO BE WRITTEN

On reflection, it's odd that the term 'civilization' – one we've not discussed much until now – ever came to be used this way in the first place. When people talk about 'early civilizations' they are mostly referring to those very same societies we've been describing in this chapter and their direct successors: Pharaonic Egypt, Inca Peru, Aztec Mexico, Han China, Imperial Rome, ancient Greece, or others of a certain scale and monumentality. All these were deeply stratified societies, held together mostly by authoritarian government, violence and the radical subordination of women. Sacrifice, as we've seen, is the shadow lurking behind this concept of civilization: the sacrifice of our three basic freedoms, and of life itself, for the sake of something always out of reach – whether that be an ideal of world order, the Mandate of Heaven or blessings from insatiable gods. Is it any wonder that in some circles the very idea of 'civilization' has fallen into disrepute? Something very basic has gone wrong here.

One problem is that we've come to assume that 'civilization' refers, in origin, simply to the habit of living in cities. Cities, in turn, were thought to imply states. But as we've seen, that is not the case historically, or even etymologically.[139] The word 'civilization' derives from Latin *civilis*, which actually refers to those qualities of political wisdom and mutual aid that permit societies to organize themselves through voluntary coalition. In other words, it originally meant the type of qualities exhibited by Andean *ayllu* associations or Basque villages, rather than Inca courtiers or Shang dynasts. If mutual aid, social co-operation, civic activism, hospitality or simply caring for others are the kind of things that really go to make civilizations, then this true history of civilization is only just starting to be written.

As we saw in Chapter Five, Marcel Mauss took some initial, furtive steps in that direction but was largely ignored; and, as he anticipated, such a history might well begin with those geographically expansive 'culture areas' or 'interaction spheres' that archaeologists can now trace back into periods far earlier than kingdoms or empires, or even cities. As we've seen, physical evidence left behind by common forms of domestic life, ritual and hospitality shows us this deep history of civilization. In some ways it's much more inspiring than monuments. Arguably, the most important findings of modern archaeology are precisely these vibrant and far-flung networks of kinship and commerce, where those who rely largely on speculation have expected to find only backward and isolated 'tribes'.

As we've been showing throughout this book, in all parts of the world small communities formed civilizations in that true sense of extended moral communities. Without permanent kings, bureaucrats or standing armies they fostered the growth of mathematical and calendrical knowledge. In some regions they pioneered metallurgy, the cultivation of olives, vines and date palms, or the invention of leavened bread and wheat beer; in others they domesticated maize and learned to extract poisons, medicines and mind-altering substances from plants. Civilizations, in this true sense, developed the major textile technologies applied to fabrics and basketry, the potter's wheel, stone industries and beadwork, the sail and maritime navigation, and so on.

A moment's reflection shows that women, their work, their concerns and innovations are at the core of this more accurate understanding of civilization. As we saw in earlier chapters, tracing the place of women in societies without writing often means using clues left, quite literally, in the fabric of material culture, such as painted ceramics that mimic both textile designs and female bodies in their forms and elaborate decorative structures. To take just two examples, it's hard to believe that the kind of complex mathematical knowledge displayed in early Mesopotamian cuneiform documents or in the layout of Peru's Chavín temples sprang fully formed from the mind of a male scribe or sculptor, like Athena from the head of Zeus. Far more likely, these represent knowledge accumulated in earlier times through concrete practices such as the solid geometry and applied calculus of weaving or beadwork.[140] What until now has passed for 'civilization' might in fact be nothing more than a gendered appropriation – by men, etching their claims in stone – of some earlier system of knowledge that had women at its centre.

We began this chapter by noting how often the expansion of

ambitious polities, and the concentration of power in a few hands, was accompanied by the marginalization of women, if not their violent subordination. This seems to be true not just of second-order regimes like Aztec Mexico and Old Kingdom Egypt but also of first-order ones like Chavín de Huántar. But what about cases where, even as societies scaled up and also took on more centralized forms of government, women and their concerns remained at the core of things? Do any such exist in history? This brings us to our final example: Minoan Crete.

Whatever was happening during the Bronze Age on Crete, the largest and most southerly of the Aegean islands, it clearly doesn't quite fit the scholarly playbook of 'state formation'. Yet the remains of what has come to be called Minoan society are too dramatic, too impressive and too close to the heart of Europe (and what was to become the classical world) to be sidelined or ignored. Indeed, in the 1970s the renowned archaeologist Colin Renfrew chose nothing less than *The Emergence of Civilisation* as the title of his important book on the prehistory of the Aegean, to the eternal confusion and annoyance of archaeologists working anywhere else.[141] Despite this high profile, and more than a century of intense fieldwork, Minoan Crete remains a kind of beautiful irritant for archaeological theory, and frankly a source of puzzlement to anyone coming at the topic from outside.

Much of our knowledge comes from the metropolis of Knossos, as well as other major centres at Phaestos, Malia and Zakros, which are usually described as 'palatial societies' that existed between 1700 and 1450 BC (the Neopalatial or 'New Palace' period).[142] Certainly, they were very impressive places at this time. Knossos, thought to have had a population of about 25,000,[143] in many ways resembles similar cities in other parts of the eastern Mediterranean, centring as it does on large palace complexes replete with industrial quarters and storage facilities, and a system of writing on clay tablets ('Linear A') which, frustratingly, has never been deciphered. The problem is that, unlike palatial societies of roughly the same age – such as those of Zimri-Lim at Mari on the Syrian Euphrates, or in Hittite Anatolia to the north, or Egypt – there is simply no clear evidence of monarchy on Minoan Crete.[144]

It's not for lack of material. We might not be able to read the writing, but Crete and the nearby island of Thera (Santorini) – where a bed of volcanic ash preserves the Minoan town of Akrotiri in splendid detail – actually furnish us with one of the most extensive bodies of pictorial art from the Bronze Age world: not just frescoes, but also ivories and detailed engravings on seals and jewellery.[145] By far the most frequent depictions of authority figures in Minoan art show adult women in boldly patterned skirts that extend over their shoulders but are open at the chest.[146] Women are regularly depicted at a larger scale than men, a sign of political superiority in the visual traditions of all neighbouring lands. They hold symbols of command, like the staff-wielding 'Mother of Mountains' who appears on seal impressions from a major shrine at Knossos; they perform fertility rites before horned altars, sit on thrones, meet together in assemblies with no male presiding and appear flanked by supernatural creatures and dangerous animals.[147] Most male depictions, on the other hand, are either of scantily clad or naked athletes (no women are depicted naked in Minoan art); or show men bringing tribute and adopting poses of subservience before female dignitaries. All this is without parallel in the highly patriarchal societies of Syria, Lebanon, Anatolia and Egypt (all regions that Cretans of the time were familiar with, since they visited them as traders and diplomats).

Scholarly interpretations of Minoan palatial art, with its array of powerful females, are somewhat perplexing. Most follow Arthur Evans, the early-twentieth-century excavator of Knossos, in identifying such figures as goddesses, or priestesses wielding no earthly power – almost as though they have no connection to the real world.[148] They tend to come up in the 'religion and ritual' sections of books on Aegean art and archaeology as opposed to 'politics', 'economics' or 'social structure' – politics, in particular, being reconstructed with almost no reference to the art at all. Others simply avoid the issue altogether, describing Minoan political life as clearly different, but ultimately impenetrable (a gendered sentiment if ever there was one). Would this keep happening if these were images of men in positions of authority? Unlikely, since the same scholars usually have no trouble identifying similar scenes that involve males – painted on the walls of Egyptian tombs, for example – or even actual representations of

Keftiu (Cretans) bringing tribute to powerful Egyptian men as reflections of real power relations.

Another puzzling bit of evidence is the nature of the wares that Minoan merchants imported from abroad. Minoans were a trading people, and the traders appear to have been mostly men. But starting in the Proto-palatial period, what they brought home from overseas had a distinctively female flavour. Egyptian sistra, cosmetic jars, figures of nursing mothers and scarab amulets do not come from the male-dominated sphere of courtly culture but the rituals of non-royal Egyptian women and the gynocentric rites of Hathor. Hathor was celebrated outside Egypt too, in temples near the Sinai turquoise mines and in maritime ports, where the horned goddess morphed into a protector of travellers. One such port was Byblos on the Lebanese coast, where an assemblage of cosmetics and amulets – almost identical to those from early Cretan tombs – was found buried as offerings in a temple. Most likely, such objects travelled along with women's cults, perhaps like the much later cults of Isis, tracking the 'official' trade of male elites. The concentration of these items within prestigious Cretan *tholos* tombs in the period just before the formation of palaces (another of those neglected 'proto-periods') suggests, at the very least, that women occupied the demand side of such long-distance exchanges.[149]

Again, this was most definitely *not* the case elsewhere. To throw things into relief, let's briefly consider the slightly later palaces of mainland Greece.

Cretan palaces were unfortified, and Minoan art makes almost no reference to war, dwelling instead on scenes of play and attention to creature comforts. All this is in marked contrast to what was happening on the Greek mainland. Walled citadels arose at Mycenae, Pylos and Tiryns around 1400 BC, and before long their rulers launched a successful takeover of Crete, occupying Knossos and assuming control of its hinterland. Compared to Knossos or Phaistos, their residences appear little more than hill forts, perched on key passes in the Peloponnese and surrounded by modest hamlets. Mycenae, the biggest, had a population of around 6,000. This is not surprising, since the palace societies of the mainland don't arise from pre-existing cities but from warrior aristocracies that produced the earlier Shaft Graves of Mycenae, with their haunting gold death masks and weaponry inlaid with scenes of male fighters and hunting bands.[150]

On to this institutional foundation – the warrior band leader and his hunting retinue – were soon added courtly finery borrowed mainly from the Cretan palaces, and a script (Linear B) adapted to write the Greek language for administration. Analysis of the Linear B tablets suggests that just a handful of literate officers did most of the administrative work themselves, personally inspecting crops and livestock, gathering taxes, distributing raw materials to artisans and supplying provisions for festivals. It was all rather limited and small-scale,[151] and a Mycenaean *wanax* (the ruler or overlord) would have exercised little true sovereignty beyond his citadel, making do with seasonal tax raids on a surrounding populace whose lives otherwise went on beyond the scope of royal surveillance.[152]

These Mycenaean overlords held court in a *megaron* or great hall, a relatively well-preserved example of which exists at Pylos. Early archaeologists were being a bit fanciful when they imagined this actually to be the palace of the Homeric king Nestor, but there is no doubt one of Homer's kings would have felt quite at home here. The *megaron* centred on a huge hearth, open to the sky; the remainder of the space, including the throne, was most likely cast in shadow. The walls bear frescoes showing a bull led to slaughter and a bard playing the lyre. The *wanax*, although not depicted, is clearly the focus of these processional scenes, which converge on his throne.[153]

We can contrast this with the 'Throne Room' of Knossos on Crete, identified as such by Arthur Evans. In this case the purported throne faces an open space, surrounded by stone benches symmetrically arranged in rows so the assembled groups could sit in comfort for long periods, each visible to all the others. Nearby was a stepped bathing chamber. There are many such 'lustral basins' (as Evans called them) in Minoan houses and palaces. Archaeologists puzzled for decades over their function, until at Akrotiri one such was found directly under a painted scene of a female initiation ceremony most likely linked to menstruation.[154] In fact, on purely architectural grounds, and notwithstanding Evans's rather desperate insistence that it 'seems better adapted for a man', the centrepiece of the Throne Room may be quite reasonably understood not as the seat of a male monarch but rather that of a council head, and its occupants more likely a succession of female councillors.

Pretty much all the available evidence from Minoan Crete suggests a system of female political rule – effectively a theocracy of some sort, governed by a college of priestesses. We might ask: why are contemporary researchers so resistant to this conclusion? One can't blame everything on the fact that proponents of 'primitive matriarchy' made exaggerated claims back in 1902. Yes, scholars tend to say that cities ruled by colleges of priestesses are unprecedented in the ethnographic or historical record. But by the same logic, one could equally point out that there is no parallel for a kingdom run by men, in which all the visual representations of authority figures are depictions of women. Something different was clearly happening on Crete.

Certainly, the way in which Minoan artists represented life attests to a profoundly different sensibility to that of Crete's neighbours on mainland Greece. In an essay called 'The Shapes of Minoan Desire', Jack Dempsey points out that erotic attention seems to be displaced from the female body on to just about every other facet of life, starting with the lithe, scantily clad figures of young men as they dart in and out of the bodies of bulls who tease them, or gyrate in sporting activities, or the naked boys represented carrying fish. It's all a world away from the stiff animal figures that populate the walls of Pylos, or indeed those of Zimri-Lim's court, let alone the scenes of brutal warfare on later Assyrian wall reliefs. In the Minoan frescoes everything merges – except, that is, for the sharply delineated figures of those leading females, who stand apart or in small groups, happily chatting with one another or admiring some spectacle. Flowers and reeds, birds, bees, dolphins, even hills and mountains are in the throes of a perpetual dance, weaving in and out of each other.

Minoan objects too bleed into one another in an extraordinary play on materials – a true 'science of the concrete' – that turns pottery into crusted shell and melds the worlds of stone, metal and clay together into a common realm of forms, each mimicking the others.

All this unfolds to the undulating rhythms of the sea, the eternal backdrop to this garden of life, and all with a remarkable absence of 'politics', in our sense, or what Dempsey calls the 'self-perpetuating, power-hungry ego'. What these scenes celebrate, as he eloquently puts it, is quite the opposite of politics: it is the 'ritually induced release from individuality, and an ecstasy of being that is overtly erotic and spiritual at the same time (*ek-stasis*, "standing beyond oneself") – a cosmos that both nurtures and ignores the individual, that vibrates with inseparable sexual energies and spiritual epiphanies'. There are no heroes in Minoan art – only players. Crete of the palaces was the realm of *Homo ludens*. Or perhaps, better said, *Femina ludens* – not to mention *Femina potens*.[155]

What we've learned in this chapter can be briefly summarized. The process usually called 'state formation' can in fact mean a bewildering number of very different things. It can mean a game of honour or chance gone terribly wrong, or the incorrigible growth of a particular ritual for feeding the dead; it can mean industrial slaughter, the appropriation by men of female knowledge, or governance by a college of priestesses. But we've also learned that when studied and compared more closely, the range of possibilities is far from limitless.

In fact, there seem to be both logical and historical constraints on the variety of ways in which power can expand its scope; these limits are the basis of our 'three principles' of sovereignty, administration and competitive politics. What we can also see, though, is that – even within these constraints – there were far more interesting things going on than we might ever have guessed by sticking to any conventional definition of 'the state'. What was really happening in the Minoan palaces? They seem to have been in some sense theatrical stages, in some sense women's initiation societies, and administrative hubs all at the same time. Were they even a regime of domination at all?

It's also important to recall the very uneven nature of the evidence we've been dealing with. What would we be saying about Minoan Crete, or Teotihuacan, or Çatalhöyük for that matter, were it not for the fact that their elaborate wall paintings happen to have been preserved? More than almost any other form of human activity, painting on walls is something people in virtually any cultural setting seem inclined to do. This has been true almost since the beginnings of humanity itself. We can hardly doubt that similar images were produced, on skins and fabrics as well as directly on walls, in any number of so-called 'early states' from which only bare stone building blocks or mud-brick enclosures now survive.

Archaeology, using a barrage of new scientific techniques, will undoubtedly reveal many more such 'lost civilizations', as it is already in the process of doing, from the deserts of Saudi Arabia or Peru to the once seemingly empty steppes of Kazakhstan and the tropical forests of Amazonia. As the evidence accumulates, year on year, for large settlements and impressive structures in previously unsuspected locations, we'd be wise to resist projecting some image of the modern nation state on to their bare surfaces, and consider what other kinds of social possibilities they might attest to.

11
Full Circle
On the historical foundations of the indigenous critique

We appear to have come a long way from where this book began, with the Wendat statesman Kandiaronk and the critique of European civilization that developed among indigenous people in North America during the seventeenth century. Now it's time to bring the story full circle. Recall how, by the eighteenth century, the indigenous critique – and the deep questions it posed about money, faith, hereditary power, women's rights and personal freedoms – was having an enormous influence on leading figures of the French Enlightenment, but also resulted in a backlash among European thinkers which produced an evolutionary framework for human history that remains broadly intact today. Portraying history as a story of material progress, that framework recast indigenous critics as innocent children of nature, whose views on freedom were a mere side effect of their uncultivated way of life and could not possibly offer a serious challenge to contemporary social thought (which came increasingly to mean just European thought).[1]

In reality, we have not strayed far at all from this starting point, because the conventional wisdom we've been challenging throughout this book – about hunter-gatherer societies, the consequences of farming, the rise of cities and states – has its genesis right there: with Turgot, Smith and the reaction against the indigenous critique. Of course, the idea that human societies evolved over time was not particularly special to the eighteenth century, or to Europe.[2] What was new in the version of world history put forward by European writers of that century was an insistence on classifying societies by means of subsistence (so that agriculture could be seen as a fundamental break in the history of human affairs); an assumption that as societies grew larger, they inevitably grew more complex; and that 'complexity' means not just a greater differentiation of functions, but also the reorganization of human societies into hierarchical ranks, governed from the top down.

This European backlash was so effective that generations of philosophers, historians, social scientists, and almost anyone else since who wishes to address the human story on a broad scale, feels secure in their knowledge of how it should properly start and where it is leading. It begins with an imaginary collection of tiny hunter-gatherer bands and ends with the current collection of capitalist nation states (or some projection of what might come after them). Anything going on in between can be considered interesting – mainly insofar as it contributed to moving us all on down that particular pathway. As we've been discovering, one consequence is that huge swathes of the human past disappear from the purview of history, or remain effectively invisible (except to the eyes of a tiny number of researchers, who rarely explain the implications of their findings to each other, let alone to anyone else).

Since the 1980s it has been commonplace for social theorists to claim we are living in a new 'post-modern' age, marked by a suspicion towards metanarratives. This claim is often used as justification for a sort of hyper-specialization: to cast one's intellectual net wider – to compare notes with colleagues in other fields, even – smacks of imposing a single, imperialistic vision of history. For this very reason, the 'idea of progress' is usually held up as a prime example of the way we no longer think about history and society. But such claims are odd, since almost everyone making them nonetheless continues to think in evolutionary terms. We could go further: thinkers who do seek to knit together the findings of specialists, to describe the course of human history on a grand scale, haven't entirely got past the biblical notion of the Garden of Eden, the Fall and subsequent inevitability of domination. Blinded by the 'just so' story of how human societies evolved, they can't even see half of what's now before their eyes.

As a result, the same portrayers of world history who profess themselves believers in freedom, democracy and women's rights continue to treat historical epochs of relative freedom, democracy and women's rights as so many 'dark ages'. Similarly, as we've seen, the concept of 'civilization' is still largely reserved for societies whose defining characteristics include high-handed autocrats, imperial conquests and the use of slave labour. Presented with undeniable cases of large and materially sophisticated societies for which evidence of such things is conspicuously lacking – ancient centres like Teotihuacan or Knossos, for example – the standard recourse is to throw up one's hands and say: who can tell what was really going on there? or insist that Ozymandias' throne room must be lurking in there somewhere, but that we simply haven't found it yet.

IN WHICH WE CONSIDER JAMES C. SCOTT'S ARGUMENTS ABOUT THE LAST 5,000 YEARS AND ASK WHETHER CURRENT GLOBAL ARRANGEMENTS WERE, IN FACT, INEVITABLE

You may object: perhaps much of human history was more complicated than we usually admit, but surely what matters is how things ended up. For at least 2,000 years, most of the world's population have been living under kings or emperors of one sort or another. Even in places where monarchy did not exist – much of Africa or Oceania, for example – we find that (at the very least) patriarchy, and often violent domination of other sorts, have been widespread. Once established, such institutions are very hard to get rid of. So your objection might run: all you're saying is that the inevitable took a little longer to happen. That doesn't make it any less inevitable.

Similarly, with farming. True, your objection might run: agriculture might not have transformed everything overnight, but surely it laid the groundwork for later systems of domination? Wasn't it really just a matter of time? Did not the very possibility of piling up large surpluses of grain, in effect, lay a trap? Wasn't it inevitable that, sooner or later, some warrior-prince like Narmer of Egypt would begin amassing stockpiles for his henchmen? And once he did, surely the game was over. Rival kingdoms and empires would quickly come into being. Some would find the means to expand; they would insist on their subjects producing more and more grain, and those subjects would grow in number, even as the number of remaining free peoples tended to remain stable. Once again, was it not just a matter of time before one of those kingdoms (or, as it turned out, a small collection of them) came up with a successful formula for world conquest – just the right combination of guns, germs and steel – and imposed its system on everybody else?

James Scott – a renowned political scientist who has devoted much of his career to understanding the role of states (and those who succeed in evading them) in human history – has a compelling description of how this agricultural trap works. The Neolithic, he suggests, began with flood-retreat agriculture, which was easy work and encouraged redistribution; the largest populations were, indeed, concentrated in deltaic environments, but the first states in the Middle East (he concentrates largely on these; and China) developed upriver, in areas with an especially strong focus on cereal agriculture – wheat, barley, millet – and relatively limited access to a range of other staples. The key to the importance of grain, Scott notes, is that it was durable, portable, easily divisible and quantifiable by bulk, and therefore an ideal medium to serve as a basis for taxation. Growing above ground – unlike, say, certain tubers or legumes – grain crops were also highly visible and amenable to appropriation. Cereal agriculture did not cause the rise of extractive states, but it was certainly predisposed to their fiscal requirements.[3]

Like money, grain allows a certain form of terrifying equivalence. Whatever the reasons why it initially became a predominant crop in a given region (as we've seen – in Egypt for example – this had much to do with changes in rituals for the dead), once this happened a permanent kingdom could always emerge. However, Scott also points out that for much of history this process turned out to be a trap for these newfound 'grain states' as well, limiting them to areas that favoured intensive agriculture, and leaving surrounding highlands, fenlands and marshes largely beyond their reach.[4] What's more, even within those confines the grain-based kingdoms were fragile, always prone to collapse under the weight of over-population, ecological devastation and the kind of endemic diseases that always seemed to result when too many humans, domesticated animals and parasites accumulated in one place.

Ultimately, though, Scott's focus isn't really on states at all: it's about the 'barbarians' – a term Scott uses for all those groups which came to surround the little islands of authoritarian-bureaucratic rule, and which existed in a largely symbiotic relation with them: some ever-shifting mix of raiding, trading and mutual avoidance. As Scott argued about the hill peoples of Southeast Asia, some of these 'barbarians' became, effectively, anarchists: organizing their lives in explicit opposition to the valley societies below, or to prevent the emergence of stratified classes in their own midst. As we've seen, such conscious rejection of bureaucratic values – another example of cultural schismogenesis – could also give rise to 'heroic societies', a hurly-burly of petty lords whose pre-eminence was founded on dramatic contests of war, feasting, boasting, duelling, games, gifts and sacrifice. Monarchy itself is likely to have started that way, on the fringes of urban-bureaucratic systems.

But to continue with Scott: barbarian monarchies remained either small-scale or, if they did expand – as was spectacularly the case under figures like Alaric, Attila, Genghis or Tamerlane – the expansion was short-lived. Throughout much of history, grain states and barbarians remained 'dark twins', locked together in an unresolvable tension, since neither could break out of their ecological niches. When the states had the upper hand, slaves and mercenaries flowed in one direction; when the barbarians were dominant, tribute flowed to appease the most dangerous warlord; or alternatively, some overlord would manage to organize an effective coalition, sweep in on the cities and either lay waste to them, or more typically, attempt to rule them, and inevitably find himself and his retinue absorbed as a new governing class. As the Mongolian adage went, 'One can conquer a kingdom on horseback, to rule it one must dismount.'

Scott, though, doesn't draw any particular conclusions. Rather, he simply remarks that while the period from about 3000 BC to AD 1600 was a fairly miserable one for the bulk of the world's farmers, it was a Golden Age for the barbarians, who reaped all the advantages of their proximity to dynastic states and empires (luxuries to loot and plunder), while themselves living comparatively easy lives. And it was usually possible for at least some of the oppressed to join their ranks. For most of history, he suggests, this is what rebellion typically looked like: defection to join the ranks of nearby barbarians. To put the matter in our own terms, while these agrarian kingdoms managed largely to abolish the freedom to ignore orders, they had a much harder time abolishing the freedom to move away. Empires were exceptional and short-lived, and even the most powerful – Roman, Han, Ming, Inca – could not prevent large-scale movements of people into and out of their spheres of control. Until around a half-millennium ago, a large proportion of the world's population still lived either beyond the tax collector's purview or within reach of some relatively straightforward means of escaping it.[5]

Yet today, in our twenty-first-century world, this is obviously no longer the case. Something did go terribly wrong – at least from the point of view of the barbarians. We no longer live in that world. But merely recognizing that it existed for so long allows us to pose a further important question. How inevitable, really, were the type of governments we have today, with their particular fusion of territorial sovereignty, intense administration and competitive politics? Was this really the necessary culmination of human history?

One problem with evolutionism is that it takes ways of life that developed in symbiotic relation with each other and reorganizes them into separate stages of human history. By the late nineteenth century, it was becoming clear that the original sequence as developed by Turgot and others – hunting, pastoralism, agriculture, then finally industrial civilization – didn't really work. Yet at the same time, the publication of Darwin's theories meant that evolutionism became entrenched as the only possible scientific approach to history – or at least the only one likely to be given credence in universities. So the search was on for more workable categories. In his 1877 *Ancient Society*, Lewis Henry Morgan proposed a series of steps from 'savagery' through 'barbarism' to 'civilization' which was widely adopted in the new field of anthropology. Meanwhile, Marxists concentrated on forms of domination, and the move out of primitive communism towards slavery, feudalism and capitalism, to be followed by socialism (then communism). All these approaches were basically unworkable, and eventually had to be thrown away as well.

Since the 1950s, a body of neo-evolutionist theory has sought to define a new version of the sequence, based on how efficiently groups harvest energy from their environment.[6] As we've seen, almost nobody today subscribes to this framework in its entirety. Indeed, whole volumes have been written taking it to task, or pointing out the many exceptions to its logic; we are 'over all that' and have 'moved on' would be the standard reaction of most anthropologists and archaeologists when confronted with such an evolutionary scheme today. But if our fields have moved on, they have done so, it seems, without putting any alternative vision in place, the result being that almost anyone who is not an archaeologist or anthropologist tends to fall back on the older scheme when they set out to think or write about world history on a large canvas. For this reason it might be useful to summarize the older scheme's basic sequence here:

Band societies: the simplest stage is still assumed to be made up of hunter-gatherers like the !Kung or Hadza, supposedly living in small mobile groups of twenty to forty individuals, without any formal political roles and minimal division of labour. Such societies are thought to be egalitarian, effectively by default.

Tribes: societies like the Nuer, Dayaks or Kayapo. Tribesmen are typically assumed to be 'horticulturalists', which is to say they farm but don't create irrigation works or use heavy equipment like ploughs; they are egalitarian, at least among those of the same age and gender; their leaders have only informal, or at least no coercive, power. 'Tribes' are typically arranged into the sort of complex lineage or totemic clan structures beloved of anthropologists. Economically, the central figures are 'big men' – such as were typically found in Melanesia – responsible for creating voluntary coalitions of contributors to sponsor rituals and feasts. Ritual or craft specialism is limited and usually part-time; tribes are numerically larger than bands, but settlements tend to be roughly of the same size and importance.

Chiefdoms: while the clans of tribal society are all, ultimately, equivalent, in chiefdoms the kinship system becomes the basis for a system of rank, with aristocrats, commoners and even slaves. The Shilluk, Natchez or Calusa are typically treated as chiefdoms; so are, say,

Polynesian kingdoms, or the lords of ancient Gaul. Intensification of production leads to a significant surplus, and classes of full-time craft and ritual specialists emerge, not to mention the chiefly families themselves. There is at least one level of settlement hierarchy (the chief's residence, and everyone else), and the main economic function of the chief is redistributive: pooling resources, often forcibly, and then doling them out to everyone, usually during spectacular feasts.

States: much as already described, these tend to be characterized by intensive cereal agriculture, a legal monopoly on the use of force, professional administration and a complex division of labour.

As many twentieth-century anthropologists pointed out at the time, this scheme doesn't really work either. In reality, 'big men' seem almost entirely confined to Melanesia. 'Indian chiefs' such as Geronimo or Sitting Bull were, in fact, tribal headmen, whose role was nothing like big men in Papua New Guinea. Most of those labelled 'chiefs' in the neo-evolutionist model, as we've already noted, look suspiciously like what we normally think of as 'kings' and may well live in fortified castles, wear ermine robes, support court jesters, have hundreds of wives and harem eunuchs. However, they rarely engage in the mass redistribution of resources, at least not in any systematic way.

The evolutionist response to such critiques was not to abandon the scheme but to fine-tune it. Perhaps chiefdoms are more predatory, evolutionists argued, but they are still fundamentally different to states. What's more, they can be subdivided between 'simple' and 'complex' chiefdoms: in the former, the chief really was just a glorified big man, still working like everyone else, with only minimal administrative assistance; in the complex version, he was backed up by at least two levels of administrative staff, allowing a genuine class structure. Finally, chiefdoms 'cycled', which is to say that the simple overlords were constantly, often quite methodically, trying to patch together tiny empires by conquering or subordinating local rivals, so as to catapult themselves towards the next stage of complexity (characterized by three levels of administrative hierarchy), or even to found states. While a few ambitious chiefs did manage to pull this off, most failed; they reached their ecological or social limits; this rankled with people; the whole jerry-built contraption collapsed, leaving it for some other aspiring dynast to begin trying to conquer the world – or at least, those parts of it considered worth conquering.

In academic circles, an odd disjuncture has developed around the use of such schemes. Most cultural anthropologists view this kind of evolutionary thinking as a sort of quaint relic from their discipline's past, which no one today could possibly take seriously; while most archaeologists only employ terms like 'tribe', 'chiefdom' or 'state' for lack of an alternative terminology. Yet almost anyone else will treat such schemes as the self-evident basis for all further discussion. Throughout this book, we have spent a good deal of time demonstrating how deceptive all this is. The reason why these ways of thinking remain in place, no matter how many times people point out their incoherence, is precisely because we find it so difficult to imagine history that isn't teleological – that is, to organize history in a way which does not imply that current arrangements were somehow inevitable.

As we have already remarked, one of the most puzzling aspects of living in history is that it's almost impossible to predict the course of future events; yet, once events have happened, it's difficult to know what it would even mean to say something else 'could' have happened. A properly historical event has, perhaps, two qualities: it could not have been predicted beforehand, but it only happens once. One does not get to fight the Battle of Gaugamela over again, to see what would have happened if Darius had actually won. Speculating what might have happened – had Alexander, say, been hit by a stray arrow, and there had never been a Ptolemaic Egypt or Seleucid Syria – is at best an idle game. It might raise profound questions – how much difference can an individual really make in history? – but nevertheless, these are questions that cannot ever be definitively answered.

The best we can do, when confronted with unique historical events or configurations such as the Persian or Hellenistic empires, is to engage in a project of comparison. This at least can give us an idea of the sort of things that *might* happen, and at best a sense of the pattern by which one thing is likely to follow another. The problem is that ever since the Iberian invasion of the Americas, and subsequent European colonial empires, we can't even really do that, because there's ultimately been just one political-economic system and it is global. If we wish, say, to assess whether the modern nation state, industrial capitalism and the spread of lunatic asylums are necessarily linked, as opposed to separate phenomena that just happen to have come together in one part of the world, there's simply no basis on which to judge.[7] All three emerged at a time when the planet was effectively a single global system and we have no other planets to compare ourselves to.

One could make the argument – many do – that for most of human history this was already the case. Eurasia and Africa already formed a single interconnected system. Certainly, people, objects and ideas did move back and forth across the Indian Ocean and the Silk Roads (or their Bronze and Iron Age precursors); as a result, dramatic political and economic changes often appeared to occur in more or less co-ordinated fashion across the Eurasian land mass. To take one famous example: almost a century ago, the German philosopher Karl Jaspers noted that all the major schools of speculative philosophy we know today seem to have emerged – apparently independently – in Greece, India and China at roughly the same time, between the eighth and third centuries BC; what's more, they emerged in precisely those cities which had recently seen the invention and widespread adoption of coined money. Jaspers called this the Axial Age, a term since expanded by others to include the period that saw the birth of all today's world religions, stretching from the Persian prophet Zoroaster (*c.*800 BC) to the coming of Islam (*c.*AD 600). Now, the core period of Jaspers's Axial Age – encompassing the lifetimes of Pythagoras, the Buddha and Confucius – corresponds not only to the invention of metal coinage and new forms of speculative thought, but also the spread of chattel slavery across Eurasia, even in places where it had barely existed before; moreover, chattel slavery would eventually fall into decline after a succession of Axial Age empires dissolved (the Maurya, Han, Parthian, Roman), along with their prevailing systems of currency.[8] Obviously, it would be wrong to say that Eurasia can be treated as just one place, and therefore to conclude that comparing how these processes unfolded in different parts of Eurasia is meaningless. Equally, it would be wrong to conclude that such patterns are universal features of human development. They might just be what happened in Eurasia.

Much of Africa, Oceania or northwestern Europe for that matter, was so tied into the great empires of this period – notably with the convergence of terrestrial and maritime trade routes around the Indian Ocean and Mediterranean in the fifth century BC, but arguably already much earlier – that it's hard to know whether they can be taken as independent points of comparison either. The only real exception were the Americas. Admittedly, even before 1492 there must have been some occasional movement back and forth between the two hemispheres (otherwise there wouldn't have been a human population in the Americas to begin with); but prior to the Iberian invasion, the Americas were not in direct or regular communication with Eurasia. They were in no sense part of the same 'world system'. This is important, because it means we do have one truly independent point of comparison (possibly even two, if we consider North and South America as separate), where it is possible to ask: does history really have to take a certain direction?

In the case of the Americas, we actually can pose questions such as: was the rise of monarchy as the world's predominant form of government inevitable? Is cereal agriculture really a trap, and can one really say that once the farming of wheat or rice or maize becomes sufficiently widespread, it's only a matter of time before some enterprising overlord seizes control of the granaries and establishes a regime of bureaucratically administered violence? And once he does, is it inevitable that others will imitate his example? Judging by the history of pre-Columbian North America, at least, the answer to all these questions is a resounding 'no'.

In fact, although archaeologists of North America use the language of 'bands', 'tribes', 'chiefdoms' and 'states,' what actually seems to have happened there defies all such assumptions. We've already seen how in the western half of the continent there was, if anything, a movement *away* from agriculture in the centuries before the European invasion; and how Plains societies often seem to have moved back and forth, over the course of any given year, between bands and something that shares at least some of the features we identify with states – in other words, between what should have been opposite ends of the scale of social evolution. Even more startling in its own way is what happened in the eastern part of the continent.

From roughly AD 1050 to 1350 there was, in what's now East St Louis, a city whose real name has been forgotten, but which is known to history as Cahokia.[9] It appears to have been the capital of what James Scott would term a classic budding 'grain state', rising magnificently and seemingly from nowhere, around the time that the Song Dynasty ruled in China and the Abbasid Caliphate in Iraq. Cahokia's population peaked at something in the order of 15,000 people; then it abruptly dissolved. Whatever Cahokia represented in the eyes of those under its sway, it seems to have ended up being overwhelmingly and resoundingly rejected by the vast majority of its people. For centuries after its demise the site where the city once stood, and hundreds of miles of river valleys around it, lay entirely devoid of human habitation: a 'vacant quarter' (rather like the Forbidden Zone in Pierre Boulle's *Planet of the Apes*), a place of ruins and bitter memories.[10]

Successor kingdoms to Cahokia sprang up to the south but then likewise crumbled. By the time Europeans arrived on the eastern seaboard of North America, 'Mississippian civilization' – as it has come to be known – was but a distant memory and the descendants of Cahokia's subjects and neighbours appear to have reorganized themselves into *polis*-sized tribal republics, in careful ecological balance with their natural environment. What had happened? Were the rulers of Cahokia and other Mississippian cities overthrown by popular uprisings, undermined by mass defection, victims of ecological catastrophe, or (more likely) some intricate mix of all three? Archaeology may one day supply more definitive answers. Until such time, what we can say with some confidence is that the societies encountered by European invaders from the sixteenth century onwards were the product of centuries of political conflict and self-conscious debate. They were, in many cases, societies in which the ability to engage in self-conscious political debate was itself considered one of the highest human values.

It is impossible to understand the devotion to individual liberty, or even the sceptical rationalism of figures like Kandiaronk, outside this larger historical context; or at least, that is what we propose to show in the rest of this chapter. Much though later European authors liked to imagine them as innocent children of nature, the indigenous populations of North America were in fact heirs to their own, long intellectual and political history – one that had taken them in a very different

Some key archaeological sites in the Mississippi River Basin and adjacent regions
Lake Superior
N
Mississippi River
Lake Michigan
Lake Erie
Missouri River
Ohio Hopewell sites
Ohio River
Cahokia
Emerald
Angel
Common Field
Kincaid
Crosno
Turk
Towosahgy
Lilbourn
Adams
Pinson
Arkansas River
Etowah
Red River
Winterville
Moundville
Lake George
Poverty Point
Gulf of Mexico
0 100 200 300 miles
0 100 200 300 400 km

direction to Eurasian philosophers and which, arguably, ended up having a profound influence on conceptions of freedom and equality, not just in Europe but everywhere else as well.

Of course, we are taught to treat such claims as inherently unlikely, even slightly preposterous. As we've seen in the case of Turgot, evolutionary theory as we know it today was largely created so as to entrench such dismissive attitudes: to make them seem natural or obvious. If the indigenous peoples of North America aren't being imagined as living in a separate time, or as vestiges of some earlier stage of human history, then they're imagined as living in an entirely separate reality ('ontology' is the currently fashionable term), a mythic consciousness fundamentally different from our own. If nothing else, it is assumed that any intellectual tradition similar to that which produced Plotinus, Shankara or Zhuang Zu can only be the product of a literary tradition in which knowledge becomes cumulative. And since North America did not produce a written tradition – or at least not the sort we are used to recognizing as such[11] – any knowledge it generated, political or otherwise, was necessarily of a different kind. Any similarity we might see to debates or positions familiar from our own intellectual tradition is typically written off as some sort of naive projection of Western categories. Real dialogue is thus impossible.

Perhaps the most straightforward way to counteract this sort of argument is by citing a text, which describes a concept the Wendat (Huron) called *Ondinnonk*, a secret desire of the soul manifested by a dream:

> Hurons believe that our souls have other desires, which are, as it were, inborn and concealed ... They believe that our soul makes these natural desires known by means of dreams, which are its language. Accordingly, when these desires are accomplished, it is satisfied; but, on the contrary, if it be not granted what it desires, it becomes angry, and not only does not give its body the good and the happiness that it wished to procure for it, but often it also revolts against the body, causing various diseases, and even death.[12]

The author goes on to explain that, in dreams, such secret desires are communicated in a kind of indirect, symbolic language, difficult to understand, and that the Wendat therefore spend a great deal of time trying to decipher the meaning of one another's dreams, or consulting specialists.

All this might seem like an oddly clumsy projection of Freudian theory, but for one thing. The text is from 1649. It was written by a certain Father Ragueneau in a *Jesuit Relation*, precisely 250 years before the appearance of the first edition of Freud's *The Interpretation of Dreams* (1899), an event which, like Einstein's theory of relativity, is widely seen as one of the founding events of twentieth-century thought. What's more, Ragueneau is not our only source. Numerous missionaries attempting to convert other Iroquoian peoples at the same time reported similar theories – which they considered absurd and obviously false (though probably, they concluded, not actually demonic) and attempted to refute, in order to bring their interlocutors around instead to the truth of Holy Scripture.

Does this mean that the community in which Kandiaronk grew up was composed of Freudians? Not exactly. There were significant differences between Freudian psychoanalysis and Iroquoian practice, most dramatically in the collective nature of the therapy. 'Dream-guessing' was often carried out by groups, and realizing the desires of the dreamer, either literally or symbolically, could involve mobilizing an entire community: Ragueneau reported that the winter months in a Wendat town were largely devoted to organizing collective feasts and dramas, literally in order to make some important man or woman's dreams come true. The point here is that it would be very unwise to dismiss such intellectual traditions as inferior – or for that matter, entirely alien – to our own.

One thing that makes the Wendat and Haudenosaunee unusual is that their traditions are so well documented: many other societies were either entirely destroyed, or reduced to traumatized remnants, long before any such records could be written down. One can only wonder what other intellectual traditions might thus have been forever lost. What we are going to do in the remainder of this chapter, then, is examine the history of the Eastern Woodlands of North America from roughly AD 200 to 1600 in exactly this light. Our aim here is to understand the local roots of the indigenous critique of European civilization, and how those roots were entangled in a history that began at Cahokia or perhaps even considerably earlier.

IN WHICH WE ASK HOW MUCH OF NORTH AMERICA CAME TO HAVE A SINGLE UNIFORM CLAN SYSTEM, AND CONSIDER THE ROLE OF THE 'HOPEWELL INTERACTION SPHERE'

Let's start with a puzzle. We've already had occasion to mention how the same basic repertoire of clan names could be found distributed more or less everywhere across Turtle Island (the indigenous name for the North American continent). There were endless local differences, but there were also consistent alliances, so that it was possible for a traveller hailing from a Bear, or Wolf or Hawk clan in what's now Georgia to travel all the way to Ontario or Arizona and find someone obliged to host them at almost any point in between. This seems all the more remarkable when one considers that literally hundreds of different languages were spoken in North America, belonging to half a dozen completely unrelated language families. It hardly seems likely that clan systems were brought over, fully fledged, with the first human arrivals from Siberia; they must have developed in more recent times. But – and here's our puzzle – considering the distances involved, it's hard to imagine how that could have happened.

As Elizabeth Tooker, doyenne of Iroquoian studies, pointed out back in the 1970s, this puzzle is all the more perplexing because it's not entirely clear if North American clans should strictly be considered 'kinship' groups at all. They are more like ritual societies, each dedicated to maintaining a spiritual relation with a different totem animal which is usually only figuratively their 'ancestor'. True, members are recruited by (matrilineal or patrilineal) descent, and fellow clan members consider one another brothers and sisters whom one therefore cannot marry. Yet nobody kept track of genealogies, and there were no ancestor cults or property claims based on descent: all clan members were, effectively, equal. There wasn't even much in the way of collective property other than certain forms of ritual knowledge, dances or chants, bundles of sacred objects and also a collection of names.

A clan typically had a fixed stock of names which were assigned to children. Some of these were chiefly names but, like the sacred paraphernalia, they were rarely directly inherited; instead, they were assigned to the most likely candidate when a title holder died. A community, moreover, was never made up of just one clan. There were usually quite a number, grouped together into two halves (or moieties), which acted as rivals and complements to one another, competing against one another in sports and burying one another's dead. The overall effect was to efface personal histories from public contexts: since names were titles, it would be as if the head of one half of the community would always be John F. Kennedy and the other always Richard Nixon. This fusing of titles and names is a peculiarly North American phenomenon. Some version of it appears almost everywhere on Turtle Island, but almost nowhere else in the world do we see anything quite like it.

Finally, Tooker notes, clans played a key role in diplomacy: not just in providing hospitality to travellers, but organizing the protocol for diplomatic missions, the paying of compensation to prevent wars, or the incorporation of prisoners, who could simply be assigned a name and thereby become a clan member in their new community – even the replacement for someone who had died in that very conflict. The system appeared to be designed to maximize people's capacity to move, individually or collectively, or for that matter to reshuffle social arrangements. Within these parameters there is an endless, almost kaleidoscopic range of possibilities. But where did this set of parameters come from in the first place? Tooker suggested it might be remnants of some long-forgotten 'trading empire', perhaps originally established by merchants from central Mexico, but the suggestion wasn't taken seriously by her fellow scholars – her essay, in fact, is hardly ever cited. There is no evidence that any such trading empire ever existed.

It seems more reasonable to assume that a ritual and diplomatic system has its origins in, well, ritual and diplomacy. The first point where we have unmistakable evidence that such a phenomenon *could* have happened – that is, where active ties developed between virtually all parts of North America – lies in what archaeologists refer to as the 'Hopewell Interaction Sphere', a network with its epicentre in the Scioto and Paint Creek river valleys of Ohio. Between roughly 100 BC and AD 500, communities participating in this network deposited treasures under burial mounds, often piled up in extraordinary quantities. The treasures included quartz-crystal arrowheads, mica and obsidian from the Appalachians, copper and silver from the Great Lakes, conch shells and shark teeth from Gulf of Mexico, grizzly-bear molars from the Rockies, meteoric iron, alligator teeth, barracuda jaws and more.[13] Most of these materials seem to have been used for the manufacture of ritual gear and magnificent costumes – including metal-sheathed pipes and mirrors – worn by shamans, priests and a host of minor officials in a complex organizational structure, the precise nature of which is fiendishly difficult to reconstruct.

Even more striking, many of these tombs were located in the vicinity of gigantic earthworks, some literally miles across. The inhabitants of the Central Ohio valley had been creating such structures since the beginning of what archaeologists called the Adena period, around 1000 BC, and earthworks do also appear in earlier 'Archaic' phases of North American history. As we've already seen in the case of Poverty Point, whoever designed them was capable of making remarkably sophisticated astronomical calculations and employed accurate systems of measurement. One would imagine such people could also marshal and deploy enormous amounts of labour – although here we must be careful. Evidence from more recent times suggests that the tradition of mound-building could have been, in some cases, a side effect of creating dancing-grounds or other flat open spaces for feasts, games and assemblies. Each year before a major ritual these spaces would be swept and flattened, and the accumulated dirt and debris piled up in the same place. Over centuries, this could obviously become a very large amount of material to be shaped. Among the Muskogee, for example, such artificial hills would be covered each year by a new mantle of red, yellow, black or white earth. This work was organized by officials on rotating duties and did not require top-down structures of command.[14]

Such is clearly not the case, however, with really large structures like Poverty Point or the Hopewell earthworks. These did not grow by slow accretion but were planned in advance. The most impressive sites are almost invariably in river valleys, typically quite close to bodies of water. They rise, literally, out of the sodden mud. As anyone who has played as a child with sand or mud (that is, pretty much anyone, including ancient Amerindians) will be aware, it's easy to make structures out of such material, but almost impossible to keep them from crumbling or washing away again in damp locations. This is where the really impressive engineering comes in. A typical Hopewell site is a complex, mathematically aligned mix of circles, squares and octagons – all made of mud. One of the largest, the Newark Earthworks in Licking County, Ohio, which apparently functioned as a lunar observatory, extends over two square miles and contains embankments more than sixteen feet tall. The only way to create stable structures of this sort – so stable that they still exist today – was by the use of ingenious building techniques, alternating layers of earth with carefully selected gravels and sand.[15] To anyone seeing them for the first time, rising above the swamps, the effect would be similar to witnessing an ice cube that refused ever to melt in the midday sun; a kind of cosmogonic miracle.

We've already mentioned how researchers calculating the maths were startled to discover that, from the Archaic phase onwards, geometric earthworks across large parts of the Americas appear to have been using the same system of measurement: one apparently based on the arrangement of cords into equilateral triangles. So the fact that people and materials were converging from far and wide upon the Hopewell mound complexes is not in itself extraordinary. Yet as archaeologists have also observed, the geometric systems characteristic of the 'Woodland peoples' who created Hopewell also mark something of a break with past custom: the introduction of a different metrical system, and a new geometry of forms.[16]

Central Ohio was just the epicentre. Sites with earthworks based on this new, Hopewellian geometrical system can be found dotted along the upper and lower reaches of the Mississippi valley. Some are the size of small towns. They might, and often did, contain meeting houses, craft workshops and charnel houses for the processing of human remains, along with crypts for the dead. A few might have had resident caretakers, though this isn't entirely clear. What is clear is that for most of the year these sites remained largely or completely empty. Only on specific ritual occasions did they come to life as theatres for elaborate ceremonies, densely populated for a week or two at a time, with people drawn from across the region and occasional visitors from very far away.

This is another of the puzzles of Hopewell. It had all the elements required to create a classic 'grain state' (as Scott would define it). The Scioto-Paint Creek bottomlands, where the largest centres were built, are so fertile they later came to be nicknamed 'Egypt' by European settlers; and at least some of the inhabitants will have been familiar with maize cultivation. But in the same way that they appear to have largely avoided this crop – except perhaps for limited, ritual purposes – they also largely avoided the valley bottoms, preferring to live in isolated homesteads scattered across the landscape and mostly on higher ground. Such homesteads often consisted of a single family; or, at most, three or four. Sometimes these tiny groups moved back and forth between summer and winter houses, pursuing a combination of hunting, fishing, foraging and cultivating local weedy crops in small garden plots; sunflowers, sumpweed, goosefoot, knotweed and maygrass, along with a smattering of vegetables.[17] Presumably people were in regular contact with their neighbours. They seem to have got on with them well enough, since there is little evidence for warfare or organized violence of any sort.[18] But they never came together to create any sort of ongoing village or town life.[19]

Monumental architecture on the scale of the Hopewell earthworks is generally assumed to imply a significant agricultural surplus, governed by chiefs or a stratum of religious leaders. Yet this isn't what was going on. Rather we find just the sort of 'play farming' familiar from our discussions in Chapter Six, as well as shamans and engineers who spent the overwhelming majority of their time with the same five or six companions, but who periodically walked out on to the stage of an extended society that encompassed much of the North American continent. It is all so strikingly different from anything we know of later Woodlands societies that it's difficult to reconstruct exactly what these settlement patterns meant in practice. If nothing else, however, this overall situation illustrates the profound irrelevance of a conventional evolutionist terminology, based on a progression from 'bands' to 'tribes' and 'chiefdoms'.

So what kind of societies were these?

One thing we can definitely say is that they were artistically brilliant. For all their modest living arrangements, Hopewellians produced one of the most sophisticated repertories of imagery in the pre-Columbian

Americas: everything from effigy pipes topped by exquisite animal carvings (used to smoke a variety of tobacco strong enough to induce trance-like states, along with other herbal concoctions); to fired earthen jars covered in elaborate designs; and small copper sheets, worn as breastplates, cut into intricate geometrical designs. Much of the imagery is evocative of shamanic ritual, vision quests and soul journeys (as we noted, there is a particular emphasis on mirrors), but also periodic festivals of the dead.

Like Chavín de Huántar in the Andes, or indeed Poverty Point, social influence derived from control over esoteric forms of knowledge. The main difference is that the Hopewell Interaction Sphere has no discernible centre, no single capital, and unlike Chavín it offers little evidence for the existence of permanent elites, priestly or otherwise. Analysis of burials reveals at least a dozen different sets of insignia, ranging perhaps from funerary priests to clan chief or diviner. Remarkably, it also appears to reveal the existence of a developed clan system, since the ancient inhabitants of central Ohio developed the historically unusual – but from an archaeologist's point of view extraordinarily convenient – habit of including bits of their totem animal – jaws, teeth, claws or talons, often fashioned into pendants or jewellery – in their tombs. All the clans most familiar from later North America – Deer, Wolf, Elk, Hawk, Snake and so on – were already represented.[20] The really striking thing is that, despite the existence of a system of offices and clans, there appears to be virtually no relation between the two. It is possible that clans sometimes 'owned' certain offices, but there is little evidence for the existence of a hereditary, ranked elite.[21]

Some suggest that much of Hopewell ritual consisted of heroic-style feasts and contests: races, games and gambling, which – if at all like later Feasts of the Dead in the American Northeast – often ended by covering great treasures beneath carefully layered strata of soil and gravel, so that nobody (except, perhaps, gods or spirits) would ever see them again.[22] Both the games and burials would, obviously, tend to militate against the accumulation of wealth – or, better put, would ensure that social differences remained largely theatrical. Indeed, even those systemic differences that can be detected seem to be an effect of the ritual system, for the Hopewell heartland appears to break down into a Tripartite Alliance, three great clusters of sites.

In the northernmost, centred on Hopewell itself, funerary assemblages focus on shamanic ritual, heroic male figures travelling between cosmic domains. In the southern, best exemplified by the Turner Site in southwest Ohio, the emphasis is on an imagery of impersonal masked figures, hilltop earth shrines and chthonic monsters. Still more remarkably, in the northern cluster all those buried with badges of office are men; in the southern, those buried with the same badges of office are just as exclusively women. (The central cluster of sites is mixed, in both respects.)[23] What's more, there was clearly some kind of systemic co-ordination between the clusters, with causeways joining them.[24]

It's informative, at this point, to compare and contrast the Hopewell Interaction Sphere with a phenomenon we discussed in the previous chapter: the 'Ubaid village societies of Mesopotamia in the fifth millennium BC. The comparison might seem a stretch, but both can be conceived as culture areas on the grandest possible scale, the first in their respective hemispheres to encompass the entire span of a great river system – the Mississippi and the Euphrates respectively – from headwaters to delta, including all the surrounding plains and coastlands.[25] The establishment of regular cultural interaction on such a scale, across sharply contrasting landscapes and environmental niches, often marks an important turning point in history. In the case of the 'Ubaid it created a certain self-conscious form of standardization, a social egalitarianism, that laid foundations for the world's first cities.[26] What happened in the case of Hopewell seems rather different.

In fact, in many ways Hopewell and 'Ubaid are polar cultural opposites. The unity of the 'Ubaid interaction sphere lay in the suppression of individual differences between people and households; in contrast, the unity of Hopewell lay in the celebration of difference. To take one example: while later North American societies would distinguish entire clans and nations by characteristic hairstyles (so it was a simple matter to distinguish a Seneca, Onondaga or Mohawk warrior at a distance), it is difficult to find two figures in Hopewell art – and there are quite a few of them – that have the same hair. Everybody appears to have been free to make a spectacle of themselves, or to obtain some dramatic role in the theatre of society, and this individual expressiveness was reflected in miniature depictions of people sporting what seem to be an endless variety of playful, idiosyncratic styles of haircut, clothing and ornamentation.[27]

Yet all this was intricately co-ordinated over large areas. Even locally, each earthwork was one element in a continuous ritual landscape. The earthworks' alignments often reference particular segments of the Hopewell calendar (such as the solstices, phases of the moon and so on), with people presumably having to move back and forth regularly between the monuments to complete a full ceremonial cycle. This is complex: one can only imagine the kind of detailed knowledge of stars, rivers and seasons that would have been required to co-ordinate people from hundreds of miles away, such that they might congregate on time for rituals in centres that lasted only for periods of five or six days at a time, over the course of a year. Let alone what it would take to actually transform such a system across the length and breadth of a continent.

In later times, Feasts of the Dead were also occasions for the 'resurrection' of names, as the titles of those who were now gone passed to the living. It may have been through some such mechanisms that Hopewell disseminated the basic structure of its clan system across North America. It's even possible that when the spectacular burials in Hopewell came to an end around AD 400, it was largely because Hopewell's work was done. The idiosyncratic nature of its ritual art, for instance, gave way to standardized versions disseminated across the continent; while great treks to fantastic, temporary capitals that rose miraculously from the mud were no longer required to establish ties between groups, who now had a shared idiom for personal diplomacy, a common set of rules for interacting with strangers.[28]

IN WHICH WE TELL THE STORY OF CAHOKIA, WHICH LOOKS LIKE IT OUGHT TO BE THE FIRST 'STATE' IN AMERICA

One of the many puzzles of Hopewell is how its social arrangements seem to anticipate much later institutions. There was a division between 'white' and 'red' clans: the first identified with summer, circular houses and peacemaking; the second with winter, square houses and warfare.[29] Most later indigenous societies had a separation between peace chiefs and war chiefs: an entirely different administration came into force in times of military conflict, then melted away as soon as matters were resolved. Some of this symbolism appears to originate in Hopewell. Archaeologists even identify certain figures as war chiefs; and yet, despite all this, there is an almost total lack of evidence for actual warfare. One possibility is that conflict took a different, more theatrical form – as in later times, when rival nations or 'enemy' moieties would often play out their hostilities through aggressive games of lacrosse.[30]

In the centuries following the decline of the Hopewell centres, roughly from AD 400 to 800, we start to see a series of familiar developments. First, some groups begin adopting maize as a staple crop and growing it in river valleys along the Mississippi floodplain. Second, actual armed conflict becomes more frequent. In at least some places, this led to populations living for longer periods around their local earthworks. Especially in the Mississippi valley and on adjacent bluffs, a pattern emerged of small towns centred on earthen pyramids and plazas, some fortified, often surrounded by extensive stretches of no-man's-land. A few came to resemble tiny kingdoms. Eventually this situation led to a veritable urban explosion with its epicentre at the site of Cahokia, which was soon to become the greatest city in the Americas north of Mexico.

Cahokia lies in an extensive floodplain along the Mississippi known as the American Bottom. It was a bounteous and fertile environment, ideal for growing maize, but still a challenging place to build a city since much of it was swampland, foggy and full of shallow pools. Charles Dickens, who once visited this place, described it as 'an unbroken slough of black mud and water'. In Mississippian cosmology, watery places like this were connected to the chaotic underworld – seen as the diametrical opposite of a precise, predictable celestial order – and it's no doubt significant that some of the first large-scale construction at Cahokia centred on a processional walkway known as the Rattlesnake Causeway, designed to rise from the surrounding waters and leading towards the surrounding ridge-top tombs (a Path of Souls, or Way of the Dead). To begin with, then, Cahokia was likely a place of pilgrimage, much like some of the Hopewell sites.[31]

Its inhabitants also shared with Hopewell the same love of games. Around AD 600, someone living at Cahokia or close by seems to have come up with the idea for chunkey, later to become one of the most popular sports in North America. Chunkey was a complex and highly co-ordinated affair in which running players tried to throw poles as close as possible to a rolling wheel or ball without actually touching it.[32] It was played at several earthwork sites that sprang up along the American Bottom: one way of holding together the increasingly disparate groups of people who came to settle there. In social terms, it had certain things in common with Mesoamerican ball games, though the rules were entirely different. It could be either a substitute for, or continuation of, war; it was tied into legend (in this case, the story of Red Horn the Morning Star who, much like the Maya hero-twins, confronted gods of the underworld); and it could become the focus of frenetic gambling, when some would even raise themselves or their families as stakes.[33]

In Cahokia and its hinterland we can chart the rise of social hierarchies through the lens of chunkey, as the game became increasingly monopolized by an exclusive elite. One sign of this is how stone chunkey discs disappear from ordinary burials, just as beautifully crafted versions of them start to appear in the richest graves. Chunkey was becoming a spectator sport, and Cahokia the sponsor of a new regional, Mississippian elite. We are not sure exactly how it happened – as an act of religious revelation, perhaps – but around AD 1050 Cahokia exploded in size, growing from a fairly modest community to a city of over six square miles, including more than 100 earthen mounds built around spacious plazas. Its original population of a few thousand was augmented by perhaps 10,000 more, coming from outside to settle in Cahokia and its satellite towns, totalling something in the order of 40,000 in the American Bottom as a whole.[34]

The main part of the city was designed and built according to a master plan in a single burst of activity. Its focus was a huge packed-earth pyramid known today as Monk's Mound, standing before an enormous plaza. In a smaller plaza to the west stood a 'woodhenge' of cypress posts marking out the sun's annual course. Some of Cahokia's pyramids were topped with palaces or temples; others with charnel houses or sweat lodges. A calculated effort was made to resettle foreign populations – or at least their most important, influential representatives – in newly designed thatch houses, arranged in neighbourhoods around smaller plazas and earthen pyramids; many had their own craft specializations or ethnic identity.[35] From the summit of Monk's Mound the city's ruling elite enjoyed powers of surveillance over these planned residential zones.[36] At the same time, existing villages and hamlets in Cahokia's hinterland were disbanded and the rural population dispersed, scattered in homesteads of one or two families.[37]

What's so striking about this pattern is its suggestion of an almost complete dismantling of any self-governing communities outside the city. For those who fell within its orbit, there was nothing much left between domestic life – lived under constant surveillance from above – and the awesome spectacle of the city itself.[38] That spectacle could be terrifying. Along with games and feasts, in the early decades of Cahokia's expansion there were mass executions and burials, carried out in public. As with fledgling kingdoms in other parts of the world, these large-scale killings were directly associated with the funerary rites of nobility; in this case, a mortuary facility centred on the paired burials of high-status males and females,[39] whose shrouded bodies were placed around a surface built up from some thousands of shell beads. Around them an earthen mound was formed, precisely oriented to an azimuth derived from the southernmost rising point of the moon. Its contents included four mass graves holding the stacked bodies of mainly young women (though one was over fifty), who were killed specifically for the occasion.[40]

Carefully sifting through the ethnographic and historical evidence, scholars have reconstructed the outlines of what Cahokia – and later kingdoms modelled on it – must have looked like. While something endured of the earlier clan organizations, the old moiety system was transformed into an opposition between nobles and commoners. The Mississippians appear to have been matrilineal, which meant that a *mico* (ruler) was not succeeded by his children but by his eldest nephew. Nobles could only marry commoners, and after several generations of such intermarriage the descendants of kings might lose their noble status entirely. So a pool of nobles-turned-commoners always existed from which warriors and administrators could be drawn. Genealogies were carefully preserved, and there was a priesthood devoted to maintaining the temples, which contained images of royal ancestors. Lastly, there was a system of titles for heroic achievement in war, which made it possible for commoners to win their way into the nobility, a status symbolized in bird-man imagery, which also invoked the prestige of competing at chunkey tournaments.[41]

Bird-man symbolism was especially marked in the smaller kingdoms – some fifty in all – that began to appear up and down the Mississippi, of which the largest are at places called Etowah, Moundville and Spiro. The rulers of these towns were often buried with what seem to be precious badges and insignia manufactured at Cahokia. Sacred images in Cahokia itself focused not so much on the hawk and falcon symbolism that appeared everywhere else as – fittingly for an increasingly prominent centre of intensive grain production – on the figure of the Corn Mother, who also appears as the Old Woman, a goddess holding a loom. During the eleventh and twelfth centuries, Mississippian sites with links of various kinds to Cahokia appear everywhere from Virginia to Minnesota, often in aggressive conflict with their neighbours. Trade routes spanning the continent were activated, the materials for new treasures pouring into the American Bottom much as they once had to Hopewell.[42]

Very little of this expansion was directly controlled from the centre. We are unlikely to be talking about an actual empire so much as an intricate ritual alliance, backed up ultimately by force – and things began to grow increasingly violent, fairly fast. Within a century of the initial urban explosion at Cahokia, in about AD 1150, a giant palisaded wall was built, though it only included some parts of the city and not others. This marked the beginning of a long and uneven process of war, destruction and depopulation. At first people seem to have fled the metropolis for the hinterlands, then ultimately abandoned the rural bottomlands entirely.[43] This same process can be observed in many of the smaller Mississippian towns. Most appear to have begun as co-operative enterprises before becoming centralized around the cult of some royal line and receiving patronage from Cahokia. Then, over the course of a century or two, they emptied out (in much the same way as the Natchez Great Village was later to do, and possibly for much the same reasons, as subjects sought freer lives elsewhere) until finally being sacked, burned or simply deserted.

Whatever happened in Cahokia, it appears to have left extremely unpleasant memories. Along with much of its bird-man mythology, the place was erased from any later oral traditions. After AD 1400 the entire fertile expanse of the American Bottom (which at the city's height had contained perhaps as many as 40,000 people), along with the territory from Cahokia up to the Ohio River, became what's referred to in the literature as the Vacant or Empty Quarter: a haunted wilderness of overgrown pyramids and housing blocks crumbling back into swamp, occasionally traversed by hunters but devoid of permanent human settlement.[44]

Scholars continue to debate the relative importance of ecological and social factors in Cahokia's collapse, just as they argue about whether or not it should be considered a 'complex chiefdom' or a 'state'.[45] In our own terms (as set out in the last chapter), what we appear to have in Cahokia is a second-order regime in which two of our three elementary forms of domination – in this case, control over violence and charismatic politics – came together in a powerful, even explosive cocktail. This is the same combination we found in the Classic Maya elite, for whom competitive sports and warfare were similarly fused; and who extended their sovereignty by bringing large populations into their orbit through organized spectacle, or by capture, or other forms of compulsion we can only guess at.

Both in Cahokia and the Classic Maya, managerial activities seem to have focused on the administration of otherworldly matters, notably in the sophistication of their ritual calendars and precise orchestration of sacred space. These, however, had real-world effects, especially in the areas of city-planning, labour mobilization, public surveillance and careful monitoring of the maize cycle.[46] Perhaps we are dealing here with attempts to create 'third-order' regimes of domination, albeit of a very different kind to modern nation states, in which control over violence and esoteric knowledge became caught up in the spiralling political competition of rival elites. This may also explain why, in both cases – Cahokia and the Maya – the collapse of such totalizing (totalitarian, even) projects, when it happened, was itself sudden, comprehensive and total.

Whatever the precise combination of factors at play, by about AD

1350 or 1400 the result was mass defection. Just as the metropolis of Cahokia was founded through its rulers' ability to bring diverse populations together, often from across long distances, in the end the descendants of those people simply walked away. The Vacant Quarter implies a self-conscious rejection of everything the city of Cahokia stood for.[47] How did it happen?

Among descendants of Cahokian subjects, migration is often framed as implying the restructuring of an entire social order, merging our three elementary freedoms into a single project of emancipation: to move away, to disobey and to build new social worlds. As we'll see, the Osage – a Siouan people who appear originally to have inhabited the region of Fort Ancient in the Middle Ohio River valley before abandoning it for the Great Plains – used the expression 'moving to a new country' as a synonym for constitutional change.[48] It is important to bear in mind that in this part of North America, populations were relatively sparse. There were extensive stretches of uninhabited territory (often marked by ruins and effigies, their builders long since forgotten), so it was not difficult for groups simply to relocate. What we would now call social movements often took the form of quite literal physical movements.

To get a sense of the kind of ideological conflicts that must have been going on, let's consider the history of the Etowah river valley, part of a region then inhabited by ancestors of the Choctaw, in Georgia and Tennessee. Around the time of Cahokia's initial take-off between AD 1000 and 1200, this area was emerging from a period of generalized warfare. Post-conflict settlement involved the creation of small towns, each with its temple-pyramids and plaza, and in every case centred on a large council house, designed as a meeting place for the entire adult community. Grave goods of the time show no indications of rank. Around 1200 the Etowah valley was for some reason abandoned; then, around half a century later, people returned to it. A burst of construction ensued, including a palace and charnel house on top of giant mounds – walled off from commoner eyes – and a royal tomb, placed directly atop the ruins of the communal council house. Burials there were accompanied by magnificent bird-man costumes and regalia apparently sent from the workshops of Cahokia itself. Smaller villages were broken up, some of their old residents moved into Etowah, and in the countryside they were replaced by the familiar pattern of scattered homesteads.[49]

Enclosed by a perimeter ditch and substantial palisade wall, the town of Etowah was at this point clearly the capital of some sort of kingdom. In 1375 someone – whether external enemies or internal rebels, we do not know – sacked Etowah and desecrated its holy places; then, after a brief and abortive attempt at reoccupation, Etowah was again entirely abandoned, as were all the towns across the region. During this period the priestly orders seem largely to vanish across much of the Southeast, to be replaced by warrior *micos*. Occasionally, these petty rulers would become paramount in a given region, but they lacked either the ritual authority or economic resources to create the kind of urban life that existed before. In about 1500 the Etowah valley fell under the sway of the kingdom of Coosa, by which time most of the original population appears to have left and moved on, leaving behind little more than a museum of earthworks for the Coosa to lord it over.[50]

Some of those who walked away concentrated around the new capitals. In 1540, a member of Hernando De Soto's expedition described the *mico* of Coosa and his core territory (a place now known, oddly enough, as Little Egypt) in the following terms:

> The cacique came out to welcome him in a carrying chair borne on the shoulders of his principal men, seated on a cushion, and covered with a robe of marten skins of the form and size of a woman's shawl. He wore a crown of feathers on his head; and around him were many Indians playing and singing. The land was very populous and had many large towns and planted fields which reached from one town to the other. It was a charming and fertile land, with good cultivated fields stretching along the rivers.[51]

In the sixteenth and seventeenth centuries, petty kingdoms of this sort appear to have been the dominant political form in much of the Southeast. Their rulers were treated with reverence and received tribute, but their rule was brittle and unstable. The Coosa *mico*'s litter, like that of his chief rival, the Lady of Cofitachequi, was carried by subordinate lords, largely because the latter couldn't be trusted not to rise up unless kept under constant surveillance. Shortly after de Soto's departure several of them did just that, causing the kingdom of Coosa to collapse. Meanwhile, outside the central towns much more egalitarian forms of communal life were taking form.

ON HOW THE COLLAPSE OF THE MISSISSIPPIAN WORLD AND REJECTION OF ITS LEGACY OPENED THE WAY TO NEW FORMS OF INDIGENOUS POLITICS AROUND THE TIME OF THE EUROPEAN INVASION

By the early eighteenth century these petty kingdoms, and the very practice of building mounds and pyramids, had almost entirely vanished from the American South and Midwest. At the edge of the prairies, for example, people living in scattered homesteads began migrating seasonally, leaving the very young and old behind in the earthwork towns and taking to extended hunting and fishing in the surrounding uplands, before finally relocating entirely. In other areas, the towns would be reduced to ceremonial centres or Natchez-style hollow courts, where the *mico* continued to be paid magnificent tokens of respect but held almost no actual power. Then finally, when those rulers were definitively gone, people would begin descending back into the valleys, but this time in communities organized on very different principles: small towns of a few hundred people, or at most 1,000 or 2,000, with egalitarian clan structures and communal council houses.

Today historians seem inclined to see these developments as in large part a reaction to the shock of war, slavery, conquest and disease introduced by European settlers. However, they appear to have been the logical culmination of processes that had been going on for centuries before that.[52]

By 1715, the year of the Yamasee War, the dismantling of petty kingdoms was complete across the entire region of former Mississippian influence, except for isolated hold-overs like the Natchez. Earthworks and homesteads were both things of the past, and the Southeast came to be divided among tribal republics, of the sort familiar from early ethnography.[53] A number of factors made this possible. The first was demographic. As we've noted, North American societies were, with few exceptions, marked by low birth rates and low population densities, which in turn facilitated mobility and made it easier for agriculturalists to shift back to a mode of subsistence more oriented to hunting, fishing and foraging; or simply to relocate entirely. Meanwhile women – who in one of Scott's 'grain states' would typically be viewed by the (male) authorities as little more than baby-making machines, and when not pregnant or nursing to be engaged in industrial tasks like spinning and weaving – took on stronger political roles.

Such details form part of the cultural background to a political struggle over the role of hereditary leadership and privileged esoteric knowledge. These battles were still being fought into relatively recent times. Consider the Nations known in the colonial period as the 'five civilized tribes' of the American Southeast: Cherokee, Chickasaw, Choctaw, Creek and Seminole. All of them exemplify this pattern, being governed by communal councils in which all had equal say and operating by a process of consensus-finding. Yet at the same time, all shared traces of the older priests, castes and princes. In some cases hereditary leadership may have persisted into the nineteenth century, straining against the wider preference for more democratic forms of government.[54]

Some see the egalitarian institutions themselves as an outcome of self-conscious social movements, centred on the summer Green Corn ceremonies.[55] In art their symbol was the looped square; architecturally this symbolic template was realized in the creation not just of council or town houses, but also square grounds for public meetings, a feature with no precedent in the old Mississippian towns and cities. Among the Cherokee we find evidence of priests claiming to be sent from the heavens with special knowledge to impart. Yet we also find stories, such as that of the Aní-Kutánî, about the existence long ago of a theocratic society governed by a hereditary caste of male priests and how they so systematically abused their power, particularly in their abuse of women, that the people rose up and massacred the lot of them.[56]

Much like the arguments Iroquoian speakers presented to Jesuit missionaries, or for that matter their theories about dreams, descriptions of daily life in these post-Mississippian townships often feel strikingly familiar – perhaps disturbingly so for anyone committed to the idea that the Age of Enlightenment was the result of a 'civilizing process' originating exclusively in Europe. Among the Creek, for instance, the post of *mico* was reduced to a facilitator of the assembly and supervisor of collective granaries. Each day the adult men of a town would gather to spend much of the day arguing about politics, in a spirit of rational debate, in conversations punctuated by the smoking of tobacco and drinking of caffeinated beverages.[57] Both tobacco and the 'black drink' had originally been drugs ingested by shamans or other spiritual virtuosos in intense and highly concentrated doses so as to produce altered states of consciousness; now, instead, they were doled out in carefully measured portions to everyone assembled. What Jesuits reported in the Northeast seems to apply here too: 'They believe that there is nothing so suitable as Tobacco to appease the passions; that is why they never attend a council without a pipe or calumet in their mouths. The smoke, they say, gives them intelligence, and enables them to see clearly through the most intricate matters.'[58]

Now, if all this sounds suspiciously reminiscent of an Enlightenment coffee-house it isn't a total coincidence. Tobacco, for example, was adopted around this period by settlers then taken back and popularized in Europe itself, and it was indeed promoted in Europe as a drug to be taken in small doses to focus the mind. Obviously, there is no direct cultural translation here. There never is. But as we have seen, indigenous North American ideas – from the advocacy of individual liberties to scepticism of revealed religion – certainly had an impact on the European Enlightenment, even though, like pipe-smoking, such ideas underwent many transformations in the process.[59] No doubt it would be too much to suggest that the Enlightenment itself had its first stirrings in seventeenth-century North America. But it's possible, perhaps, to imagine some future non-Eurocentric history where such a suggestion would not be treated as almost by definition outrageous and absurd.

HOW THE OSAGE CAME TO EMBODY THE PRINCIPLE OF SELF-CONSTITUTION, LATER TO BE CELEBRATED IN MONTESQUIEU'S *THE SPIRIT OF THE LAWS*

Clearly, evolutionist categories only confuse the issue here. Arguing about whether Hopewellians were 'bands', 'tribes' or 'chiefdoms', or indeed whether Cahokia was a 'complex chiefdom' or a 'state', tells us virtually nothing. Insofar as we can speak of 'states' and 'chiefdoms' at all, in the case of Native North America the state-making project seems to come first, virtually out of nowhere, and the chiefdoms observed by de Soto and his successors appear to be little more than the rubble left behind by its downfall.

There must be more interesting and useful questions to ask of the past, and the categories we've been developing in this book suggest what some of these might be. As we've seen, an important feature in much of the Americas is the relationship between esoteric and bureaucratic knowledge. On the surface, the two might not have much to do with one another. It is easy enough to see how brute force can take institutional form in sovereignty, or as the assertion of charisma in a competitive political field. The path from knowledge, as a general form of domination, to administrative power might seem more circuitous. Does the kind of esoteric knowledge we encounter at Chavín, often founded in hallucinogenic experience, really have anything in common with the accounting methods of the later Inca? It seems highly unlikely – until, that is, we recall that even in much more recent times, qualifications to enter bureaucracies are typically based on some form of knowledge that has virtually nothing to do with actual administration. It's only important because it's obscure. Hence in tenth-century China or eighteenth-century Germany, aspiring civil servants had to pass exams on proficiency in literary classics, written in archaic or even dead languages, just as today they will have had to pass exams on rational choice theory or the philosophy of Jacques Derrida. The arts of administration are really only learned later on and through more traditional means: by practice, apprenticeship or informal mentoring.

Similarly, those who designed the great construction projects of Poverty Point or Hopewell were clearly drawing on esoteric knowledge of some sort – astronomical, mythic, numerological – which was contiguous with the practical knowledge of maths, engineering and construction, not to mention techniques of organizing and monitoring human labour (even voluntary labour) which were required to realize those designs. Over the long term of pre-Columbian history, this particular sort of knowledge always seems to lie at the core of systems of domination that periodically emerged. Hopewell is a perfect example, since the heroic games that accompanied ceremonial projects were not really the basis for systematic domination at all.[60] Cahokia, on the other hand, appears to represent a self-conscious effort to turn that style of administrative esoterica into a basis for sovereignty; the gradual transformation of geometric earthworks, designed on cosmic principles, into actual fortifications being only the most obvious indication. In the end it didn't work. Political power retreated back into heroic theatre, if in a decidedly more violent form.

Even more strikingly, however, the very principle of esoteric knowledge came increasingly to be challenged.

What we saw in Hopewell was a kind of 'reformation', in the same sense that the European Reformation of the sixteenth century involved a fundamental reorientation of access to the sacred – albeit one which had knock-on effects in just about every other aspect of social life, from the organization of work to the nature of politics. In Europe these battles played out over the medium of scripture: the translation of the Bible from obscure ancient languages into regional vernaculars, and its release from the closed sanctuary of the High Faith into mass dissemination via the printing press. In the pre-Columbian Americas, the equivalent media revolution focused instead on the (quite literal) reformation of mathematical principles underlying the creation of complex geometrical earthworks which captured the sacred in spatial form.

In both cases, such reformations determined who could and could not partake of a sacred power encapsulated in stories and myths, encoded on the one hand as complex layers of scripture (the Old and New Testaments and other holy books), and on the other as a network of landscape monuments, just as complex in their own way. Indeed, there is every reason to think that the images of chthonic and other beings frozen in ancient earthworks were testaments of a sort. They were mnemonic schemes that prompted the recollection and re-enactment of exploits carried out by founding ancestors at the beginning of days and magnified in monumental form, to be witnessed by the powers dwelling 'on high'. While European clergy burned incense to form a sentient bond with the invisible (a distant echo of biblical animal sacrifice), Hopewell peoples lit tobacco in their effigy pipes, sending smoke up towards the heavens.

Here we begin to comprehend what it might have meant actually to stop creating such monuments entirely, or to repurpose drugs like tobacco towards collective, rational debate. Of course, this does not necessarily imply a systematic, Enlightenment-style rejection of esoteric knowledge. It could also mean the democratization of such knowledge – or at least the transformation of what had once been a theocratic elite into a kind of oligarchy. We find an excellent example of this in the history of the Osage.

A nation of the Great Plains, the Osage are directly descended from Mississippianized Fort Ancient people, and much of their ritual and mythology can be traced back directly to their Midwestern origins.[61] The Osage were doubly fortunate. First, because they succeeded in taking advantage of a strategic position on the Missouri River to ally with the French government and thereby maintain their independence, even creating something of a trading empire from 1678 to 1803. Second, because the ethnographer who documented their ancient traditions in the first decades of the twentieth century, Francis La Flesche, was himself a native speaker of Omaha (a closely related language), and therefore appears to have been unusually capable and receptive. As a result we have a much better sense of how Osage elders thought about their own traditions than is the case for most other Plains societies.

Let us begin with a map of a typical Osage summer village. Osage communities typically moved between three seasonal locations: permanent villages of multi-family lodge houses made up of perhaps 2,000 people; summer camps; and camps for the annual midwinter bison hunts. The basic village pattern was a circle divided into two exogamous moieties, sky and earth, with twenty-four clans in all, each of which had to be represented in any settlement or camp, just as at least one representative of each had to be present for any major ritual. The system was initially based on a tripartite division: seven clans each designated Sky People, Earth People and Water People, with the last two grouped together as the earth moiety in relation to sky, making twenty-one; then over time this was expanded when clans were added to become 7+2 (sky, *Tsizhu*) against 7+7+1 (earth, *Honga*), giving twenty-four in total.

At this point you may well be wondering how, precisely, did it ever come about that people arranged themselves in such intricate patterns? Who exactly decided that each of the twenty-four clans would

Arrangement of different clans (1-5)
in an Osage village

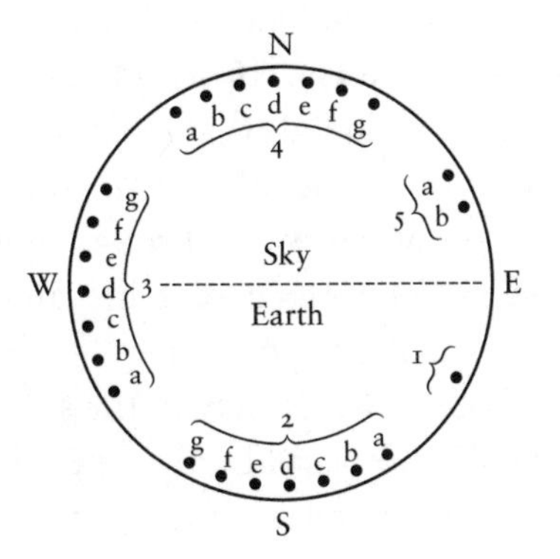

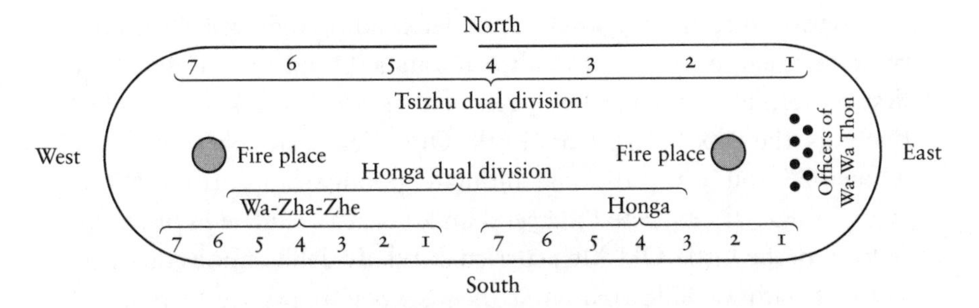

How representatives of the same clans arranged
themselves inside a lodge for a major ritual

be represented in every village, and how did they orchestrate things so it would happen? In the case of the Osage we actually have something of an answer, since Osage history was remembered essentially as a series of constitutional crises in which the elders of the community gradually worked out exactly this arrangement.

The history, according to La Flesche, is difficult to piece together because it is distributed among the clans. Or, to be more exact, a bare-bones version of the story, full of cryptic allusions, is known to everyone; but each clan also has its own history and stock of secret knowledge, whereby the true meaning of certain aspects of the story is revealed over the course of seven levels of initiation. The real story then can be said to be broken into 168 pieces – arguably 336, since each revelation contained two parts: a political history and an accompanying philosophical reflection on what that history reveals about the forces responsible for dynamic aspects of the visible world that caused the stars to move, plants to grow, and so forth.

Records, La Flesche observed, had also been kept of particular discussions in which various results of this study of nature were debated and discussed. Osage concluded that this force was ultimately unknowable and gave it the name *Wakonda*, which could alternately be translated as 'God' or 'Mystery'.[62] Through lengthy investigation, La Flesche notes, elders determined that life and motion was produced by the interaction of two principles – sky and earth – and therefore they divided their own society in the same way, arranging it so that men from one division could only take wives from the other. A village was a model of the universe, and as such a form of 'supplication' to its animating power.[63]

Initiation through the levels of understanding required a substantial investment of time and wealth, and most Osage only attained the first or second tier. Those who reached the top were known collectively as the *Nohozhinga* or 'Little Old Men' (though some were women),[64] and were also the ultimate political authorities. While every Osage was expected to spend an hour after sunrise in prayerful reflection, the Little-Old-Men carried out daily deliberations on questions of natural philosophy and their specific relevance to political issues of the day. They also kept a history of the most important discussions.[65] La Flesche explains that, periodically, particularly perplexing questions would come up: either about the nature of the visible universe, or about the application of these understandings to human affairs. At this point it was customary for two elders to retreat to a secluded spot in the wilderness and carry out a vigil for four to seven days, to 'search their minds', before returning with a report on their conclusions.

The *Nohozhinga* were the body that met daily to discuss affairs of state.[66] While larger assemblies could be called to ratify decisions, they were the effective government. In this sense one could say that the Osage were a theocracy, though it would be more accurate, perhaps, to say there was no difference between officials, priests and philosophers. All were title-bearing officials, including the 'soldiers' assigned to help chiefs enforce their decisions, while 'Protectors of the Land' assigned to hunt down and kill outsiders who poached game were also religious figures. As for the history: it begins in mythic terms, as an 'allegorical fable', then rapidly turns into a story about institutional reform.

In the beginning, the three main divisions – Sky People, Earth People and Water People – descended into the world and set out in search of its indigenous inhabitants. When they located these inhabitants, they were discovered to be in a repulsive state: living amid filth, bones and carrion, feeding on offal, rotting flesh, even each other. Despite this more-than-Hobbesian situation, the Isolated Earth People (as they came to be known) were also powerful sorcerers, capable of using the four winds to destroy life everywhere. Only the chief of the Water division had the courage to enter their village, negotiate with their leader and convince his people to abandon their murderous and unsanitary ways. In the end, he persuaded the Isolated Earth People to join them in a federation – to 'move to a new country', free from the pollution of decaying corpses. This is how the circular village plan was first conceived, with the one-time wizards placed opposite Water, at the eastern door, where they were in charge of the House of Mystery, used for all peaceful rituals, and where all children were brought to be named. The Bear clan of the Earth division was put in charge of an opposite House of Mystery, responsible for rituals concerning war. The problem was that the Isolated Earth People, while no longer murderous, did not prove particularly effective allies either. Before long everything had descended into continual strife and feuding, until the Water division demanded another 'move to a new country', which initiated, among other things, an elaborate process of constitutional reform, making declarations of war impossible without the acquiescence of every clan. This too proved problematic over time, since it meant that if an external enemy entered the country, at least a week was required to organize a military response. Eventually it became necessary yet again to 'move to another country', which this time involved the creation of a new, decentralized clan-by-clan system of military authority. This in turn led to a new crisis and round of reforms: in this case, the separation of civil and military affairs with the creation of a hereditary peace chief for each division, their houses placed on the east and west extremes of the village, and various subordinate officials, as well as a parallel structure with responsibility for all five major Osage villages.

We will not linger over the details. But two elements of the story deserve emphasis. The first is that the narrative sets off from the neutralization of arbitrary power: the taming of the Isolated Earth People's leader – the chief sorcerer, who abuses his deadly knowledge – by according him some central position in a new system of alliances. This is a common story among the descendants of groups that had formerly come under the influence of Mississippian civilization. In the process of co-opting their leader, the destructive ritual knowledge once held by the Isolated Earth People was, eventually, distributed to everyone, along with elaborate checks and balances concerning its use. The second is that even the Osage, who ascribed key roles to sacred knowledge in their political affairs, in no sense saw their social structure as something given from on high but rather as a series of legal and intellectual discoveries – even breakthroughs.

This last point is critical, because – as outlined earlier – we are used to imagining that the very notion of a people self-consciously creating their own institutional arrangements is largely a product of the Enlightenment. Obviously, the idea that nations could be effectively created by great lawmakers such as Solon of Athens, Lycurgus in Sparta or Zoroaster in Persia, and that their national character was in some sense a product of that institutional structure, was a familiar one in antiquity. But we are generally taught to think of the French political philosopher Charles-Louis de Secondat, Baron de Montesquieu as the first to build an explicit and systematic body of theory based on the principle of institutional reform with his book *The Spirit of the Laws* (1748). By doing so, it's widely believed, he effectively created modern politics. The Founding Fathers of the United States, all avid readers of Montesquieu, were consciously trying to put his theories into practice when they attempted to create a constitution that would preserve the spirit of individual liberty, and spoke of the results as a 'government of laws and not of men'.

As it turns out, precisely this sort of thinking was commonplace in North America well before European settlers appeared on the scene. It might not be a coincidence, in fact, that in 1725 a French explorer named Bourgmont brought an Osage and Missouria delegation across the Atlantic to Paris, around the time Lahontan's works were at the height of their popularity. It was traditional at the time to organize a series of public events around such 'savage' diplomats and arrange private meetings with prominent European intellectuals. We don't know whom specifically they met with, but Montesquieu was indeed in Paris at the time, and already working on such subjects. As one historian of the Osage notes, it is hard to imagine Montesquieu would *not* have attended. At any rate, the chapters in *The Spirit of the Laws* which speculate on the modes of savage government seem an almost exact reproduction of what Montesquieu would likely have heard from them, albeit framed by an artificial distinction between those who do or don't cultivate the land.[67]

The connections may well run deeper than we think.

IN WHICH WE RETURN TO IROQUOIA, AND CONSIDER THE POLITICAL PHILOSOPHIES LIKELY TO HAVE BEEN FAMILIAR TO KANDIARONK IN HIS YOUTH

We have come full circle. The case of North America not only throws conventional evolutionary schemes into chaos; it also clearly demonstrates that it's simply not true to say that if one falls into the trap of 'state formation' there's no getting out. Whatever happened in

Cahokia, the backlash against it was so severe that it set forth repercussions we are still feeling today.

What we are suggesting is that indigenous doctrines of individual liberty, mutual aid and political equality, which made such an impression on French Enlightenment thinkers, were neither (as many of them supposed) the way all humans can be expected to behave in a State of Nature. Nor were they (as many anthropologists now assume) simply the way the cultural cookie happened to crumble in that particular part of the world. This is not to say there is no truth whatsoever in either of these positions. As we've said before, there are certain freedoms – to move, to disobey, to rearrange social ties – that tend to be taken for granted by anyone who has not been specifically trained into obedience (as anyone reading this book, for instance, is likely to have been). Still, the societies that European settlers encountered, and the ideals expressed by thinkers like Kandiaronk, only really make sense as the product of a specific political history: a history in which questions of hereditary power, revealed religion, personal freedom and the independence of women were still very much matters of self-conscious debate, and in which the overall direction, for the last three centuries at least, had been explicitly anti-authoritarian.

East St Louis is, of course, a long way from Montreal, and no one to our knowledge has ever suggested that Iroquoian-speaking peoples of the Great Lakes region were ever, themselves, directly under Mississippian rule. So it would be going a bit too far to suggest that the views recorded by men like Lahontan were, in any literal sense, the ideology that overthrew Mississippian civilization. Still, a careful review of oral traditions, historical accounts and the ethnographic record shows that those who framed what we call the 'indigenous critique' of European civilization were not only keenly aware of alternative political possibilities, but for the most part saw their own social orders as self-conscious creations, designed as a barrier against all that Cahokia might have represented – or indeed, all those qualities they were later to find so objectionable in the French.

Let us start with the available oral traditions. These are unfortunately somewhat limited. During the late sixteenth and early seventeenth centuries, Iroquoia was divided between a number of shifting political coalitions and confederacies of which the most prominent were the Wendat (Huron), based in what's now Quebec; the Five Nations or Haudenosaunee (often referred to as 'League Iroquois'), distributed across what's now upstate New York; and an Ontario-based confederation that the French referred to as the 'Neutrals'. The Wendat referred to this last as Attiwandaronk – which literally means 'those whose speech is not quite right'. We don't actually know what these Neutrals called themselves (clearly it wasn't that); but according to early accounts, they were by far the most numerous and powerful, at least until their society was devastated by famine and disease in the 1630s and 1640s. Afterwards the survivors were absorbed by the Seneca, given names and thus incorporated in one or other Seneca clan.

A similar fate befell the Wendat Confederation, whose power had been decisively broken in the year Kandiaronk was born, 1649, when they were scattered or absorbed during the notorious 'Beaver Wars'. In Kandiaronk's own lifetime the remaining Wendat were leading a fairly precarious existence: partly driven north towards Quebec; partly under the protection of a French fort in a place called Michilimackinac, near Lake Michigan. Kandiaronk himself spent much of his life trying to put the confederation's pieces back together and, according to oral histories at least, attempting to found a coalition that would unite the warring nations against the invaders. In this he failed. As a result, we don't actually know the stories told by members of any of these other great confederacies about the origins of their political institutions. By the time oral histories began to be written down in the nineteenth century, only the Haudenosaunee remained.

We do, however, have numerous versions of the foundation of the League of Five Nations (the Seneca, Oneida, Onondaga, Cayuga and Mohawk), an epic known as the *Gayanashagowa*. What is most remarkable about this epic, in the present context at least, is the degree to which it represents political institutions as self-conscious human creations. Certainly, the story contains magical elements. In a certain sense, the main characters – Deganawideh the Peacemaker, Jigonsaseh the Mother of Nations and so forth – are reincarnations of characters from the creation myth. But what comes through most strongly in the text is its representation of a social problem with a social solution: a breakdown of relationships in which the country is plunged into chaos and revenge, spiralling to a point where social order has dissolved away and where the powerful have become literal cannibals. Most powerful of all is Adodarhoh (Tadodaho), who is represented as a witch, deformed, monstrous and capable of commanding others to do his bidding.

The narrative centres on a hero, Deganawideh the Peacemaker, who appears from what is later to be the Attiwandaronk (Neutral) territory to the northwest, determined to put an end to this chaotic state of affairs. He wins to his cause first the Jigonsaseh, a woman famous for standing outside all quarrels (he finds her hosting and feeding war parties from all sides of the conflict); and then Hiawatha, one of Adodarhoh's cannibal henchmen. Together they set about winning over the people of each nation to agree on creating a formal structure for heading off disputes and creating peace. Hence the system of titles, nested councils, consensus-finding, condolence rituals and the prominent role of female elders in formulating policy. In the story, the very last to be won over is Adodarhoh himself, who is gradually healed of his deformities and turned into a human being. In the end the laws of the League are 'spoken into' belts of *wampum*, which serves as its constitution; the records are transferred to the keeping of Adodarhoh; and, his work finished, the Peacemaker vanishes from the earth.

Since Haudenosaunee names are passed on like titles, there has continued to be an Adodarhoh, just as there is also still a Jigonsaseh and Hiawatha, to this day. Forty-nine sachems, delegated to convey the decisions of their nation's councils, continue to meet regularly. These meetings always begin with a rite of 'condolence', in which they wipe away the grief and rage caused by the memory of anyone who died in the interim, to clear their minds to go about the business of establishing peace (the fiftieth, the Peacemaker himself, is always represented by an empty place). This federal system was the peak of a complex apparatus of subordinate councils, male and female, all with carefully designated powers – but none with actual powers of compulsion.

In its essence, the story is not so different from the founding of the Osage social order: a terrifying witch is brought back into society, and in the process transformed into a peacemaker. The main difference is that, in this case, Adodarhoh is quite explicitly a ruler, one invested with power of command:

> South of the Onondaga town lived an evil-minded man. His lodge was in a swale and his nest was made of bulrushes. His body was distorted by seven crooks and his long tangled locks were adorned by withering living serpents. Moreover, this monster was a devourer of raw meat, even of human flesh. He was also a master of wizardry and by his magic he destroyed men but he could not be destroyed. Adodarhoh was the name of the evil man.
>
> Notwithstanding the evil character of Adodarhoh the people of Onondaga, the Nation of Many Hills, obeyed his commands and though it cost many lives they satisfied his insane whims, so much did they fear him and his sorcery.[68]

It is an anthropological commonplace that if you want to get a sense of a society's ultimate values it is best to look at what they consider to be the worst sort of behaviour; and that the best way to get a sense of what they consider to be the worst possible behaviour is by examining ideas about witches. For the Haudenosaunee, the giving of orders is represented as being almost as serious an outrage as the eating of human flesh.[69]

Representing Adodarhoh as a king might seem surprising, since there seems no reason to think that, before the arrival of Europeans, either the Five Nations or any of their immediate neighbours had any immediate experience of arbitrary command. This raises precisely the question often directed against arguments[70] that indigenous institutions of chiefship were in fact designed to prevent any danger of states emerging: how could so many societies be organizing their entire political system around heading off something (i.e. 'the state') that they had never experienced? The straightforward response is that most of the narratives were gathered in the nineteenth century, by which time any indigenous American was likely to have had long and bitter experience of the United States government: men in uniforms carrying legal briefs, issuing arbitrary commands and much more besides. So perhaps this element was added to these narratives later?

Anything is possible, of course, but this strikes us as unlikely.[71]

Even in more recent times, the danger of being accused of witchcraft was deployed against office holders to ensure that none could accumulate any appreciable advantage over their fellows – particularly in wealth. Here we have to return to the Iroquoian theory about dreams as repressed desires, mentioned earlier in the chapter. One interesting twist of this theory is that it was considered the responsibility of others to realize a fellow community member's dream: even if one dreamed of appropriating a neighbour's possession, it could only be refused at the risk of endangering their health. To do so was considered beyond awkward; almost socially impossible. Even if one did it would cause outraged gossip, and very possibly bloody revenge: if somebody was thought to have died because somebody else refused to grant a soul wish, his or her relatives might retaliate physically, or by supernatural means.[72]

Any member of an Iroquoian society given an order would have fiercely resisted it as a threat to their personal autonomy – but the one exception to this norm was, precisely, dreams.[73] One Huron-Wendat chief gave away his prized European cat, which he had carried by canoe all the way from Quebec, to a woman who dreamed she could only be cured by owning it (Iroquoians also feared becoming the victims of witchcraft practised consciously or unconsciously by people who envied them). Dreams were treated as if they were commands, delivered either by one's own soul or possibly, in the case of a particularly vivid or portentous dream, by some greater spirit. The spirit might be the Creator or some other spirit, perhaps entirely unknown. Dreamers could become prophets – if only, usually, for a relatively brief period of time.[74] During that time, however, their orders had to be obeyed. (Needless to say, there were few more terrible crimes than to falsify a dream.)

In other words, the image of the witch was at the centre of a complex of ideas that had everything to do with unconscious desire, including the unconscious desire to dominate, and the need both to realize it and to keep it under control.

How did all this come about, historically?

The exact time and circumstance of the League of Five Nations' creation is unclear; dates have been proposed ranging from AD 1142 to sometime around 1650.[75] No doubt the creation of such confederacies was an ongoing process; and surely, like almost all historical epics, the *Gayanashagowa* patches together elements, many historically accurate, others less so, drawn from different periods of time. What we know from the archaeological record is that Iroquoian society as it existed in the seventeenth century began to take form around the same time as the heyday of Cahokia.

By around AD 1100 maize was being cultivated in Ontario, in what later became Attiwandaronk (Neutral) territory. Over the next several centuries, the 'three sisters' (corn, beans and squash) became ever more important in local diets – though Iroquoians were careful to balance the new crops with older traditions of hunting, fishing and foraging. The key period seems to be what's called the Late Owasco phase, from AD 1230 to 1375, when people began to move away from their previous settlements (and from their earlier patterns of seasonal mobility) along waterways, settling in palisaded towns occupied all year around in which longhouses, presumably based in matrilineal clans, became the predominant form of dwelling. Many of these towns were quite substantial, containing as many as 2,000 inhabitants (that is, something approaching a quarter of the population of central Cahokia).[76]

References to cannibalism in the *Gayanashagowa* epic are not pure fantasy: endemic warfare and the torture and ceremonial sacrifice of war prisoners are sporadically documented from AD 1050. Some contemporary Haudenosaunee scholars think the myth refers to an actual conflict between political ideologies within Iroquoian societies at the time; turning especially on the importance of women, and agriculture, against defenders of an older male-dominated order where prestige was entirely based in war and hunting.[77] (If so, it would not look so very different from the kind of ideological divergence we've suggested might have been taking place in the Middle East during the early phases of the Neolithic.)[78] Some kind of compromise between these two positions appears to have been reached around the eleventh century AD, one result of which was a stabilizing of population at a modest level. Population numbers increased fairly quickly for two or three centuries after the widespread adoption of maize, squash and beans, but by the fifteenth century they had levelled off. The Jesuits later reported how Iroquoian women were careful to space their births, setting optimal population to the fish and game capacities of the region, not its potential agricultural productivity. In this way the cultural emphasis on male hunting actually reinforced the power and autonomy of Iroquoian women, who maintained their own councils and officials and whose power in local affairs at least was clearly greater than that of their men.[79]

In the period spanning the twelfth to fourteenth centuries, neither the Wendat Confederacy nor Haudenosaunee show much evidence of having extensive contact or even much trade with the Mississippians, whose main presence in the Northeast was in the Fort Ancient region along the Ohio River and the nearby Monongahela valley. This is not true, however, of the Attiwandaronk. By AD 1300, much of the Ontario area was indeed under Mississippian influence. It is doubtful, but not totally inconceivable, that there were migrations from the Cahokian heartland.[80] Even if there weren't, the Attiwandaronk appear to have been monopolizing trade to the south and through it to the Chesapeake Bay and beyond, leaving the Wendat and Haudenosaunee to form relations with Algonkian peoples to their north and east. The sixteenth century saw a sharp increase in Mississippian influences in Ontario, including various cult objects and ceremonial regalia, and even large numbers of chunkey stones of the same style that also appear at Fort Ancient.

Archaeologists refer to all this as 'Mississippianization', and it is accompanied by strong evidence for a renewed burst of trade at least as far as Delaware culminating in, among other things, the arrival of enormous quantities of shells and shell beads derived from the mid-Atlantic seaboard from around 1610 onwards, to be piled up in Attiwandaronk tombs. By that time, the Attiwandaronk population was several times larger than any of the neighbouring confederations, Wendat, Haudenosaunee, let alone the Erie, Petun, Wenro or other small rivals; and its capital, Ounotisaston, was then among the largest settlements in the Northeast. (Scholars, predictably, argue about whether the Neutrals could thus qualify as a 'simple chiefdom' rather than a mere 'tribe'.)

Certainly, the Jesuits who visited the region before Attiwandaronk society was, effectively, destroyed by plagues and famines were unanimous in insisting that its constitution was fundamentally different from that of its neighbours. We will probably never have the means to reconstruct precisely how. For instance, the French referred to the

Attiwandaronk as 'the Neutral Nation' largely because they took no part in the near-constant conflicts between the various nations making up the Wendat and Haudenosaunee, but instead allowed war parties from both sides free passage through their territories. This echoes the behaviour attributed to the Jigonsaseh, Mother of Nations, the highest-ranking woman official among the later Haudenosaunee, in their national epic, who was indeed said to have been of Attiwandaronk origin. But at the same time the Attiwandaronk were in no sense neutral in their relations with many of their western and southern neighbours.

Indeed, according to the Recollect Father Joseph de la Roche Daillon, in 1627 the Attiwandaronk were dominated by a warlord named Tsouharissen, 'the chief of the greatest credit and authority that has ever been in all these nations, because he is not only chief of his town, but of all those of his nation . . . It is unexampled in the other nations to have a chief so absolute. He acquired this honour and power by his courage, and by having been many times at war against seventeen nations who are their enemies.'[81] When he was away at war, in fact, the federal council (in all other Iroquoian societies the ultimate authority) could make no important decisions. Tsouharissen seems to have been something at least very like a king.

What was the relation of Tsouharissen and the Jigonsaseh, a figure who came to exemplify principles of reconciliation that are in many ways precisely the opposite of kingship and self-aggrandizement? We don't know. The only source we have for details of Tsouharissen's life is very much contested, an oral history purporting to be the testimony of Tsouharissen's third wife, passed down three centuries to the present day.[82] Almost all historians discount it – but that isn't necessarily an absolute disqualification. At any rate, according to the account Tsouharissen was a child prodigy, a brilliant student of esoteric knowledge. The story of his existence reached a certain Cherokee priest, who travelled to become his tutor; he found a great crystal which he said marked him as a reincarnation of the Sun and fought many wars and married four times. But when he decided to hand on his mantle to the daughter of his youngest Tuscarora wife, a similar child prodigy, disaster struck. So infuriated by this plan was his senior (Attiwandaronk) wife, of the highest-ranking Turtle clan, that she ambushed and killed the daughter, whose mother took her own life in despair. Tsouharissen, in a rage, massacred the culprit's entire lineage, including his own heirs, thus effectively destroying any possibility of dynastic succession.

As we say, we have no idea how much credit to give the story: but we do know that its broad outlines reflect realities. Attiwandaronk at that time did indeed have regular connections with nations as far away as the Cherokee; while the problems of how to square esoteric knowledge with democratic institutions, or even the difficulties faced by powerful men trying to establish dynasties when descent was organized according to matrilineal clans with no internal ranking, would have been familiar issues in North America at that time. Tsouharissen definitely existed, and he did apparently try to translate his success as a warrior into centralized power. We also know it ultimately came to nothing. We just don't know if it really came to nothing in this particular way.

By the time Baron de Lahontan was serving with the French army in Canada, and Kandiaronk was holding forth on questions of political theory at his periodic dinners with Governor Frontenac, the Attiwandaronk no longer existed. Still, the events surrounding Tsouharissen's life were likely to have been familiar to Kandiaronk, as they would have been vivid childhood memories for many of the elders known to him in his formative years. The Jigonsaseh, Mother of Nations, for instance, was still very much alive, the last Attiwandaronk holder of the title having been incorporated in the Wolf clan of the Seneca in 1650. She remained established in her traditional seat, a fortress called Kienuka overlooking the Niagara gorge.[83] Either that Jigonsaseh – or, more likely, her successor – was still there in 1687, when Louis XIV decided to put an end to the ongoing threat the Five Nations posed to French settlement by sending a seasoned military commander, the Marquis de Denonville, as governor, with orders to use whatever force necessary to drive the Nations from what is now upstate New York.

We have a report on what happened from Lahontan's own memoirs. Feigning interest in a peace settlement, Denonville invited the League council, as a body, to negotiate terms in a place called Fort Frontenac (after the former governor). Some 200 delegates arrived, including all the permanent officers of the confederation and many from the women's councils as well. Summarily arresting them, Denonville shipped them off to France to serve as galley slaves. Then, taking advantage of the resulting confusion, he ordered his men to invade the Five Nations' territory. (Lahontan, who strongly disapproved of the proceedings, got himself into trouble for trying to intervene and stop some underlings from casually torturing the prisoners – he was ordered away, but in the end spared further sanction after protesting that he had been drunk. Some years later, in a different context, an order was put out for his arrest on grounds of insubordination, and he had to flee to Amsterdam.)[84]

The Jigonsaseh, however, had chosen not to attend Denonville's meeting. The arrest of the entire Grand Council left her the highest-ranking League official. Since in such an emergency there was no time to raise new chiefs, she and the remaining clan mothers themselves raised an army. Many of those recruited, it is reported, were themselves Seneca women. As it turned out, the Jigonsaseh was a far superior military tactician to Denonville. After routing the invading French troops near Victor, New York, her forces were at the point of entering Montreal when the French government sued for peace, agreeing to dismantle Fort Niagara and return the surviving galley slaves.[85] When Lahontan later notes that, like Kandiaronk, 'those who had been galley slaves in France' were highly critical of French institutions, he is referring largely to those taken prisoner on this occasion – or more specifically the dozen or so, out of the original 200, who made it back alive.

In such a lethal context, why draw attention to the depredations of a self-appointed lord such as Tsouharissen? What his example demonstrates, we suggest, is that even within indigenous society, the political question was never definitively settled. Certainly, the overall direction, in the wake of Cahokia, was a broad movement away from overlords of any sort and towards constitutional structures carefully worked out to distribute power in such a way that they would never return. But the possibility that they might always lurked in the background. Other paradigms of governance existed, and ambitious men – or women – could, if occasion allowed, appeal to them. After her defeat of Denonville the Jigonsaseh appears to have demobilized her army and returned to the process of selecting new officials to reconstitute the Great Council. If she had chosen to act otherwise, however, precedents were available.

It was precisely this combination of such conflicting ideological possibilities – and, of course, the Iroquoian penchant for prolonged political argument – that lay behind what we have called the indigenous critique of European society. It would be impossible to understand the origins of its particular emphasis on individual liberty, for instance, outside that context. Those ideas about liberty had a profound impact on the world. In other words, not only did indigenous North Americans manage almost entirely to sidestep the evolutionary trap that we assume must always lead, eventually, from agriculture to the rise of some all-powerful state or empire; but in doing so they developed political sensibilities that were ultimately to have a deep influence on Enlightenment thinkers and, through them, are still with us today.

In this sense, at least, the Wendat won the argument. It would be impossible for a European today, or anyone, really – whatever they actually thought – to take a position like that of the seventeenth-century Jesuits and simply declare themselves opposed to the very principle of human freedom.

12
Conclusion
The dawn of everything

This book began with an appeal to ask better questions. We started out by observing that to inquire after the origins of inequality necessarily means creating a myth, a fall from grace, a technological transposition of the first chapters of the Book of Genesis – which, in most contemporary versions, takes the form of a mythical narrative stripped of any prospect of redemption. In these accounts, the best we humans can hope for is some modest tinkering with our inherently squalid condition – and hopefully, dramatic action to prevent any looming, absolute disaster. The only other theory on offer to date has been to assume that there were no origins of inequality, because humans are naturally somewhat thuggish creatures and our beginnings were a miserable, violent affair; in which case 'progress' or 'civilization' – driven forward, largely, by our own selfish and competitive nature – was itself redemptive. This view is extremely popular among billionaires but holds little appeal to anyone else, including scientists, who are keenly aware that it isn't in accord with the facts.

It's hardly surprising, perhaps, that most people feel a spontaneous affinity with the tragic version of the story, and not just because of its scriptural roots. The more rosy, optimistic narrative – whereby the progress of Western civilization inevitably makes everyone happier, wealthier and more secure – has at least one obvious disadvantage. It fails to explain why that civilization did not simply spread of its own accord; that is, why European powers should have been obliged to spend the last 500 or so years aiming guns at people's heads in order to force them to adopt it. (Also, if being in a 'savage' state was so inherently miserable, why so many of those same Westerners, given an informed choice, were so eager to defect to it at the earliest opportunity.) During the nineteenth-century heyday of European imperialism, everyone seemed more keenly aware of this. While we remember that age as one of naive faith in 'the inevitable march of progress', liberal, Turgot-style progress was actually never really the dominant narrative in Victorian social theory, let alone political thought.

In fact, European statesmen and intellectuals of that time were just as likely to be obsessed with the dangers of decadence and disintegration. Many were overt racists who held that most humans are not capable of progress, and therefore looked forward to their physical extermination. Even those who did not share such views tended to feel that Enlightenment schemes for improving the human condition had been catastrophically naive. Social theory, as we know it today, emerged largely from the ranks of such reactionary thinkers, who – looking back over their shoulders at the turbulent consequences of the French Revolution – were less concerned with disasters being visited on peoples overseas than on growing misery and public unrest at home. As a result, the social sciences were conceived and organized around two core questions: (1) what had gone wrong with the project of Enlightenment, with the unity of scientific and moral progress, and with schemes for the improvement of human society? And: (2) why is it that well-meaning attempts to fix society's problems so often end up making things even worse?

Why, these conservative thinkers asked, did it prove so difficult for Enlightenment revolutionaries to put their ideas into practice? Why couldn't we just imagine a more rational social order and then legislate it into existence? Why did the passion for liberty, equality and fraternity end up producing the Terror? There must surely be some underlying reasons.

If nothing else, these preoccupations help to explain the continued relevance of an otherwise not particularly successful eighteenth-century Swiss musician named Jean-Jacques Rousseau. Those primarily concerned with the first question saw him as the first to ask it in a quintessentially modern way. Those mainly concerned with the second were able to represent him as the ultimate clueless villain, a simple-minded revolutionary who felt that the established order, being irrational, could simply be brushed aside. Many held Rousseau personally responsible for the guillotine. By contrast, few nowadays read the 'traditionalists' of the nineteenth century, but they're actually important since it is they, not the Enlightenment *philosophes*, who are really responsible for modern social theory. It's long been recognized that almost all the great issues of modern social science – tradition, solidarity, authority, status, alienation, the sacred – were first raised in the works of men like the theocratic Vicomte de Bonald, the monarchist Comte de Maistre, or the Whig politician and philosopher Edmund Burke as examples of the kind of stubborn social realities which they felt that Enlightenment thinkers, and Rousseau in particular, had refused to take seriously, with (they insisted) disastrous results.

These nineteenth-century debates between radicals and reactionaries never really ended; they keep resurfacing in different forms. Nowadays, for instance, those on the right are more likely to see themselves as defenders of Enlightenment values, and those on the left its most ardent critics. But over the course of the argument all parties have come to agree on one key point: that there was indeed something called 'the Enlightenment', that it marked a fundamental break in human history, and that the American and French Revolutions were in some sense the result of this rupture. The Enlightenment is seen as introducing a possibility that had simply not existed before: that of self-conscious projects for reshaping society in accord with some rational ideal. That is, of genuine revolutionary politics. Obviously, insurrections and visionary movements had existed before the eighteenth century. No one could deny that. But such pre-Enlightenment social movements could now largely be dismissed as so many examples of people insisting on a return to certain 'ancient ways' (that they had often just made up), or else claiming to act on a vision from God (or the local equivalent).

Pre-Enlightenment societies, or so this argument goes, were 'traditional' societies, founded on community, status, authority and the sacred. They were societies in which human beings did not ultimately act for themselves, individually or collectively. Rather, they were slaves of custom; or, at best, agents of inexorable social forces which they projected on to the cosmos in the form of gods, ancestors or other supernatural powers. Supposedly, only modern, post-Enlightenment people had the capacity to self-consciously intervene in history and change its course; on this everyone suddenly seemed to agree, no matter how virulently they might disagree about whether it was a good idea to do so.

All this might seem a bit of a caricature, and only a minority of authors were willing to state matters quite so bluntly. Yet most modern thinkers have clearly found it bizarre to attribute self-conscious social projects or historical designs to people of earlier epochs. Generally, such 'non-modern' folk were considered too simple-minded (not having achieved 'social complexity'); or to be living in a kind of mystical dreamworld; or, at best, were thought to be simply adapting themselves to their environment at an appropriate level of technology. Anthropology, it must be confessed, did not play a stellar role here.

For much of the twentieth century, anthropologists tended to describe the societies they studied in ahistorical terms, as living in a kind of eternal present. Some of this was an effect of the colonial situation under which much ethnographic research was carried out. The British Empire, for instance, maintained a system of indirect rule in various parts of Africa, India and the Middle East where local institutions like royal courts, earth shrines, associations of clan elders, men's houses and the like were maintained in place, indeed fixed by legislation. Major political change – forming a political party, say, or leading a prophetic movement – was in turn entirely illegal, and anyone who tried to do such things was likely to be put in prison. This obviously made it easier to describe the people anthropologists studied as having a way of life that was timeless and unchanging.

Since historical events are by definition unpredictable, it seemed more scientific to study those phenomena one could in fact predict: the things that kept happening, over and over, in roughly the same way. In a Senegalese or Burmese village this might mean describing the daily round, seasonal cycles, rites of passage, patterns of dynastic succession, or the growing and splitting of villages, always emphasizing how the same structure ultimately endured. Anthropologists wrote this way because they considered themselves scientists ('structural-functionalists', in the jargon of the day). In doing so they made it much easier for those reading their descriptions to imagine that the people being studied were quite the opposite of scientists: that they were trapped in a mythological universe where nothing changed and very little really happened. When Mircea Eliade, the great Romanian historian of religion, proposed that 'traditional' societies lived in 'cyclical time', innocent of history, he was simply drawing the obvious conclusion. As a matter of fact, he went even further.

In traditional societies, according to Eliade, everything important has already happened. All the great founding gestures go back to mythic times, the *illo tempore*,[1] the dawn of everything, when animals could talk or turn into humans, sky and earth were not yet separated, and it was possible to create genuinely new things (marriage, or cooking, or war). People living in this mental world, he felt, saw their own actions as simply repeating the creative gestures of gods and ancestors in less powerful ways, or as invoking primordial powers through ritual. According to Eliade, historical events thus tended to merge into archetypes. If anyone in what he considered a traditional society does do something remarkable – establishes or destroys a city, creates a unique piece of music – the deed will eventually end up being attributed to some mythic figure anyway. The alternative notion, that history is actually going somewhere (the Last Days, Judgment, Redemption), is what Eliade referred to as 'linear time', in which historical events take on significance in relation to the future, not just the past.

And this 'linear' sense of time, Eliade insisted, was a relatively recent innovation in human thought, one with catastrophic social and psychological consequences. In his view, embracing the notion that events unfold in cumulative sequences, as opposed to recapitulating some deeper pattern, rendered us less able to weather the vicissitudes of war, injustice and misfortune, plunging us instead into an age of unprecedented anxiety and, ultimately, nihilism. The political implications of this position were, to say the least, unsettling. Eliade himself had been close to the fascist Iron Guard in his student days, and his basic argument was that the 'terror of history' (as he sometimes called it) was introduced by Judaism and the Old Testament – which he saw as paving the way for the further disasters of Enlightenment thought. Being Jewish, the authors of the present book don't particularly appreciate the suggestion that we are somehow to blame for everything that went wrong in history. Still, for present purposes, what's startling is that anyone ever took this sort of argument seriously.

Imagine we tried applying Eliade's distinction between 'historical' and 'traditional' societies to the full scope of the human past, on the sort of scale we've been covering in the preceding chapters. Would this not have to mean that most of history's great discoveries – for example the first weaving of fabrics, or the first navigations of the Pacific Ocean, or the invention of metallurgy – were made by people who didn't believe in discovery or in history? This seems unlikely. The only alternative would be to argue that most human societies only became 'traditional' more recently: perhaps they each eventually found a state of equilibrium, settled into it and all came up with a shared ideological framework to justify their newfound condition. Which would mean there actually *was* some kind of previous *illo tempore* or time of creation, when all humans were capable of thinking and acting in the kind of highly creative ways we now consider quintessentially modern; one of their major achievements apparently being to find a way of abolishing most future prospects of innovation.

Both positions are, self-evidently, quite absurd.

Why are we entertaining such ideas? Why does it seem so odd, even counter-intuitive, to imagine people of the remote past as making their own history (even if not under conditions of their own choosing)? Part of the answer no doubt lies in how we have come to define science itself, and social science in particular.

Social science has been largely a study of the ways in which human beings are not free: the way that our actions and understandings might be said to be determined by forces outside our control. Any account which appears to show human beings collectively shaping their own destiny, or even expressing freedom for its own sake, will likely be written off as illusory, awaiting 'real' scientific explanation; or if none is forthcoming (why *do* people dance?), as outside the scope of social theory entirely. This is one reason why most 'big histories' place such a strong focus on technology. Dividing up the human past according to the primary material from which tools and weapons were made (Stone Age, Bronze Age, Iron Age) or else describing it as a series of revolutionary breakthroughs (Agricultural Revolution, Urban Revolution, Industrial Revolution), they then assume that the technologies themselves largely determine the shape that human societies will take for centuries to come – or at least until the next abrupt and unexpected breakthrough comes along to change everything again.

Now, we are hardly about to deny that technologies play an important role in shaping society. Obviously, technologies are important: each new invention opens up social possibilities that had not existed before. At the same time, it's very easy to overstate the importance of new technologies in setting the overall direction of social change. To take an obvious example, the fact that Teotihuacanos or Tlaxcalteca employed stone tools to build and maintain their cities, while the inhabitants of Mohenjo-daro or Knossos used metal, seems to have made surprisingly little difference to those cities' internal organization or even size. Nor does our evidence support the notion that major innovations always occur in sudden, revolutionary bursts, transforming everything in their wake. (This, as you'll recall, was one of the main points to emerge from the two chapters we devoted to the origins of farming.)

Nobody, of course, claims that the beginnings of agriculture were anything quite like, say, the invention of the steam-powered loom or the electric light bulb. We can be fairly certain there was no Neolithic equivalent of Edmund Cartwright or Thomas Edison, who came up with the conceptual breakthrough that set everything in motion. Still, it often seems difficult for contemporary writers to resist the idea that some sort of similarly dramatic break with the past must have occurred. In fact, as we've seen, what actually took place was nothing like that. Instead of some male genius realizing his solitary vision, innovation in Neolithic societies was based on a collective body of knowledge accumulated over centuries, largely by women, in an endless series of apparently humble but in fact enormously significant discoveries. Many of those Neolithic discoveries had the cumulative effect of reshaping everyday life every bit as profoundly as the automatic loom or lightbulb.

Every time we sit down to breakfast, we are likely to be benefiting from a dozen such prehistoric inventions. Who was the first person to figure out that you could make bread rise by the addition of those microorganisms we call yeasts? We have no idea, but we can be almost certain she was a woman and would most likely not be considered 'white' if she tried to immigrate to a European country today; and we definitely know her achievement continues to enrich the lives of billions of people. What we also know is that such discoveries were, again, based on centuries of accumulated knowledge and experimentation – recall how the basic principles of agriculture were known long before anyone applied them systematically – and that the results of such experiments were often preserved and transmitted through ritual, games and forms of play (or even more, perhaps, at the point where ritual, games and play shade into each other).

'Gardens of Adonis' are a fitting symbol here. Knowledge about the nutritious properties and growth cycles of what would later become staple crops, feeding vast populations – wheat, rice, corn – was initially maintained through ritual play farming of exactly this sort. Nor was this pattern of discovery limited to crops. Ceramics were first invented, long before the Neolithic, to make figurines, miniature models of animals and other subjects, and only later cooking and storage vessels. Mining is first attested as a way of obtaining minerals to be used as pigments, with the extraction of metals for industrial use coming only much later. Mesoamerican societies never employed wheeled transport; but we know they were familiar with spokes, wheels and axles since they made toy versions of them for children. Greek scientists famously came up with the principle of the steam engine, but only employed it to make temple doors that appeared to open of their own accord, or similar theatrical illusions. Chinese scientists, equally famously, first employed gunpowder for fireworks.

For most of history, then, the zone of ritual play constituted both a scientific laboratory and, for any given society, a repertory of knowledge and techniques which might or might not be applied to pragmatic problems. Recall, for example, the 'Little Old Men' of the Osage and how they combined research and speculation on the principles of nature with the management and periodic reform of their constitutional order; how they saw these as ultimately the same project and kept careful (oral) records of their deliberations. Did the Neolithic town of Çatalhöyük or the Tripolye mega-sites host similar colleges of 'Little Old Women'? We cannot know for certain, but it strikes us as quite likely, given the shared rhythms of social and technical innovation that we observe in each case and the attention to female themes in their art and ritual. If we are trying to frame more interesting questions to ask of history, this might be one: is there a positive correlation between what is usually called 'gender equality' (which might better be termed, simply, 'women's freedom') and the degree of innovation in a given society?

Choosing to describe history the other way round, as a series of abrupt technological revolutions, each followed by long periods when we were prisoners of our own creations, has consequences. Ultimately it is a way of representing our species as decidedly less thoughtful, less creative, less free than we actually turn out to have been. It means *not* describing history as a continual series of new ideas and innovations, technical or otherwise, during which different communities made collective decisions about which technologies they saw fit to apply to everyday purposes, and which to keep confined to the domain of experimentation or ritual play. What is true of technological creativity is, of course, even more true of social creativity. One of the most striking patterns we discovered while researching this book – indeed, one of the patterns that felt most like a genuine breakthrough to us – was how, time and again in human history, that zone of ritual play has also acted as a site of social experimentation – even, in some ways, as an encyclopaedia of social possibilities.

We are not the first to suggest this. In the mid twentieth century, a British anthropologist named A. M. Hocart proposed that monarchy and institutions of government were originally derived from rituals designed to channel powers of life from the cosmos into human society. He even suggested at one point that 'the first kings must have been dead kings',[2] and that individuals so honoured only really became sacred rulers at their funerals. Hocart was considered an oddball by his fellow anthropologists and never managed to secure a permanent job at a major university. Many accused him of being unscientific, just engaging in idle speculation. Ironically, as we've seen, it is the results of contemporary archaeological science that now oblige us to start taking his speculations seriously. To the astonishment of many, but much as Hocart predicted, the Upper Palaeolithic really has produced evidence of grand burials, carefully staged for individuals who indeed seem to have attracted spectacular riches and honours, largely in death.

The principle doesn't just apply to monarchy or aristocracy, but to other institutions as well. We have made the case that private property first appears as a concept in sacred contexts, as do police functions and powers of command, along with (in later times) a whole panoply of formal democratic procedures, like election and sortition, which were eventually deployed to limit such powers.

Here is where things get complicated. To say that, for most of human history, the ritual year served as a kind of compendium of social possibilities (as it did in the European Middle Ages, for instance, when hierarchical pageants alternated with rambunctious carnivals), doesn't really do the matter justice. This is because festivals are already seen as extraordinary, somewhat unreal, or at the very least as departures from the everyday order. Whereas, in fact, the evidence we have from Palaeolithic times onwards suggests that many – perhaps even most – people did not merely imagine or enact different social orders at different times of year, but actually lived in them for extended periods of time. The contrast with our present situation could not be more stark. Nowadays, most of us find it increasingly difficult even to picture what an alternative economic or social order would be like. Our distant ancestors seem, by contrast, to have moved regularly back and forth between them.

If something did go terribly wrong in human history – and given the current state of the world, it's hard to deny something did – then perhaps it began to go wrong precisely when people started losing that freedom to imagine and enact other forms of social existence, to such a degree that some now feel this particular type of freedom hardly even existed, or was barely exercised, for the greater part of human history. Even those few anthropologists, such as Pierre Clastres and later Christopher Boehm, who argue that humans were always able to imagine alternative social possibilities, conclude – rather oddly – that for roughly 95 per cent of our species' history those same humans recoiled in horror from all possible social worlds but one: the small-scale society of equals. Our only dreams were nightmares: terrible visions of hierarchy, domination and the state. In fact, as we've seen, this is clearly not the case.

The example of Eastern Woodlands societies in North America, explored in our last chapter, suggests a more useful way to frame the problem. We might ask why, for example, it proved possible for their ancestors to turn their backs on the legacy of Cahokia, with its overweening lords and priests, and to reorganize themselves into free republics; yet when their French interlocutors effectively tried to follow suit and rid themselves of their own ancient hierarchies, the result seemed so disastrous. No doubt there are quite a number of reasons. But for us, the key point to remember is that we are not talking here about 'freedom' as an abstract ideal or formal principle (as in 'Liberty, Equality and Fraternity!').[3] Over the course of these chapters we have instead talked about basic forms of social liberty which one might actually put into practice: (1) the freedom to move away or relocate from one's surroundings; (2) the freedom to ignore or disobey commands issued by others; and (3) the freedom to shape entirely new social realities, or shift back and forth between different ones.

What we can now see is that the first two freedoms – to relocate, and to disobey commands – often acted as a kind of scaffolding for the third, more creative one. Let us clarify some of the ways in which this 'propping-up' of the third freedom actually worked. As long as the first two freedoms were taken for granted, as they were in many North American societies when Europeans first encountered them, the only kings that could exist were always, in the last resort, play kings. If they overstepped the line, their erstwhile subjects could always ignore them or move someplace else. The same would go for any other hierarchy of offices or system of authority. Similarly, a police force that operated for only three months of the year, and whose membership rotated annually, was in a certain sense a play police force – which makes it slightly less bizarre that their members were sometimes recruited directly from the ranks of ritual clowns.[4]

It's clear that something about human societies really has changed here, and quite profoundly. The three basic freedoms have gradually receded, to the point where a majority of people living today can barely comprehend what it might be like to live in a social order based on them.

How did it happen? How did we get stuck? And just how stuck are we really?

*

'There is no way out of the imagined order,' writes Yuval Noah Harari in his book *Sapiens*. 'When we break down our prison walls and run towards freedom', he goes on, 'we are in fact running into the more spacious exercise yard of a bigger prison.'[5] As we saw in our first chapter, he is not alone in reaching this conclusion. Most people who write history on a grand scale seem to have decided that, as a species, we are well and truly stuck and there is really no escape from the institutional cages we've made for ourselves. Harari, once again echoing Rousseau, seems to have captured the prevailing mood.

We'll come back to this point, but for now we want to think a bit further about this first question: how did it happen? To some degree this must remain a matter for speculation. Asking the right questions may eventually sharpen our understanding, but for now the material at our disposal, especially for the early phases of the process, is still too sparse and ambiguous to provide definitive answers. The most we can offer are some preliminary suggestions, or points of departure, based on the arguments presented in this book; and perhaps we can also begin to see more clearly where others since the time of Rousseau have been going wrong.

One important factor would seem to be the gradual division of human societies into what are sometimes referred to as 'culture areas'; that is, the process by which neighbouring groups began defining themselves against each other and, typically, exaggerating their differences. Identity came to be seen as a value in itself, setting in motion processes of cultural schismogenesis. As we saw in the case of Californian foragers and their aristocratic neighbours on the Northwest Coast, such acts of cultural refusal could also be self-conscious acts of political contestation, marking the boundary (in this case) between societies where inter-group warfare, competitive feasting and household bondage were rejected – as in those parts of Aboriginal California closest to the Northwest Coast – and where they were accepted, even celebrated, as quintessential features of social life. Archaeologists, taking a longer view, see a proliferation of such regional culture areas, especially from the end of the last Ice Age on, but are often at a loss to explain why they emerged or what constitutes a boundary between them.

Still, this appears to have been an epochal development. Recall, for example, how post-Ice Age hunter-gatherers, especially in coastal or woodland regions, were enjoying something of a Golden Age. There appear to have been all sorts of local experiments, reflected in a proliferation of opulent burials and monumental architecture, the social functions of which often remain enigmatic: from shell-built 'amphitheatres' along the Gulf of Mexico to the great storehouses of Sannai Maruyama in Jōmon Japan, or the so-called 'Giants' Churches' of the Bothnian Sea. It is among such Mesolithic populations that we often find not just the multiplication of distinct culture areas, but also the first clear archaeological indications of communities divided into permanent ranks, sometimes accompanied by interpersonal violence, even warfare. In some cases this may already have meant the stratification of households into aristocrats, commoners and slaves. In others, quite different forms of hierarchy may have taken root. Some appear to have become, effectively, fixed in place.

The role of warfare warrants further discussion here, because violence is often the route by which forms of play take on more permanent features. For example, the kingdoms of the Natchez or Shilluk might have been largely theatrical affairs, their rulers unable to issue orders that would be obeyed even a mile or two away; but if someone was arbitrarily killed as part of a theatrical display, that person remained definitively dead even after the performance was over. It's an almost absurdly obvious point to make, but it matters. Play kings cease to be play kings precisely when they start killing people; which perhaps also helps to explain the excesses of ritually sanctioned violence that so often ensued during transitions from one state to the other. The same is true of warfare. As Elaine Scarry points out, two communities might choose to resolve a dispute by partaking in a contest, and often they do; but the ultimate difference between war (or 'contests of injuring', as she puts it) and most other kinds of contest is that anyone killed or disfigured in a war remains so, even after the contest ends.[6]

Still, we must be cautious. While human beings have always been capable of physically attacking one another (and it's difficult to find examples of societies where no one ever attacks anyone else, under any circumstances), there's no actual reason to assume that war has always existed. Technically, war refers not just to organized violence but to a kind of contest between two clearly demarcated sides. As

Raymond Kelly has adroitly pointed out, it's based on a logical principle that's by no means natural or self-evident, which states that major violence involves two teams, and any member of one team treats all members of the other as equal targets. Kelly calls this the principle of 'social substitutability'[7] – that is, if a Hatfield kills a McCoy and the McCoys retaliate, it doesn't have to be against the actual murderer; any Hatfield is fair game. In the same way, if there is a war between France and Germany, any French soldier can kill any German soldier, and vice versa. The murder of entire populations is simply taking this same logic one step further. There is nothing particularly primordial about such arrangements; certainly, there is no reason to believe they are in any sense hardwired into the human psyche. On the contrary, it's almost invariably necessary to employ some combination of ritual, drugs and psychological techniques to convince people, even adolescent males, to kill and injure each other in such systematic yet indiscriminate ways.

It would seem that for most of human history, no one saw much reason to do such things; or if they did, it was rare. Systematic studies of the Palaeolithic record offer little evidence of warfare in this specific sense.[8] Moreover, since war was always something of a game, it's not entirely surprising that it has manifested itself in sometimes more theatrical and sometimes more deadly variations. Ethnography provides plenty of examples of what could best be described as play war: either with non-deadly weapons or, more often, battles involving thousands on each side where the number of casualties after a day's 'fighting' amount to perhaps two or three. Even in Homeric-style warfare, most participants were basically there as an audience while individual heroes taunted, jeered and occasionally threw javelins or shot arrows at one another, or engaged in duels. At the other extreme, as we've seen, there is an increasing amount of archaeological evidence for outright massacres, such as those that took place among Neolithic village dwellers in central Europe after the end of the last Ice Age.

What strikes us is just how uneven such evidence is. Periods of intense inter-group violence alternate with periods of peace, often lasting centuries, in which there is little or no evidence for destructive conflict of any kind. War did not become a constant of human life after the adoption of farming; indeed, long periods of time exist in which it appears to have been successfully abolished. Yet it had a stubborn tendency to reappear, if only many generations later. At this point another new question comes into focus. Was there a relationship between external warfare and the internal loss of freedoms that opened the way, first to systems of ranking and then later on to large-scale systems of domination, like those we discussed in the later chapters of this book: the first dynastic kingdoms and empires, such as those of the Maya, Shang or Inca? And if so, how direct was this correlation? One thing we've learned is that it's a mistake to begin answering such questions by assuming that these ancient polities were simply archaic versions of our modern states.

The state, as we know it today, results from a distinct combination of elements – sovereignty, bureaucracy and a competitive political field – which have entirely separate origins. In our thought experiment of two chapters ago, we showed how those elements map directly on to basic forms of social power which can operate at any scale of human interaction, from the family or household all the way up to the Roman Empire or the super-kingdom of Tawantinsuyu. Sovereignty, bureaucracy and politics are magnifications of elementary types of domination, grounded respectively in the use of violence, knowledge and charisma. Ancient political systems – especially those, such as the Olmec or Chavín de Huántar, that elude definition in terms of 'chiefdoms' and 'states' – can often be understood better in terms of how they developed one axis of social power to an extraordinary degree (e.g. charismatic political contests and spectacles in the Olmec case, or control of esoteric knowledge in Chavín). These are what we termed 'first-order regimes'.

Where two axes of power were developed and formalized into a single system of domination we can begin to talk of 'second-order regimes'. The architects of Egypt's Old Kingdom, for example, armed the principle of sovereignty with a bureaucracy and managed to extend it across a large territory. By contrast, the rulers of ancient Mesopotamian city-states made no direct claims to sovereignty, which for them resided properly in heaven. When they engaged in wars over land or irrigation systems, it was only as secondary agents of the gods. Instead they combined charismatic competition with a highly developed administrative order. The Classic Maya were different again, confining administrative activities largely to the monitoring of cosmic affairs, while basing their earthly power on a potent fusion of sovereignty and inter-dynastic politics.

Insofar as these and other polities commonly regarded as 'early states' (Shang China, for instance) really share any common features, they seem to lie in altogether different areas – which brings us back to the question of warfare, and the loss of freedoms within society. All of them deployed spectacular violence at the pinnacle of the system (whether that violence was conceived as a direct extension of royal sovereignty or carried out at the behest of divinities); and all to some degree modelled their centres of power – the court or palace – on the organization of patriarchal households. Is this merely a coincidence? On reflection, the same combination of features can be found in most later kingdoms or empires, such as the Han, Aztec or Roman. In each case there was a close connection between the patriarchal household and military might. But why exactly should this be the case?

The question has proved difficult to answer in all but superficial terms, partly because our own intellectual traditions oblige us to use what is, in effect, imperial language to do so; and the language already implies an explanation, even a justification, for much of what we are really trying to account for here. That is why, in the course of this book, we sometimes felt the need to develop our own, more neutral (dare we say scientific?) list of baseline human freedoms and forms of domination; because existing debates almost invariably begin with terms derived from Roman Law, and for a number of reasons this is problematic.

The Roman Law conception of natural freedom is essentially based on the power of the individual (by implication, a male head of household) to dispose of his property as he sees fit. In Roman Law property isn't even exactly a right, since rights are negotiated with others and involve mutual obligations; it's simply power – the blunt reality that someone in possession of a thing can do anything he wants with it, except that which is limited 'by force or law'. This formulation has some peculiarities that jurists have struggled with ever since, as it implies freedom is essentially a state of primordial exception to the legal order. It also implies that property is not a set of understandings between people over who gets to use or look after things, but rather a relation between a person and an object characterized by absolute power. What does it mean to say one has the natural right to do anything one wants with a hand grenade, say, except those things one isn't allowed to do? Who would come up with such an odd formulation?

An answer is suggested by the West Indian sociologist Orlando Patterson, who points out that Roman Law conceptions of property (and hence of freedom) essentially trace back to slave law.[9] The reason it is possible to imagine property as a relationship of domination between a person and a thing is because, in Roman Law, the power of the master rendered the slave a thing (*res*, meaning an object), not a person with social rights or legal obligations to anyone else. Property law, in turn, was largely about the complicated situations that might arise as a result. It is important to recall, for a moment, who these Roman jurists actually were that laid down the basis for our current legal order – our theories of justice, the language of contract and torts, the distinction of public and private and so forth. While they spent their public lives making sober judgments as magistrates, they lived their private lives in households where they not only had near-total authority over their wives, children and other dependants, but also had all their needs taken care of by dozens, perhaps hundreds of slaves.

Slaves trimmed their hair, carried their towels, fed their pets, repaired their sandals, played music at their dinner parties and instructed their children in history and maths. At the same time, in terms of legal theory these slaves were classified as captive foreigners who, conquered in battle, had forfeited rights of any kind. As a result, the Roman jurist was free to rape, torture, mutilate or kill any of them at any time and in any way he had a mind to, without the matter being considered anything other than a private affair. (Only under the reign of Tiberius were any restrictions imposed on what a master could do to a slave, and what this meant was simply that permission from a local magistrate had to be obtained before a slave could be ripped apart by wild animals; other forms of execution could still be imposed at the owner's whim.) On the one hand, freedom and liberty were private affairs; on the other, private life was marked by the absolute power of the patriarch over conquered people who were considered his private property.[10]

The fact that most Roman slaves were not prisoners of war, in the literal sense, doesn't really make much difference here. What's important is that their legal status was defined in those terms. What is both striking and revealing, for our present purposes, is how in Roman jurisprudence the logic of war – which dictates that enemies are interchangeable, and if they surrendered they could either be killed or rendered 'socially dead', sold as commodities – and, therefore, the potential for arbitrary violence was inserted into the most intimate sphere of social relations, including the relations of care that made domestic life possible. Thinking back to examples like the 'capturing societies' of Amazonia or the process by which dynastic power took root in ancient Egypt, we can begin to see how important that particular nexus of violence and care has been. Rome took the entanglement to new extremes, and its legacy still shapes our basic concepts of social structure.

Our very word 'family' shares a root with the Latin *famulus*, meaning 'house slave', via *familia*, which originally referred to everyone under the domestic authority of a single *paterfamilias* or male head of household. *Domus*, the Latin word for 'household', in turn gives us not only 'domestic' and 'domesticated' but *dominium*, which was the technical term for the emperor's sovereignty as well as a citizen's power over private property. Through that we arrive at (literally, 'familiar') notions of what it means to be 'dominant', to possess 'dominion' and to 'dominate'. Let us follow this line of thought a little further.

We've seen how, in various parts of the world, direct evidence of warfare and massacres – including the carrying-off of captives – can be detected long before the appearance of kingdoms or empires. Much harder to ascertain, for such early periods of history, is what happened to captive enemies: were they killed, incorporated or left suspended somewhere in between? As we learned from various Amerindian cases, things may not always be entirely clear-cut. There were often multiple possibilities. It's instructive, in this context, to return one last time to the case of the Wendat in the age of Kandiaronk, since this was one society that seemed determined to avoid ambiguity in such matters.

In certain ways Wendat, and Iroquoian societies in general around that time, were extraordinarily warlike. There appear to have been bloody rivalries fought out in many northern parts of the Eastern Woodlands even before European settlers began supplying indigenous factions with muskets, resulting in the 'Beaver Wars'. The early Jesuits were often appalled by what they saw, but they also noted that the ostensible reasons for wars were entirely different from those they were used to. All Wendat wars were, in fact, 'mourning wars', carried out to assuage the grief felt by close relatives of someone who had been killed. Typically, a war party would strike against traditional enemies, bringing back a few scalps and a small number of prisoners. Captive women and children would be adopted. The fate of men was largely up to the mourners, particularly the women, and appeared to outsiders at least to be entirely arbitrary. If the mourners felt it appropriate a male captive might be given a name, even the name of the original victim. The captive enemy would henceforth become that other person and, after a few years' trial period, be treated as a full member of society. If for any reason that did not happen, however, he suffered a very different fate. For a male warrior taken prisoner, the only alternative to full adoption into Wendat society was excruciating death by torture.

Jesuits found the details shocking and fascinating. What they observed, sometimes at first hand, was a slow, public and highly theatrical use of violence. True, they conceded, the Wendat torture of captives was no more cruel than the kind directed against enemies of the state back home in France. What seems to have really appalled them, however, was not so much the whipping, boiling, branding, cutting-up – even in some cases cooking and eating – of the enemy, so much as the fact that almost everyone in a Wendat village or town took part, even women and children. The suffering might go on for days, with the victim periodically resuscitated only to endure further ordeals, and it was very much a communal affair.[11] The violence seems all the more extraordinary once we recall how these same Wendat societies refused to spank children, directly punish thieves or murderers, or take any measure against their own members that smacked of arbitrary authority. In virtually all other areas of social life they were renowned for solving their problems through calm and reasoned debate.

Now, it would be easy to make an argument that repressed aggression must be vented in one way or another, so that orgies of communal torture are simply the necessary flipside of a non-violent community; and some contemporary scholars do make this point. But it doesn't really work. In fact, Iroquoia seems to be precisely one of those regions of North America where violence flared up only during certain specific historical periods and then largely disappeared in others. In what archaeologists term the 'Middle Woodland' phase, for instance, between 100 BC and AD 500 – corresponding roughly to the heyday of the Hopewell civilization – there seems to have been general peace.[12] Later on, signs of endemic warfare reappear. Clearly, at some points in their history people living in this region found effective ways to ensure that vendettas didn't escalate into a spiral of retaliation or actual warfare (the Haudenosaunee story of the Great Law of the Peace seems to be about precisely such a moment); at other times, the system broke down and the possibility of sadistic cruelty returned.

What, then, was the meaning of these theatres of violence? One way to approach the question is to compare them with what was happening in Europe around the same time. As the Quebecois historian Denys Delâge points out, Wendat who visited France were equally appalled by the tortures exhibited during public punishments and executions, but what struck *them* as most remarkable is that 'the French whipped, hanged, and put to death men *from among themselves*', rather than external enemies. The point is telling, as in seventeenth-century Europe, Delâge notes,

> . . . almost all punishment, including the death penalty, involved severe physical suffering: wearing an iron collar, being whipped, having a hand cut off, or being branded . . . It was a ritual that manifested power in a conspicuous way, thereby revealing the existence of an internal war. The sovereign incarnated a superior power that transcended his subjects, one that they were compelled to recognise . . . While Amerindian cannibal rituals showed the desire to take over the strength and courage of the alien so as to combat him better, the European ritual revealed the existence of a dissymmetry, an irrevocable imbalance of power.[13]

Wendat punitive actions against war captives (those not taken in for adoption) required the community to become a single body, unified by its capacity for violence. In France, by contrast, 'the people' were unified as potential victims of the king's violence. But the contrasts run deeper still.

As a Wendat traveller observed of the French system, anyone – guilty or innocent – *might* end up being made a public example. Among the Wendat themselves, however, violence was firmly excluded from the realm of family and household. A captive warrior might either be treated with loving care and affection or be the object of the worst treatment imaginable. No middle ground existed. Prisoner sacrifice was not merely about reinforcing the solidarity of the group but also proclaimed the internal sanctity of the family and the domestic realm as spaces of female governance where violence, politics and rule by command did not belong. Wendat households, in other words, were defined in exactly opposite terms to the Roman *familia*.

In this particular respect, French society under the *Ancien Régime* presents a rather similar picture to imperial Rome – at least, when both are placed in the light of the Wendat example. In both cases, household and kingdom shared a common model of subordination. Each was made in the other's image, with the patriarchal family serving as a template for the absolute power of kings, and vice versa.[14] Children were to be submissive to their parents, wives to husbands, and subjects to rulers whose authority came from God. In each case the superior party was expected to inflict stern chastisement when he considered it appropriate: that is, to exercise violence with impunity. All this, moreover, was assumed to be bound up with feelings of love and affection. Ultimately, the house of the Bourbon monarchs – like the palace of an Egyptian pharaoh, Roman emperor, Aztec *tlatoani* or Sapa Inca – was not merely a structure of domination but also one of care, where a small army of courtiers laboured night and day to attend to the king's every physical need and prevent him, as much as was humanly possible, from ever feeling anything but divine.

In all these cases, the bonds of violence and care extended downwards as well as upwards. We can do no better than put it in words made famous by King James I of England in *The True Law of Free Monarchies* (1598):

> As the father, of his fatherly duty, is to care for the nourishing, education, and virtuous government of his children; even so is the King bound to care for all his subjects . . . As the father's wrath and correction on any of his children that offendeth, ought to be a fatherly chastisement seasoned with pity, so long as there is any hope of amendment in them; so ought the King towards any of his lieges that offend in that measure . . . As the father's chief joy ought to be in procuring his children's welfare, rejoicing in their weal, sorrowing and pitying at their evil, to hazard for their safety . . . so ought a good Prince think of his People.

Public torture, in seventeenth-century Europe, created searing, unforgettable spectacles of pain and suffering in order to convey the message that a system in which husbands could brutalize wives, and parents beat children, was ultimately a form of love. Wendat torture, in the same period of history, created searing, unforgettable spectacles of pain and suffering in order to make clear that no form of physical chastisement should ever be countenanced inside a community or household. Violence and care, in the Wendat case, were to be entirely separated. Seen in this light, the distinctive features of Wendat prisoner torture come into focus.

It seems to us that this connection – or better perhaps, confusion – between care and domination is utterly critical to the larger question of how we lost the ability freely to recreate ourselves by recreating our relations with one another. It is critical, that is, to understanding how we got stuck, and why these days we can hardly envisage our own past or future as anything other than a transition from smaller to larger cages.

In the course of writing this book, we have tried to strike a certain balance. It would be intuitive for an archaeologist and an anthropologist, immersed in our subject matter, to take on all the scholarly views about, say, Stonehenge, the 'Uruk expansion' or Iroquoian social organization and explain our preference for one interpretation over another, or venture a different one. This is how the search for truth is normally conducted in the academy. But had we tried to outline or refute every existing interpretation of the material we covered, this book would have been two or three times the size, and likely would have left the reader with a sense that the authors are engaged in a constant battle with demons who were in fact two inches tall. So instead we have tried to map out what we think really happened, and to point out the flaws in other scholars' arguments only insofar as they seemed to reflect more widespread misconceptions.

Perhaps the most stubborn misconception we've been tackling has to do with scale. It does seem to be received wisdom in many quarters, academic and otherwise, that structures of domination are the inevitable result of populations scaling up by orders of magnitude; that is, that a necessary correspondence exists between social and spatial hierarchies. Time and again we found ourselves confronted with writing which simply assumes that the larger and more densely populated the social group, the more 'complex' the system needed to keep it organized. Complexity, in turn, is still often used as a synonym for hierarchy. Hierarchy, in turn, is used as a euphemism for chains of command (the 'origins of the state'), which mean that as soon as large numbers of people decided to live in one place or join a common project, they must necessarily abandon the second freedom – to refuse orders – and replace it with legal mechanisms for, say, beating or locking up those who don't do as they're told.

As we've seen, none of these assumptions are theoretically essential, and history tends not to bear them out. Carole Crumley, an anthropologist and expert on Iron Age Europe, has been pointing this out for years: complex systems don't have to be organized top-down, either in the natural or in the social world. That we tend to assume otherwise probably tells us more about ourselves than the people or phenomena that we're studying.[15] Neither is she alone in making this point. But more often than not, such observations have fallen on deaf ears.

It's probably time to start listening, because 'exceptions' are fast beginning to outnumber the rules. Take cities. It was once assumed that the rise of urban life marked some kind of historical turnstile, whereby everyone who passed through had to permanently surrender their basic freedoms and submit to the rule of faceless administrators, stern priests, paternalistic kings or warrior-politicians – simply to avert chaos (or cognitive overload). To view human history through such a lens today is really not all that different from taking on the mantle of a modern-day King James, since the overall effect is to portray the violence and inequalities of modern society as somehow arising naturally from structures of rational management and paternalistic care: structures designed for human populations who, we are asked to believe, became suddenly incapable of organizing themselves once their numbers expanded above a certain threshold.

Not only do such views lack a sound basis in human psychology. They are also difficult to reconcile with archaeological evidence of how cities actually began in many parts of the world: as civic experiments on a grand scale, which frequently lacked the expected features of administrative hierarchy and authoritarian rule. We do not possess an adequate terminology for these early cities. To call them 'egalitarian', as we've seen, could mean quite a number of different things. It might imply an urban parliament and co-ordinated projects of social housing, as with some pre-Columbian centres in the Americas; or the self-organizing of autonomous households into neighbourhoods and citizens' assemblies, as with prehistoric mega-sites north of the Black Sea; or, perhaps, the introduction of some explicit notion of equality based on principles of uniformity and sameness, as in Uruk-period Mesopotamia.

None of this variability is surprising once we recall what preceded cities in each region. That was not, in fact, rudimentary or isolated groups, but far-flung networks of societies, spanning diverse ecologies, with people, plants, animals, drugs, objects of value, songs and ideas moving between them in endlessly intricate ways. While the individual units were demographically small, especially at certain times of year, they were typically organized into loose coalitions or confederacies. At the very least, these were simply the logical outcome of our first freedom: to move away from one's home, knowing one will be received and cared for, even valued, in some distant place. At most they were examples of 'amphictyony', in which some kind of formal organization was put in charge of the care and maintenance of sacred places. It seems that Marcel Mauss had a point when he argued that we should reserve the term 'civilization' for great hospitality zones such as these. Of course, we are used to thinking of 'civilization' as something that originates in cities – but, armed with new knowledge, it seems more realistic to put things the other way round and to imagine the first cities as one of those great regional confederacies, compressed into a small space.

Of course, monarchy, warrior aristocracies or other forms of stratification could also take hold in urban contexts, and often did. When this happened the consequences were dramatic. Still, the mere existence of large human settlements in no way caused these phenomena, and certainly didn't make them inevitable. For the origins of these structures of domination we must look elsewhere. Hereditary aristocracies were just as likely to exist among demographically small or modest-sized groups, such as the 'heroic societies' of the Anatolian highlands, which took form on the margins of the first Mesopotamian cities and traded extensively with them. Insofar as we have evidence for the inception of monarchy as a permanent institution it seems to lie precisely there, and not in cities. In other parts of the world, some urban populations ventured partway down the road towards monarchy, only to turn back. Such was the case at Teotihuacan in the Valley of Mexico, where the city's population – having raised the Pyramids of the Sun and Moon – then abandoned such aggrandizing projects and embarked instead on a prodigious programme of social housing, providing multi-family apartments for its residents.

Elsewhere, early cities followed the opposite trajectory, starting with neighbourhood councils and popular assemblies and ending up being ruled by warlike dynasts, who then had to maintain an uneasy coexistence with older institutions of urban governance. Something along these lines took place in Early Dynastic Mesopotamia, after the Uruk period: here again the convergence between systems of violence and systems of care seems critical. Sumerian temples had always organized their economic existence around the nurturing and feeding of the gods, embodied in their cult statues, which became surrounded by a whole industry and bureaucracy of welfare. Even more crucially, temples were charitable institutions. Widows, orphans, runaways, those exiled from their kin groups or other support networks would take refuge there: at Uruk, for example, in the Temple of Inanna, protective goddess of the city, overlooking the great courtyard of the city's assembly.

The first charismatic war-kings attached themselves to such spaces, quite literally moving in next door to the residence of the city's leading deity. In such ways, Sumerian monarchs were able to insert themselves into institutional spaces once reserved for the care of the gods, and thus removed from the realm of ordinary human relationships. This makes sense because kings, as the Malagasy proverb puts it, 'have no relatives' – or they shouldn't, since they are rulers equally of all their subjects. Slaves too have no kin; they are severed from all prior attachments. In either case, the only recognized social relationships such individuals possess are those based on power and domination. In structural terms, and as against almost everyone else in society, kings and slaves effectively inhabit the same ground. The difference lies in which end of the power spectrum they happen to occupy.

We also know that needy individuals, taken into such temple institutions, were supplied with regular rations and put to work on the temple's lands and in its workshops. The very first factories – or, at least, the very first we are aware of in history – were charitable institutions of this kind, where temple bureaucrats would supply women with wool to spin and weave, supervise the disposal of the product (much of it traded with upland groups in exchange for wood, stone and metal, unavailable in the river valleys), and provide them with carefully apportioned rations. All this was already true long before the appearance of kings. As persons dedicated to the gods, these women must originally have had a certain dignity, even a sacred status; but already by the time of the first written documents, the situation seems to have grown more complicated.

By then, some of those working in Sumerian temples were also war captives, or even slaves, who were similarly bereft of family support. Over time, and perhaps as a result, the status of widows and orphans also appears to have been downgraded, until the temple institutions came to resemble something more like a Victorian poorhouse. How, we might then ask, did the degradation of women working in the temple factories affect the status of women more generally? If nothing else, it must have made the prospect of fleeing an abusive domestic arrangement far more daunting. Loss of the first freedom meant, increasingly, loss of the second. Loss of the second meant effacement of the third. If a woman in such a situation attempted to create a new cult, a new temple, a new vision of social relations she would instantly be marked as a subversive, a revolutionary; if she attracted followers she might well find herself confronted by military force.

All this brings into focus another question. Does this newly established nexus between external violence and internal care – between the most impersonal and the most intimate of human relations – mark the point where everything begins to get confused? Is this an example of how relations that were once flexible and negotiable ended up getting fixed in place: an example, in other words, of how we effectively got stuck? If there is a particular story we should be telling, a big question we should be asking of human history (instead of the 'origins of social inequality'), is it precisely this: how did we find ourselves stuck in just one form of social reality, and how did relations based ultimately on violence and domination come to be normalized within it?

Perhaps the scholar who most closely approached this question in the last century was an anthropologist and poet named Franz Steiner, who died in 1952. Steiner led a fascinating if tragic life. A brilliant polymath born to a Jewish family in Bohemia, he later lived with an Arab family in Jerusalem until expelled by the British authorities, conducted fieldwork in the Carpathians and was twice forced by the Nazis to flee the continent, ending his career – ironically enough – in the south of England. Most of his immediate family were killed at Birkenau. Legend has it that he completed 800 pages of a monumental doctoral dissertation on the comparative sociology of slavery, only to have the suitcase containing his drafts and research notes stolen on a train. He was friends with, and a romantic rival to, Elias Canetti, another Jewish exile at Oxford and a successful suitor to the novelist Iris Murdoch – although two days after she'd accepted his proposal of marriage, Steiner died of a heart attack. He was forty-three.

The shorter version of Steiner's doctoral work, which does survive, focuses on what he calls 'pre-servile institutions'. Poignantly, given his own life story, it is a study of what happens in different cultural and historical situations to people who become unmoored: those expelled from their clans for some debt or fault; castaways, criminals, runaways. It can be read as a history of how refugees such as himself were first welcomed, treated as almost sacred beings, then gradually degraded and exploited, again much like the women working in the

Sumerian temple factories. In essence, the story told by Steiner appears to be precisely about the collapse of what we would term the first basic freedom (to move away or relocate), and how this paved the way for the loss of the second (the freedom to disobey). It also leads us back to a point we made earlier about the progressive division of the human social universe into smaller and smaller units, beginning with the appearance of 'culture areas' (a fascination of ethnologists in the central European tradition, in which Steiner first trained).

What happens, Steiner asked, when expectations that make freedom of movement possible – the norms of hospitality and asylum, civility and shelter – erode? Why does this so often appear to be a catalyst for situations where some people can exert arbitrary power over others? Steiner worked his way in careful detail through cases ranging from the Amazonian Huitoto and East African Safwa to the Tibeto-Burman Lushai. Along the journey he suggested one possible answer to the question that had so puzzled Robert Lowie, and later Clastres: if stateless societies do regularly organize themselves in such a way that chiefs have no coercive power, then how did top-down forms of organization ever come into the world to begin with? You'll recall how both Lowie and Clastres were driven to the same conclusion: that they must have been the product of religious revelation. Steiner provided an alternative route. Perhaps, he suggested, it all goes back to charity.

In Amazonian societies, not only orphans but also widows, the mad, disabled or deformed – if they had no one else to look after them – were allowed to take refuge in the chief's residence, where they received a share of communal meals. To these were occasionally added war captives, especially children taken in raiding expeditions. Among the Safwa or Lushai, runaways, debtors, criminals or others needing protection held the same status as those who surrendered in battle. All became members of the chief's retinue, and the younger males often took on the role of police-like enforcers. How much power the chief actually had over his retainers – Steiner uses the Roman Law term *potestas*, which denotes among other things a father's power of arbitrary command over his dependants and their property – would vary, depending how easy it was for wards to run away and find refuge elsewhere, or to maintain at least some ties with relatives, clans or outsiders willing to stand up for them. How far such henchmen could be relied on to enforce the chief's will also varied; but the sheer potential was important.

In all such cases, the process of giving refuge did generally lead to the transformation of basic domestic arrangements, especially as captured women were incorporated, further reinforcing the *potestas* of fathers. It is possible to detect something of this logic in almost all historically documented royal courts, which invariably attracted those considered freakish or detached. There seems to have been no region of the world, from China to the Andes, where courtly societies did not host such obviously distinctive individuals; and few monarchs who did not also claim to be the protectors of widows and orphans. One could easily imagine something along these lines was already happening in certain hunter-gatherer communities during much earlier periods of history. The physically anomalous individuals accorded lavish burials in the last Ice Age must also have been the focus of much caring attention while alive. No doubt there are sequences of development linking such practices to later royal courts – we've caught glimpses of them, as in Predynastic Egypt – even if we are still unable to reconstruct most of the links.

Steiner may not have foregrounded the issue, but his observations are directly relevant to debates about the origins of patriarchy. Feminist anthropologists have long argued for a connection between external (largely male) violence and the transformation of women's status in the home. In archaeological and historical terms, we are only just beginning to gather together enough material to begin understanding how that process actually worked.

The research that culminated in this book began almost a decade ago, essentially as a form of play. We pursued it at first, it would be fair to say, in a spirit of mild defiance towards our more 'serious' academic responsibilities. Mainly we were just curious about how the new archaeological evidence that had been building up for the last thirty years might change our notions of early human history, especially the parts bound up with debates on the origins of social inequality. Before long, though, we realized that what we were doing was potentially important, because hardly anyone else in our fields seemed to be doing this work of synthesis. Often, we found ourselves searching in vain for books that we assumed must exist but, it turns out, simply didn't – for instance, compendia of early cities that lacked top-down governance, or accounts of how democratic decision-making was conducted in Africa or the Americas, or comparisons of what we've called 'heroic societies'. The literature is riddled with absences.

We eventually came to realize that this reluctance to synthesize was not simply a product of reticence on the part of highly specialized scholars, although this is certainly a factor. To some degree it was simply the lack of an appropriate language. What, for instance, does one even call a 'city lacking top-down structures of governance'? At the moment there is no commonly accepted term. Dare one call it a 'democracy'? A 'republic'? Such words (like 'civilization') are so freighted with historical baggage that most archaeologists and anthropologists instinctively recoil from them, and historians tend to limit their use to Europe. Does one, then, call it an 'egalitarian city'? Probably not, since to evoke such a term is to invite the obvious demand for proof that the city was 'really' egalitarian – which usually means, in practice, showing that no element of structural inequality existed in any aspect of its inhabitants' lives, including households and religious arrangements. Since such evidence will rarely, if ever, be forthcoming, the conclusion would have to be that these are not really egalitarian cities after all.

By the same logic, one might easily conclude there aren't really any 'egalitarian societies', except possibly certain very small foraging bands. Many researchers in the field of evolutionary anthropology do, in fact, make precisely this argument. But ultimately the result of this kind of thinking is to lump together all 'non-egalitarian' cities or indeed all 'non-egalitarian societies', which is a little like saying there's no meaningful difference between a hippie commune and a biker gang, since neither are entirely non-violent. All this achieves, at the end of the day, is to leave us literally at a loss for words when confronted with certain major aspects of human history. We fall strangely mute in the face of any kind of evidence for humans doing something other than 'rushing headlong for their chains'. Sensing a sea change in the evidence of the past, we decided to approach things the other way round.

What this meant, in practice, was reversing a lot of polarities. It meant ditching the language of 'equality' and 'inequality', unless there was explicit evidence that ideologies of social equality were actually present on the ground. It meant asking, for instance, what happens if we accord significance to the 5,000 years in which cereal domestication did *not* lead to the emergence of pampered aristocracies, standing armies or debt peonage, rather than just the 5,000 in which it did? What happens if we treat the rejection of urban life, or of slavery, in certain times and places as something just as significant as the emergence of those same phenomena in others? In the process, we often found ourselves surprised. We'd never have guessed, for instance, that slavery was most likely abolished multiple times in history in multiple places; and that very possibly the same is true of war. Obviously, such abolitions are rarely definitive. Still, the periods in which free or relatively free societies existed are hardly insignificant. In fact, if you bracket the Eurasian Iron Age (which is effectively what we have been doing here), they represent the vast majority of human social experience.

Social theorists have a tendency to write about the past as if everything that happened could have been predicted beforehand. This is somewhat dishonest, since we're all aware that when we actually try to predict the future we almost invariably get it wrong – and this is just as true of social theorists as anybody else. Nonetheless, it's hard to resist the temptation to write and think as if the current state of the world, in the early twenty-first century, is the inevitable outcome of the last 10,000 years of history, while in reality, of course, we have little or no idea what the world will be like even in 2075, let alone 2150.

Who knows? Perhaps if our species does endure, and we one day look backwards from this as yet unknowable future, aspects of the remote past that now seem like anomalies – say, bureaucracies that work on a community scale; cities governed by neighbourhood councils; systems of government where women hold a preponderance of formal positions; or forms of land management based on care-taking rather than ownership and extraction – will seem like the really significant breakthroughs, and great stone pyramids or statues more like historical curiosities. What if we were to take that approach now and look at, say, Minoan Crete or Hopewell not as random bumps on a road that leads inexorably to states and empires, but as alternative possibilities: roads not taken?

After all, those things really did exist, even if our habitual ways of looking at the past seem designed to put them at the margins rather than at the centre of things. Much of this book has been devoted to recalibrating those scales; to reminding us that people did actually live in those ways, often for many centuries, even millennia. In some ways, such a perspective might seem even more tragic than our standard narrative of civilization as the inevitable fall from grace. It means we *could* have been living under radically different conceptions of what human society is actually about. It means that mass enslavement, genocide, prison camps, even patriarchy or regimes of wage labour never had to happen. But on the other hand it also suggests that, even now, the possibilities for human intervention are far greater than we're inclined to think.

We began this book with a quote which refers to the Greek notion of *kairos* as one of those occasional moments in a society's history when its frames of reference undergo a shift – a metamorphosis of the fundamental principles and symbols, when the lines between myth and history, science and magic become blurred – and, therefore, real change is possible. Philosophers sometimes like to speak of 'the Event' – a political revolution, a scientific discovery, an artistic masterpiece – that is, a breakthrough which reveals aspects of reality that had previously been unimaginable but, once seen, can never be unseen. If so, *kairos* is the kind of time in which Events are prone to happen.

Societies around the world appear to be cascading towards such a point. This is particularly true of those which, since the First World War, have been in the habit of calling themselves 'Western'. On the one hand, fundamental breakthroughs in the physical sciences, or even artistic expression, no longer seem to occur with anything like the regularity people came to expect in the late nineteenth and early twentieth centuries. Yet at the same time, our scientific means of understanding the past, not just our species' past but that of our planet, has been advancing with dizzying speed. Scientists in 2020 are not (as readers of mid-twentieth-century science fiction might have hoped) encountering alien civilizations in distant star systems; but they are encountering radically different forms of society under their own feet, some forgotten and newly rediscovered, others more familiar, but now understood in entirely new ways.

In developing the scientific means to know our own past, we have exposed the mythical substructure of our 'social science' – what once appeared unassailable axioms, the stable points around which our self-knowledge is organized, are scattering like mice. What is the purpose of all this new knowledge, if not to reshape our conceptions of who we are and what we might yet become? If not, in other words, to rediscover the meaning of our third basic freedom: the freedom to create new and different forms of social reality?

Myth in itself is not the problem here. It shouldn't be mistaken for bad or infantile science. Just as all societies have their science, all societies have their myths. Myth is the way in which human societies give structure and meaning to experience. But the larger mythic structures of history we've been deploying for the last several centuries simply don't work any more; they are impossible to reconcile with the evidence now before our eyes, and the structures and meanings they encourage are tawdry, shop-worn and politically disastrous.

No doubt, for a while at least, very little will change. Whole fields of knowledge – not to mention university chairs and departments, scientific journals, prestigious research grants, libraries, databases, school curricula and the like – have been designed to fit the old structures and the old questions. Max Planck once remarked that new scientific truths don't replace old ones by convincing established scientists that they were wrong; they do so because proponents of the older theory eventually die, and generations that follow find the new truths and theories to be familiar, obvious even. We are optimists. We like to think it will not take that long.

In fact, we have already taken a first step. We can see more clearly now what is going on when, for example, a study that is rigorous in every other respect begins from the unexamined assumption that there was some 'original' form of human society; that its nature was fundamentally good or evil; that a time before inequality and political awareness existed; that something happened to change all this; that 'civilization' and 'complexity' always come at the price of human freedoms; that participatory democracy is natural in small groups but cannot possibly scale up to anything like a city or a nation state.

We know, now, that we are in the presence of myths.

Notes

1. FAREWELL TO HUMANITY'S CHILDHOOD

1. To take one example, Ian Morris's (2015) *Foragers, Farmers, and Fossil Fuels: How Human Values Evolve* sets itself the ambitious challenge of finding a uniform measure of inequality applicable across the entire span of human history, by translating the 'values' of Ice Age hunter-gatherers and Neolithic farmers into terms familiar to modern-day economists, and then using those to establish Gini coefficients (i.e. formal inequality rates). It's a laudable experiment, but one that quickly leads to some very odd conclusions. For instance, in a 2015 piece for the *New York Times*, Morris estimated the income of a Palaeolithic hunter-gatherer at $1.10 a day, pegged to 1990 currency values. Where does this figure come from? Presumably it has something to do with the calorific value of daily food intake. But if we're comparing this to daily incomes today, wouldn't we also have to factor in all the other things Palaeolithic foragers got for free, but which we ourselves would expect to pay for: free security, free dispute resolution, free primary education, free care of the elderly, free medicine, not to mention entertainment costs, music, storytelling and religious services? Even when it comes to food, we must consider quality: after all, we're talking about 100 per cent organic free-range produce here, washed down with purest natural spring water. Much contemporary income goes to mortgages and rents. But consider the camping fees for prime Palaeolithic locations along the Dordogne or the Vézère, not to mention the high-end evening classes in naturalistic rock-painting and ivory-carving – and all those fur coats. Surely all this must cost wildly in excess of $1.10 a day. As we'll see in Chapter Four, it's not for nothing that anthropologists sometimes refer to foragers as 'the original affluent society'. Such a life today would not come cheap. Admittedly, this is all a bit silly, but that's really our point: if one reduces world history to Gini coefficients, silly things will, necessarily, follow.

2. Fukuyama 2011: 43, 53–4.
3. Diamond 2012: 10–15.
4. Fukuyama 2011: 48.
5. Diamond 2012: 11.
6. In the case of Fukuyama and Diamond one can, at least, note that they were never trained in the relevant disciplines (the first is a political scientist, the other has a PhD on the physiology of the gall bladder). Still, even when anthropologists, archaeologists and historians try their hand at 'big-picture' narratives, they have an odd tendency to end up with some similarly minor variation on Rousseau. Flannery and Marcus's (2012) *The Creation of Inequality: How our Prehistoric Ancestors Set the Stage for Monarchy, Slavery, and Empire*, for example, offers all sorts of interesting insights into how inequality *might* emerge in human societies, but their overall framing of human history remains explicitly wedded to Rousseau's second *Discourse*, concluding that humanity's best hope of a more egalitarian future is to 'put hunters and gatherers in charge'. Walter Scheidel's more economically informed study, *The Great Leveller: Violence and the History of Inequality from the Stone Age to the Twenty-First Century* (2017), concludes – just as dismally – that there's really nothing we can do about inequality: civilization invariably puts in charge a small elite who grab more and more of the pie, and the only thing that has ever been successful in dislodging them is catastrophe in the form of war, plague, mass conscription, wholesale suffering and death. Half-measures never work. So if you don't want to go back to living in a cave, or die in a nuclear holocaust (which presumably also ends up with the survivors in caves), you're just going to have to accept the existence of Warren Buffett and Bill Gates.
7. Rousseau 1984 [1754]: 78.
8. As articulated by Judith Shklar (1964), the renowned Harvard political theorist.
9. Rousseau 1984 [1754].: 122.
10. As a matter of fact, Rousseau, unlike Hobbes, was not a fatalist. For Hobbes, all things large and small in history were to be understood as the unfolding of forces set in motion by God, which are ultimately beyond the capacity of humans to control (see Hunter 1989). Even a tailor making a garment is entering, from his first stitch, into a flow of historical entanglements that he is powerless to resist and of which he is largely unaware; his precise actions are tiny links in a great chain of causality that is the very fabric of human history, and – in this rather extreme metaphysics of entanglement – to suggest that he might have been doing these things some alternative way is to deny the whole,

irreversible course of world history. For Rousseau, by contrast, what humans make, they could always unmake, or at least do differently. We could free ourselves from the chains that bind us; it just wasn't going to be easy (see, again, Shklar 1964 for a classic discussion of this aspect of Rousseau's thought).

11. Pinker 2012: 39, 43.
12. If a trace of impatience can be detected in our presentation, the reason is this: so many contemporary authors seem to enjoy imagining themselves as modern-day counterparts to the great social philosophers of the Enlightenment, men like Hobbes and Rousseau, playing out the same grand dialogue but with a more accurate cast of characters. That dialogue in turn is drawn from the empirical findings of social scientists, including archaeologists and anthropologists like ourselves. Yet in fact the quality of their empirical generalizations is hardly better; in some ways it's probably worse. At some point, you have to take the toys back from the children.
13. Pinker 2012; 2018.
14. Pinker 2012: 42.
15. Tilley 2015.
16. Formicola 2007.
17. Margaret Mead did this once, when she suggested that the first sign of 'civilization' in human history was not tool use but a 15,000-year-old skeleton that showed signs of having healed from a broken femur. It takes six weeks, she noted, to recover from such an injury; most animals with broken femurs simply die because their companions abandon them; one of the things that makes humans so unusual is precisely that we take care of one another in such situations.
18. Below, n.21. As others point out, Yanomami tend to sleep together six to even ten people in the same bed. This requires a degree of good-natured mutual accommodation of which few contemporary social theorists would be capable. If they were really anything like the 'fierce savages' of undergraduate caricature, there would be no Yanomami as they'd all have long since killed each other for snoring.
19. In reality, far from being pristine exemplars of our 'ancestral condition', the Yanomami in the 1960s to 1980s, when Chagnon conducted fieldwork among them, had been exposed to decades of European incursions, intensified by the discovery of gold on their lands. Over that period, Yanomami populations were decimated by epidemics of infectious diseases introduced by missionaries, prospectors, anthropologists and government agents; see Kopenawa and Albert 2013: 2–3.

20. Chagnon 1988.
21. Some were about the statistics Chagnon presented, and his claim that men who achieved a state of ritual purity (*unokai*) obtained more wives and offspring than others. A key issue here, which Chagnon never entirely cleared up, is that *unokai* status was not reserved for men who had killed; it could also be achieved, for example, by shooting an arrow into the corpse of an enemy already slain, or indeed by causing death through non-physical means, such as sorcery. Others pointed out that most *unokai* were on the older side of the age spectrum, and some held the status of village headmen: both would have ensured more offspring, with no direct relationship to warfare. Still others pointed out a logical flaw in Chagnon's suggestion that homicide acted both as deterrent to further killing (the *unokai* having earned a fierce reputation), and at the same time kept in motion a cycle of revenge killings on the part of embittered kinsmen: a kind of 'war of all against all'. Criticisms of Chagnon: Albert 1989; Ferguson 1989; and see Chagnon 1990 for a response.
22. Geertz 2001. Academics are very prone to a phenomenon called 'schismogenesis', which we will be exploring at various points in this book.
23. The framers of the US Constitution, for example, were quite explicitly anti-democratic and made clear in their own public statements that they designed the Federal Government in large part to head off the risk of 'democracy' breaking out in one of the former colonies (they were particularly worried about Pennsylvania). Meanwhile, actual direct democratic decision-making had been practised regularly in various parts of Africa or Amazonia, or for that matter in Russian or French peasant assemblies, for thousands of years; see Graeber 2007b.
24. For example, one would not have to waste one's time coming up with convoluted reasons why, say, forms of decision-making that look like democracy outside Europe are not 'really' democracy, philosophical arguments about nature that take rigorous logical form are not 'really' science, etc.
25. Chagnon (1998: 990) chose to end his famous *Science* paper with an anecdote to this very effect: 'A particularly acute insight into the power of law to thwart killing for revenge was provided to me by a young Yanomamö man in 1987. He had been taught Spanish by missionaries and sent to the territorial capital for training in practical nursing. There he discovered police and laws. He excitedly told me that he had visited the town's largest *pata* [the territorial governor] and urged him to make law and police available to his people so that they would not have to engage any longer in their wars of revenge and have to live in constant fear.'

26. Pinker 2012: 54.
27. As recounted by Valero to Ettore Biocca and published in 1965 under the latter's authorship.
28. For which, see the evidence compiled in a (1977) thesis by J. N. Heard: 'The Assimilation of Captives on the American Frontier in the Eighteenth and Nineteenth Centuries'.
29. In his (1782) *Letters from an American Farmer* J. Hector St John de Crèvecoeur noted how parents, at the end of a war, would visit Indian towns to reclaim their children: 'To their inexpressible sorrow, they found them so completely Indianized, that many knew them no longer, and those whose more advanced ages permitted them to recollect their fathers and mothers, absolutely refused to follow them, and ran to their adopted parents for protection against the effusions of love their unhappy real parents lavished upon them.' (cited in Heard 1977: 55–6, who also notes Crèvecoeur's conclusion that the Indians must possess a 'social bond singularly captivating, and far superior to anything to be boasted of among us'.)
30. Franklin 1961 [1753]: 481–3.
31. 'Alas! Alas!' wrote James Willard Schultz – an eighteen-year-old from a prominent New York family who married into the Blackfoot, remaining with them until they were driven on to a reservation – 'Why could not this simple life have continued? Why must the ... swarms of settlers have invaded that wonderful land, and robbed its lords of all that made life worth living? They knew not care, nor hunger, nor want of any kind. From my window here, I hear the roar of the great city, and see the crowds hurrying by ... "bound to the wheel" and there is no escape from it except by death. And this is civilization! I, for one, maintain that there is no ... happiness in it. The Indians of the plains ... alone knew what was perfect content and happiness, and that, we are told, is the chief end and aim of men – to be free from want, and worry, and care. Civilization will never furnish it, except to the very, very few.' (Schultz 1935: 46; see also Heard 1977: 42)
32. See Heard 1977: 44, with references.
33. For example, the Wendat ('Huron') societies of Northeastern North America in the seventeenth century – to which we turn in the next chapter – of whom Trigger (1976: 62) notes that: 'Relations of friendship and material reciprocity were extended beyond the Huron confederacy in the form of trading arrangements. In the historic period, trade was a source not only of luxury goods but of meat and skins which were vital to a population that had outstripped the resources of its

nearby hunting territory. Important as these goods were, however, foreign trade was not merely an economic activity. It was embedded in a network of social relations that were, fundamentally, *extensions of the friendly relationships* that existed within the Huron confederacy.' (Our emphasis.) For a general anthropological survey of 'archaic trade' the classic source remains Servet 1981; 1982. Most contemporary archaeologists are well aware of this literature, but tend to get caught up in debates over the difference between 'trade' and 'gift exchange', while assuming that the ultimate point of both is to enhance somebody's status, either by profit, or by prestige, or both. Most will also acknowledge that there is something inherently valuable, even cosmologically significant, in the phenomenon of travel, the experience of remote places or the acquisition of exotic materials; but in the last resort, much of this too seems to come down to questions of status or prestige, as if no other possible motivation might exist for people interacting over long distances; for some further discussion of the issues see Wengrow 2010b.

34. On 'dream economies' among the Iroquois see Graeber 2001: 145–9.
35. Following Charles Hudson's (1976: 89–91) interpretation of Cabeza de Vaca's account.
36. DeBoer 2001.

2. WICKED LIBERTY

1. In his (2009) *Europe Through Arab Eyes, 1578–1727*, Nabil Matar considers the relative lack of interest in Frankish Europe among medieval Muslim writers, and possible reasons for it (especially, pp. 6–18).
2. Many examples of this tendency are discussed in David Allen Harvey's (2012) *The French Enlightenment and its Others*.
3. A notorious example was that of Christian Wolff, the most famous German philosopher in the period between Leibniz and Kant – he too was a Sinophile and lectured on the superiority of Chinese modes of government, with the ultimate effect that an envious colleague denounced him to the authorities, a warrant was issued for his arrest and he was forced to flee for his life.
4. Some classic statements, especially concerning North America, are to be found in: Chinard 1913; Healy 1958; Berkhofer 1978a, 1978b; Dickason 1984; McGregor 1988; Cro 1990; Pagden 1993; Sayre 1997; Franks 2002.
5. For example, Grinde 1977; Johansen 1982, 1998; Sioui 1992; Levy 1996; Tooker 1988; 1990; and cf. Graeber 2007b. The literature, however,

focuses around the impact of Native ideas on American colonists, and has become bogged down in an argument about the specific 'influence' of the Haudenosaunee political confederation on the American Constitution. The original argument was actually much broader, suggesting that European settlers in the Americas only came to think of themselves as 'Americans' (rather than English or French or Dutch) when they began to adopt certain elements of Native American standards and sensibilities, from the indulgent treatment of children to ideals of republican self-governance.

6. Alfani and Frigeni 2016.
7. The best English-language source on these debates is Pagden 1986.
8. One of Rousseau's rivals in the essay contest, the Marquis d'Argenson, who also failed to win a prize, made precisely this argument: monarchy allowed the truest equality, he argued, and absolutist monarchy most of all, since everyone is equal before the absolute power of the king.
9. Lovejoy and Boas (1935) compile and provide commentary on all the relevant texts.
10. As Barbara Alice Mann suggests to us (in personal communication), bourgeois women may have especially appreciated the *Jesuit Relations* because it allowed them to read about discussions of women's sexual freedom in a form that was entirely acceptable to the Church.
11. David Allen Harvey (2012: 75–6), for instance, places Lahontan's *Dialogues* (to which we shortly turn) in a literary class with works by Diderot and Rousseau, writers who had little if any direct experience of Native American peoples but invoked them as a 'discursively constructed Other with which to interrogate European customs and civilization'. See also Pagden 1983; 1993.
12. It rarely seems to occur to anyone that (1) there are only so many logical arguments one *can* make, and intelligent people in similar circumstances will come up with similar rhetorical approaches, and (2) it is likely that European writers trained in the classics would be especially impressed by arguments that reminded them of ones they already knew from Greek or Roman rhetoric. Obviously, such accounts do not provide a direct window on to the original conversations, but to insist that they bear no relation at all seems equally absurd.
13. Technically, the Huron were a confederation of Iroquoian speakers that existed at the time the French arrived, but later scattered under attacks from the Haudenosaunee to the south and then reformed as the Wyandot or Wendat, along with refugees from the Petun and Neutral confederations. Their contemporary descendants prefer Wendat (pronounced

'Wen-dot'), noting that 'Huron' was originally an insult, meaning (depending on the source) either 'pig-haired' or 'malodorous'. Sources at the time regularly use 'Huron', and while we have followed Barbara Mann's usage in changing it to 'Wendat' when quoting from indigenous speakers like Kandiaronk, we have maintained it in European sources.

14. Biard 1611: 173 –4, cited in Ellingson 2001: 51.
15. The Recollects were a branch of the Franciscan Order, who took vows of poverty and were among the first missionaries dispatched to New France.
16. Sagard 1939 [1632]: 192.
17. Ibid.: 88–9.
18. Wallace 1958; cf. also Graeber 2001, Chapter Five.
19. *The Jesuit Relations and Allied Documents: Travels and Explorations of the Jesuit Missionaries in New France 1610–1791*, ed. Reuben Gold Thwaites, and henceforth: *JR* 6: 109–10/241. The phrase 'captain' is used indiscriminately in the French sources for any male in a position of authority, whether that person be a simple headman of a band or village, or the holder of an official rank in the Wendat or Haudenosaunee Confederation.
20. *JR* 28: 47.
21. *JR* 28: 48–9, cf. *JR* 10: 211–21.
22. *JR* 28: 49–50. Here's a different Jesuit father, returning to the donkey theme again: 'There is nothing so difficult as to control the tribes of America. All these barbarians have the law of wild asses, – they are born, live, and die in a liberty without restraint; they do not know what is meant by bridle or bit. With them, to conquer one's passions is considered a great joke, while to give free rein to the senses is a lofty philosophy. The Law of our Lord is far removed from this dissoluteness; it gives us boundaries and prescribes limits, outside of which we cannot step without offending God and reason.' (*JR* 12:191–2).
23. *JR* 5: 175.
24. *JR* 33: 49.
25. *JR* 28: 61–2.
26. *JR* 15:155, also in Franks 2002: 4; cf. Blackburn 2000: 68.
27. They were also unevenly accepted. Most Jesuits still subscribed to the old Renaissance doctrine that 'savages' had once been of a higher level of grace and of civilization, and had degenerated (Blackburn 2000: 69).
28. A comprehensive review of the literature by Ellingson (2001) finds the view that European observers regularly romanticized those they considered savages to be entirely unfounded; even the most positive accounts tended to be fairly nuanced, recognizing both virtues and vices.

29. So according to some sources at the time, and Wendat oral traditions (Steckley 1981).
30. The official histories claim that he converted at the very end of his life, and it's true he was buried as a Christian in Notre-Dame Church in Montreal, but Mann argues convincingly that the story of the deathbed conversion and burial is likely to have been a mere political ploy on the part of the missionaries (Mann 2001: 53).
31. Chinard 1931; Allan 1966; Richter 1972; Betts 1984: 129–36; Ouellet 1990, 1995; White 1991; Basile 1997; Sayre 1997; Muthu 2003: 25–9; Pinette 2006; but for a significant exception see Hall 2003: 160 ff.
32. Sioui 1972, 1992, 1999; Steckley 1981, 2014: 56–62; Mann 2001.
33. Mann 2001: 55.
34. Ibid.: 57–61.
35. 1704: 106–7. Cited references are to the 1735 English edition of *Dialogues*, but the translation in this instance is a combination of that, Mann's (2001: 67–8), and our own. Subsequent translations are our own, based on the 1735 edition.
36. 'Assuming he is so powerful and great, how likely is it that such an unknowable being would have made himself into a man, dwelt in misery, and died in infamy, just to work off the sin of some ignoble creature who was as far beneath him as a fly is beneath the sun and the stars? Where does that leave his infinite power? What good would it do him, and what use would he make of it? For my part, it seems to me that to believe in a debasement of this nature is to doubt the unimaginable sweep of his omnipotence, while making extravagant presumptions about ourselves.' (cited in Mann 2001: 66)
37. Bateson 1935; 1936.
38. Sahlins 1999: 402, 414.
39. Allan 1966: 95.
40. Ouellet 1995: 328. After a hiatus, another spate of similar plays with Indian heroes were produced in the 1760s: *La Jeune Indienne* (1764) by Chamfort and *Le Huron* (1768) by Marmontel.
41. See Harvey (2012) for a good recent summary of the impact of foreign perspectives, real and imagined, on social thinking in the French Enlightenment.
42. The expression is Pagden's (1983).
43. So, Etienne 1876; cf. Kavanagh 1994. In 1752, just around the time de Graffigny's second edition appeared, a former soldier, spy and theatre director named Jean Henri Maubert de Gouvest also released a novel called *Lettres Iroquois*, the correspondence of an imaginary Iroquois traveller named Igli, which was also hugely successful.

44. 'Without gold, it is impossible to acquire a part of this earth which nature has given in common to all men. Without possessing what they call property, it is impossible to have gold, and by an inconsistency which is an outrage to natural common sense, and which exasperates one's reason, this haughty nation, following an empty code of honour entirely of its own invention, considers it a disgrace to receive from anyone other than the sovereign whatever is necessary to sustain one's life and position.' (de Graffigny 2009 [1747]: 58).
45. Meek 1976: 70–71. Turgot was writing on the eve of the Industrial Revolution. Later evolutionists would simply replace 'industrial' with 'commercial'. No pastoral society actually existed in the New World, but somehow early evolutionists never seemed to consider this a problem.
46. It is to be noted that the question is framed in traditional terms: the arts and sciences are assumed not to progress, but rather still to be in the process of being restored to their former (presumably ancient) glory. It was only over the course of the next decade that notions of progress became widely accepted.
47. This is the third footnote of the *Discourse on the Arts & Sciences*, sometimes referred to as 'The First Discourse'. Montaigne's essay 'On Cannibals', written in 1580, appears to be the first to consider indigenous American perspectives on European societies, with Tupinamba visitors questioning the arbitrariness of royal authority and wondering why the homeless did not burn down the mansions of the rich. The fact that so many societies appeared to maintain peace and social order without coercive institutions or even, it seemed, formal institutions of government of any kind caught the attention of European observers from very early on. Leibniz, for instance, who, as we've seen, had long been promoting Chinese models of bureaucracy as the embodiment of rational statecraft, felt this was what was really significant in Lahontan's testimony: the possibility that statecraft might not be required at all (Ouellet 1995: 323).
48. Rousseau 1984 [1754]: 109.
49. Rousseau described himself as an avid reader of travelogues and does cite Lebeau, who is basically summarizing Lahontan, as well as *l'Arlequin sauvage* (Allan 1966: 97–8; Muthu 2003: 12–13, 25–8; Pagden 1983: 33). It's extremely unlikely that Rousseau had not read Lahontan in the original, though even if he hadn't it would just mean that he had come by the same arguments second-hand.
50. Other examples: 'The cultivation of the earth necessarily brought about its distribution; and property, once recognised, gave rise to the first rules

of justice; for, to secure each man his own, it had to be possible for each to have something. Besides, as men began to look forward to the future, and all had something to lose, every one had reason to apprehend that reprisals would follow any injury he might do to another.' Compare that passage to Kandiaronk's argument, cited above, that the Wendat intentionally avoided divisions of wealth because they had no desire to create a coercive legal system. Montesquieu made the same point in discussing the Osage, noting that 'the division of lands is what principally increases the civil code. Among nations where they have not made this division there are very few civil laws' – an observation which seems to have been derived partly from Montesquieu's conversation with members of an Osage delegation that visited Paris in 1725 (Burns 2004: 362).

51. See Graeber 2011: 203–7.
52. Rousseau himself had fled home at an early age, writing to his Swiss watchmaker father that he aspired to live 'without the help of others'.
53. Barruel 1799: 104. The quote is from an anti-Illuminati tract, claiming to be the 'Code of the Illuminati', and this whole discourse is so shrouded in rumour and accusation that we can't even be entirely sure our sources didn't just make it up; but in a way it hardly matters, since the main point is that the right wing saw Rousseauian ideas as inspiring leftist revolutionary activity.
54. It is not entirely clear whether 'Illuminism' as it came to be called was a revolutionary doctrine at all, since Weishaupt himself later denied it – after the society was banned and he was himself driven from Bavaria – and characterized it as purely reformist; but his enemies of course insisted these protests were disingenuous.
55. The key difference is that Rousseau sees progress as undermining an essentially benevolent human nature, while classic conservative thought tends to see it as having undermined traditional mores and forms of authority which had previously been able to contain the less benevolent aspects of human nature.
56. Certainly, there *is* a tendency, in all this literature, when introduced to unfamiliar societies, to treat them alternately as entirely good or entirely evil. Columbus was already doing this in the 1490s. All we're saying here is that this does not mean that nothing they ever said had any bearing on the actual perspectives of those they encountered.
57. Chinard 1913: 186, translation following Ellingson 2001: 383. A similar passage: 'Rebel against all constraints, all laws, all hierarchies, the baron Lahontan and his American savage are anarchists properly speaking. *The Dialogues with a Savage* are neither a political treatise nor a learned

dissertation, they are the clarion call of a revolutionary journalist; Lahontan opens the way not just for Jean-Jacques Rousseau, but for Father Duchesne and the modern socialist revolutionaries, and all that just ten years before the death of Louis XIV.' (1913: 185, translation ours).

58. Ellingson 2001: 383.
59. The construction 'our own' of course presumes that Native Americans don't read books, or those that do don't matter.
60. Chinard 1913: 214.
61. 'His imagination paints no pictures; his heart makes no demands on him. His few wants are so readily supplied, and he is so far from having the knowledge which is needful to make him want more, that he can have neither foresight nor curiosity . . . His soul, which nothing disturbs, is wholly wrapped up in the feeling of its present existence, without any idea of the future, however near at hand; while his projects, as limited as his views, hardly extend to the close of day. Such, even at present, is the extent of the native Caribbean's foresight: he will improvidently sell you his cotton-bed in the morning, and come crying in the evening to buy it again, not having foreseen he would want it again the next night.' (Rousseau 1984 [1754]: 90).
62. 'Fraternity' might seem the odd man out here, at least insofar as Native American influences go – though a case can be made that it echoes the responsibility for mutual aid and support which American observers so often remarked on. Montesquieu in *The Spirit of the Laws* makes a great point of the sense of fraternal commitment among the Osage, and his book was a powerful influence on the political theorists of both the American and French Revolutions; as we'll see in Chapter Eleven, Montesquieu himself appears to have met with an Osage delegation visiting Paris and his observations may be based on direct communication with them (Burns 2004: 38, 362).
63. In the sense that women controlled land and its produce and also most other productive resources, but men controlled most of the important political offices.

3. UNFREEZING THE ICE AGE

1. The authoritative account, well into the nineteenth century, was that of James Ussher, Archbishop of Armagh, first published in 1650, though it is important to note that none other than Sir Isaac Newton proposed an alternative calculation, suggesting the actual date was 3988 BC.

2. The phrase we owe to Thomas Trautmann's (1992) account of this 'time revolution'. While the field of anthropology came into existence during the 'decade of Darwin' (i.e. between the publication of *The Origin of Species* in 1859 and *The Descent of Man* in 1871), it was not actually Darwinism but archaeological excavations that established the timescale of human prehistory as we know it. Geology paved the way, replacing the biblically inspired view of earth's genesis as a series of rapid titanic upheavals with a more mechanistic and gradual account of our planet's origins. More detailed studies of the early development of scientific prehistory, and how fossil evidence and stone tools were first fitted into this expanded chronology of life on earth, can be found in Schnapp 1993 and Trigger 2006.
3. The key findings are summarized in Scerri et al. 2018. For an accessible account see also Scerri's feature article in *New Scientist*, published online (25 April 2018) as 'Origin of our species: why humans were once so much more diverse'.
4. The Sahara seems to have acted as a kind of turnstile for human evolution, periodically turning green and then dry again with the cyclical advance/retreat of monsoon rains, opening and shutting the gates of interaction between northern and southern parts of the African continent (see Scerri 2017).
5. Geneticists assume, reasonably enough, that a fair amount of genetic admixture did take place.
6. Green et al. 2010; Reich et al. 2010. Fossil evidence tells us that the first expansions of modern humans out of Africa began as far back as 210,000 years ago (Harvati et al. 2019), but these were often tentative and quite short-lived, at least until the more decisive radiations of our species began around 60,000 BC.
7. Recent and historical hunter-gatherers, as we shall see, present an enormous range of possibilities, from assertively egalitarian groups like the Ju/ 'hoansi of the Kalahari, the Mbendjele BaYaka of Congo or the Agta in the Philippines to assertively hierarchical ones like the populations of the Canadian Northwest Coast, the Calusa of Florida Keys or the forest-dwelling Guaicurú of Paraguay (these latter groups, far from being egalitarian, are known to have traditionally kept slaves and lived in ranked societies). Holding up any particular subset of recent foragers as representatives of 'early human society' is essentially a matter of picking cherries.
8. Hrdy 2009.
9. Will, Conard and Tryon 2019, with further references.
10. For important reviews and critiques of the 'Human Revolution' idea see McBrearty and Brooks 2000; Mellars et al. 2007.

11. The term 'Venus figurine' is still widely used, but has links to scientific racism in the nineteenth and early twentieth centuries, when direct comparisons were drawn between prehistoric images and the anatomy of modern individuals considered living specimens of humanity in its 'primitive' forms. A tragic example is the life story of Sara Baartman, a Khoikhoi woman who was exhibited around Europe as a 'freak' owing to her large buttocks under the stage name 'Hottentot Venus'. See Cook 2015.
12. Renfrew 2007.
13. The case against European exceptionalism was laid out by Sally McBrearty and Alison Brooks in a key (2000) publication; and has since been supplemented by discoveries in South Asia (James and Petraglia 2005) and Africa (Deino et al. 2018).
14. Shipton et al. 2018.
15. Aubert et al. 2018.
16. Conceivably this included the making of cave art; Hoffmann et al. 2018.
17. Recent efforts to estimate the overall human population at the start of the Upper Palaeolithic (known as the Aurignacian period) suggest a mean figure of just 1,500 people for the whole of western and central Europe, which is remarkably low; Schmidt and Zimmermann 2019.
18. For the relationship between demographic density and enhanced cultural transmission in Upper Palaeolithic Europe see the (2009) arguments of Powell, Shennan and Thomas.
19. This is obviously only a last resort and usually extreme measures are employed to ensure it's really called for: in rural Madagascar, for instance, when police were effectively absent, the usual rule was that one could only lynch such a person if his parents gave permission first – which was usually effective as a way to simply drive the person out of town. (D. Graeber, field observation.)
20. Boehm 1999: 3–4.
21. Initially, but as it turns out wrongly thought to be a boy and girl; for new genetic evidence on this point see Sikora et al. 2017.
22. Again, modern genetic studies of the group burial at Dolní Věstonice have confirmed the male identity of all three burials, which was previously in doubt; Mittnik 2016.
23. Evidence from these various sites is usefully summarized and evaluated in Pettitt 2011, with further references; and see also Wengrow and Graeber 2015.
24. See e.g. White 1999; Vanhaeren and D'Errico 2005. Inheritance is hardly the only possible explanation for the association of wealth with children: in many societies where wealth circulates freely (for instance,

where it's socially impossible to refuse a request to hand over one's necklace or bracelet to an admirer), a lot of ornament ends up festooning children to keep it out of circulation. If elaborate ornaments were buried in part to take them out of circulation, so as not to create invidious distinctions, burying them with children might be the ideal way to accomplish this.

25. Schmidt 2006; and for a convenient digest see also https://www.dainst.blog/the-tepe-telegrams/
26. As ventured by Haklay and Gopher 2020, based on geometrical regularities and correspondences found among the ground plans of some three large enclosures; but doubts remain, as their study does not take into account the complex and dynamic construction sequences that lie behind the enclosures, and compares building phases that are not strictly contemporaneous.
27. Acemoğlu and Robinson 2009: 679; and see also Dietrich et al. 2019; Flannery and Marcus 2012: 128–31.
28. For the monumental character of mammoth structures in their Ice Age settings see Soffer 1985; Iakovleva 2015: 325, 333. As we note below, current research by Mikhail Sablin, Natasha Reynolds and colleagues shows that the term 'mammoth houses' or 'dwellings' may well be misleading in some cases; in fact, the precise functions of these impressive structures may have varied considerably across regions and periods (see also Pryor et al. 2020). For the massive wooden enclosures as evidence of large, seasonal gatherings see Zheltova 2015.
29. Sablin, Reynolds, Iltsevich and Germonpré (manuscript in preparation; made available to us by courtesy of Natasha Reynolds).
30. Ibid.
31. In fact, even small children are typically far more imaginative than this, and as we all know spend a considerable part of their time constructing alternative roles and symbolic worlds to inhabit. Robert L. Kelly, in his magisterial survey of the 'foraging spectrum', offered a clear statement of the problem concerning the stereotyping of forager populations, urging a study of 'hunter-gatherer prehistory in terms other than broad typological contrasts such as generalized versus specialized, simple versus complex, storing versus non-storing, or immediate versus delayed return' (2013: 274). Still, we note that in the main part of his study Kelly himself maintains just such a broad dichotomy between 'egalitarian' and 'non-egalitarian' hunter-gatherers as distinct types of society with supposedly fixed internal characteristics (tabulated as a binary contrast between 'simple versus complex' forms; Kelly 2013: 242, table 9-1).

32. The British historian Keith Thomas, for instance, compiled a whole list of casual rejections of Christianity from medieval and Renaissance English sources. 'The Bishop of Exeter complained in 1600 that in his diocese it was "a matter very common to dispute whether there be a God or not" . . . In Essex a husband-man of Bradwell-near-the-Sea was said to "hold his opinion that all things cometh by nature, and does affirm this as an atheist" . . . At Wing, Rutland, in 1633 Richard Sharpe was accused of saying "there is no God and that he hath no soul to save". From Durham in 1635 came the case of Brian Walker who, when asked if he did not fear God, retorted that, "I do not believe there is either God or Devil; neither will I believe anything but what I see": as an alternative to the Bible he commended "the book called Chaucer"' (1978: 202). The difference of course is that while expressing such opinions among the Winnebago might make you a figure of fun, under the government of Queen Elizabeth or King James it could get you into serious trouble – as evidenced by the fact that we know most of these people because of trial documents.
33. Beidelman 1971: 391–2. The account assumes that prophets are male but there are documented cases of female prophets as well. Douglas Johnson (1997) provides the definitive history of Nuer prophets in the early twentieth century.
34. Lévi-Strauss 1967 [1944]: 61.
35. Lee and Devore 1968: 11. It's worth noting, perhaps, that Lévi-Strauss offered a forlorn epilogue to *Man the Hunter*, which is not read any more.
36. Formicola (2007) surveys the evidence; and see also Trinkaus 2018; Trinkaus and Buzhilova 2018.
37. This is the general pattern (Pettitt 2011). Of course, it is not completely universal – the Romito dwarf, for example, does not seem to have been buried with grave goods.
38. Archaeologists have observed close spatial associations between large Upper Palaeolithic (Magdalenian) aggregation sites in the French Périgord and natural choke points or 'bottlenecks' along the Dordogne and Vézère such as fords or meanders: ideal locations for intercepting herds of reindeer on their seasonal migrations (White 1985). In northern Spain, the famous cave sites of Altamira and Castillo have long been identified as aggregation locales based on their topographical location and the preponderance of seasonal resources like deer, ibex and shellfish among the animal remains found there (Straus 1977). On the periglacial 'mammoth steppe' of Central Russia, spectacularly large settlements like

Mezhirich and Mezin – with their mammoth-bone dwellings, fixed storage pits and abundant evidence of art and trade – were aligned on major river systems (Dnepr and Desna), which also channelled the annual north–south movements of steppe bison, horse, reindeer and mammoth (Soffer 1985). Similarly, the Pavlov Hills of southern Moravia, where Dolní Věstonice is located, once formed part of a narrow belt of forest-steppe, bridging the non-glaciated zones of eastern and western Europe (see contributions by Jiří Svoboda, in Roebroeks et al. 2000). Year-round habitation was certainly possible in some of these locations, but population densities are still likely to have fluctuated markedly between seasons. Recently, archaeologists have begun using more fine-grained analytical techniques – like the microscopic study of growth patterns in animal teeth and antlers, as well as measuring geochemical proxies of seasonal variation such as stable isotope ratios in animal remains – to determine the migration patterns and diets of hunted game (for a useful survey see Prendergast et al. 2018).

39. Lang et al. 2013; L. Dietrich et al. 2019 (grinding slabs, stone bowls, hand pounders, pestles and mortars are all found in impressive numbers at Göbekli Tepe); and see also O. Dietrich et al. 2012.
40. Parker Pearson (2012) provides a detailed survey and interpretation of the archaeology of Stonehenge, including the results of recent fieldwork. The argument for Neolithic aristocracy is based on close analysis and dating of human remains associated with different phases in the construction of Stonehenge, which prove consistent with the idea that the first stone circle was linked to a high-status cemetery, where the cremated remains of a nuclear family were placed around the start of the third millennium BC. Subsequent removals and rebuildings, including the incorporation of massive sarsen stones, were apparently linked to ongoing mortuary rituals, as the same family's lineage presumably expanded in size and status over a period of centuries.
41. For the rejection of cereal-farming in prehistoric Britain during periods of megalithic construction see Stevens and Fuller 2012; for the seasonality of midwinter meat-feasting at Durrington Walls, as detected from tooth remains, see Wright et al. 2014.
42. Viner et al. 2010; Madgwick et al. 2019.
43. Of course, humans are not alone in this. Non-human primates, like chimpanzees and bonobos, also vary the size and structure of their groups on a seasonal basis according to the changing distribution of edible resources in what primatologists call 'fission-fusion' systems (Dunbar 1988). So too, in fact, do all sorts of other gregarious animals. But what Mauss was

talking about and what we're considering here is categorically different from this. Uniquely, for humans such alternations also involve corresponding changes in moral, legal and ritual organization. Not just strategic alliances, but entire systems of roles and institutions are liable to be periodically disassembled and reconstructed, allowing for more or less concentrated ways of living at different times of year.

44. Mauss and Beuchat 1979 [1904–5]. It's worth noting that politics wasn't the aspect of seasonal variations they themselves chose to emphasize, being more concerned with the contrast between secular and ceremonial arrangements and the effects this had on the self-consciousness of the group. E.g. 'Winter is a season when Eskimo society is highly concentrated and in a state of continual excitement and hyperactivity. Because individuals are brought into close contact with one another, their social interactions become more frequent, more continuous and coherent; ideas are exchanged; feelings are mutually revived and reinforced. By its existence and constant activity, the group becomes more aware of itself and assumes a more prominent place in the consciousness of individuals.' (p. 76)
45. It's surely no coincidence that so much of Kwakiutl art plays visually on the relation of 'name,' 'person' and 'role' – relations laid open to scrutiny by their seasonal practices (Lévi-Strauss 1982).
46. Lowie 1948: 18.
47. 'One does not find in these Plains military societies the germs of law and of the state. One finds that the germs have germinated and grown up. They are comparable, not antecedent, to our modern state and what would appear to be the important problem for study is not the investigation of how one grew out of the other, but what they have in common which might throw light on the nature of law and of the state.' (Provinse 1937: 365)
48. Much of the rest of Lowie's essay focuses on the role of chiefs, arguing that the power of political leaders over the 'anarchic' societies of the Americas was so carefully circumscribed as to exclude the emergence of permanent structures of coercion. Insofar as indigenous states developed there, he concluded, it could only have been through the power of prophecy: the promise of a better world, with religious figures claiming authority directly from the gods. A generation later, Pierre Clastres made almost exactly the same argument in his 1974 essay, *Society Against the State*. He follows Lowie so closely that he can only have been directly inspired. While Lowie is now largely forgotten, Clastres is remembered for arguing that stateless societies do not represent an evolutionary stage, innocent of higher organization, but are based on the self-conscious and principled

rejection of coercive authority. Interestingly, the one element not carried over from Lowie to Clastres is that of seasonal variations in modes of authority; and this is despite the fact that Clastres himself focused largely on Amazonian societies, which did in fact have very different structures at different times of year (see Maybury-Lewis ed. 1979). A common and logical objection to Clastres's argument, which remains hugely influential, is to ask how Amazonian societies could have consciously organized themselves against the emergence of forms of authority they'd never actually experienced. It seems to us that bringing seasonal variations back into the debate goes a significant way to resolving this dilemma.

49. Seasonal kings or lords like 'John Barleycorn' – a variant of the sacred ruler, destined to end his tenure and be killed each year at harvest time – are stock figures of British folklore to this day, but there is little agreement on how much further back they go beyond their earliest written mentions in the sixteenth century AD. The ubiquity of such 'temporary kings' in European, African, Indian, and Greco-Roman myth and legend was the subject of Book III in James Frazer's *The Golden Bough*, which he called 'The Dying God'.
50. Perhaps one reason why the published paper (Lowie 1948) has been forgotten is because of the distinctly uninspiring title: 'Some aspects of political organisation among the American Aborigines'.
51. Knight 1991.
52. Discussed further by D. Graeber in 'Notes on the politics of divine kingship: Or, elements for an archaeology of sovereignty', in Graeber and Sahlins 2017, Chapter Seven.
53. On the 'carnivalesque' the classic text is Bakhtin's *Rabelais and His World* (1940).
54. This is hardly the place to go into detail about the history of these debates, but it's interesting to observe that they emerge directly from Mauss's research on seasonality, which he carried out in co-ordination with his uncle, Émile Durkheim, who is considered the founder of French sociology in the same sense that Mauss is of French anthropology. In 1912, in *The Elementary Forms of Religious Life*, Durkheim relied upon Mauss's research on indigenous Australian societies to contrast what they described as the ordinary economic existence of Australian bands – concerned mostly with getting food – with the 'effervescence' of their seasonal gatherings, called *corroboree*. It was in the excitement of *corroboree*, he argued, that the power to create society appeared to them, as if it were an alien force projected into totemic spirits and their emblems. This was the first formulation of the basic problematic that almost all theorists have been forced to

grapple with ever since: that rituals are simultaneously moments where social structure is manifested and moments of 'anti-structure' in which new social forms can pop up. British social anthropology, which took its initial theoretical inspiration primarily from Durkheim, worked through the problem in various ways (notably the work of Edmund Leach, Victor Turner or Mary Douglas). The most sophisticated, and to our minds compelling, proposals for how to resolve the dilemma are currently Maurice Bloch's (2008) notion of the 'transcendental' versus 'transactional' realms; and Seligman et al.'s (2008) argument that ritual creates a 'subjunctive' or 'as if' domain of order, consciously set apart from a reality that is always seen in a contrasting light, as fragmented and chaotic. Ritual creates a world which is marked off as standing apart from ordinary life, but is also where essentially imaginary, ongoing institutions (like clans, empires, etc.) exist and are maintained.

55. As Peter Burke (2009: 283–5) notes, the idea that rituals of rebellion were simply 'safety valves' or ways of allowing common folk to 'let off steam' is first documented only two years after the invention of the steam engine – the favoured metaphor before that had been to let off the pressure in a wine cask. At the same time, though, medieval authorities were keenly aware of the fact that most peasant revolts or urban insurrections would begin precisely during such ritual moments. This ambivalence appears again and again. Rousseau already considered the popular festival to embody the spirit of revolution. Such ideas were later developed in Roger Caillois's seminal essay on 'the festival', written for George Bataille's Collège de Sociologie (transl. 2001 [1939]). It went through two drafts, the first holding forth the festival as a model of revolutionary social liberation, the second holding it forth as a harbinger of fascism.

4. FREE PEOPLE, THE ORIGIN OF CULTURES, AND THE ADVENT OF PRIVATE PROPERTY

1. Or at least broadly similar in form and function: specialists in prehistoric stone tool analysis, of course, spend a great deal of time differentiating specific 'industries' on the basis of fine-grained analysis, but even those who find themselves more 'splitters' than 'lumpers' would not deny the broad similarities of Upper Palaeolithic traditions – the Aurignacian, Gravettian, Solutrean, Magdalenian, Hamburgian and so on – over very impressive geographical spans. For some recent discussion of the issues see Reynolds and Riede 2019.
2. Schmidt and Zimmerman 2019.

3. Bird et al. 2019; see also Hill et al. 2011.
4. This was one reason for the North Americans' famous development of sign language. It is interesting that in either case, one is dealing with systems of totemic clans: raising the question of whether such systems are themselves typically forms of long-distance organization (cf. Tooker 1971). If nothing else, the common stereotype that 'primitive' peoples saw anyone outside their particular local group only as enemies appears to be entirely groundless.
5. Jordan et al. 2016; also Clarke 1978; Sherratt 2004.
6. We'll see examples in the next chapter.
7. One can agree, for instance, with the arguments of James C. Scott (2017), that an affinity exists between grain economies and the interests of predatory elites imposing their authority through taxation, raiding and tribute (grain being an eminently visible, quantifiable, appropriable and storable resource). Nowhere, however, does Scott make the naive claim that taking up cereal-farming will in every case produce a state: he simply points out that, for these very pragmatic reasons, a majority of successful states and empires have chosen to promote – and often enforce – the production of a small number of grain crops among their subject populations, while similarly discouraging the pursuit of more chaotic, fluid and thus unmanageable forms of subsistence, such as nomadic pastoralism, garden cultivation or seasonal hunting and gathering. We will return to these issues in later chapters.
8. For basic texts: Woodburn 1982, 1988, 2005.
9. Leacock 1978; for a more extended argument, Gardner 1991.
10. *JR* 33: 49. When Lallemant says the Wendat had never known what it means to forbid something he presumably means by human law: they were, no doubt, familiar with ritual prohibitions of one sort or another.
11. By this we mean that their power was largely theatrical – though of course they also played a critical advisory role.
12. The way we are using the term here somewhat echoes Amartya Sen (2001) and Martha Nussbaum's (2011) 'Capability Approach' to social welfare, which also speaks of 'substantive freedoms' as the ability to take part in economic or political activity, live to old age etc.; but we actually arrived at the term independently.
13. Gough 1971; see also Sharon Hutchinson 1996 for the full implications for women's autonomy, taking matters down to post-colonial times.
14. Evans-Pritchard 1940: 182.
15. It is intriguing to note, in this regard, that all human languages have an imperative form; there are no people, even in radically anti-authoritarian

societies like the Hadza, who are entirely unfamiliar with the idea of a command. Yet at the same time, many societies clearly arrange things in such a way that no one can give another person orders systematically.

16. It must be recalled in this context that Turgot was writing in the mid eighteenth century, so most of the criteria we use nowadays to justify the superiority of 'Western civilization' (a concept that did not exist at the time) clearly would not apply: European standards of hygiene and public health, for example, were appalling, much worse than prevailed among 'primitive' peoples of the time; Europe had no democratic institutions to appeal to, its legal systems were barbaric by world standards (e.g. Europeans were still imprisoning heretics and burning witches, something which happened almost nowhere else); standards of living and even de facto wage levels were lower than in India, or China, or under the Ottoman Empire or Safavid Persia until perhaps the 1830s.
17. The proposal that medieval European peasants worked fewer hours overall than contemporary American office workers was first, famously, made by the American sociologist Juliet Schor in *The Overworked American* (1991). It has been contested but appears to have stood the test of time.
18. It was based, in fact, on his own brief contribution to the *Man the Hunter* symposium two years before. The original essay has been reprinted in various editions of Sahlins's collected essays under the overall title *Stone Age Economics* (most recently, Sahlins 2017).
19. The key studies relied on by Sahlins are gathered in Lee and Devore 1968. Earlier ethnographic work, by contrast, was hardly ever supported by statistical data.
20. Braidwood 1957: 22.
21. The concept of a Neolithic Revolution, now more often called the Agricultural Revolution, was introduced by the Australian prehistorian V. Gordon Childe in the 1930s, who identified the origins of farming as the first of three major revolutions in human civilization, the second being the Urban Revolution and the third the Industrial. See Childe 1936.
22. As we've seen in Chapter One, people do in fact still make this kind of claim quite routinely, but they do so in flagrant disregard of the evidence presented by Sahlins, Lee, Devore, Turnbull and many others, almost as if none of this research had ever been published.
23. This reading of Augustine is actually derived from Sahlins's own later work (1996, 2008). At the time, of course, all this could only be informed speculation. Now new discoveries about the evolving relationship

between people and crops are forcing us to revisit his thesis, as we'll see in Chapters Six and Seven.

24. Sahlins 2017 [1968]: 36–7.
25. Codere 1950:19.
26. Oddly enough, Poverty Point is actually situated almost exactly midway between the Bayou Macon Wildlife Management Area and the Black Bear Golf Club.
27. We quote here from Kidder's (2018) summary article. For a more extended, if somewhat idiosyncratic, account of Poverty Point archaeology see Gibson 2000; and for a wide-ranging assessment, Sassaman 2005.
28. As Lowie (1928) demonstrated, in more recent Amerindian societies it was usually ownership of these 'incorporeal' goods (which he compared to our patents and copyrights) that unlocked rights of usufruct over land and resources, rather than direct ownership of territory.
29. Clark 2004.
30. Gibson and Carr 2004: 7, here citing Sahlins's 'original affluent society' on the matter of 'simple, ordinary foragers'.
31. For which see also Sassaman 2005: 341–5; 2010: 56 ff.; Sassaman and Heckenberger 2004.
32. A special issue of the Society for American Archaeology's magazine contains useful discussion of 'Archaic' shell-mound cultures in various parts of North America; see Sassaman (ed.) 2008. For evidence of prehistoric coastal fortifications, trade and warfare in British Columbia see Angelbeck and Grier 2012; Ritchie et al. 2016.
33. Sannai Maruyama, the largest and most impressive Jōmon site, was occupied between 3900 and 2300 BC, and lies in the Aomori Prefecture in northern Japan. Habu and Fawcett (2008) provide a lively account of the site's discovery, reception and contemporary interpretation. For broader discussions of Jōmon material culture, settlement patterns and uses of the environment see Takahashi and Hosoya 2003; Habu 2004; Kobayashi 2004; Matsui and Kanehara 2006; Crema 2013. It's worth noting that the ancient Jōmon have been infiltrating modern consciousness in other ways too: the distinctive 'rope pattern' aesthetics of their highly crafted ceramics provided the graphic template for one of Nintendo's most popular video games, *The Legend of Zelda: Breath of the Wild*. Jōmon seems quite at home in the digital age.
34. In Europe, the term 'Mesolithic' refers to the history of fisher-hunter-gatherers after the Ice Age, including their first encounters with farming populations, which we'll discuss in Chapter Seven. Some consider the

Finnish 'Giant's Churches' to have had a defensive function (Sipilä and Lahelma 2006), while others note their astronomical alignments and possible role in signifying the division of the year into four seasons, as per the much later medieval Nordic calendar. For the dating and analysis of the so-called Shigir Idol see Zhilin et al. 2018. And for Mesolithic burial traditions in Karelia and Europe's Atlantic seaboard see Jacobs 1995; Schulting 1996.

35. Sassaman (ed.) 2008.
36. English edition of Lahontan (1735), p. 113.
37. Tully 1994. Locke's position was repudiated by Chief Justice Marshall in 1823, in the case *Johnson and Grahame's Lesee* v. *McIntosh*. But in some countries the related principle of *terra nullius* ('land belonging to no one') was revoked only much more recently, in Australia as recently as the 1992 'Mabo Decision', which ruled that Aboriginal and Torres Strait islanders did after all have their own distinct forms of land tenure before British colonization.
38. This is the argument of *Dark Emu* by Bruce Pascoe (2014); whether or not one accepts this technical definition of farming, the strength of the evidence he presents is overwhelming, to show that indigenous populations were routinely working, cultivating and enhancing their territories, and had been for millennia.
39. Of course, the existence of past inequalities and exploitation doesn't in any way weaken claims to title by indigenous groups, unless one wants to argue that only groups living in some imaginary State of Nature are worthy of legal compensation.
40. Marquardt 1987: 98.
41. Frank Cushing, of the Bureau of American Ethnology, was among the first to embark on a systematic study of the remains of Calusa society, which dwindled into obsolescence in the seventeenth and eighteenth centuries. Cushing, even with the rudimentary archaeological methods of his time, reached conclusions that have been borne out by later research: 'The development of the Key Dwellers in this direction, is attested by every Key ruin – little or great – built so long ago, yet enduring the storms that have since played havoc with the mainland; is mutely yet even more eloquently attested by every great group of the shell mounds on these Keys built for the chief's houses and temples; by every lengthy canal built from materials of slow and laborious accumulation from the depths of the sea. Therefore, to my mind, there can be no question that the executive, rather than the social side of government was developed among these ancient Key Dwellers to an almost disproportionate degree; to a

degree which led not only to the establishment among them of totemic priests and headmen, as among the Pueblos, but to more than this – to the development of a favoured class, and of chieftains even in civil life little short of regal in power and tenure of office.' (Cushing 1896: 413; and for more recent accounts see Widmer 1988; Santos-Granero 2009).

42. For a summary of the evidence on Calusa subsistence and its socio-economic implications see Widmer 1988: 261–76.
43. Flew 1989.
44. Trouillot 2003.
45. Consider the reaction of Otto von Kotzebue, commander of a Russian ship called the *Rurik*, on first catching sight of the Sacramento River in November 1824: 'The many rivers flowing through this fruitful country will be of greatest use to future settlers. The low ground is exactly adapted to the cultivation of rice; and the higher, from the extraordinary strength of the soil, would yield the finest wheat-harvests. The vine might be cultivated here to great advantage. All along the banks of the river grapes grow wild, in as much profusion as the rankest weeds: the clusters were large; and the grapes, though small, very sweet, and agreeably flavoured. We often ate them in considerable quantities, and sustained no inconvenience from them. The Indians also ate them voraciously.' Cited in Lightfoot and Parrish 2009: 59.
46. Nabokov 1996: 1.
47. In Florida we find stone tools together with mastodon bones at least 14,000 years old (Halligan et al. 2016). Evidence for early coastal penetration into the Americas along the so-called 'kelp highway' is presented by Erlandson et al. 2007.
48. In a now classic discussion, Bailey and Milner (2002) laid out a powerful case for the central role of coastal hunter-gatherers in the evolution of human societies, between the Late Pleistocene and the Mid-Holocene, noting how changing sea levels have grossly distorted our conventional picture of early human demography, submerging a greater part of the evidence. The Tågerup promontory in western Scania, Sweden – and the wider region of southern Scandinavia – offer excellent examples of large scale and longevity in Mesolithic settlement, and for every such ancient coastal landscape that survives we must surely imagine hundreds more, long hidden below the waves (Larsson 1990; Karsten and Knarrström 2013).
49. For a more detailed analysis of the Natchez divine kingship see Graeber and Sahlins 2017: 390–95. We only know that the Great Sun's power was so limited because when the French and English were competing for allies they found that each Natchez village adopted its own, often contradictory,

foreign policy, regardless of what the Great Sun told them to do. If the Spanish had limited their dealings to the court, they might well have missed out entirely on this side of things.

50. Woodburn 2005: 26 (our emphasis). Nor, we should add, is it difficult to find other examples of free societies (for instance, in aboriginal California or Tierra del Fuego), where no adult would ever presume to give another a direct order, but where the one exception is during ritual masquerades when gods, spirits and ancestors who impose laws and punish infractions are presumed to be, somehow, present; see also Loeb 1927; also Sahlins's essay on the 'original political society' which opens Graeber and Sahlins 2017.
51. See Turnbull 1985 for a description.
52. Women have to pretend they don't know it's really their own brothers and husbands, and so forth. No one quite knows whether the women really know (it seems they almost certainly do), whether the men really know the women know, whether the women know the men know they know, and so forth ...
53. This is why, as MacPherson – our principal source here – notes in his *Political Theory of Possessive Individualism* (1962), negative rights make so much better sense to us than positive rights – that is, despite the fact that the UN Human Rights Charter guarantees everyone jobs and livelihood as basic human rights, no government is ever accused of a human rights abuse for throwing people out of work or removing subsidies on basic foodstuffs, even if it causes widespread hunger; but only for 'trespass' on their persons.
54. Consider here the way that indigenous land claims almost invariably involve invoking some notion of the sacred: sacred mountains, sacred precincts, earth mothers, ancestral burial grounds and so forth. This is precisely by way of opposing the prevailing ideology, where what is ultimately sacred is the freedom afforded by being able to make absolute, exclusive property claims.
55. Lowie 1928.
56. Walens 1981: 56–8 provides an elaborate analysis of Kwakiutl feast dishes, which are both corporeal and incorporeal property at the same time since they can die and be reincarnated.
57. Lowie 1928: 557; see also Zedeño 2008.
58. Fausto 2008; see also Costa 2017.
59. Costa and Fausto 2019: 204.
60. Durkheim 1915, Book Two, Chapter One: 'The Principal Totemic Beliefs: the Totem as Name and Emblem'; see also Lévi-Strauss 1966: 237–44.

61. Strehlow 1947: 99–100.
62. As in so many examples of what we are calling 'free societies', maternal nurture sought to inculcate a sense of autonomy and independence; while male nurture – because the trials and ordeals of the Australian initiation ceremonies were, indeed, meant to complete a process of 'growing up' – was designed to ensure that, in those contexts at least, exactly the opposite instincts came to the fore. In this connection it's worth noting that a considerable literature exists, starting from Barry, Child and Bacon (1959), which suggests, as Gardner put it, that 'while non-foragers tend to push children towards obedience and responsibility, foragers tend to press for self-reliance, independence and individual achievement' (1971: 543).

5. MANY SEASONS AGO

1. Indigenous population figures are highly contested, but there is agreement that the Pacific littoral was among the most densely populated regions of aboriginal North America; see Denevan 1992; Lightfoot and Parrish 2009.
2. In Kroeber's magisterial *Handbook of the Indians of California*, he at one point remarks that 'Agriculture had only touched one periphery of the State, the Colorado River bottom, although the seed-using and fairly sedentary habits of virtually all the other tribes would have made possible the taking over of the art with relatively little change of mode of life. Evidently planting is a more fundamental innovation to people used to depending on nature than it seems to those who have once acquired the practice' (1925: 815), though elsewhere he duly acknowledges that a number of California peoples – 'the Yurok, Hupa, and probably Wintun and Maidu' – did in fact plant and grow tobacco (ibid: 826). So planting could not have been such a conceptual innovation after all. As Bettinger noted more recently, 'that agriculture never managed to spread to California was not due to isolation. California was always in more or less direct communication with agriculturalists, whose products occasionally turn up in archaeological sites' (2015: 28). He argues that Californians simply developed a 'superior adaptation' to the local environment; though this does not explain the systematic nature of the rejection.
3. Hayden 1990.
4. We still see this mindset to this day, of course: witness the endless fascination of journalists with the idea that somewhere on earth there must be some group of humans that could be said to have lived in untouched isolation since the Stone Age. In fact, no such groups exist.

5. This was admittedly not the only way to organize displays: most US museums before Boas organized objects by types: beadwork, canoes, masks, etc.
6. 'Ethnology' is today a minor sub-branch of anthropology, but in the early twentieth century it was regarded as the highest form of synthesis, bringing together the findings of hundreds of micro-studies to compare and analyse the connections and divergences among human societies.
7. This is surely understandable. Exponents of scientific racism took theories like the 'Hamitic hypothesis' to new extremes, notably followers of the Austrian-German 'culture-circle school' (*Kulturkreislehre*), but equally many contemporary writings by French, Russian, British and American scholars. One particular interest of the culture-circle school of ethnology were the origins of monotheism, long considered a unique and seminal contribution of Jewish culture to Europe. The idea of studying an extraordinary variety of 'herding cultures', 'shepherds' and 'cattle keepers' was at least partly to show that there was nothing special about the religious achievements of the ancient Israelites, and that monotheistic beliefs about 'High Gods' were quite likely to crop up in almost any tribal society that spent much of its time moving with animals through arid or steppe landscapes. Published debates on this matter in the mid twentieth century could fill a small library, starting with Wilhelm Schmidt's twelve-volume *Der Ursprung der Gottesidee* (*The Origin of the Idea of God*, 1912–55).
8. Wissler 1927: 885.
9. Hence E. B. Tylor, the founder of British anthropology, wrote that 'though cat's-cradle is now known over all western Europe, I find no record of it at all ancient in our part of the world. It is known in South-east Asia, and the most plausible explanation seems to be that this is its centre of origin, whence it migrated westward into Europe, and eastward and southward through Polynesia and into Australia.' (1879: 26). A JSTOR search for 'string figures' in anthropology journals between 1880 and 1940 yields 212 results, and forty-two essays with 'string figures' in the title.
10. Collected in Mauss 1968–9, and now also compiled and translated into English, with commentary and historical context, in Schlanger 2006.
11. Since the 1930s and 1940s anthropologists turned first to structural-functionalist paradigms, then later to others that focused more on cultural meanings, but in either case concluded that the historical origins of customs is not a particularly interesting question since it tells you almost nothing about what the meaning of the custom is today.
12. See Dumont 1992: 196.

13. This was, in some ways, closer to the kind of approach advocated in today's research on how culture spreads, although the ultimate causes now tend to be sought in universal factors of human cognition (e.g. Sperber 2005).
14. Mauss, in Schlanger 2006: 44, and see also pp. 69, 137.
15. For a general overview and history of Northwest Coast peoples, ecology and material culture see Ames and Maschner 1999; for the equivalent in Aboriginal California see Lightfoot and Parrish 2009.
16. E.g. Hayden 2014.
17. Such broad-brush distinctions based on food preferences and resource availability were at the foundation of 'culture areas' when these were first defined by Clark Wissler and others in the early twentieth century. In *The American Indian* (1922), Wissler actually first defined 'food areas' and then subdivided them into 'culture areas'. For more recent and critical views of these broad ecological classifications see Moss 1993; Grier 2017. It's worth noting that the presence or absence of slavery never factors into the 'culture areas' described in Wissler's influential book *The American Indian* (admittedly, chattel slavery was an unusual institution among the indigenous societies of North America, but it did exist).
18. Actually, he was speaking of a cluster of related peoples, primarily the Yurok, Karuk and Hupa, who shared very similar cultural and social institutions even though they spoke entirely unrelated languages. In the anthropological literature, the Yurok have often come to stand in for Californians in general (just as 'Kwakiutl' have come to stand in for all Northwest Coast peoples), which is unfortunate since, as we'll see, they were in some ways quite unusual.
19. Goldschmidt 1951: 506–8. In fact, all this was unusual even for California: as we'll see, most Californian societies used shell money, but a man or woman's wealth was ritually destroyed at death.
20. Benedict 1934: 156–95. The comparison between Northwest Coast societies and the noble households of medieval Europe was explored by Claude Lévi-Strauss in a piece which is most famous for its definition of 'house societies', and is reprinted as part of his collected essays under the title 'Anthropology and Myth' (Lévi-Strauss 1987: 151; and see also Lévi-Strauss 1982 [1976]).
21. Hajda (2005) provides a fine-grained discussion of the different forms of slavery on the lower Columbia River and further north on the Northwest Coast, and how these developed in the early period of European contact (1792–1830). But she does not go into the wider contrast with indigenous societies south of Cape Mendocino, which rejected slavery altogether.
22. Sahlins 2004: 69.

23. Goldschmidt 1951: 513.
24. Drucker 1951: 131.
25. Elias 1969.
26. See Boas and Hunt 1905; Codere 1950. Ethnographers in the early twentieth century certainly regarded the occasional introduction of such practices into northern Californian societies as highly exotic and anomalous, as in Leslie Spier's (1930) discussion of the Klamath, who took up slaving and limited aspects of *potlatch* after their adoption of the horse.
27. Powers 1877: 408; Vayda 1967; Goldschmidt and Driver 1940.
28. See, especially, Blackburn 1976: 230–35.
29. Chase-Dunn and Mann 1998: 143–4. Napoleon Chagnon (1970: 17–18) went so far as to argue that 'it was functionally necessary for the Yurok to "desire" *dentalia* [i.e. money], but only if they were obtained from their neighbours. The social prestige involved with obtaining wealth in this fashion effected a more stable adaptation to the distribution of resources by allowing trade to be the alternative to raid in times of local insufficiency.'
30. See Donald 1997.
31. Ames 2008; cf. Coupland, Steward and Patton 2010.
32. The case for some form of social stratification in this early period was convincingly laid out in numerous pioneering works by the archaeologist Kenneth Ames (e.g. Ames 2001).
33. Arnold 1995; Ames 2008; Angelbeck and Grier 2012.
34. Santos-Granero 2009.
35. Patterson 1982; and thus, Goldman on Kwakiutl slaves: 'captive aliens, they had no kinship connections with their new homes, and no genuine ties any longer with their original tribes and villages. As persons violently torn loose from their roots, slaves existed in a state equivalent to being dead. Being on the margins of death they were by Kwakiutl standards the proper sacrificial victims for cannibalistic feasts.' (1975: 54)
36. Patterson 1982; Meillassoux 1996.
37. Santos-Granero 2009: 42–4.
38. See Wolf 1982: 79–82.
39. According to Santos-Granero, who has carefully compiled information about what raiders actually said they were doing.
40. Fausto 1999.
41. Santos-Granero 2009: 156. This does not appear to be a mere analogy: slaves in most Amazonian societies that kept them seem to have had the same formal status as pets; pets in turn were, as we already observed, seen as the paradigm for property more generally in much of Amazonia (see also Costa 2017). For instance, despite the kind treatment, a man or

woman's dogs, horses, parrots and slaves were typically all ritually sacrificed on their death (Santos-Granero, op. cit.: 192–4).

42. See also Graeber 2006. It is interesting in this connection that the Guaicurú, though they captured farmers, did not set those they took as slaves to work planting or tending crops, but integrated them into their own foraging lifestyle.
43. Powers 1877: 69.
44. They'd been among the first on the Pacific littoral to succumb to diseases introduced by traders and settlers. Combined with genocidal attacks, this caused the Chetco and nearby groups to suffer almost total demographic collapse in the nineteenth century. As a result, there are no detailed accounts of these groups to compare with the two major 'culture areas' of California and the Northwest Coast, which lie to either side of their former territories. Indeed, this complex subsector of the coast, between the Eel River and the mouth of the Columbia River, posed significant problems of classification for scholars seeking to delineate the boundaries of those culture areas, and the issue of their affiliation remains contentious today. See Kroeber 1939; Jorgensen 1980; Donald 2003.
45. The historicity of First Nations oral narratives concerning ancient migrations and wars on the Northwest Coast has been the subject of an innovative study which combines archaeology with the statistical modelling of demographic shifts that can be scientifically dated back to periods well over a millennium into the past. Its authors conclude that the 'Indigenous oral record has now been subjected to extremely rigorous testing. Our result – that the [in this case] Tsimshian oral record is correct (properly not disproved) in its accounting of events from over 1,000 years ago – is a major milestone in the evaluation of the validity of Indigenous oral traditions.' (Edinborough et al. 2017: 12440)
46. We cannot know how common such cautionary tales were because they are not the kind of stories early observers were likely to have recorded (this particular tale survived only because Chase believed the Wogies might have been shipwrecked Japanese!).
47. Spott and Kroeber 1942: 232.
48. Intriguingly, in some parts of California reliance on acorns as a dietary staple can be traced back some four millennia, long before the intensive exploitation of fish. See Tushingham and Bettinger 2013.
49. On the Northwest Coast, bulk harvesting of salmon and other anadromous species extends back to 2000 BC and remained a cornerstone of aboriginal economy until recent times. See Ames and Maschner 1999.
50. Suttles 1968.

51. Turner and Loewen 1998.
52. Take, for example, Joaquin Miller's (1873: 373–4) description in his *Life Amongst the Modocs, Unwritten History*: 'Here we passed groves of magnificent oak. Their trunks are five and six feet in diameter, and the boughs were then covered with acorns and fairly matted with mistletoe. Coming down to the banks of the Pit river, we heard the songs and shouts of Indian girls gathering acorns. They were up in the oaks and half covered with mistletoe. They would beat off the acorns with sticks, or cut off the little branches with tomahawks, and the older squaws gathered them from the ground, and threw them over their shoulders in baskets borne by a strap around the forehead.' (He then goes off on an excursus about how Indian girls have exquisitely small and attractive feet, despite not wearing tight European-style shoes, thus exposing a 'popular delusion' among overweening mothers in frontier communities of the time: feet could be free, and still elegant.)
53. As Bettinger puts it, the acorn is 'so very back-loaded that its capture as stores represents little saved time . . . with correspondingly less potential for developing inequality, likewise for attracting raiders or developing organizational means to defend or retaliate' (2015: 233). His argument basically seems to be that what the remote ancestors of the Maidu, Pomo, Miwok, Wintu and other Californian groups sacrificed in short-term nutritional value they gained over the long term in food security.
54. Much of what we've presented in the preceding paragraphs is based on a more detailed argument by Tushingham and Bettinger (2013), but the basis of their approach – including the suggestion that forager slavery is rooted in the seasonal exploitation of aquatic resources – can be found in publications going all the way back to Herman Nieboer's *Slavery as an Industrial System* (1900).
55. For a general reconstruction of traditional raiding practices on the Northwest Coast, and further discussion, see Donald 1997.
56. Drucker 1951: 279.
57. Golla 1987: 94.
58. Compare Ames 2001; 2008. Slaves could and often did attempt to escape as well, often successfully – especially when a number of slaves from the same community were held in the same place (see e.g. Swadesh 1948: 80).
59. There appears to have been something of a transitional zone on the lower reaches of the Columbia River where chattel slavery dwindled into various forms of peonage, while beyond stretched a largely slave-free zone (Hajda 2005); and for other limited exceptions see Kroeber 1925: 308–20; Powers 1877: 254–75; and Spier 1930).

60. MacLeod 1928; Mitchell 1985; Donald 1997.
61. Kroeber 1925: 49. Macleod (1929: 102) was unconvinced of this point, noting the existence of similar legal mechanisms among Tlingit and other Northwest Coast groups, which did not prevent the 'subjection of foreign groups, tribute taking, and enslavement of captives'. Yet all sources concur that the only real slaves in Northwest California were debt-slaves, and that even these were few in number (cf. Bettinger 2015: 171). If not Kroeber's then some other mechanism for the suppression of chattel slavery must have been at work.
62. Donald 1997: 124–6.
63. Goldschmidt 1951: 514.
64. Brightman 1999.
65. Boas 1966: 172; cf. Goldman 1975: 102.
66. Kan 2001.
67. Lévi-Strauss 1982.
68. Garth 1976: 338.
69. Buckley 2002: 117; cf. Kroeber 1925: 40, 107.
70. 'The northwest is perhaps also the only part of California that knew slavery. This institution rested wholly upon an economic basis here. The Chumash may have held slaves; but precise information is lacking. The Colorado River tribes kept women captives from motives of sentiment, but did not exploit their labor.' (Kroeber 1925: 834)
71. Loeb 1926: 195; Du Bois 1935: 66; Goldschmidt 1951: 340–41; Bettinger 2015: 198. Bettinger notes that (archaeologically visible) inequalities of wealth steadily declined after the introduction of dentalium in central California, and argues that the overall effect of the introduction of money appears to have been to limit debt relations, and thus reduce overall dependency and 'inequality.'
72. Pilling 1989; Lesure 1998.
73. While captives taken in war were quickly redeemed, it seems that unlike in other parts of California, where tribal divisions assumed collective responsibility for doing so, here it was up to the individual family. Debt peonage seems to have resulted from inability to pay. Bettinger (2015: 171) suggests that this nexus of debt and warfare may partly explain the demographic fragmentation of Northwest Californian groups and break-up of collective groups to begin with, which were never very strong (totemic clans, for instance, were absent), but did exist further south. One early source (Waterman 1903: 201) adds that killers unable to pay compensation, but not forced into peonage, became a disgrace to their communities and retreated into isolation, often remaining there even after

settling their debts. The overall situation did come to look a bit like a class system as men of inherited wealth often initiated wars, directed the peace-making ceremonies that followed, and then managed the resulting debt arrangements – in the course of which one class of poorer household would fall into marginal status, its members scattering across the land-scape and dissolving into patrilineal bands, while another concentrated as dependants around the victors. However, unlike the situation on the Northwest Coast, the degree to which grandees could compel their 'slaves' to work was decidedly limited (Spott and Kroeber 1942: 149–53).

74. As further argued in Wengrow and Graeber (2018), with subsequent comments by regional experts in the archaeology and anthropology of West Coast foragers and their descendants, and authors' response.

6. GARDENS OF ADONIS

1. *Phaedrus* 276B.
2. For the former opinion, see Detienne 1994, and for the latter, Piccaluga 1977.
3. From the children's story *Where the Wild Things Are* (1963).
4. Mellaart 1967.
5. Our understanding of the site largely follows that developed by its recent excavator, Ian Hodder, except that we have emphasized the importance of seasonal variations in social structure to a greater degree. See Hodder 2006; and for further information, images and databases see also www.catalhoyuk.com, with references to multiple field reports, sections of which are also referred to below.
6. Meskell et al. 2008.
7. See, for example, Gimbutas 1982. More recent studies make the point that Gimbutas's publications often inflated the frequency of female forms within Neolithic figurine assemblages, which on closer inspection contain a more balanced proportion of clearly female, clearly male, mixed or simply unsexed forms (e.g. Bailey 2017).
8. Charlene Spretnak (2011) discusses the successive waves of criticism levelled at Gimbutas and provides further references.
9. The key publication on the genomics of steppe migration is Haak et al. 2015. Shortly after these findings were published, the eminent prehistorian Colin Renfrew delivered a lecture at the University of Chicago entitled 'Marija Rediviva [Marija Born Again]: DNA and Indo-European Origins'. He suggested that Gimbutas's '*kurgan* hypothesis' had been 'magnificently vindicated' by the findings of ancient DNA, which

suggest links between the dispersal of Indo-European languages and the westward spread of the Yamnaya cultural complex from the steppe north of the Black Sea in the late fourth and early third millennia BC. It's worth noting that these findings contradict Renfrew's own (1987) hypothesis, that Indo-European languages originated in the region of Anatolia and spread, some millennia earlier, with the dispersal of Neolithic farming cultures. Other archaeologists, however, feel that the genomic data is still too coarse to permit talk of large-scale migrations, let alone establish links between biological inheritance, material culture and the spread of languages (for a detailed critique see Furholt 2018).

10. See here Sanday's *Women at the Center* (2002). Sanday notes that Gimbutas rejects the term 'matriarchy' because she sees it as a mirror image of patriarchy, and therefore that it would imply autocratic rule or political dominance of women, and therefore prefers 'matric'. Sanday notes that Minangkabau themselves use the English term 'matriarchate', employing it in a different sense (ibid.: 230–37).
11. See Hodder 2003; 2004; 2006, plate 22. For the most recently discovered statuette of an (elderly?) female in limestone see also Chris Kark's short but informative item in *Stanford News*, 'Archaeologists from Stanford find an 8,000-year-old "goddess figurine" in central Turkey' (2016), including comments from key researchers.
12. For the occurrence in these regions of likely masked figurines, and the connections between figurines and other Neolithic depictions of masked human forms, see e.g. Belcher 2014: *passim*; Bánffy 2017.
13. Hodder 2006: 210, with further references.
14. Hodder and Cessford 2004.
15. Çatalhöyük actually comprises two main archaeological mounds. Everything we've been talking about so far applies to the early 'East Mound', while the 'West Mound' relates mainly to later periods of prehistory, beyond the scope of our discussion here.
16. Matthews 2005.
17. See Fairbairn et al. 2006.
18. Of the kind we discussed in Chapter Three.
19. Bogaard et al. 2014, with further references.
20. Arbuckle 2013; Arbuckle and Makarewicz 2009.
21. See Scheffler 2003.
22. In terms of environmental history, the upland regions of the Fertile Crescent fall within the Irano-Turanian bioclimatic zone. Current reconstructions suggest that the establishment of deciduous woodlands in this region did not follow directly from the onset of warmer and wetter conditions at the

beginning of the Holocene, but was to a significant degree a product of landscape-management strategies carried out initially by forager populations, and subsequently by cultivators and herders (Asouti and Kabukcu 2014).

23. Based on the analysis of carbonized residues of wood found in archaeological sites, Asouti et al. (2015) reconstruct a moister environment for this region in the Early Holocene, with considerably more tree cover than is apparent today, especially along and adjacent to the Jordan Rift Valley. Towards the Mediterranean coast these lowland regions acted as refugia for wood- and grassland species, which survived continuously through the Last Glacial Maximum and into the Early Holocene.
24. Prehistorians have experimented with all kinds of different ways of classifying the Fertile Crescent into 'culture areas' or 'interaction spheres' corresponding to the main distinctions of the Late and Epi-Palaeolithic era, and the Early (or Pre-Pottery) Neolithic. The history of these various classifications is reviewed and evaluated by Asouti (2006). Here we follow the distinctions outlined by Sherratt (2007), which are based on the correlation of broad ecological and cultural patterns rather than isolated (and fairly arbitrary) categories of archaeological data, such as different ways of manufacturing stone tools and weapons. Sherratt's classification also has the advantage of avoiding teleological tendencies found in some other studies, which assume all evidence of cultural complexity (such as substantial settlements and architecture) must be in some way related to the development of food production; in other words, he allows for such developments within foraging societies that had no strong investment in the domestication of plants and animals.
25. For craft specialization in Early Neolithic communities see Wright et al. 2008; and, in general, Asouti and Fuller 2013.
26. Sherratt 1999.
27. Willcox 2005; 2007.
28. For western Iran and Iraqi Kurdistan see Zeder and Hesse 2000; and for eastern Anatolia see Peters et al. 2017.
29. Asouti and Fuller 2013: 314–23, 326–8.
30. Harari 2014: 80.
31. Hillman and Davies 1990.
32. Maeda et al. 2016.
33. The initial growth of permanent villages – between 11,000 and 9500 BC – may have had much to do with a temporary return of glacial weather conditions (known as the 'Younger Dryas' episode) after the end of the last Ice Age obliging foragers in the lowland parts of the

Fertile Crescent to commit to well-watered locations (Moore and Hillman 1992).

34. This conclusion is based on a combination of genetic and botanical data from samples recovered in archaeological excavations, as explained further below; and for a summary, see Fuller 2010; Fuller and Allaby 2010.
35. See Willcox et al. 2008; Willcox 2012.
36. Fuller 2007; 2010; Asouti and Fuller 2013, with further references.
37. Cf. Scott 2017: 72.
38. In fact, they did not even wish it on their slaves.
39. As proposed by Fuller 2010: 10; see also Fuller et al. 2010.
40. The significance of flood-recession cultivation for the origins of farming was first pointed out in a seminal (1980) article by Andrew Sherratt; republished and updated in Sherratt 1997.
41. Cultivation systems of this kind have been pursued up until recent times in rural India and Pakistan, and also in the American Southwest. As one geographer observed of Pueblo cultivation in New Mexico: 'The kinds of places suitable for farming under the system . . . have existed from the time of prehistoric settlement; but cultivation, by its disturbance of the surface, leads to washing and channelling, which temporarily or permanently ruin a field. Thus, at the same site the best places to plant are limited in area and changeable in position. The Indians of the present day, like their prehistoric ancestors, hardly disturb the ground, as they do not plough but merely insert the seed in a hole made with planting stick . . . Even with the use of their methods fields must be periodically abandoned for later reoccupation. One of the principal causes of such shifts in location lies in the habits of ephemeral streams in the stage of alluviation.' (Bryan 1929: 452)
42. For which see Sanday 1981, especially her Chapter Two: 'Scripts for Male Dominance'.
43. Diamond 1987.
44. See Murdock 1937; Murdock and Provost 1973.
45. Owen 1994; 1996.
46. Barber 1991; 1994.
47. Soffer et al. 2000.
48. Lévi-Strauss 1966: 269.
49. See MacCormack and Strathern (eds) 1980.
50. We find ourselves reminded of Silvia Federici's *Caliban and the Witch* (1998), where she showed how, in Europe, such 'magical' approaches to production came to be associated not just with women but with witchcraft. Federici argues that the elimination of such attitudes was essential

for the establishment of modern (male-dominated) science, and also for the growth of capitalist wage labour: 'This is how we must read the attack against witchcraft and against that magical view of the world which, despite the best efforts of the Church, had continued to prevail on a popular level through the Middle Ages ... In this perspective ... every element – herbs, plants, metals, and most of all the human body – hid virtues and powers peculiar to it ... From palmistry to divination, from the use of charms to sympathetic healing, magic opened a vast number of possibilities ... Eradicating these practices was a necessary condition for the capitalist rationalisation of work, since magic appeared as an illicit form of power and an instrument *to obtain what one wanted without work* ... "Magic kills industry," lamented Francis Bacon, admitting that nothing repelled him so much as the assumption that one could obtain results with a few idle experiments, rather than with the sweat of one's brow.' (Federici 1998: 142)

51. Lévi-Strauss 1966: 15.
52. Wengrow 1998; 2003; and for the evolution of Neolithic counting devices, and its relationship to the invention of script, see also Schmandt-Besserat 1992.
53. Vidale 2013.
54. Schmidt 2006; Köksal-Schmidt and Schmidt 2010; Notroff et al. 2016. A stone figure known to archaeologists as the 'gift bearer' also carries a human head to some unknown destination. Images of many of these sculptures and other finds from Göbekli Tepe can be found at https://www.dainst.blog/the-tepe-telegrams/
55. Schmidt 1998; Stordeur 2000, fig. 6.1, 2.
56. O. Dietrich et al. 2019.
57. Which is not to say that such evidence for violent conflict is entirely lacking either. The largest preserved sample of human remains from any Early Holocene site in the Middle East comes from the site of Körtik Tepe, which lies northeast of Göbekli Tepe on a bank of the Upper Tigris, twelve miles from the modern town of Batman, firmly within the upland sector of the Fertile Crescent. Remains of more than 800 individuals have been recovered from the site, of which the 446 individuals so far analysed reveal high rates of skeletal trauma (of 269 skulls, some 34.2 per cent showed signs of cranial injury, including two female skulls with penetrative depressed fractures; post-cranial injuries were found on over 20 per cent of the individuals studied, including three cases of healed parry fractures to the forearm). Given the absence of other signs of warfare, this evidence has been explained – perhaps not altogether persuasively – in

terms of interpersonal violence within a community of settled hunter-fisher-gatherers living in a region of abundant wild resources. A significant number of the human remains recovered from Körtik Tepe were subject to post-mortem modification, including the presence of cut marks on human crania, although none of these can be definitively linked to scalping or the taking of trophy heads (as reported by Erdal 2015).

58. Faunal and botanical remains from Çayönü Tepesi reveal a flexible economy which underwent numerous changes over a period of some 3,000 years. In the earlier phases of the site, which concern us here, its inhabitants made extensive use of wild pulses and nuts, as well as peas, lentil and bitter vetch, with smaller amounts of wild cereals. Cultivation of at least some of these crops is likely, but there is no clear evidence of plant domestication until the later phases of the site. Animal remains suggest its inhabitants pursued mixed herding and hunting strategies which varied over many centuries, including at times a strong reliance on pigs and wild boar, as well as sheep, cattle, gazelle and red deer, and also smaller game such as hares (see Pearson et al. 2013, with further references).
59. For the House of Skulls and the analysis of associated human remains at Çayönü Tepesi see Özbek 1988; 1992; Schirmer 1990; Wood 1992; also Kornienko's (2015) broader review of evidence for ritual violence in the northern part of the Fertile Crescent. Isotopic analysis of human remains indicates that the individuals whose remains came to be stored in the Skull Building had significantly different diets to those buried elsewhere on the site, which might indicate local differences of status (Pearson et al. 2013), or conceivably the incorporation of outsiders into local mortuary rituals.
60. Allsen (2016) provides a sweeping account of the practical and symbolic relationships between hunting and monarchy in Eurasian history, which are remarkably consistent from the Middle East to India, Central Asia and China, and from antiquity down to the time of the British Raj.
61. Gresky et al. 2017.
62. No doubt these contrasts could also be found within the societies themselves; the key difference lies in how these various styles of technical activity were valued, and which were selected as the basis of artistic and ritual systems. For the overall absence of gender hierarchy in Early Neolithic societies of the southern Levant, and evidence for women's participation on equal terms in ritual and economic activities, see also Peterson 2016.
63. Kuijt 1996; Croucher 2012; Slon et al. 2014.
64. Santana 2012.

65. Confronted with objects they can't explain, archaeologists often turn to ethnographic analogies. Among the cases considered here is that of the Iatmul of the Sepik River in Papua New Guinea, a people who practised skull decoration until quite recent times. For the Iatmul, the making of skull portraits was intimately linked to head-hunting. Generally, it began by taking enemy heads in warfare, and was exclusively performed by men. The head of a defeated enemy was honoured by decorating it with clay, pigments, hair and shell. Once transformed, it was then cared for and 'fed' alongside other skulls in a special men's house (Silverman 2001: 117 ff.). This case is important, because it shows how ancestor veneration and the violent taking of heads in conflict may, in some cases, form part of the same ritual system. In 2008 the social anthropologist Alain Testart published an article in the journal *Paléorient* arguing that similar things must have been going on among Neolithic societies in the Middle East, and archaeologists had missed the obvious connection between skull portraits and head-hunting. That prompted an outpouring of responses from archaeologists in the same journal, many indignant, pointing out the lack of evidence for warfare among those same communities, and even proposing that skull-modelling was a ritual strategy for promoting peaceful and egalitarian relations among Neolithic villagers (as first argued by Kuijt 1996). What we are suggesting here is that both parties to the debate were, in a sense, correct; but that they were really talking at cross purposes; or rather about different sides of the same coin. On the one hand, we should acknowledge mounting evidence that predatory violence (including the display of trophy skulls) was at least ritually and symbolically important among foragers in the northern (upland) part of the Fertile Crescent. Equally, we might consider if skull portraits (or 'plastered skulls') represent an inversion of such values in the more southerly (lowland) parts of the region. Not everything has to fit the same model, just because it was happening at the same time, and in this case just the opposite seems likely to be true.
66. Clarke 1973: 11.

7. THE ECOLOGY OF FREEDOM

1. For spectacular developments in stone vessel and bead production in the upper valley of the Tigris see Özkaya and Coşkun 2009.
2. Elinor Ostrom (1990) offers a range of field studies and historical examples, as well as formal economic models for the collective management of shared natural resources; but the basic point was already widely noted in earlier studies, some of which we cite below.

3. Georgescu-Roegen 1976: 215.
4. Periodic repartition of land at the local level was also discussed in O'Curry's *On the Manners and Customs of the Ancient Irish* (1873), and in Baden-Powell's famous treatise on *The Indian Village Community* (1896). See, more recently, Enajero 2015.
5. Palestinian villages under Ottoman and British rule practised annual redistribution of communal grazing and farming lands under *masha'a* tenure, whereby owners of adjacent plots pooled resources to complete labour-intensive tasks such as ploughing, seeding, weeding and harvesting, responding to annual fluctuations in rainfall (Atran 1986). In Bali, irrigated rice-farming traditionally operated through a system of elected 'water committees'. Local representatives attend temple meetings, where access to land and water is negotiated annually on a consensual basis (Lansing 1991). Other examples of sustainable land management under some form of communal organization can be found in the recent histories of Sri Lanka (Pfaffenberger 1998), and also Japan, for instance (Brown 2006).
6. Fuller 2010, with further references.
7. Diamond 1997: 178.
8. Bettinger and Baumhoff 1982; Bettinger 2015: 21–8.
9. For a review of how such processes played out in various parts of the world see Fuller and Lucas 2017. None of this is to deny the fact that crops frequently 'got around' various parts of the Old World, and often on a surprising scale, as with the westward transfer of Chinese millets to the Indus, mirrored by the eastward dissemination of western/central Asian bread wheat to China around 2000 BC. But efforts (notably by Jones et al. 2011) to characterize such early crop transfers as precursors to the 'Columbian exchange' of the sixteenth century AD (see below) are misplaced, since they ignore a number of important contrasts. These are spelled out by Boivin, Fuller and Crowther (2012), who note that early crop transfers in Eurasia took place within a 'long-term, slow-growing network of connections and exchanges' over many millennia, often initially in small and experimental quantities, and driven not by centres of urban expansion but by highly mobile and often small-scale intermediary groups such as the mounted pastoralists of the Eurasian steppe or the maritime nomads of the Indian Ocean. It was precisely this slow millennial history of cultural exchange and gene-flow between Eurasian species that prevented the kind of major ecological ruptures which occurred once those same species were unleashed upon the Americas and Oceania.
10. Crosby 1972; 1986.
11. See Richerson, Boyd and Bettinger 2001.

12. Recent estimates for the population of the Americas before Europeans landed in 1492 are around 60 million. The figure of 50 million lost hectares of arable land is calculated on the basis of a land use per capita model, in the key study by Koch et al. 2019.
13. For eustatic changes in sea levels at the Late Pleistocene to Holocene transition see Day et al. 2012; Pennington et al. 2016; and for the role of anthropogenic activities in altering terrestrial species' distributions over the same time period Richerson et al. 2001; Boivin et al. 2016.
14. See Bailey and Milner 2002; Bailey and Flemming 2008; Marean 2014.
15. Boivin et al. 2016, with further references.
16. Clarke's (1978) *Mesolithic Europe: The Economic Basis* remains a foundational study of these processes; for a more up-to-date (and global) overview see Mithen 2003; and also Rowley-Conwy 2001; Straus et al. (eds) 1990.
17. Bookchin 1982. In adopting the title of Bookchin's landmark volume on social ecology, we cannot follow his own ideas about human prehistory or the origins of agriculture, which are based on information that is now many decades out of date. We do, however, find much to learn much from his basic insight: that human engagements with the biosphere are always strongly conditioned by the types of social relationships and social systems that people form among themselves. The mutual differentiation of forager ecologies on the American West Coast, discussed in Chapter Five, would be another excellent example of the same principle.
18. As the anthropologist Eric Wolf once put it.
19. Bruce Smith (2001) discusses the whole phenomenon under the rubric of 'low level food production', which he takes to describe economies that occupy 'the vast and diverse middle ground between hunting-fishing-foraging and agriculture'.
20. Wild et al. 2004; Schulting and Fibiger (eds) 2012; Meyer et al. 2015; see also Teschler-Nicola et al. 1996.
21. For a broad account of the spread of Neolithic farming to Europe, understood through behavioural ecology and theories of cultural evolution, see Shennan 2018.
22. Coudart 1998; Jeunesse 1997; Kerig 2003; van der Velde 1990.
23. Shennan et al. 2013; and see also Shennan 2009; Shennan and Edinborough 2006.
24. Haak et al. 2005; 2010; Larson et al. 2007; Lipson et al. 2017.
25. Zvelebil 2006; and for evidence of status differences marked by wealth in Mesolithic cemeteries see O'Shea and Zvelebil (1984) on cemeteries from the region of Karelia; Nilsson Stutz (2010) on southern Scandinavia; and Schulting (1996) on the Breton coast.

26. Kashina and Zhulnikov 2011; Veil et al. 2012.
27. Schulting and Richards 2001.
28. Golitko and Keeley (2007) envisage hostile encounters between Neolithic farmers and more established Mesolithic populations, noting that fortified villages tend to cluster around the fringes of Neolithic colonization.
29. Wengrow 2006, Chapters Two to Three; Kuper and Kroepelin 2006.
30. Wengrow et al. 2014, with further references.
31. See Spriggs 1997; Sheppard 2011.
32. Denham et al. 2003; Golson et al. (eds) 2017; see also Yen 1995.
33. See Spriggs 1995; the Lapita habit of leapfrogging established populations into empty spaces may be partially confirmed by the findings of ancient DNA, for which: Skoglund et al. 2016.
34. Kirch 1990; Kononenko et al. 2016. Gell (1993) provides a systematic, comparative study of regional traditions of body art and tattooing in more recent Polynesian societies, and their social and conceptual permutations.
35. Holdaway and Wendrich (eds) 2017.
36. As we noted, Lapita is associated with the dispersal of Austronesian languages. Correlations between the spread of farming and language also seem likely for Nilotic cultures (and the much later Bantu expansion from western to southern Africa). For a general consideration of these and other cases of language-farming dispersal see Bellwood 2005; Bellwood and Renfrew (eds) 2002. An association between Indo-European and the spread of Early Neolithic farming in Europe is now considered unlikely (see Haak et al. 2015, with further references).
37. Capriles 2019.
38. Fausto 1999; Costa 2017.
39. Descola 1994; 2005.
40. Roosevelt 2013.
41. Hornborg 2005; Hornborg and Hill (eds) 2011.
42. 'Complex' being the operative word here – the indigenous arts of Amazonia are incredibly rich and diverse, with many regional and ethnic variations. Analysts have, nonetheless, found similar principles at work in visual culture over surprisingly large regions. For a Brazilian perspective see Lagrou 2009.
43. Erickson 2008; Heckenberger and Neves 2009; Heckenberger et al. 2008; Pärssinen et al. 2009.
44. Lombardo et al. 2020.
45. Piperno 2011; Clement et al. 2015; see also Fausto and Neves 2018.
46. Arroyo-Kalin 2010; Schmidt et al. 2014; Woods et al. (eds) 2009.
47. Scott 2009.

48. Smith 2001.
49. Evidence derives from archaeological sites in the valley of Mexico's Rio Balsas; Ranere et al. 2009.
50. Smith 2006.
51. Fuller 2007: 911–15.
52. Redding 1998. Such 'flirtations' probably took the form of selective herd management with husbandry limited to females, allowing the males to roam wild.
53. In the archaeological phase termed Pre-Pottery Neolithic C (PPNC).
54. Colledge et al. 2004; 2005. It's important to note that a decline in crop diversity may have commenced *within* the Fertile Crescent, at roughly the time when the Neolithic farming package was carried north and west towards Europe, via Turkey and the Balkans. By around 7000 BC (the end of the Late Pre-Pottery Neolithic B period) average crop diversity at sites in the Fertile Crescent had dropped from ten or eleven original founder crops to a mere five or six. Interestingly, what followed in this region (during the PPNC period) was a downturn in population, associated with the abandonment of large villages and the beginning of a more dispersed pattern of human settlement.
55. See also Bogaard 2005.

8. IMAGINARY CITIES

1. E.g. Dunbar 1996; 2010.
2. Dunbar 1996: 69–71. The cognitive basis of Dunbar's Number is inferred from comparative studies of non-human primates, which suggest a correlation between neocortex size and group size in various species of monkeys and apes (Dunbar 2002). The significance of those findings for primate studies is not in question here, only whether they can be extended in any simple or direct way to our own species.
3. Bird et al. 2019; see also Hill et al. 2011; Migliano et al. 2017.
4. Sikora et al. 2017.
5. Bloch 2013.
6. Anderson 1991.
7. See Bird et al. 2019; and compare Bloch 2008.
8. Fischer 1977: 454.
9. See especially Childe 1950.
10. We hope to treat the rich African material, outside ancient Egypt, more fully in future work, along with many other valuable cases that could

not be included here, such as the Pueblo traditions of the American Southwest, to name but one. For important existing discussions of African material, which bear out a number of our observations about the decentralized and self-organizing nature of early cities, see e.g. S. McIntosh 2009; R. McIntosh 2005.

11. Most archaeologists are generally happy to call any densely inhabited settlement over around 150 hectares, or certainly over 200 hectares, in size a 'city' (see, for example, Fletcher 1995).
12. Fleming 2009: *passim*.
13. For direct evidence of in-migration to Teotihuacan, based on isotopic studies of human remains, see White et al. 2008; for similar evidence at Harappa see Valentine et al. 2015. For a general discussion of neighbourhoods and their role in the formation of early cities, Smith 2015.
14. Plunket and Uruñuela 2006.
15. Day et al. 2007; Pennington et al. 2016.
16. See Pournelle 2003.
17. Sherratt 1997; Styring et al. 2017.
18. See Pournelle 2003; Scott 2017.
19. For China see Underhill et al. 2008; for Peru see Shady Solis, Haas and Creamer 2001.
20. Inomata et al. 2020. They key site here is in Tabasco State, and goes by the name Aguada Fénix. Dated between 1000 and 800 BC, it's now recognized as the 'oldest monumental construction ever found in the Maya area and the largest in the entire pre-Hispanic history of the region'. Aguada Fénix is by no means an outlier. Massive architectural features, implying communal labour on the scale of ancient Egyptian pyramids, have now been found at numerous sites in the Maya lowlands, many centuries before the inception of Classic Maya kingship. Mostly these comprise not pyramids but earthen platforms of staggering proportions and horizontal extent, carefully laid out in roughly E-shaped formations; their function remains unclear, as most of these sites were revealed by remote sensing (using LiDAR technology) and are yet to be excavated on any scale.
21. Anthony 2007.
22. Much of this research (published exclusively in Russian) was cutting-edge by the standards of the time, including aerial photography, subsurface prospection and careful excavation. For summaries and descriptions in English see Videiko 1996; Menotti and Korvin-Piotrovskiy 2012.
23. Shumilovskikh, Novenko and Giesecke 2017. What distinguishes these soils, in physical terms, is their high humus content and capacity for storing moisture.

24. Anthony 2007: 160–74.
25. To get a sense of relative scale, consider that just this vacant centre of a mega-site alone could have contained a large Neolithic town such as Çatalhöyük more than twice over.
26. Scientific dating shows that some of the largest known mega-sites were contemporaneous; Müller et al. 2016: 167–8.
27. Ohlrau et al. 2016; Shumilovskikh, Novenko and Giesecke 2017.
28. Nebbia et al. (2018) present evidence in support of this extreme seasonal model, while leaving room for other possibilities.
29. The people of the mega-sites had a tradition of deliberately burning their houses, which complicates matters for modern analysts, trying to ascertain how much of each site was in use simultaneously. It's not known why this burning was done (for ritual purposes, or hygiene, or both?). Did it take place routinely within settlements, so part of the mega-site was living and growing, with the other part lingering on as a sort of 'house-cemetery'? Ordinarily, careful modelling of high-precision radiocarbon dates would allow archaeologists to resolve such issues. Frustratingly, in this case, an anomaly in the calibration curve for the fourth millennium BC is preventing them from doing so.
30. Kirleis and Dal Corso 2016.
31. Chapman and Gaydarska 2003; Manzura 2005.
32. One should also allow for different answers, varying from one mega-site to another. For example, some of them, such as Maidenetske and Nebelivka, mobilized their populations to dig perimeter ditches, marking out a garden space between the outer circuit of houses and the edge of the settlement. Others, such as Taljanky, did not. It is worth stressing that these ditches cannot possibly have functioned as fortifications or defences of any kind – they were shallow, with frequent gaps so that people could come and go. It's worth stressing this, because earlier scholarship often viewed the mega-sites as 'refuge towns' formed for the defence of a local population, a view that has now been largely abandoned in the absence of any clear evidence for warfare or other forms of conflict (see Chapman 2010; Chapman, Gaydarska and Hale 2016).
33. Bailey 2010; Lazarovici 2010.
34. As John Chapman and colleagues show, there is nothing in these assembly houses to suggest they housed a political or religious upper class: 'Those expecting the architectural and artefactual reflections of a hierarchical society with elites ruling over thousands of inhabitants in the Trypillia mega-site will be disappointed.' (Chapman, Gaydarska and Hale 2016: 120). Aside from their scale, and sometimes an accentuated

entranceway, these buildings are similar in their furnishings to ordinary dwellings, except for the interesting absence of installations for the preparation and storage of food. They have 'none of the depositional characteristics of a ritual or administrative centre' (ibid.), and do not seem to have been permanently occupied on any scale, which supports the idea that they were used for periodic, perhaps seasonal gatherings.

35. Chapman, Gaydarska and Hale 2016.
36. The Basque system of settlement organization is described by Marcia Ascher in Chapter Five of her book *Mathematics Elsewhere* (2004). We cannot do justice to the subtleties of Ascher's account here or the mathematical insight she brings, and refer interested readers to her study and to the original ethnographic material she relies on (Ott 1981).
37. Ascher 2004: 130.
38. As one of their leading excavators, the prehistorian Johannes Müller (2016: 304) puts it: 'The new and unique character of spatial organization in Late Trypillia [or 'Tripolye'] mega-sites displays some insights into human and group behaviour which might still be relevant for us today. Both the ability of non-literate societies to agglomerate in huge population groups under rural conditions of production, distribution, and consumption and their ability to avoid unnecessary social pyramids and instead practice a more public structure of decision making, reminds us of our own possibilities and abilities.'
39. *Heartland of Cities* was the title of a landmark archaeological survey and analysis of the central Mesopotamian floodplain by Robert McCormick Adams (1981).
40. The marshes of southern Iraq are home to the Ma'dān (sometimes called Marsh Arabs), best known to Europeans through the writings of Wilfred Thesiger. The marshes were systematically drained by Saddam Hussein's Ba'ath government in an act of political retribution, leading to the mass displacement of the indigenous population, and enormous damage to this ancient habitat. Since 2003 there have been sustained and partially successful efforts to reconstitute the marshes and their ancestral communities and ways of life.
41. Oates et al. 2007. Key evidence is in Syria, where military conflict has interrupted archaeological work at sites like Tell Brak, on the Khabur River (a major tributary of the Euphrates). Archaeologists call these grasslands in northern Mesopotamia the 'dry-farming' zone, because agriculture based on rainfall was possible there. The contrast is with southern Mesopotamia, an arid zone, where irrigation from the major rivers was mandatory for cereal-farming.

42. These mounds are the great material accretions of human life and death known by the Arabic word *tell*, built up through successive foundation and collapse of mud-brick architecture over tens or often hundreds of generations.
43. For a survey of 'the Sumerian world' see Crawford (ed.) 2013.
44. This also fitted rather well with British colonial concerns in the modern region they called 'Mesopotamia' which were based on a policy of elevating (and occasionally creating) local monarchies favourable to their own interests (see Cannadine 2001).
45. See Dalley 2000.
46. Wengrow 2010: 131–6; Steinkeller 2015. Scribes sometimes used another word (*bala*) – meaning 'term' or 'cycle' – to refer to corvée labour and also the succession of royal dynasties, but this is a later development. It is interesting to compare the whole phenomenon with the Malagasy *fanompoana* or 'service', a theoretically unlimited labour duty owed to the monarchy; in this case the monarch's own family was exempt, but there are similar accounts of the absolute equality of everyone who came together to dig earth on royal projects and the cheerful enthusiasm with which they did so (Graeber 2007a: 265–7).
47. Steinkeller 2015: 149–50.
48. Written evidence from various periods of Mesopotamian history shows that rulers quite routinely proclaimed debt amnesties on jubilees and other festive occasions, wiping the slate clean for their subjects and allowing them to resume a productive civic life. Redemption of accrued debts, either by royal proclamation or in 'years of forgiveness', made good fiscal sense. It was a mechanism for restoring balance to the economy of Mesopotamian cities, and by releasing debtors and their kin from servitude it allowed them to continue living productive civic lives (see Graeber 2011; Hudson 2018).
49. Women were citizens and owned land. Some of the earliest stone monuments from anywhere in Mesopotamia record transactions between male and female owners, who appear as legal parties on an equal footing. Women also held high rank in temples, and female royals trained as scribes. If their husbands fell into debt they could become acting heads of households. Women also formed the backbone of Mesopotamia's prolific textile industry, which financed its foreign trade ventures. They worked in temples or other large institutions, often under the supervision of other women, who received land allotments in similar proportions to men. Some women were independent financial operators, issuing credit to other women; see, in general, Zagarell 1986; van de Mieroop 1989; Wright 2007; Asher-Greve

2012. Some of the earliest documentation on these matters comes from Girsu, in the city-state of Lagash, around the middle of the third millennium BC. It comprises some 1,800 cuneiform texts derived mostly from an institution named 'the House of the Woman' and later called 'the House of the Goddess Baba', for which see Karahashi 2016.

50. Chattel slavery, the keeping of slaves as property in private households, was so deeply rooted in the economy and society of classical Greece that many feel justified in defining Greek cities as 'slave societies'. We find no obvious equivalent to this in ancient Mesopotamia. Temples and palaces held prisoners of war and debt defaulters as slaves or semi-free workers, who performed manual tasks such as grain-grinding or porterage all year round for food rations and owned no land of their own. Even then, they formed only a minority of the workforce in the public sector. Outright chattel slavery also existed, but played no comparably central role in the Mesopotamian economy; see Gelb 1973; Powell (ed.) 1987; Steinkeller and Hudson (eds) 2015: *passim.*
51. Jacobsen 1943; see also Postgate 1992: 80–81.
52. Barjamovic 2004: 50 n.7.
53. Fleming 2004.
54. As John Wills (1970) noted long ago, something of the conduct of assemblies is likely preserved in the speeches ascribed to gods and goddesses in Mesopotamian myth. The deities too convene to sit in assemblies, where they exhibit skills of rhetoric, persuasive speech, logical argumentation and occasional sophistry.
55. Barjamovic 2004: 52.
56. One such 'urban village', as Nicholas Postgate (1992: 81–2) terms it, is documented in a tablet recovered from the city of Eshnunna in the Diyala valley, which lists Amorites 'living in the city' according to their wards, designated by the names of male family heads and their sons.
57. See e.g. Van de Mieroop 1999, especially p. 123.
58. Ibid. 160–61.
59. Stone and Zimansky 1995: 123.
60. Fleming 2009: 1–2.
61. Fleming (2009: 197–9) notes the 'tradition [at Urkesh] of a powerful collective balance to leadership by kings may be the inheritance of a long urban history', and that the council of elders cannot possibly be construed as part of the king's own circle of advisors. It was rather an 'entirely independent political force' of some antiquity, a collective form of urban leadership, which 'cannot be regarded as a minor player in a primarily monarchic framework'.

62. To reconstruct early urban political systems in Mesopotamia, Jacobsen relied especially on the story of 'Gilgamesh and Agga', a brief epic composition about the war between Uruk and Kish, which describes a city council divided into two chambers.
63. Hence population estimates for the fourth millennium BC city are based almost entirely on topographical surveys and distributions of surface finds (see Nissen 2002).
64. Nissen, Damerow and Englund 1993.
65. Englund 1998: 32–41; Nissen 2002. A significant number of the monumental structures on the Eanna complex are spectacularly enlarged versions of a common household type (the so-called 'tripartite house' form) which is ubiquitous in villages of the preceding 'Ubaid period of the fifth millennium BC. Specialists debate whether some of these buildings might have been private palaces rather than temples, but in fact they don't resemble later palaces *or* temples very much. In essence, they are up-scaled versions of traditional house forms, where meetings of large numbers of people probably took place in the idiom of an extended family under the patronage of a deity-in-residence (Wengrow 1998; Ur 2014). The first compelling examples of palace architecture in cities of the southern Mesopotamian alluvium come only centuries later, in the Early Dynastic period (Moorey 1964).
66. See Crüsemann et al. (eds) 2019 for a magnificent survey of Uruk's architectural development over the ages; although we note that their interpretation plays down those aspects of urban planning we would see as clearly relating to civic participation (especially with regard to the early phases of the Eanna sanctuary they tend to assume, even in the absence of written evidence, that any sort of grand architectural project must necessarily have been intended to establish the exclusivity of a ruling elite).
67. Among them are early copies of the so-called 'Titles and Professions List', which was widely reproduced in later times and includes (among other things) terms for various kinds of judges, mayors, priests, chairs of 'the assembly', ambassadors, messengers, overseers of flocks, groves, fields and farming equipment, and also of potting and metalworking. Nissen, Damerow and Englund (1993: 110–11) review the immense difficulties of extracting any kind of social history from such documents, which depends on finding corroborations between particular terms and their recurrence in functional administrative texts of the same period, and even then is somewhat tendentious.
68. Though we should also note that, at least by Old Babylonian times (*c.*2000–1500 BC), much scribal instruction also went on in private households.

69. Englund 1988.
70. Bartash 2015. There is a possibility some were already slaves or war captives at this time (Englund 2009), and as we'll see, this becomes much more commonplace later; indeed, it is possible that what was originally a charitable organization gradually transformed as captives were added to the mix. For the demographic composition of the temple workforce in the Uruk period see also Liverani 1998.
71. Another aspect of quality control in urban temples was the use of cylinder seals. These tiny, near-indestructible carved stones are our main source of knowledge for about 3,000 years of image-making in the Middle East, from the time of the first cities to the Persian Empire (*c.*3500–500 BC). They had many functions, and were not simply 'art objects'. In fact, cylinder seals were among the earliest devices for mechanically reproducing complex images, done by rolling the seal on to a strip or block of clay to make raised figures and signs appear, so they stand at the beginning of print media. They were impressed on inscribed clay tablets, but also marked clay stoppers of jars containing food and drink. In this way, tiny images of people, animals, monsters, gods and so on were made to guard and authorize the contents, which distinguished the otherwise standard products of temple and later palace workshops and guaranteed their authenticity as they passed among unknown parties (see Wengrow 2008).
72. Some Assyriologists once believed this sphere encompassed almost everything: that the first Mesopotamian cities were 'temple states' governed on the basis of 'theocratic socialism'. This thesis has been convincingly refuted; see Foster 1981. We don't really know what economic life was like outside the area administered by the temples; we just know that the temples administered a certain portion of the economy, but not all, and that they had nothing like political sovereignty.
73. On the Uruk Vase the figure of the goddess, probably Inanna, is larger than the males who march towards her. The only exception is the figure who approaches her directly, at the head of the parade, which is mostly lost due to a break in the vessel but is most likely the same standard male figure that appears on cylinder seals and other monuments of the time with his characteristic beard, hair gathered into a chignon, and long woven garments. It is impossible to tell what status this male figure refers to, or if it was occupied on a hereditary or rotating basis. The goddess wears a long robe, which almost completely disguises the contours of her body, while the smaller male figures appear nude, and arguably sexualised (Wengrow 1998: 792; Bahrani 2002).
74. See Yoffee 1995; Van de Mieroop 2013: 283–4.

75. See Algaze 1993. There is no hint of these colonies in the administrative correspondence of the mother-city (and writing was hardly used in the colonies themselves).
76. In essence, these were the sacred origins of what we now call commodity branding; see Wengrow 2008.
77. See Frangipane 2012.
78. Helwig 2012.
79. Frangipane 2006; Hassett and Sağlamtimur 2018.
80. Treherne 1995: 129.
81. Among the more remarkable finds from the Early Bronze Age cemetery at Başur Höyük in eastern Anatolia is an early set of sculpted gaming pieces.
82. Largely as predicted, in fact, by Andrew Sherratt (1996); and see also Wengrow 2011. Where urban and upland societies converged, a third element emerged which resembles neither the tribal aristocracies nor the more egalitarian cities. Archaeologists know this other element as the Kura-Araxes or Transcaucasian culture, but it has proved hard to define in terms of settlement types, which vary widely within it. For archaeologists, what identifies the Transcaucasian culture above all is its highly burnished pottery, which achieved a remarkable distribution extending south from the Caucasus as far as the Jordan valley. Over such considerable distances, methods for making pottery and other distinctive craft products stayed remarkably constant, suggesting to some the migration of artisans, and perhaps even whole communities, to settle in remote locations. Such diaspora groups seem to have been widely involved in the circulation and working of metals, especially copper. They carried with them certain other distinctive practices such as the use of portable cooking hobs, sometimes decorated with faces, which supported lidded pots used to prepare a cuisine based on stews and casseroles: a somewhat eccentric practice in regions where roasting and baking food in fixed ovens was an age-old practice going back to Neolithic times (see Wilkinson 2014, with further references).
83. Recent work attributes the eventual decline of the Indus civilization to changes in the flood regime of the major river systems, prompted by alterations in the monsoon cycle. This is most evident in the drying-up of the Ghaggar-Hakra, once a major course of the Indus, and a shift of human settlement to more easily watered areas where the Indus meets the rivers of Punjab, or to parts of the Indo-Gangetic plain which still fell within the catchment of the monsoon belt; Giosan et al. 2012.

84. For a review of the debates see Green (2020), who develops an argument that the Indus civilization was a case of egalitarian cities, but along rather different lines to our own.
85. For general overviews of the Indus civilization, and further description of the major sites, see Kenoyer 1998; Possehl 2002; Ratnagar 2016.
86. For an overview of the Indus valley's far-flung commercial and cultural contacts in the Bronze Age see Ratnagar 2004; Wright 2010.
87. For the Indus script in general see Possehl 1996; for the Dholavira street-sign, Subramanian 2010; and for the function of Indus seals, Frenez 2018.
88. See Jansen 1993.
89. Wright 2010: 107–10.
90. See Rissman 1988.
91. Kenoyer 1992; H. M.-L. Miller 2000; Vidale 2000.
92. 'The Indus Civilization is something of a faceless sociocultural system. Individuals, even prominent ones, do not readily emerge from the archaeological record, as they do in Mesopotamia and Dynastic Egypt, for example. There are no clear signs of kingship in the form of sculpture or palaces. There is no evidence for a state bureaucracy or the other trappings of "stateness".' (Possehl 2002: 6)
93. Daniel Miller's (1985) perceptive discussion of these points remains important.
94. As discussed by, among others, Lamberg-Karlovsky 1999. It is sometimes objected that viewing the Bronze Age civilization of the Indus valley through the lens of caste means painting an artificially 'timeless' picture of South Asian societies, and thus slipping into 'orientalist' tropes, because the earliest written mention of the caste system and its basic social distinctions or *varnas* occurs only around a millennium later, in the hymns of the *Rig Veda*. In many ways, it's a puzzling – and to some extent self-defeating – objection, because it only makes sense if one assumes that a social system based on caste principles cannot itself evolve, in the same way that, say, class or feudal systems undergo important structural transformations over time. There are, certainly, those who have explicitly taken this position (most famously, Dumont 1972). Obviously, however, that is not the position we are taking here; nor do we see any continuity in this context between caste, language and racial identity (another false equation, which has hampered these kinds of discussions in the past).
95. On this point see Vidale's important (2010) reassessment of Mohenjo-daro and its archaeological record.

96. The general scarcity of weapons from Harappan sites remains striking; but as Corke (2005) points out, in other Bronze Age civilizations (e.g. Egypt, China, Mesopotamia) weaponry tends to be found in burials rather than settlements; so – he reasons – the visibility of weapons and warfare in the Indus valley may be greatly reduced by an overall lack of funerary remains. As he also points out, though, there is no evidence that weapons were used as symbols of authority (by contrast with Mesopotamia, for instance) or in any way formed 'a significant part of elite identity' in the Indus civilization. What is definitely absent is the *glorification* of weapons and the kind of people who employ them.
97. Obviously, it's partly just the desire to preserve the credit for having 'invented' democracy for something called 'the West'. Part of the explanation might also lie in the fact that academia itself is organized in an extremely hierarchical fashion, and most scholars therefore have little or no experience of making democratic decisions themselves, and find it hard to imagine anyone else doing so as a result.
98. Gombrich 1988: 49–50, 110 ff. See also Muhlenberg and Paine 1996: 35–6.
99. As with all such cases, just about everything on the topic of early Indian 'democracy' is contested. The earliest literary sources, the Vedas, assume a society that's entirely rural, and that monarchy is the only possible form of government – though some Indian scholars detect traces of earlier democratic institutions (Sharma 1968); however, by the time of Buddha in the fifth century BC the Ganges valley was home to a host of city-states, small republics and confederations, many of which (the *gana-sangha*) appear to have been governed by assemblies made up of all male members of the warrior caste. Greek travellers like Megasthenes were perfectly willing to describe them as democracies, since Greek democracies were basically the same thing, but contemporary scholars debate how democratic they really were. The entire discussion seems to be premised on the assumption that 'democracy' was some sort of remarkable historical breakthrough, rather than a habit of self-governance that would have been available in any historical period (see, for example, Sharan 1983; Thapar 1984; our thanks to Matthew Milligan for guiding us to relevant source material, although he bears no responsibility for the use we've made of it).
100. On the *seka* principle see Geertz and Geertz 1978; Warren 1993.
101. Lansing 1991.
102. As argued in Wengrow 2015.
103. Possehl 2002: *passim*; Vidale 2010.

104. Independent cities were only entirely abolished in Europe in the seventeenth and eighteenth centuries, as part of the creation of the modern nation state. European empires, and the creation of the modern interstate system in the twentieth century, succeeded in wiping out any traces of them in other parts of the world.
105. Bagley 1999.
106. Steinke and Ching 2014.
107. Interestingly, some of the smallest are in Henan itself, the heartland of the later named dynasties. The town of Wangchenggang, associated with the Xia Dynasty – the semi-legendary precursor to the Shang – has a total walled area of around thirty hectares; see Liu and Chen 2012: 222.
108. Ibid.: *passim*; Renfrew and Liu 2018.
109. Some scholars initially suggested that the Longshan period was an age of high shamanism, an appeal to the later myth of Pan Gu, who prised heaven and earth apart in such a way that only those with spiritual powers could journey between them. Others at first related it to classical legends of *wan guo*, the period of Ten Thousand States, before power was localized to the Xia, Shang and Zhou dynasties; see Chang 1999.
110. Jaang et al. 2018.
111. He 2013: 269.
112. Ibid.
113. He 2018.

9. HIDING IN PLAIN SIGHT

1. The precise location of Aztlán is unknown. Various lines of evidence suggest that populations speaking Nahuatl (the language of the Mexica/Aztec) were dispersed among both urban and rural settings before their southward migration. Most likely they were present, alongside a range of other ethnic and linguistic groups, in the Toltec capital of Tula, which lies north of the Basin of Mexico (Smith 1984).
2. So-called for the founding political union of three city-states: Tenochtitlan, Texcoco and Tlacopan.
3. Mexica kings claimed partial descent from the Toltec rulers of a city called Culhuacan, where they sojourned in the course of their migrations, whence the ethnonym Culhua-Mexica; see Sahlins 2017.
4. Stuart 2000.
5. See Taube 1986; 1992.
6. Published estimates range as high as 200,000 and drop down to as low as 75,000 people (Millon 1976: 212), but the most thorough reconstruction

to date (by Smith et al. 2019) rounds off at 100,000 and relates to the Xolalpan-Metepec phases of the city's occupation, between *c.*AD 350 and 600. At that time, much of the population – both rich and poor – lived in fine masonry apartment blocks, as we'll go on to discuss.

7. In fact, it's quite likely some form of writing system was used at Teotihuacan, but all we can see of it are isolated signs, or small groups, repeated on wall paintings and pots where they caption human figures. Perhaps one day they will yield answers to some of the burning questions about the society that built Teotihuacan, but for the moment they remain largely inscrutable. Scholars can't even tell yet, with any degree of confidence, if the signs name individuals or groups, or perhaps places of origin; for recent and sometimes contradictory discussions see Taube 2000; Headrick 2007; Domenici 2018. It is quite possible, of course, that the residents of Teotihuacan produced more extensive inscriptions on media that have not survived, such as the ephemeral reed or bark paper (*amatl*) later used by Aztec scribes.
8. Other immigrants came to Teotihuacan from as far as Veracruz and Oaxaca, forming their own residential quarters and nurturing their traditional crafts. We should probably imagine at least some of the city's many districts as so many 'Chiapas-towns', 'Yucatán-towns', and so on; see Manzanilla 2015.
9. For the cosmological and political significance of ball-courts in Classic Maya cities see Miller and Houston 1987.
10. See Taube 1986.
11. It is worth noting, in passing, that rather similar arguments were made by the archaeologist and art historian Henri Frankfort (1948; 1951) with regard to the emergence of Egypt and Mesopotamia as parallel, but in some ways opposite, types of civilization; see also Wengrow 2010.
12. Pasztory 1988: 50; and see also Pasztory 1992; 1997.
13. Millon 1976; 1988: 112; see also Cowgill 1997: 155–6; and for more recent arguments, on similar lines, see Froese, Gershenson and Manzanilla 2014.
14. Sharer 2003; Ashmore 2015.
15. See Stuart 2000; Braswell (ed.) 2003; Martin 2001; and for the recent and unprecedented discovery of Maya wall paintings at Teotihuacan see Sugiyama et al. 2019.
16. For Captain Cook as Lono see Sahlins 1985. Hernan Cortés attempted something similar in 1519 after some interpreted him as the second coming of Quetzalcoatl, the once and future king of the Aztecs, though most contemporary historians have concluded he and Moctezuma were really

playing a game where none took the attribution particularly seriously. For other examples, and the general phenomenon of 'stranger-kings', see also Sahlins 2008; Graeber and Sahlins 2017.

17. Based on chemical analysis of human remains from an adult male found in the Hunal Tomb at Copan, which suggest an origin for its occupant – identified as the dynastic founder K'inich Yax K'uk Mo' – in the central Petén region (Buikstra et al. 2004).
18. Cf. Cowgill 2013. Some later conquistadors played a similar role, such as the notorious Nuño Beltrán de Guzmán (*c.*1490–1558), who started his career in the Spanish court as a bodyguard of Charles V before going on to found cities in northwest Mexico, where he ruled as founder-tyrant.
19. Parallels with the fifth century AD seem striking, but again no scholarly consensus exists on how to interpret the evidence connecting these two Tollans of Chichén Itzá and Tula (see Kowalski and Kristan-Graham 2017).
20. Millon 1964.
21. Plunket and Uruñuela 2005; see also Nichols 2016.
22. Froese, Gershenson and Manzanilla 2014.
23. Carballo et al. (2019: 109) note that domestic architecture from this early phase of Teotihuacan's expansion is very poorly understood. Such traces as have been found suggest irregular and unimposing dwellings, erected on posts rather than stone foundations. See also Smith et al. 2017.
24. See Manzanilla 2017.
25. Arguably, the whole affair has a strong millenarian flavour when set against the backdrop of mass displacements and the loss of former homes to natural disasters; cf. Paulinyi 1981: 334.
26. Pasztory 1997: 73–138; and for more up-to-date accounts of the various construction phases, with associated radiocarbon determinations, see S. Sugiyama and Castro 2007; N. Sugiyama et al. 2013.
27. Sugiyama 2005. And for detailed studies of humans remains and their origins see also White, Price and Longstaffe 2007; White et al. 2002.
28. Cowgill 1997: 155.
29. See Cowgill 2015: 145–6.
30. Sugiyama and Castro 2007. Froese et al. (2014: 3) note that the Pyramids of the Sun and Moon may have been considered as 'large-scale public goods on a continuum with the constructions of large-scale housing for most of the population'.
31. Carballo et al. 2019; cf. Smith et al. 2017.
32. Pasztory (1992: 287) observed, 'No other common people in Mesoamerican history lived in such houses', though as we shall see, the case of social housing at Teotihuacan is not as isolated as once thought.

33. See Manzanilla 1993; 1996.
34. Millon 1976: 215.
35. Manzanilla 1993.
36. Froese et al. 2014: 4–5; cf. Headrick 2007: 105–6, fig. 6.3; Arnauld, Manzanilla and Smith (eds) 2012. A significant number of these larger three-temple complexes lie at various points along the Way of the Dead, while others are distributed among the residential zones of the city.
37. Giveaways include dizzying colour contrasts, fractal arrangements of organic forms that merge into one another, and intense geometrical patterning, bordering on kaleidoscopic.
38. Most famously, in the murals from the apartment compounds of Tepantitla district, which also show ball games played in open civic spaces rather than courts (as we discuss further in Chapter Ten). See Uriarte 2006; and also Froese et al. 2014: 9–10.
39. Isolated elements of glyphic writing may complicate this picture, designating specific groups or individuals, although on what criteria exactly is still unknown; Domenici 2018.
40. Manzanilla 2015.
41. Domenici (2018: 50–51), drawing on the work of Richard Blanton (1998; Blanton et al. 1996), proposes a plausible sequence of developments whereby tensions grew between civic responsibilities and the interests of largely self-governing neighbourhoods. Some form of privatization is envisaged, undermining the collective ethos or 'corporate ideology' of earlier times.
42. As pointed out by the historian Zoltán Paulinyi (1981: 315–16).
43. Mann 2005: 124.
44. For an important but still quite isolated exception see the works by Lane F. Fargher and colleagues cited below.
45. Cortés 1928 [1520]: 51.
46. For which, see Isaac 1983.
47. Crosby 1986; Diamond 1997.
48. In sixteenth-century Mexican city-states (or *altepetl*), these urban wards called *calpolli* enjoyed considerable autonomy. *Calpolli* were ideally organized into symmetrical sets, with reciprocal rights and duties. The city as a whole worked on the basis of each *calpolli* taking its turn to fulfil the obligations of municipal government, rendering its share of tribute, workers for corvée service and personnel to staff the higher ranks of political office, including – in the case of royal cities – the office of *tlatoani* (king, or literally 'speaker'). Special land allotments often went along with official roles, to support the incumbent administrator,

and had to be surrendered at the end of a term. This opened up positions of authority to those who lacked hereditary estates. *Calpolli* also existed outside cities – in rural settlements and small towns – where they may have corresponded more closely to extended kin units; within cities they were often defined administratively by their shared responsibilities for delivering tithes, taxes and corvée, but also sometimes along ethnic or occupational lines, or in terms of shared religious duties, or even origin myths. Rather like the English term 'neighbourhood', *calpolli* seems to have become a nebulous concept in the modern scholarly literature, potentially covering an enormous variety of social forms and units; see Lockhart 1985; Fargher et al. 2010; Smith 2012: 135–6 and *passim*.

49. For the literary context of Cervantes de Salazar's writings on New Spain see González González 2014; and also Fargher, Heredia Espinoza and Blanton 2010: 236, with further references.
50. Nuttall 1921: 67.
51. Ibid. 88–9.
52. If any of this seems somewhat unlikely, we would ask the reader to consider that the 1585 manuscript of Diego Muñoz Camargo's remarkable *Historia de Tlaxcala*, which in fact comprises three sections – one textual, in Spanish, and two pictographic, in Spanish and Náhuatl – effectively vanished from sight for around two centuries and was not registered at all in the comprehensive survey of Mesoamerican manuscripts undertaken in 1975. It eventually resurfaced in a collection bequeathed to the University of Glasgow by Dr William Hunter in the eighteenth century, and a facsimile edition was produced only in 1981.
53. Courtesy of Biblioteca Virtual Universal, Buenos Aires, the reader can find a digital edition of Cervantes de Salazar's text, *Crónica de la Nueva España*, at: http://www.cervantesvirtual.com/obra-visor/cronica-de-la-nueva-espana--o/html/
54. Xicotencatl the Younger, or Xicotencatl Axayacatl, was initially cast as a traitor in both Spanish colonial and Tlaxcalteca accounts; according to Ross Hassig (2001) his rehabilitation and reputation as an indigenous fighter against the Spanish only took place after Mexico declared independence.
55. We paraphrase here from the Spanish. We are not aware of any authorized translation of these words into English. Incidentally, Xicotencatl the Elder was quite right about all this: it didn't take long after the conquest of Tenochtitlan for Tlaxcala to lose its privileges and exemptions with the Spanish Crown, reducing its populace to just another source of tribute.

56. Hassig (2001: 30–32) provides a summary of the standard account, drawing mainly on Bernal Díaz del Castillo; he also considers possible factors behind the Spanish execution of Xicotencatl the Younger, who died by hanging at the age of thirty-seven.
57. The possibility that Kandiaronk, whom the Jesuits considered to rank among the smartest people that ever lived, might himself have learned about some of Lucian's best lines in his conversations with the French, been impressed, and deployed variations of them in later debate is one that seems utterly inconceivable to such scholars.
58. See Lockhart, Berdan and Anderson 1986; and for Nahuatl traditions of direct speech and political rhetoric, also Lockhart 1985: 474.
59. MacLachlan (1991: xii and n.12) is fairly typical in this regard when commenting on the 'remarkable adjustment' of Tlaxcalteca members of the council to (supposedly) European mores, which he attributes almost entirely to native self-interest under conditions of imperial domination.
60. For a useful discussion of shifting scholarly opinion on such matters, Lockhart (1985) remains valuable.
61. As, for instance, with academic responses to the so-called 'Influence Debate', which we touch on in a later chapter, triggered by the proposal that Haudenosaunee federal structures (the Six Nations of the Iroquois) might have been one model for the US Constitution.
62. Motolinía 1914 [1541]: 227. Even if we can't always establish direct links between the surviving texts of de Salazar, Motolinía and other chroniclers, it seems safe to assume that by the 1540s there would have been any number of bilingual Nahuatl and Spanish speakers in large centres like Tlaxcala, exchanging stories about the deeds and sayings of their notable recent ancestors.
63. Gibson 1952; and see also Fargher, Heredia Espinoza and Blanton 2010: 238–9.
64. On Chichimec see also Sahlins 2017, with further references.
65. Balsera 2008.
66. Fargher et al. 2011.

10. WHY THE STATE HAS NO ORIGIN

1. Lévi-Strauss (1987) referred to Northwest Coast societies as 'house' societies, that is, ones where kinship was organized around noble households, which were precisely the holders of titles and heirloom treasures (as well as slaves, and the loyalty of retainers). This arrangement seems typical of heroic societies more generally; the palace at Arslantepe, which we

described in Chapter Eight, is most likely just a more elaborate version of the same thing. There is a direct line from here to what Weber called 'patrimonial' and 'prebendal' forms of governance, where entire kingdoms or even empires are imagined as extensions of a single royal household.

2. This again is easy to observe in activist groups, or any group self-consciously trying to maintain equality between members. In the absence of formal powers, informal cliques that gain disproportionate power almost invariably do so through privileged access to one or another form of information. If self-conscious efforts are made to pre-empt this, and make sure everyone has equal access to important information, then all that's left is individual charisma.
3. This definition held sway for a long time in Europe. It is why medieval England could begin holding elections to select parliamentary representatives as far back as the thirteenth century; but it never occurred to anyone that this had something to do with 'democracy' (a term which, at the time, was held in near-total disrepute). It was only much more recently, in the late nineteenth century, when men like Tom Paine came up with the idea of 'representative democracy' that the right to weigh in on spectacular competitions among the political elite came to be seen as the essence of political freedom, rather than its antithesis.
4. Definitions that ignore sovereignty have little currency. One could argue, hypothetically, that the essence of 'statehood' is a system of governance with at least three tiers of administrative hierarchy, staffed by professional bureaucrats. But by that definition we would have to define the European Union, UNESCO and the IMF as 'states', and this would be silly. They are not states by any common definition, precisely because they lack sovereignty and make no claim to it.
5. Which is not, of course, to say that they didn't make grandiose claims to territorial sovereignty; just that careful analysis of the ancient written and archaeological sources shows these claims to be generally hollow; see Richardson 2012.
6. On 'Early Bronze Age urbanism and its periphery' in western Eurasia see also Sherratt 1997: 457–70; and more generally, Scott's (2017: 219–56) reflections on 'The golden age of the barbarians'.
7. This pattern much resembles Weber's famous notion of 'the routinization of charisma', where the vision of a 'religious virtuoso', whose charismatic quality was based explicitly on presenting a total break with traditional ideas and practices, is gradually bureaucratized in subsequent generations. Weber argued that this was the key to understanding the internal dynamics of religious change.

8. Nash 1978: 356, citing Soustelle (1962), citing Bernardino de Sahagún's *Historia general de las cosas de Nueva España*.
9. Dodds Pennock (2017: 152–3) discusses a revealing episode in 1427, when Aztec visitors to a Tepanec banquet were made to dress as women on the orders of the Maxtla (the Tepanec ruler) in order to humiliate both them and their own ruler, who had of late failed to avenge the rape of Aztec women by Tepanec in the market at Coyoacan; things came full circle two years later, when Aztec armies entered Atzcapotzalco and sacrificed Maxtla to the gods.
10. As reported, for example, in the memoirs of Bernal Díaz (in Maudslay's translation), see among others the section on *Complaints of Montezuma's tyranny*: 'but they [the local chiefs] said that Montezuma's tax gatherers carried off their wives and daughters if they were handsome and ravished them, and this they did throughout the land where the Totonac language was spoken.' See also Townsend 2006; Gómez-Cano 2010: 156.
11. Dodds Pennock (2008) situates the public practice of religious violence within broader Aztec notions of gender, vitality and sacrifice; and see also Clendinnen 1991.
12. See Wolf 1999: 133–96; Smith 2012.
13. For general overviews of the Inca Empire and its archaeological remains see Morris and van Hagen 2011; D'Altroy 2015.
14. Murra 1982.
15. *Ayllu*, as we will discuss again later in the chapter, were land-holding groups, bound by ties of descent that cut across households. Their original function was to manage the redistribution of labour within, and sometimes between, villages, so no household was left to fend for itself. The kind of tasks usually taken on by *ayllu* corporations were routinely necessary, but fell beyond the capacity of a typical nuclear household: such things as clearing fields, harvesting crops, managing canals and reservoirs, porterage or fixing bridges and other buildings. Importantly, the *ayllu* organization also acted as a support system for families who found themselves unable to obtain the basic material requirements of lifecycle rituals – *chicha* for funerals, houses for newly married couples and so on. See Murra 1956; Godoy 1986; Salomon 2004.
16. Gose 1996; 2016.
17. See Kolata 1992; 1997.
18. Silverblatt 1987; and cf. Gose 2000.
19. Urton and Brezine 2005.
20. Hyland 2016.
21. Hyland 2017.

22. Clendinnen 1987.
23. Maya writings from the early colonial period, such as the books of *Chilam Balam*, almost invariably treat the Spanish not as the actual government but as irritating interlopers, and rival factions of Maya nobility – engaged in ongoing struggles for influence that the supposed conquerors appear to have been entirely unaware of – as still constituting the real government (Edmonson 1982).
24. Just how much else remains to be discovered is highlighted by new (LiDAR) techniques for mapping tropical landscapes, which recently led experts to triple their estimates for the Classic Maya population; see Canuto et al. 2018.
25. See Martin and Grube 2000; Martin 2020.
26. For a tentative reconstruction of how Maya rulership evolved out of earlier forms of shamanic power see Freidel and Schele 1988.
27. In the absence of definitive evidence, theories of collapse have tended to follow the political concerns of their day. During the Cold War, many Euro-American Mayanists seemed to assume some kind of class conflict or peasant revolution; since the 1990s there has been more of a tendency to focus on ecological crises of one sort or another as the main causal factor.
28. Ringle 2004; see also Lincoln 1994. These reconstructions remain hotly debated (see Braswell (ed.) 2012), but if broadly correct, even in outline, they would correspond to what Graeber and Sahlins (2017) describe as a shift from 'divine' to 'sacred' forms of kingship, or even 'adverse sacralisation'.
29. Kowaleski 2003.
30. And for *K'iche* parallels see Frauke Sachse: 'The Martial Dynasties – the Postclassic in the Maya Highlands', in Grube et al. (eds) 2001: 356–71.
31. Kubler 1962.
32. Kroeber (1944: 761) began his grand conclusion as follows: 'I see no evidence of any true law in the phenomena dealt with; nothing cyclical, regularly repetitive, or necessary. There is nothing to show either that every culture must develop patterns within which a florescence of quality is possible, or that, having once so flowered, it must wither without chance of revival.' Neither did he find any necessary relation between cultural achievement and systems of government.
33. In continental Europe, there's an entire category of scholarship known as 'proto-history' which describes the study of peoples like the Scythians, Thracians or Celts, who briefly break into the light of history through

the writings of Greek or Roman colonizers, only to fizzle back out again when the literate gaze turns elsewhere.

34. In their additional cultic role as the 'god's hand' the wives of Amun – such as Amenirdis I and Shepenupet II – were also obliged to assist the male creator-god in acts of cosmic masturbation; so, in ritual terms, she was as subordinate to a male principal as one could possibly imagine, while in reality running a good portion of Upper Egypt's economy and calling political shots at court. Judging by the grand locations of their funerary chapels at Karnak and Medinet Habu, the combination made for some very effective *realpolitik*.
35. See John Taylor's chapter on 'The Third Intermediate Period' in Shaw (ed.) 2000: 330–69, especially 360–62; also Ayad 2009.
36. Schneider 2008: 184.
37. In *The Oxford History of Ancient Egypt* (Shaw ed. 2000), for instance, the relevant chapter is called 'Middle Kingdom Renaissance (*c.*2055–1650 BC).'
38. For a useful summary see Pool 2007.
39. Rosenwig 2017. Again, this picture is liable to change quite dramatically with the application of LiDAR survey techniques in the provinces of Tabasco and Veracruz, which is already under way at the time of writing.
40. See Rosenwig 2010.
41. Attention to individual differences and personal aesthetics is also evident in a second major category of Olmec sculpture, most abundantly documented at San Lorenzo. It depicts human figures with unusual or anomalous features, including images of hunchbacks, dwarfs, lepers and possibly also images based on people's observations of miscarried embryos; see Tate 2012.
42. See Drucker 1981; Clark 1997; Hill and Clark 2001.
43. See Miller and Houston 1987.
44. Hill and Clark 2001. It's of more than passing interest, in this context, that Teotihuacan – governed on more collective principles than Olmec, Maya or Aztec cities – had no such arena for the official staging of ball games. Excluding a public ball-court from the municipal plan must surely have been a deliberate choice, since many of Teotihuacan's occupants would have been familiar with such spectacles, and as we saw in Chapter Nine, nearly everything else in the city centre was laid out with exacting foresight and precision. When ball games do appear at Teotihuacan, it's in a different context. So very different, that one begins to suspect some conscious inversion of ideas that were canonical in the

surrounding kingdoms of Oaxaca and the Maya lowlands (recalling that people moved regularly between these regions and were familiar with the practices of their neighbours).

The evidence comes from domestic wall paintings in one of Teotihuacan's well-appointed housing estates, known as Tepantitla. Gods are depicted, but also the earliest known images of people playing ball games with their feet, hands and sticks – something on the lines of soccer, basketball and hockey (see Uriarte 2006). All this was in violation of aristocratic norms. The scenes have a street setting, with large numbers of participants all shown at the same scale. Associated with these scenes is a recurrent symbolism of water lilies, a powerful hallucinogen. Perhaps what we are seeing here is something peculiar to Teotihuacan; or perhaps we are glimpsing something of the games played by ordinary folk throughout Mesoamerica, a side of life that is largely invisible to us in more stratified polities.

45. Clendinnen 1991: 144.
46. In this respect, Wilk's (2004) stimulating comparison between the dynamics of the Olmec horizon and the cultural/political impact of modern beauty pageants, such as Miss World and Miss Universe, seems very apposite. Geertz coined the phrase 'theatre state' (1980) to describe Balinese kingdoms, where, he suggested, the entire apparatus of tribute basically existed for the purpose of organizing spectacular rituals, rather than the other way round. His argument has some notable weaknesses – especially as seen from the perspective of Balinese women – but the analogy may still be helpful, especially when one considers the original role of those famous Balinese cock fights (familiar to any first-year anthropology student); they were initially promoted and staged by royal courts as a way of putting people into debt, which not infrequently led to one's wife and children being handed over to the palace for use as slaves or concubines, or for onward sale abroad (Graeber 2011: 157–8, 413 n.88).
47. As we saw in Chapter Eight.
48. See Conklin and Quilter (eds) 2008.
49. See Isbell 2008.
50. See Quilter 2002; Castillo Butters 2005.
51. Cf. Weismantel 2013.
52. These are precisely the sort of highly complex images studied by the anthropologist Carlo Severi (2015) in his classic analysis of the 'chimera principle'.
53. Burger 2011; Torres 2008. The stone carvings at Chavín de Huántar seem mostly concerned with making permanent what were inherently

ephemeral experiences of altered states of consciousness. Animal motifs typical of Chavín art – such as felines, snakes and crested eagles – actually occur up to 1,000 years earlier on cotton textiles and in beadwork, which already circulated widely between the highlands and the coast. Interestingly, more fully preserved textiles from later periods show that, even at the height of Chavín's power, coastal societies were approaching their deities in explicitly female forms (Burger 1993). At Chavín de Huántar itself, women appear to be absent from the surviving repertory of figural sculpture.

54. Rick 2017.
55. See Burger 2008.
56. See Weismantel 2013.
57. For a more detailed discussion of the divine kingship of the Natchez, with full references, see Graeber's chapter 'Notes on the Politics of Divine Kingship', in Graeber and Sahlins 2017: 390–98.
58. Cited in Graeber, ibid. p. 394.
59. Lorenz 1997.
60. See Gerth and Wright Mills (eds) 1946: pp. 233–4.
61. Brown 1990: 3, quoting John Swanton's *Indian Tribes of the Lower Mississippi Valley and Adjacent Coast of the Gulf of Mexico* (1911) (Bureau of American Ethnology, Bulletin 43).
62. For a good summary of such royal 'exploits' see de Heusch 1982; the king most famous for gunning down his own subjects was the Ganda King Mutesa, who was trying to impress David Livingstone after the latter presented him with the gift of a rifle, but it's by no means a unique event: see Simonse 1992; 2005.
63. Graeber and Sahlins 2017: 129.
64. Crazzolara 1951: 139.
65. Reported in Diedrich Westermann's *Shilluk People: their Language and Folklore* (1911). Philadelphia: Board of Foreign Missions of the United Presbyterian Church of North America, p. 175.
66. Graeber and Sahlins 2017: 96, 100–101, 130.
67. We will be considering such possibilities further in the next chapter.
68. Actually, we are being disingenuous here. This is not just a thought experiment: the remains of the Great Village – now known to archaeologists as the Fatherland Site, in Adams County – were in fact excavated, notably by Stu Neitzel in a few intermitted seasons of fieldwork, during the 1960s and early 1970s. In the centuries since its abandonment, what was left of the site had been covered by up to ten feet of colluvial mud deposited by the St Catherine Creek, which first had to be removed with

heavy machinery (bulldozers), playing havoc with the archaeological remains below and obliterating key evidence. What Neitzel (1965; 1972) reported accords in broad outline with what we have just described; no doubt, more careful and modern techniques could do a lot better in terms of archaeological reconstruction (cf. Brown 1990).

69. In fact, early excavations in the vicinity of Mound C, the likely location of the Natchez temple, did turn up more than twenty burials with grave goods including objects of French as well as local manufacture; but their excavation was poorly conducted, with no systematic documentation, and they likely date to the very final period of temple use, just before the building was razed, and when the power of the Great Sun was no doubt already much diminished (see Brown 1990: 3; Neitzel 1965, reporting finds made by Moreau B. C. Chambers in 1930).
70. Egyptologists refer to the First and Second Dynasties as Egypt's 'Early Dynastic' period while the 'Old Kingdom' – somewhat confusingly – begins only in the Third Dynasty.
71. See Dickson 2006; Morris 2007; Campbell (ed.) 2014; Graeber and Sahlins 2017: 443–4, with further references.
72. For the latter possibility, and a review of earlier interpretations, see Moorey (1977); but for an alternative view, which sees them as true royal burials, see Marchesi 2004.
73. Campbell 2014.
74. Cf. Campbell 2009.
75. Although perhaps not the only tombs, since Egypt's earliest rulers may occasionally have split their ancestors' bodies up, burying them in more than one location to distribute their mortuary cult as widely as possible; see Wengrow 1996: 226–8.
76. Wengrow 1996: 245–58; Bestock 2008; see also Morris 2007; 2014.
77. Macy Roth 2002.
78. Maurice Bloch (2008) has observed, in a similar vein, that early states almost invariably involve an explosion of spectacular and often apparently random violence, and that the final result of such states is to 'disorganize' the ritual life of ordinary households in a way that, somehow, can never be put back to how it was even if those states collapse. It's from this dilemma, he argues, that the phenomenon of universalizing religion emerges.
79. One effect of this was to create a series of 'no-man's-lands' around Egypt's territorial borders. For example, the political separation of Egypt from once closely related lands and peoples in Sudan seems to have involved the depopulation of territories on Egypt's newly established

southern boundary, and the dismantling of a former apparatus of chiefly power within Nubia: the so-called A-Group, as archaeologists call it. This took place in an act of violent domination, commemorated in a rock carving at Gebel Sheikh Suleiman on the Second Cataract. So, in effect, we have a kind of symmetry between extremes of ritual killing at the centre of the new Egyptian polity (on the occasion of a ruler's demise) and the foundational violence taking place, or commemorated, on its territorial frontiers; Baines 2003; Wengrow 2006: *passim*.

80. For which see Lehner 2015.

81. Wengrow et al. 2014.

82. Jones et al. 2014. Neolithic burials were usually located in the arid margins of the Nile valley (areas dry enough to afford a certain amount of natural preservation for the corpse), and sometimes further into adjacent desert lands; they seem not to have had any sort of durable superstructures, but were often laid out in large cemeteries, and other lines of evidence show that people remembered, revisited and reused the same locations over a period of generations; see Wengrow 2006: 41–71; Wengrow et al. 2014.

83. Indeed, Egyptologists had long noted certain elements of later kingship showing up in art far 'too early' – for instance, the famous Red Crown of Lower Egypt appears on a piece of pottery dated almost 1,000 years before the Red and White Crowns were combined to become an official symbol of Egyptian unity; the standard motif of a king wielding a mace to smite his foes crops up in a painted tomb at Hierakonpolis, 500 years before the Narmer Palette, and so on. See Baines 1995 for further examples and references.

84. Recent Nilotic peoples have tended to be strictly patrilineal; this, actually, does not entirely exclude women from taking on prominent positions, but generally they do so by playing the part of men. Among the Nuer, for instance, a 'bull' or village leader with no male heir can simply declare his daughter a man, and she might well take over his position, even marry a woman and be recognized as father of her children. It's probably no coincidence that in Egyptian history as well, women who took on dominant positions often did so by declaring themselves, effectively, males (a notable exception to this being the god's wives of Amun, whom we discussed earlier in this chapter).

85. See Wengrow 2006: Chapters One, Four and Five; Kemp 2006; Teeter (ed.) 2011. Population estimates for these 'proto-kingdoms' remain highly speculative due to the inaccessibility of ancient living quarters and the burial of large areas of prehistoric settlement beneath modern field systems and floodplains.

86. See Friedman 2008; 2011.

87. See Wengrow 2006: 92–8.
88. Ibid.: 142–6.
89. Integration of large-scale *chicha* consumption into lifecycle rituals was not actually an Inca innovation – it traces back to the expansion of Tiwanaku, midway between Chavín (with its very different ritual comestibles) and the Inca; see Goldstein 2003.
90. See Murra 1956: 20–37.
91. Wengrow 2006: 95, 160–63, 239–45, with further references.
92. Lehner 2015.
93. See also Roth 1991.
94. Symbolic and likely also practical associations between monumental architecture and the activities of ships' crews are also suggested for the later Bronze Age stone temples of Byblos (Jbeil) in Lebanon, a port town with close trading and cultural links to Egypt (see Wengrow 2010b: 156); and ethnographic descriptions of how team-skills transfer from boat-handling to the manipulation of heavy stone-work can be found, for instance, in John Layard's classic ethnography of a Melanesian island, *Stone Men of Malekula* (1942). London: Chatto and Windus.
95. The production line analogy is inspired by Lewis Mumford on the 'megamachine', where he famously argued that the first complex machines were in fact made of people. The 'rationalization' of labour typical of the factory system was, as scholars like Eric Williams long ago suggested, really pioneered on slave plantations in the seventeenth and eighteenth centuries, but others have recently pointed out that ships around that time, both merchant and military, seem to have been another major zone of experimentation, since being on board such vessels was one of the few circumstances where large numbers of people were assigned tasks entirely under a single overseer's command.
96. As pointed out by feminist theorists (e.g. Noddings 1984).
97. It is worth recalling here that in the tombs of some of Egypt's highest-ranking officials, during the Old Kingdom, we find among their most important titles not just military, bureaucratic and religious offices but also duties such as 'Beloved Acquaintance of the King', 'Overseer of the Palace Manicurists', and so on (Strudwick 1985)
98. Compare Baines 1997; 2003; Kolata 1997.
99. For the different forms of Egyptian and Mesopotamian kingship see Frankfort 1948; Wengrow 2010a; for rare exceptions to this pattern, in which Mesopotamian kings appear to have claimed divine or near-divine status, see the contributions by Piotr Michalowski and Irene Winter in Brisch (ed.) 2008, both stressing the exceptional and ambivalent nature of such claims.

100. This situation persisted even in later Mesopotamian history: when Hammurabi erected a stela with his famous law code in the eighteenth century BC, this might have seemed the quintessential sovereign act, decreeing how violence could and could not be used within the king's territories, creating a new order out of nothing; but in fact most of these grand edicts appear never to have been systematically enforced. Babylonian subjects continued to use the same complex patchwork of traditional legal codes and practices they had before. Moreover, as the decorative scheme of the stela makes clear, Hammurabi is acting on the authority of the sun-god Shamash; see Yoffee 2005: 104–12.
101. And here we can draw a further contrast with Mesopotamia, where administration was an established feature of earthly government, but the cosmos – far from being predictably organized – was inhabited by gods whose actions (like those of the biblical Yahweh) often came in the form of unexpected interventions, and frequently chaotic ruptures in human affairs; Jacobsen 1976.
102. Other examples of regimes where sovereignty and competitive politics dominated the earthly sphere, and administrative hierarchies were projected on to the universe, might include many South Asian societies, which exhibit a similar fascination with cosmic cycles, and medieval Europe, where the Church and its image of angelic hierarchies seems to have preserved a memory of the old legal-bureaucratic order of ancient Rome.
103. Martin and Grube 2000: 20; Martin 2020.
104. See Bagley 1999.
105. Shaughnessy 1989.
106. Appeal to divination is limited in Egypt before the New Kingdom, and had an ambivalent role in Inca systems of government. As Gose (1996: 2) explains, in the Inca case oracular performances were actually at odds with the personal authority of living kings. They centred instead on the mummified bodies of royal ancestors or their statue equivalents, which provided one of the few venues for expressing subaltern (and potentially subversive) views, in a manner that did not challenge the assumption of the ruler's absolute sovereignty and supreme authority. In a similar way, in Renaissance times, to take a king or queen's horoscope was often considered to be an act of treason. Maya kings used bloodletting and stone-tossing as forms of divination, but they do not seem to have been of central importance to affairs of state.
107. Yuan and Flad 2005. Outside the sphere of literacy, divination with animal parts was also widely practised.
108. See Keightley 1999.

109. Shaughnessy 1999.
110. Shang China might well be considered the paradigm for what the anthropologist Stanley Tambiah (1973) has described as 'galactic polities', also the most common form in later Southeast Asian history, where sovereignty concentrated at the centre and then attenuated outwards, focusing in some places, fading in others to the point where, at the edges, certain rulers or nobles might actually claim to be part of empires – even distant descendants of the founders of empires – whose current ruler had never heard of them. We can contrast this sort of outward proliferation of sovereignty with another kind of macro-political pattern, emerging first in the Middle East and then gradually across much of Eurasia, where diametrically opposed notions of what actually constitutes 'government' would face off against each other in dynamic tension, creating the great frontier zones that separated bureaucratic regimes (whether in China, India or Rome) from the heroic politics of nomadic peoples which threatened constantly to overwhelm them; for which see Lattimore 1962; Scott 2017.
111. The clearest illustration of this royal theme from the Old Kingdom is to be found in the reliefs surviving from Sahure's mortuary temple at Abusir; see Baines 1997.
112. Baines 1999.
113. See Seidlmayer 1990; Moreno García 2014.
114. Translation as per Seidlmayer 1990: 118–21. Especially striking, in this regard, are the nomarch's claims to keep his people not just healthy, but also provided with the basic necessities for a full social life: the resources to sustain a family, conduct proper funerals, and the guarantee that one would not be cut loose from one's local moorings or condemned to live as a refugee.
115. Dunbar 1996: 102; Diamond 2012: 11. This assumption is enshrined both in the kind of 'scalar stress' theories we discussed at the start of Chapter Eight, and also in a certain strand of evolutionary psychology (see also Dunbar 2010), which argues that bureaucracy supplies the administrative solution to problems of information storage and management, arising when societies scale up beyond a certain threshold of face-to-face interaction. Bureaucracy, according to such theories, acts as a kind of 'external symbolic storage' when the innate capacities of the human mind for storing and recalling information (e.g. relating to the flow of commodities or labour) are overtaxed. So far as we are aware, there is absolutely no empirical evidence to support this hypothetical but deeply ingrained reconstruction of the 'origins of bureaucracy'.

116. Interestingly, the archival records of the nineteenth-century Merina kingdom of Madagascar are much the same: the kingdom was conceived in patrimonial terms as a royal household, and each descent group's true nature was defined by the service they performed for the king, who was often imagined as a child, with the people as his or her nursemaids. The records go into endless and exact detail concerning every item that passed in and out of the royal household to maintain the ruler, but are otherwise almost silent on economic affairs (see Graeber, 'The People as Nursemaids of the King,' in Graeber and Sahlins 2017).
117. Akkermans (ed.) 1996.
118. See Schmandt-Besserat 1992.
119. Oddly, very few such seals have been found at Tell Sabi Abyad itself, perhaps because they were made of materials that have not survived such as wood; miniature stamp seals made of stone, and of the kind we are referring to here, are quite ubiquitous on other northern Mesopotamian sites of the same ('Late Neolithic' or 'Halaf') period.
120. Akkermans and Verhoeven 1995; Wengrow 1996.
121. The suggestion – sometimes voiced – that all this was a result of part of the population being absent from the village during the herding season, when they took their flocks to graze on nearby hillsides (a practice called transhumance), is probably much too simplistic. It also doesn't make a great deal of sense – there were still the elderly, spouses, siblings and offspring left in the village to look after property or report problems.
122. See Wengrow 2010a: Chapter Four.
123. Wengrow 1998: 790–92; 2001.
124. This is of course somewhat ironic, since archaeologists working within older frameworks of social evolution had long assumed that 'Ubaid societies must have been organized into some sort of 'complex chiefdoms', simply because they are chronologically located between the earliest farming settlements and the earliest cities (which, in turn, are assumed to usher in the 'the birth of the state'). The logical circularity of such arguments will by now be very obvious, as will their lack of fit to the archaeological evidence for the periods in question.
125. Murra 1956: 156.
126. See, especially, Salomon 2004. It may be observed in passing that market systems in medieval villages in England appear to have worked in much the same way, though less formally: the vast majority of transactions were on credit, and every six months to a year a collective accounting was held in an effort to cancel all debts and credits back to zero (Graeber 2011: 327).

127. Salomon 2004: 269; Hyland 2016. It is possible that string-based counting techniques may also have been used in prehistoric Mesopotamia, in conjunction with clay symbols and stamp seals, as demonstrated by evidence for the suspension of perforated lumps of clay, shaped into regular forms and sometimes bearing impressed signs or symbols (Wengrow 1998: 787).
128. See Wernke 2006: 180–81, with further references.
129. For the mode of organization and labour tribute schedules see Hyland 2016.
130. John Victor Murra in his magisterial thesis on *The Economic Organization of the Inca State* (1956), cites Spanish sources who tell of local misanthropes and misfits elevated to new positions of authority; of neighbours turning against each other; and debtors uprooted from their villages – though one can never be sure how much of this was a result of the *conquista* itself; see also Rowe 1982.
131. For which see *Debt: The First 5000 Years*, by one of the authors of the present volume (Graeber 2011); and also Hudson 2018.
132. Von Dassow 2011: 208.
133. Murra (1956: 228) concludes that the illusion of the Inca state as socialistic derives from 'ascribing to the state what was actually an *ayllu* function'. 'Security for the incapacitated was provided by an age-old, pre-Incaic system of automatic access to community assets and surpluses as well as reciprocal labour services,' he goes on: 'There may have been some state relief in the case of major frost and drought; the references to this are late and very few compared to the hundreds that describe the use of reserves for military, court, church and administrative purposes.' Probably, this is overstated, since the Inca also inherited administrative structures and an apparatus of social welfare from some of the kingdoms they conquered, so the reality must surely have varied from place to place (S. Rockefeller, personal communication).
134. See, for example, Richardson (2012) on Mesopotamia, and Schrakamp (2010) on the seasonal dimensions of early dynastic military organization; Tuerenhout (2002) on seasonal warfare among the Classic Maya; and for further examples and discussion see contributions to Neumann et al. (eds) (2014); Meller and Schefik (2015).
135. James Scott (2017: 15) makes a similar observation at the beginning of his book *Against the Grain*: 'In a good part of the world, the state, even when it was robust, was a seasonal institution. Until very recently, during the annual monsoons in Southeast Asia, the state's ability to project its power shrank back virtually to its palace walls. Despite the state's

self-image and its centrality in most standard histories, it is important to recognize that for thousands of years after its first appearance, it was not a constant but a variable, and a very wobbly one at that in the life of much of humanity.'

136. Buoyed, perhaps, by the illusion common to so many who claim arbitrary power, that the fact that you are able to kill your subjects is somehow equivalent to having given them life.
137. In a brilliant and under-appreciated book called *Domination and the Arts of Resistance* (1990), James Scott makes the point that whenever one group has overwhelming power over another, as when a community is divided between lords and serfs, masters and slaves, high-caste and untouchable, both sides tend to end up acting as if they were conspiring to falsify the historical record. That is: there will always be an 'official version' of reality – say, that plantation owners are benevolent paternal figures who only have the best interest of their slaves at heart – which no one, neither masters or slaves, actually believes, and which they are likely to treat as self-evidently ridiculous when 'offstage' and speaking only to each other, but which the dominant group insists subordinates play along with, particularly at anything that might be considered a public event. In a way, this is the purest expression of power: the ability to force the dominated to pretend, effectively, that two plus two is five. Or that the pharaoh is a god. As a result, the version of reality that tends to be preserved for history and posterity is precisely that 'official transcript'.
138. Abrams 1977.
139. See also Wengrow 2010a.
140. As pointed out, for example, by Mary Harris (1997); recall here also that the centralized knowledge systems of Chavín, the Classic Maya and other pre-Columbian polities may well have rested on a continent-wide system of mathematics, originally calculated with the aid of strings and cords, and hence grounded ultimately in fabric technologies (Clark 2004); and that the invention of cuneiform mathematics in cities was preceded by some thousands of years of sophisticated weaving technologies in villages, echoes of which are preserved in the forms and decoration of prehistoric ceramic traditions throughout Mesopotamia (Wengrow 2001).
141. Renfrew 1972.
142. A chronological scheme, widely used by archaeologists for the entire island, starts with the 'Pre-palatial', moving on to 'Proto-palatial', 'Neo-palatial' and so on.
143. Whitelaw 2004.
144. See Davis 1995.

145. Preziosi and Hitchcock 1999.
146. We should perhaps mention here Arthur Evans's notorious identification of a 'priest-king' among the painted figures he uncovered at Knossos at the turn of the twentieth century (for which see S. Sherratt 2000). In fact, the various bits of decorated wall relief Evans used to piece this image together came from different archaeological strata, and probably never belonged to a single figure (he himself thought this to begin with, but changed his mind). Even the gender of the priest-king is now questioned by art historians. But the more basic question is why anybody would want to seize on a single, possibly male figure with an impressive plumed hat as evidence for kingship, when the overwhelming majority of Minoan pictorial art is pointing us in another direction altogether. We will follow it in a moment. But in short, the priest-king of Knossos is no better contender for a throne than his similarly named and similarly isolated counterpart at the earlier Bronze Age city of Mohenjo-daro in the Indus valley, whom we encountered in Chapter Eight.
147. Younger 2016.
148. 'It is certain', Evans wrote, 'that, however much the male element had asserted itself in the domain of government by the great days of Minoan Civilization, the stamp of Religion still continued to reflect the older matriarchal stage of social development.' (cited in Schoep 2018: 21)
149. For a detailed discussion of early Egyptian imports to Crete, via Lebanon, and their likely association with women's rituals see Wengrow 2010b.
150. Voutsaki 1997.
151. Palaces only called in tribute on certain specific goods – such as flax, wool and metals – which were converted into a still more specific range of items in palace workshops, mainly fine textiles, chariots, weapons and scented oils. Other major industries, such as pottery manufacture, are entirely missing from the administrative records. See Whitelaw 2001.
152. See Bennett 2001; S. Sherratt 2001.
153. Kilian 1988.
154. Rehak 2002.
155. Groenewegen-Frankfort 1951.

11. FULL CIRCLE

1. Uncultivated, quite literally, for as Montesquieu put it, perhaps most succinctly: 'These people enjoy great liberty; for as they do not cultivate the earth, they are not fixed: they are wanderers and vagabonds ...'

(*Spirit of the Laws*, 18: 14 – 'Of the political State of the People who do not cultivate the Land').

2. Lovejoy and Boas 1965.
3. Scott 2017: 129–30.
4. Ibid.: 135.
5. Ibid.: 253.
6. Sahlins and Service 1960.
7. The closest we have to a historical comparison is economic: the Socialist Bloc, which existed from roughly 1917 to 1991 and at its peak encompassed a fair amount of the world's land mass and population, is often treated as a (failed) experiment in this sense. But some would argue that it was never really independent from the larger capitalist world system, but simply a subdivision of state capitalism.
8. We are drawing here on examples discussed at greater length in *Debt: The First 5000 Years* (Graeber 2011: Chapter Nine), which describes these coordinated changes largely in terms of the alternation between physical (gold and silver) currency and various forms of abstract (intangible) credit money.
9. For a general overview of Cahokia see Pauketat 2009.
10. Williams 1990.
11. Severi (2015) discusses evidence for the use of pictographic writing systems among the indigenous peoples of North America, and why their characterization as 'oral' societies is misleading in many ways.
12. *JR* (1645–6) 30: 47; see also Delâge 1993: 74.
13. Carr et al. 2008.
14. Knight 2001; 2006.
15. Sherwood and Kidder 2011.
16. 'The Archaic measuring unit appears to have survived into the Adena period . . . but Woodland peoples employed a different system of measurement and geometric forms . . . derived, at least in part, from Formative Mesoamerica . . . The system used a shorter measuring cord (1.544 m) for the Standard Unit and its permutations, but it otherwise preserved many of the traditional counts and arithmetic . . . Also, reliance on triangles was replaced by the use of square grids, and circles and squares, at various SMU [Standard Macro Unit] increments, as is so evident in Hopewell earthwork.' (Clark 2004: 205, with further references)
17. Yerkes 2005: 245.
18. Specialists have come to refer to this as the *Pax Hopewelliana*; for a general review, as well some occasional exceptions in the form of trophy skulls, see Seeman 1988.
19. See Carr and Case (eds) 2005; Case and Carr (eds) 2008: *passim*.

20. Contemporary scholars list at least nine clans among the 'Tripartite Alliance' of sodalities in the central Hopewell region: Bear, Canine, Feline, Raptor, Raccoon, Elk, Beaver, Nonraptorial Bird and Fox. These correspond roughly to the largest clans documented among Central Algonkian peoples who still inhabit the region (Carr 2005; Thomas et al. 2005: 339–40).
21. As one might imagine, this has been a matter of some debate, but we follow here the extensively documented views put forward in Carr and Case (eds) 2005, with further detailed references.
22. DeBoer 1997: 232: 'I view Hopewell earthwork sites as ceremonial centres, places where various activities, including mortuary rituals and other activities such as feasts, causeway-directed foot races and other "games," as well as dances and gambling were conducted periodically in the absence of large permanent populations resident at the centres themselves.' On burial: Seeman 1979.
23. See Coon (2009) on the north/south distinction; he also notes that in the south, burials are mostly collective and undifferentiated, and treasures are buried apart from bodies, not identified with specific individuals. The art shows figures in costume, dressed as monsters, rather than individuals wearing headdresses as at Hopewell. All this suggests a more self-consciously egalitarian, or at least anti-heroic, ideology in the south. On the pairing of shamanic and earth shrine sites see DeBoer 1997; on gender and office see Field, Goldberg and Lee 2005; Rodrigues 2005. Carr (in Carr and Case 2005: 112) speculates that the north/south division might reflect a distinction between the ancestors of later, patrilineal Great Lakes Algonkians and the matrilineal Southeastern societies (Cree, Cherokee, Choctaw, etc); but the pattern reflected in tombs seems far more radical: aside from some male mortuary priests, all major office holders in the south appear to be women. Rodrigues's (2005) analysis of skeletal remains suggests even more surprising differences in the south, where 'women also participated in maintenance and subsistence activities more commonly done by Native American men, including flint knapping and running possibly involved in hunting. Inversely, men shared in processing plant foods, stereotypically associated with women.' (Carr, in Case and Carr eds 2008: 248). It is rather surprising that these findings have not been more broadly discussed.
24. On the causeway: Lepper 1997.
25. As we discussed in Chapter Four, the monumental hunter-gatherer centre of Poverty Point on the Lower Mississippi drew in objects and materials from a similarly wide region almost 2,000 years earlier, and it

may well have disseminated various forms of intangible goods and knowledge far and wide in return; but Poverty Point had a different character to Hopewell, tightly focused on a single centre of gravity, and less clearly marked by the spread of social institutions such as burial rituals or settlement patterns.

26. Between 3500 and 3200 BC a cultural spread of similar scope, but very different character, also preceded the emergence of the first large territorial kingdom in Egypt; this is often referred to in the literature as a 'cultural unification' that preceded political unification, although in fact much of that unification between the valley and delta of the Nile seems to have been confined to the sphere of funerary rituals and their associated forms of personal display (Wengrow 2006: 38, 89).
27. Seeman 2004: 58–61.
28. For a more detailed argument on these lines see Braun (1986).
29. DeBoer 1997.
30. Hudson 1976: *passim*. Residents of New York City might be interested to know that Broadway was originally an Indian road, and that Astor Place, where it begins, was the shared lacrosse field for the three nations that occupied Manhattan.
31. On Rattlesnake Causeway and mound and the origins of Cahokia see Baires 2014; 2015; on Cahokia's beginnings as place of pilgrimage, Skousen 2016.
32. Chunkey appears to have been modelled on a popular children's game called Hoop and Pole. On the origins of chunkey and its later role see DeBoer 1993; Pauketat 2009: Chapter Four.
33. A later observer of the Choctaw recorded: 'Their favorite game of chunké . . . they play from morning till night, with an unwearied application, and they bet high; here you may see a savage come and bring all his skins, stake them and lose them; next his pipe, his beads, trinkets and ornaments; at last his blankets and other garments, and even all their arms, and after all it is not uncommon for them to go home, borrow a gun and shoot themselves.' (Romans, cited in Swanton 1931: 156–7). At the time of European contact, such extreme sports seem to have acted as a levelling mechanism, since few stayed on top for long and even those who sold themselves don't seem to have remained that way for very long.
34. Pauketat 2009: 20. The literature on Cahokia is vast. In addition to the general overviews we have already cited see also Alt 2018; Byers 2006; Emerson 1997a; Fowler 1997; Milner 1998; Pauketat 1994; 2004; and essays in Emerson and Lewis 1991; Pauketat and Emerson (eds) 1997; for the environmental context, Benson et al. 2009; Woods 2004.

35. Emerson et al. 2018.
36. Smith 1992: 17.
37. Emerson 1997a; 1997b: 187; cf. Alt 2018. Pauketat et al. (2015: 446) refer to this as a process of 'ruralization'.
38. Betzenhauser and Pauketat 2019. As Emerson notes (1997b), between AD 1050 and 1200, surveillance was also extended into the countryside through the establishment of what he terms 'civic nodes', which seem to have performed a mixture of ritual and managerial functions; see also Pauketat et al. 2015: 446–7.
39. Originally identified as the central burial of two males, flanked by retainers; but see now Emerson et al. 2016 for the true complexity of this deposit, which lies within the tumulus known to archaeologists as Mound 72, some way south of the Great Plaza.
40. Fowler et al. 1999 reported the mass graves as all-female, but in fact the picture is again more complex; see now Ambrose et al. 2003; Thompson et al. 2015.
41. Knight 1986; 1989; Knight et al. 2011; Pauketat 2009: Chapter Four; for other possible readings of the bird-man symbolism see Emerson et al. 2016.
42. Emerson 2007; 2012.
43. The precise reasons behind Cahokia's collapse are hotly contested; for a range of views see Emerson and Headman 2014; Kelly 2008, with further references.
44. See Cobb and Butler 2002.
45. For a range of views on the latter issue, compare Holt 2009; Pauketat 2007; and see also Milner 1998.
46. For which, at Cahokia, see Smith 1992; Pauketat 2013.
47. Cf. Pauketat et al. 2015: 452.
48. La Flesche 1921: 62–3; Rollings 1992: 28; Edwards 2010: 17.
49. See King 2003.
50. King 2003; 2004; 2007; Cobb and King 2005.
51. In Clayton et al. (eds) 1993: 92–3.
52. As argued in Ethridge 2010.
53. Ibid.: 33–7, 74–7. The indigenous forms of republican government that emerged in the Southeast during the eighteenth century also presumed a certain relationship with nature, but this was in no sense one of harmony. Ultimately, it was a relation of war. Plants were human allies and animals were enemies; killing prey without following correct ritual formula was a violation of the laws of war, which would cause animals to send disease into human communities by way of revenge. Yet at the same

time the business of hunting tended to be understood, especially by men, as representing a certain ideal of individual freedom.

54. Ibid.: 82–3.
55. The argument was put forward by Waskelov and Dumas but never published; it's cited and discussed in Ethridge (2010: 83–4) and Stern (2017: 33), though in our opinion it makes the whole case backwards, seeing the 'creation of new coalescent communities [and] . . . emergence of a more egalitarian, consensus-based social structure' in the face of disasters caused by the European invasion, and only then the emergence of a new cosmology whose symbol was the looped square, representing the council-ground as the universe, as a sort of adaptation to this 'new reality'. But how could self-conscious ideals of egalitarianism have emerged and been adopted at all, except through some sort of cosmological expression?
56. Fogelson 1984. There were, as Fogelson notes, and as we'll soon see, Cherokee priests in the seventeenth century, though they were gradually replaced by individual curers. It's hard not to see the legend as to some degree reflecting real historical events: Etowah, for instance, was in what later became Cherokee territory.
57. Coffee itself was first cultivated either in Ethiopia or Yemen; the American equivalent was called 'the Black Drink' and traces back to at least Hopewell times, when it was used in intense doses for ritual purposes (Hudson 1979; Crown et al. 2012). On Creek daily gatherings: Hahn 2004; Fairbanks 1979.
58. Brebeuf in *JR* 10: 219.
59. Certainly, the emerging soft drug regime in Europe – which was in many ways also the foundation of the emerging world economy of the time (founded, first on the spice trade, then on the drug, arms and slave trades), was quite different, since it focused so much on new regimes of work. While in the Middle Ages almost everyone had consumed mild intoxicants like wine or beer on a daily basis, the new regime saw a division between mild drugs meant to facilitate work – coffee and tea, especially used as vehicles for sugar, along with tobacco – and hard liquor for the weekends (see various contributions in Goodman, Lovejoy and Sherratt eds 1995).
60. In our terms, it's not even clear that Adena-Hopewell was a 'first-order regime of domination'; in most respects, as we've indicated, it seems closer to the kind of grand hospitality zones, culture areas, interaction spheres, or civilizations we've encountered so many times before in other parts of the world.
61. See Kehoe (2007) for an extensive comparison of ethno-historical data on the Osage with Cahokian archaeology (also Hall 1997). Their exact

relationship with Cahokia is not archaeologically clear, however, and Robert Cook (2017) supplies the most recent breakdown of their origins in Fort Ancient, a Mississippianized region of central Ohio whose population seems to have interacted with the Cahokian heartland (see especially, pp. 141–2, 162–3).

62. La Flesche 1930: 530; Rollings 1992: 29–30; Bailey and La Flesche 1995: 60–62.
63. La Flesche 1921: 51.
64. Rollings 1992: 38; Edwards 2010.
65. La Flesche (1921: 48–9) writes: 'In the course of this study of the Osage tribe, covering a number of years, it was learned from some of the older members of the *Nohozhinga* of the present day that, aside from the formulated rites handed down by the men of the olden days who had delved into the mysteries of nature and of life, stories also came down in traditional form telling of the manner in which these seers conducted their deliberations. The story that seemed most to impress the Nohozhinga of today is the one telling of how those men, those students of nature, gradually drifted into an organized association that became known by the name *Nohozhinga*, Little-Old-Men. As time went on this association found a home in the house of a man who had won, by his kindness and hospitality, the affection of his people ... Since that time it has been regarded by prominent men as an honor to entertain them.'
66. La Flesche 1939: 34.
67. See above, n.1; and also Burns 2004: 37–8, 362. Burns himself is of partly Osage descent and was brought up as Osage. We find it striking how regularly indigenous authors are open to the possibility that such dialogues were two-way, and how quickly European historians, or Americans of European descent, dismiss any such suggestions as preposterous and effectively shut them down.
68. Parker 1916: 17; italics ours. It's interesting to note that some early sources, like Josiah Clark, refer to the later figure of Adodharoh as 'the king', though alternatively as 'principal civil affairs officer of the confederacy' (in Henige 1999: 134–5).
69. It is worthy of note here that Arthur Parker describes Iroquois witches of his day as essentially those who have the power to turn themselves into monstrous beasts, and at the same time to bend others to their will through telepathic commands (Parker 1912: 27–8 n.2; cf. Smith 1888; Dennis 1993: 90–94; Shimony 1961: 261–88; 1970; Tooker 1964: 117–20). Mann also emphasizes the political nature of the designation: 'the closest Iroquoian thought comes to European witchcraft is a general

disgust with anyone who uses [charms] in an underhanded way, to trick another person into behaviour that is neither voluntary nor self-directed' (Mann 2000: 318; cf. Graeber 1996 for a similar case in Madagascar, also focusing on love magic).

70. We are thinking here, particularly, of the arguments made by Robert Lowie and Pierre Clastres, discussed at various points in our earlier chapters.
71. For instance, the Haudenosaunee also claimed they were descended from a population of escaped serfs, subjugated by a numerically superior enemy whom they called the Adirondaks ('Barkeaters') (Holm 2002: 160). Subjugation and insurrection were in no sense entirely alien concepts here.
72. Trigger 1990: 136–7.
73. Ibid.: 137.
74. Fremin in Wallace (1958: 235): 'The Iroquois have, properly speaking, only a single Divinity – the dream. To it they render their submission, and follow all its orders with the utmost exactness. The Tsonnontouens [Seneca] are more attached to this superstition than any of the others; their Religion in this respect becomes even a matter of scruple; whatever it be that they think they have done in their dreams, they believe themselves absolutely obliged to execute at the earliest moment. The other nations content themselves with observing those of their dreams which are the most important; but this people, which has the reputation of living more religiously than its neighbors, would think itself guilty of a great crime if it failed in its observance of a single dream. The people think only of that, they talk about nothing else, and all their cabins are filled with their dreams . . . Some have been known to go as far as Quebec, traveling a hundred and fifty leagues, for the sake of getting a dog, that they had dreamed of buying there . . .'. Wallace argues that this is a direct psychological consequence of the stoicism and importance of personal freedom and autonomy in Iroquoian societies. See also Blau 1963; Graeber 2001: 136–9.
75. The earlier dates are in reference to an eclipse mentioned in the foundation text (Mann and Fields 1997; cf. Henige 1999; Snow 1991; Atkins 2002; Starna 2008).
76. For the general state of archaeological understanding see Tuck 1978; Bamann et al. 1992; Engelbrecht 2003; Birch 2015. On the Ontario inception of maize cultivation: Johansen and Mann 2000: 119–20.
77. Mann and Fields 1997: 122–3; Johansen and Mann 2000: 278–9.
78. See Chapter Six.
79. Morgan 1851; Beauchamp 1907; Fenton 1949; 1998; Tooker 1978; on the role of women specifically: Brown 1970; Tooker 1984; Mann 1997; 1998; 2000.

80. Jamieson 1992: 74.
81. In Noble 1985: 133, cf. 1978: 161. There is some debate over how seriously to take the missionary claims about Tsouharissen: Trigger (1985: 223) for instance insists he was simply an unusually prominent war chief, but the preponderance of anthropological opinion seems weighted towards seeing the Neutral as a 'simple chiefdom'.
82. Noble 1985: 134–42.
83. Parker 1919: 16, 30–32.
84. Lahontan 1990 [1703]: 122–4.
85. The story is told in some detail in Mann 2000: 146–52.

12. CONCLUSION

1. Sometimes he also used the phrase *illud tempus*; see, among many other works, Eliade 1959.
2. Hocart 1954: 77; see also Hocart 1969 [1927]; 1970 [1936].
3. On reflection, many of what we consider to be quintessential freedoms – such as 'freedom of speech' or 'the pursuit of happiness' – are not really *social* freedoms at all. You can be free to say whatever you like, but if nobody cares or listens, it hardly matters. Equally, you can be as happy as you like, but if that happiness comes at the price of another's misery, it hardly amounts to much either. Arguably, the things often quoted as quintessential freedoms are based on the very illusion created by Rousseau in his second *Discourse*: the illusion of a human life that is solitary.
4. For which see Graeber and Sahlins 2017: *passim*.
5. Harari 2014: 133.
6. Scarry 1985.
7. Kelly 2000.
8. See Haas and Piscitelli 2013.
9. Patterson 1982.
10. For further discussion see Graeber 2011: 198–201 and the sources cited there.
11. One might imagine these public torments as wild or disorderly in their conduct; but in fact the preparation of a prisoner for sacrifice was one of the few occasions on which an office holder might issue commands for calm and orderly behaviour, as well as forbidding sexual intercourse. For all of the above see Trigger 1976: 68–75.
12. For a period of perhaps five centuries or more, human remains across the whole of Eastern North America display remarkably little evidence of traumatic injuries, scalping or other forms of interpersonal violence

(Milner et al. 2013). Evidence for interpersonal violence and warfare exists in both earlier and later periods, the most famous later examples being a mass grave excavated at Crow Creek and an Oneota village cemetery with extensive evidence of trauma, both dating to around 700 years ago. Such evidence accounts for perhaps a few decades or more of social history – a century at most – and is fairly localized. There is absolutely no reason to believe that the entire region somehow existed in a Hobbesian state for millennia, as contemporary theorists of violence blandly assume.

13. Delâge 1993: 65–6.
14. See Merrick 1991.
15. In a (1995) article that has undoubtedly been influential, but still not nearly as influential as it deserves, Crumley pointed out the need for alternatives to hierarchical models of social complexity in archaeological interpretation. As she noted, the archaeological record is full of evidence for the development of social and ecological systems that were complex and highly structured, just not according to hierarchical principles. 'Heterarchy' – the umbrella term she introduced for those other types of systems – was borrowed from cognitive science. Many of the societies we've focused on in this book – from Upper Palaeolithic mammoth hunters to the shifting coalitions and confederacies of sixteenth-century Iroquoia – could be described in these terms (had we chosen to adopt the language of systems theory), on the basis that power was dispersed or distributed in flexible ways across different elements of society, or at different scales of integration, or indeed across different times of year within the same society.

Bibliography

Abrams, Philip. 1977. 'Notes on the difficulty of studying the State.' *Journal of Historical Sociology* 1 (1): 58–89.

Acemoğlu, Daron and James Robinson. 2009. 'Foundations of societal inequality.' *Science* 326: 678–9.

Adams, Robert McCormick. 1981. *Heartland of Cities: Surveys of Ancient Settlement and Land Use on the Central Floodplain of the Euphrates*. Chicago and London: University of Chicago Press.

Akkermans, Peter M. M. G. (ed.). 1996. *Tell Sabi Abyad: Late Neolithic Settlement. Report on the Excavations of the University of Amsterdam (1988) and the National Museum of Antiquities Leiden (1991–1993) in Syria*. Istanbul: Nederlands Historisch-Archaeologisch Instituut te Istanbul.

Akkermans, Peter M. M. G. and Mark Verhoeven. 1995. 'An image of complexity: the burnt village at Late Neolithic Sabi Abyad, Syria.' *American Journal of Archaeology* 99 (1): 5–32.

Albert, Bruce. 1989. 'Yanomami "violence": inclusive fitness or ethnographer's representation?' *Current Anthropology* 30 (5): 637–40.

Alfani, Guido and Roberta Frigeni. 2016. 'Inequality (un)perceived: the emergence of a discourse on economic inequality from the Middle Ages to the age of Revolution.' *Journal of European Economic History* 45 (1): 21–66.

Algaze, Guillermo. 1993. *The Uruk World System: The Dynamics of Expansion of Early Mesopotamian Civilization*. Chicago: University of Chicago Press.

Allan, Peter. 1966. 'Baron Lahontan.' Master's thesis, University of British Columbia.

Allsen, Thomas T. 2016. *The Royal Hunt in Eurasian History*. Philadelphia: University of Pennsylvania Press.

Alt, Susan. M. 2018. *Cahokia's Complexities*. Tuscaloosa: University of Alabama Press.

Ambrose, Stanley H., Jane Buikstra and Harold W. Krueger (2003). 'Status and gender differences in diet at Mound 72, Cahokia, revealed by isotopic analysis of bone.' *Journal of Anthropological Archaeology* 22 (3): 217–26.

Ames, Kenneth M. 1995. 'Chiefly power and household production on the Northwest Coast.' In T. Douglas Price and Gary M. Feinman (eds), *Foundations of Social Inequality*. New York: Plenum Press, pp. 155–87.

—. 2001. 'Slaves, chiefs and labour on the northern Northwest Coast.' *World Archaeology* 33 (1): 1–17.

—. 2008. 'Slavery, household production and demography on the southern Northwest Coast: cables, tacking and ropewalks.' In Catherine M. Cameron (ed.), *Invisible Citizens: Captives and their Consequences*. Salt Lake City: University of Utah Press, pp. 138–58.

Ames, Kenneth M. and Herbert D. G. Maschner. 1999. *Peoples of the Northwest Coast*. London: Thames and Hudson.

Anderson, Benedict. 1991. *Imagined Communities: Reflections on the Origin and Spread of Nationalism*. London: Verso.

Angelbeck, Bill and Colin Grier. 2012. 'Anarchism and the archaeology of anarchic societies: resistance to centralization in the Coast Salish Region of the Pacific Northwest Coast.' *Current Anthropology* 53 (5): 547–87.

Anthony, David. W. 2007. *The Horse, the Wheel, and Language: How Bronze-Age Riders from the Steppes Shaped the Modern World*. Princeton, NJ and Oxford: Princeton University Press.

—. (ed.) 2010. *The Lost World of Old Europe: the Danube Valley 5000–3500 BC*. Princeton, NJ and Oxford: Princeton University Press.

Arbuckle, Benjamin S. 2013. 'The late adoption of cattle and pig husbandry in Neolithic Central Turkey.' *Journal of Archaeological Science* 40: 1805–15.

Arbuckle, Benjamin S. and Cheryl Makarewicz. 2009. 'The early management of cattle (*Bos taurus*) in Neolithic Central Anatolia.' *Antiquity* 83 (321): 669–86.

Arnauld, Charlotte M., Linda Manzanilla and Michael E. Smith (eds). 2012. *The Neighborhood as a Social and Spatial Unit in Mesoamerican Cities*. Tucson: University of Arizona Press.

Arnold, Jeanne E. 1995. 'Transportation, innovation and social complexity among maritime hunter-gatherer societies.' *American Anthropologist* 97 (4): 733–47.

Arroyo-Kalin, Manuel. 2010. 'The Amazonian Formative: crop domestication and anthropogenic soils.' *Diversity* 2: 473–504.

Ascher, Marcia. 2004. *Mathematics Elsewhere: An Exploration of Ideas Across Cultures*. Princeton, NJ: Princeton University Press.

Asher-Greve, Julia M. 2013. 'Women and agency: a survey from Late Uruk to the end of Ur III.' In Crawford (ed.), pp. 359–77.

Ashmore, Wendy. 2015. 'Contingent acts of remembrance: royal ancestors of Classic Maya Copan and Quirigua.' *Ancient Mesoamerica* 26: 213–31.

Asouti, Eleni. 2006. 'Beyond the Pre-Pottery Neolithic-B interaction sphere.' *Journal of World Prehistory* 20: 87–126.

Asouti, Eleni and Dorian Q. Fuller. 2013. 'A contextual approach to the emergence of agriculture in Southwest Asia: reconstructing early Neolithic plant-food production.' *Current Anthropology* 54 (3): 299–345.

Asouti, Eleni and Ceren Kabukcu. 2014. 'Holocene semi-arid oak woodlands in the Irano-Anatolian region of Southwest Asia: natural or anthropogenic?' *Quaternary Science Reviews* 90: 158–82.

Asouti, Eleni et al. 2015. 'Early Holocene woodland vegetation and human impacts in the arid zone of the southern Levant.' *The Holocene* 25 (10): 1565–80.

Atkins, Sandra Erin. 2002. 'The Formation of the League of the Haudenosaunee (Iroquois): Interpreting the Archaeological Record through the Oral Narrative Gayanashagowa.' Master's thesis, Trent University.

Atran, Scott. 1986. 'Hamula organisation and Masha'a tenure in Palestine.' *Man* (N.S.) 21 (2): 271–95.

Aubert, M. et al. 2014. 'Pleistocene cave art from Sulawesi, Indonesia.' *Nature* 514: 223–7.

—. 2018. 'Palaeolithic cave art in Borneo.' *Nature* 564: 254–7.

—. 2019. 'Earliest hunting scene in prehistoric art.' *Nature* 576: 442–5.

Ayad, Mariam F. 2009. *God's Wife, God's Servant: The God's Wife of Amun (c.740–525 BC)*. London and New York: Routledge.

Baden-Powell, Baden Henry. 1896. *The Indian Village Community*. London, New York, Bombay: Longmans, Green and Co.

Bagley, Robert. 1999. 'Shang archaeology.' In Loewe and Shaughnessy (eds), pp. 124–231.

Bahrani, Zeinab. 2002. 'Performativity and the image: narrative, representation and the Uruk vase.' In E. Ehrenberg (ed.), *Leaving No Stones Unturned: Essays on the Ancient Near East and Egypt in Honor of Donald P. Hansen*. Winona Lake, Indiana: Eisenbrauns, pp. 15–22.

Bailey, Douglass W. 2010. 'The figurines of Old Europe.' In Anthony (ed.), pp. 112–27.

—. 2017. 'Southeast European Neolithic figurines: beyond context, interpretation, and meaning.' In Insoll (ed.), pp. 823–50.

Bailey, Garrick and Francis La Flesche. 1995. *The Osages and the Invisible World. From the Works of Francis La Flesche*. Norman and London: University of Oklahoma Press.

Bailey, Geoff N. and Nicky J. Milner. 2002. 'Coastal hunter-gatherers and social evolution: marginal or central?' In *Before Farming: The Archaeology of Old World Hunter-Gatherers* 3–4 (1): 1–15.

Bailey, Geoff N. and Nicholas C. Flemming. 2008. 'Archaeology of the continental shelf: marine resources, submerged landscapes and underwater archaeology.' *Quaternary Science Reviews* 27: 2153–65.

Baines, John. 1995. 'Origins of Egyptian kingship.' In D. O'Connor and D. Silverman (eds), *Ancient Egyptian Kingship*. Leiden, New York and Cologne: Brill, pp. 95–156.

—. 1997. 'Kingship before literature: the world of the king in the Old Kingdom.' In R. Gundlach and C. Raedler (eds), *Selbstverständnis und Realität: Akten des Symposiums zur ägyptischen Königsideologie Mainz 15–17.6.1995*. Wiesbaden: Harrassowitz, pp. 125–86.

—. 1999. 'Forerunners of narrative biographies.' In A. Leahy and J. Tait (eds), *Studies on Ancient Egypt in Honour of H.S. Smith*. London: Egypt Exploration Society, pp. 23–37.

—. 2003. 'Early definitions of the Egyptian world and its surroundings.' In T. F. Potts, M. Roaf and D. Stein (eds), *Culture through Objects: Ancient Near Eastern Studies in Honour of P .R. S. Moorey*. Oxford: Griffith Institute, pp. 27–57.

Baires, Sarah E. 2014. 'Cahokia's Rattlesnake Causeway.' *Midcontinental Journal of Archaeology* 39 (2): 145–62.

—. 2015. 'The role of water in the emergence of the pre-Columbian Native American City.' *Wiley Interdisciplinary Reviews* 2 (5): 489–503.

Bakhtin, Mikhail M. (transl. H. Iswolsky). 1993 [1940]. *Rabelais and His World*. Bloomington: Indiana University Press.

Balsera, Viviana Díaz. 2008. 'Celebrating the rise of a new sun: the Tlaxcalans conquer Jerusalem in 1539.' *Estudios de cultura Náhuatl* 39: 311–30.

Bamann, Susan et al. 1992. 'Iroquoian archaeology.' *Annual Review of Anthropology* 21: 435–60.

Bánffy, Eszter. 2017. 'Neolithic Eastern and Central Europe.' In Insoll (ed.), pp. 705–28.

Barber, Elizabeth. J. W. 1991. *Prehistoric Textiles*. Princeton, NJ: Princeton University Press.

—. *Women's Work: The First 20,000 Years*. New York: W. W. Norton.

Barjamovic, Gojko. 2003. 'Civic institutions and self-government in Southern Mesopotamia in the mid-first millennium BC.' In J. G. Dercksen (ed.), *Assyria and Beyond: Studies Presented to M. T. Larsen*. Leiden, Istanbul: NINO, pp. 47–98.

Barruel, Abbé. 1799. *Memoirs Illustrating the History of Jacobinism*, vol. 3: *The Anti-Social Conspiracy*. New York: Isaac Collins.

Barry, Herbert, Irvin. L. Child and Margaret. K. Bacon. 1959. 'Relation of child training to subsistence economy.' *American Anthropologist* 61: 51–63.

Bartash, Vitali. 2015. 'Children in institutional households of Late Uruk period Mesopotamia.' *Zeitschrift für Assyriologie* 105 (2): 131–8.

Basile, Paola. 1997. 'Lahontan et l'évolution moderne du mythe du "bon sauvage".' Master's thesis, McGill University.

Bateson, Gregory. 1935. 'Culture Contact and Schismogenesis.' *Man* 35: 178–83.

—. 1936. *Naven. A Survey of the Problems Suggested by a Composite Picture of the Culture of a New Guinea Tribe Drawn from Three Points of View.* Cambridge: Cambridge University Press.

Bean, Lowell J. and Thomas C. Blackburn. 1976. *Native Californians: A Theoretical Retrospective*. Socorro, NM: Ballena Press.

Beauchamp, William M. 1907. *Civil, Religious, and Mourning Councils and Ceremonies of Adoption of the New York Indians*. New York State Museum Bulletin 113. Albany, NY: New York State Education Department.

Beidelman, Thomas. O. 1971. 'Nuer priests and prophets: charisma, authority and power among the Nuer.' In T. O. Beidelman (ed.), *The Translation of Culture: Essays to E.E. Evans-Pritchard*. London: Tavistock, pp. 375–415.

Belcher, Ellen. 2014. 'Embodiment of the Halaf: Sixth Millennium Figurines in Northern Mesopotamia.' PhD dissertation, Columbia University, New York.

Bell, Ellen E., Marcello Canuto and Robert J. Sharer (eds). 2004. *Understanding Early Classic Copan*. Philadelphia: University of Pennsylvania Museum, pp. 191–212.

Bellwood, Peter. 2005. *First Farmers: The Origins of Agricultural Societies.* Malden, MA and Oxford: Blackwell.

Bellwood, Peter and Colin Renfrew (eds). 2002. *Explaining the Farming/Language Dispersal Hypothesis*. Cambridge: McDonald Institute for Archaeological Research.

Benedict, Ruth. 1934. *Patterns of Culture*. London: Routledge.

Bennett, John. 2001. 'Agency and bureaucracy: thoughts on the nature and extent of administration in Bronze Age Pylos.' In Voutsaki and Killen (eds), pp. 25–37.

Benson, Larry, Timothy R. Pauketat and Edwin Cook. 2009. 'Cahokia's boom and bust in the context of climate change.' *American Antiquity* 74: 467–83.

Berkhofer, Robert F. 1978a. 'White conceptions of Indians.' In William C. Sturtevant and Bruce G. Trigger (eds), *Handbook of North American Indians*, vol. 15: *Northeast*. Washington: Smithsonian Institution Press, pp. 522–47.

—. 1978b. *The White Man's Indian: Images of the American Indian from Columbus to the Present.* New York: Knopf.

Berrin, Kathleen (ed.). 1988. *Feathered Serpents and Flowering Trees: Reconstructing the Murals of Teotihuacan.* San Francisco: Fine Ars Museums of San Francisco.

Berrin, Kathleen and Esther Pasztory (eds). 1993. *Teotihuacan: Art from the City of the Gods.* London: Thames and Hudson.

Bestock, Laurel D. 2008. 'The Early Dynastic funerary enclosures of Abydos.' *Archéo-Nil* 18: 43–59.

Bettinger, Robert L. 2015. *Orderly Anarchy: Sociopolitical Evolution in Aboriginal California.* Berkeley: University of California Press.

Bettinger, Robert L. and Martin A. Baumhoff. 1982. 'The Numic spread: Great Basin cultures in competition.' *American Antiquity* 47: 485–503.

Betts, Christopher J. 1984. 'Early Deism in France: from the so-called "Deistes" of Lyon (1564) to Voltaire's "Lettres Philosophiques" (1734).' *International Archives of the History of Ideas.* Leiden: Martinus Nijhoff Publishers.

Betzenhauser, Alleen and Timothy R. Pauketat. 2019. 'Elements of Cahokian neighborhoods.' *Archaeological Papers of the American Anthropological Association* 30: 133–47.

Biocca, Ettore and Helena Valero. 1965. *Yanoáma: dal racconto di una donna rapita dagli Indi.* Bari: Leonardo da Vinci.

Birch, Jennifer. 2015. 'Current research on the historical development of Northern Iroquoian societies.' *Journal of Archaeological Research* 23: 263–323.

Bird, Douglas W. et al. 2019. 'Variability in the organization and size of hunter-gatherer groups: foragers do not live in small-scale societies.' *Journal of Human Evolution* 131: 96–108.

Blackburn, Carole. 2000. *Harvest of Souls: The Jesuit Missions and Colonialism in North America, 1632–1650.* Montreal and Kingston: McGill-Queen's University Press.

Blackburn, Thomas C. 1976. 'Ceremonial integration and social interaction in Aboriginal California.' In Bean and Blackburn, pp. 225–44.

Blanton, Richard E. 1998. 'Beyond centralization: steps toward a theory of egalitarian behaviour in archaic states.' In G. A. Feinman and J. Marcus (eds), *Archaic States.* Santa Fe: School of American Research, pp. 135–72.

Blanton, Richard E., Gary Feinman, Stephen A. Kowalewski and Peter N. Peregrine. 1996. 'A dual-processual theory for the evolution of Mesoamerican civilization.' *Current Anthropology* 37 (1): 1–14.

Blanton, Richard and Lane Fargher. 2008. *Collective Action in the Formation of Pre-Modern States.* New York: Springer.

Blau, Harold. 1963. 'Dream guessing: a comparative analysis.' *Ethnohistory* 10: 233–49.

Bloch, Maurice. 1977. 'The past and the present in the present.' *Man* (N.S.) 12 (2): 278–92.

—. 2008. 'Why religion is nothing special but is central.' *Philosophical Transactions of the Royal Society B* 363: 2055–61.

—. 2013. *In and Out of Each Other's Bodies: Theory of Mind, Evolution, Truth, and the Nature of the Social.* Boulder, CO: Paradigm.

Boas, Franz and George Hunt. 1905. *Kwakiutl Texts.* Publications of the Jesup North Pacific Expedition, vol. 3. Leiden: Brill.

Boehm, Christopher. 1999. *Hierarchy in the Forest: The Evolution of Egalitarian Behaviour.* Cambridge, MA: Harvard University Press.

Bogaard, Amy. 2005. '"Garden agriculture" and the nature of early farming in Europe and the Near East.' *World Archaeology* 37 (2): 177–96.

Bogaard, Amy et al. 2014. 'Locating land use at Neolithic Çatalhöyük, Turkey: the implications of 87SR/86SR signatures in plants and sheep tooth sequences.' *Archaeometry* 56 (5): 860–77.

Boivin, Nicole, Dorian Q. Fuller and Alison Crowther. 2012. 'Old World globalization and the Columbian exchange: comparison and contrast.' *World Archaeology* 44 (3): 452–69.

Boivin, Nicole et al. 2016. 'Ecological consequences of human niche construction: examining long-term anthropogenic shaping of global species distributions.' *Proceedings of the National Academy of Sciences* 113: 6388–96.

Bookchin, Murray 1982. *The Ecology of Freedom.* Palo Alto: Cheshire Books.

—1992. *Urbanization Without Cities: The Rise and Decline of Citizenship.* Montreal and New York: Black Rose Books.

Bowles, Samuel and Jung-Kyoo Choi. 2013. 'Coevolution of farming and private property during the early Holocene.' *Proceedings of the National Academy of Sciences* 110 (22): 8830–35.

Braidwood, Robert. 1957. *Prehistoric Men.* Chicago: Natural History Museum Press.

Braswell, Geoffrey E. (ed.). 2003. *The Maya and Teotihuacan: Reinterpreting Early Classic Interaction.* Austin: University of Texas Press.

—. 2012. *The Ancient Maya of Mexico: Reinterpreting the Past of the Northern Maya Lowlands.* Sheffield: Equinox.

Braun, David P. 1986. 'Midwestern Hopewellian exchange and supralocation interaction.' In Colin Renfrew and John. F. Cherry (eds), *Peer Polity Interaction and Socio-Political Change.* Cambridge and New York: Cambridge University Press, pp. 117–26.

Brightman, Robert. 1999. 'Traditions of subversion and the subversion of tradition: cultural criticism in Maidu clown performances.' *American Anthropologist* 101 (2): 272–87.

Brisch, Nicole (ed.). 2008. *Religion and Power: Divine Kingship in the Ancient World and Beyond.* Chicago: Chicago University Press.

Broodbank, Cyprian. 2014. *The Making of the Middle Sea: A History of the Mediterranean from the Beginning to the Emergence of the Classical World.* London: Thames and Hudson.

Brown, James A. 1990. 'Archaeology confronts history at the Natchez temple.' *Southwestern Archaeology* 9 (1): 1–10.

Brown, Judith K. 1970. 'Economic organization and the position of women among the Iroquois.' *Ethnohistory* 17 (3–4): 151–67.

Brown, Philip C. 2006. 'Arable land as commons: land reallocation in early modern Japan.' *Social Science History* 30 (3): 431–61.

Bryan, Kirk. 1929. 'Flood-water farming.' *Geographical Review* 19 (3): 444–56.

Buckley, Thomas. 2002. *Standing Ground: Yurok Indian Spirituality, 1850–1990.* Berkeley: University of California Press.

Buikstra, Jane E. et al. 2004. 'Tombs from the Copan Acropolis: a life history approach.' In E. E. Bell et al. (eds.), pp. 185–205.

Burger, Richard. L. 2003. 'The Chavin Horizon: chimera or socioeconomic metamorphosis.' In Don S. Rice (ed.), *Latin American Horizons.* Washington: Dumbarton Oaks, pp. 41–82.

—. 2008. 'Chavín de Huántar and its sphere of influence.' In H. Silverman and W. Isbell (eds), *Handbook of South American Archaeology.* New York: Springer, pp. 681–703.

—. 2011. 'What kind of hallucinogenic snuff was used at Chavín de Huántar? An iconographic identification.' *Journal of Andean Archaeology* 31 (2): 123–40.

Burke, Peter. 2009. *Popular Culture in Early Modern Europe.* Farnham, Surrey: Ashgate.

Burns, Louis F. 2004. *A History of the Osage People.* Tuscaloosa: University of Alabama Press.

Byers, A. Martin. 2006. *Cahokia: A World Renewal Cult Heterarchy.* Gainsville: University Press of Florida.

Caillois, R. (transl. M. Barash). 2001 [1939]. *Man and the Sacred.* Glencoe: University of Illinois Press.

Campbell, Roderick. 2009. 'Towards a networks and boundaries approach to early complex polities: the Late Shang Case.' *Current Anthropology* 50 (6): 821–48.

—. 2014. 'Transformations of violence: on humanity and inhumanity in early China.' In Campbell (ed.), pp. 94–118.

—. (ed.) 2014. *Violence and Civilization: Studies of Social Violence in History and Prehistory*. Oxford: Oxbow.

Canetti, Elias. 1962. *Crowds and Power*. London: Gollancz.

Cannadine, David. 2001. *Ornamentalism: How the British Saw their Empire*. London: Penguin.

Canuto, Marcello et al. 2018. 'Ancient lowland Maya complexity as revealed by airborne laser scanning of northern Guatemala.' *Science* 361 (6409): eaau0137.

Capriles, José. 2019. 'Persistent Early to Middle Holocene tropical foraging in southwestern Amazonia.' *Science Advances* 5 (4): eaav5449.

Carballo, David M. et al. 2019. 'New research at Teotihuacan's Tlajinga district, 2012–2015.' *Ancient Mesoamerica* 30: 95–113.

Carbonell, Eudald and Marina Mosquera. 2006. 'The emergence of a symbolic behaviour: the sepulchral pit of Sima de los Huesos, Sierra de Atapuerca, Burgos, Spain.' *Comptes Rendus Palevol* 5 (1–2): 155–60.

Carr, Christopher. 2005. 'The tripartite ceremonial alliance among Scioto Hopewellian communities and the question of social ranking.' In Case and Carr (eds), pp. 258–338.

Carr, Christopher and D. Troy Case. 2005. *Gathering Hopewell: Society, Ritual, and Ritual Interaction*. New York: Kluwer Academic.

Carr, Christopher et al. 2008. 'The functions and meanings of Ohio Hopewell ceremonial artifacts in ethnohistorical perspective.' In Case and Carr (eds), pp. 501–21.

Case, D. Troy and Christopher Carr (eds), 2008. *The Scioto Hopewell and their Neighbors: Bioarchaeological Documentation and Cultural Understanding*. Berlin: Springer.

Castillo Butters, Luis Jaime. 2005. 'Las Señoras de San José de Moro: Rituales funerarios de mujeres de élite en la costa norte del Perú.' In *Divina y humana. La mujer en los antiguos Perú y México*. Lima: Ministerio de Educación, pp. 18–29.

Cervantes de Salazar, Franciso. 1914. *Crónica de la Nueva España*. Madrid: The Hispanic Society of America.

Chadwick, H. M. 1926. *The Heroic Age*. Cambridge: Cambridge University Press.

Chagnon, Napoleon. 1968. *Yanomamö: The Fierce People*. New York; London: Holt, Rinehart and Winston.

—. 1970. 'Ecological and adaptive aspects of California shell money.' *UCLA Archaeological Survey Annual Report* 12:1–25.

—. 1988. 'Life histories, blood revenge, and warfare in a tribal population.' *Science* 239 (4843): 985–92.

—. 1990. 'Reply to Albert.' *Current Anthropology* 31 (1): 49–53.

Chang, Kwang-chih. 1999. 'China on the eve of the historical period.' In Loewe and Shaughnessy (eds), pp. 37–73.

Chapman, John. 2010. 'Houses, households, villages, and proto-cities in Southeastern Europe.' In Anthony (ed.), pp. 74–89.

Chapman, John and Bisserka Gaydarska. 2003. 'The provision of salt to Tripolye mega-Sites.' In Aleksey Korvin-Piotrovsky, Vladimir Kruts and Sergei M. Rizhov (eds), *Tripolye Settlements-Giants*. Kiev: Institute of Archaeology, pp. 203–11.

Chapman, John, Bisserka Gaydarska and Duncan Hale. 2016. 'Nebelivka: assembly houses, ditches, and social structure.' In Müller et al. (eds), pp. 117–32.

Charles, Douglas and Jane E. Buikstra (eds). 2006. *Recreating Hopewell*. Gainesville: University Press of Florida.

Charlevoix, Pirre Francois Xavier de. 1944 [1744]. *Journal d'un voyage fait par ordre du roi dans l'Amérique septentrionnale*. Édition critique par Pierre Berthiaume. 2 vols. Bibliothèque du Nouveau Monde. Montreal: Les Presses de l'Université de Montréal.

Chase, Alexander W. 1873. 'Indian mounds and relics on the coast of Oregon.' *American Journal of Science and Arts* 7 (31): 26–32.

Chase-Dunn, Christopher K. and Kelly Marie Mann. 1998. *The Wintu and Their Neighbors: A Very Small World System*. Tucson: University of Arizona Press.

Chiappelli, Fredi (ed.). 1976. *First Images of America: The Impact of the New World on the Old*. Berkeley: University of California Press.

Childe, V. G. 1936. *Man Makes Himself*. London: Watts.

—. 1950. 'The urban revolution.' *Town Planning Review* 21: 3–17.

Chinard, Gilbert. 1911. *L'Exotisme Américain dans la littérature française au XVIe siècle*. Paris: Hachette.

—. 1913. *L'Amérique et le rêve exotique dans la littérature française au XVIIe et au XVIIIe siècle*. Paris: Hachette.

—. (ed.) 1931. 'Introduction.' *Dialogues curieux entre l'auteur et un sauvage de bons sens qui a voyagé, et Mémoires de l'Amérique septentrionale by Lahontan, Louis Armand de Lom d'Arce*. Baltimore: Johns Hopkins University Press.

Christie, Agatha. 1936. *Murder in Mesopotamia*. London: Collins.

Clark, John E. 1997. 'The arts of government in Early Mesoamerica.' *Annual Review of Anthropology* 26: 211–34.

—. 2004. 'Surrounding the sacred: geometry and design of early mound groups as meaning and function.' In Jon L. Gibson and Philip J. Carr (eds), *Signs of Power: The Rise of Complexity in the Southeast*. Tuscaloosa: University of Alabama Press, pp. 162–213.

Clarke, David. L. 1973. 'Archaeology: the loss of innocence.' *Antiquity* 43: 6–18.

—. 1978. *Mesolithic Europe: The Economic Basis*. London: Duckworth.

Clastres, Pierre. 1987 [1974]. *Society Against the State: Essays in Political Anthropology*. New York: Zone Books.

Clayton, Lawrence A., Vernon J. Knight and Edward C. Moore. 1993. *The De Soto Chronicles: The Expedition of Hernando de Soto to North America in 1539–1543*. Tuscaloosa: University of Alabama Press.

Clement, Charles R. et al. 2015. 'The domestication of Amazonia before European conquest.' *Proceedings of the Royal Society B* 282: 20150813.

Clendinnen, Inga. 1987. *Ambivalent Conquests: Maya and Spaniard in Yucatan, 1517–1570*. Cambridge: Cambridge University Press.

—. 1991. *Aztecs: An Interpretation*. Cambridge: Cambridge University Press.

Cobb, Charles R. and Brian M. Butler. 2002. 'The Vacant Quarter revisited: Late Mississippian abandonment of the Lower Ohio Valley.' *American Antiquity* 67 (4): 625–41.

Cobb, Charles R. and Adam King. 2005. 'Re-Inventing Mississippian tradition at Etowah, Georgia.' *Journal of Archaeological Method and Theory* 12 (3): 167–93.

Codere, Helen. 1950. *Fighting with Property: A Study of Kwakiutl Potlatching and Warfare, 1792–1930*. New York: J. J. Augustin.

Colas, Pierre. R. 2011. 'Writing in space: glottographic and semasiographic notation at Teotihuacan.' *Ancient Mesoamerica* 22 (1): 13–25.

Colledge, Sue, James Conolly and Stephen Shennan. 2004. 'Archaeobotanical evidence for the spread of farming in the eastern Mediterranean.' *Current Anthropology* 45 (4): 35–58.

—. 2005. 'The evolution of Neolithic farming from SW Asian Origins to NW European limits.' *European Journal of Archaeology* 8 (2): 137–56.

Colledge, Sue and James Conolly (eds). 2007. *The Origins and Spread of Domestic Plants in Southwest Asia and Europe*. Walnut Creek, CA: Left Coast Press.

Conklin, William J. and Jeffrey Quilter (eds). 2008. *Chavín: Art, Architecture, and Culture*. Los Angeles: Cotsen Institute of Archaeology.

Cook, Jill. 2015. 'Was bedeutet ein Name? Ein Rückblick auf die Ursprünge, Geschichte und Unangemessenheit des Begriffs *Venusfigur*.' *Zeitschrift für niedersächsische Archäologie* 66: 43–72.

Cook, Robert A. 2017. *Continuity and Change in the Native American Village: Multicultural Origins and Descendants of the Fort Ancient Culture.* Cambridge: Cambridge University Press.

Coon, Matthew S. 2009. 'Variation in Ohio Hopewell political economies.' *American Antiquity* 74 (1): 49–76.

Cork, Edward. 2005. 'Peaceful Harappans? Reviewing the evidence for the absence of warfare in the Indus Civilization of north-west India and Pakistan (*c.*2500–1900 BC).' *Antiquity* 79 (304): 411–23.

Cortés, Hernando. 1928. *Five Letters, 1519–1526*. London: Routledge.

Costa, Luiz. 2017. *The Owners of Kinship: Asymmetrical Relations in Indigenous Amazonia*. Chicago: HAU Books.

Costa, Luiz and Carlos Fausto. 2019. 'The enemy, the unwilling guest, and the jaguar host: an Amazonian story.' *L'Homme* (231–2): 195–226.

Coudart, Anick. 1998. *Architecture et Société Néolithique*. Paris: Éditions de la Maison des Sciences de l'Homme.

Coupland, Gary, Kathlyn Stewart and Katherine Patton. 2010. 'Do you ever get tired of salmon? Evidence for extreme salmon specialization at Prince Rupert Harbour, British Columbia.' *Journal of Anthropological Archaeology* 29: 189–207.

Cowgill, George L. 1997. 'State and society at Teotihuacan, Mexico.' *Annual Review of Anthropology* 26: 129–61.

—. 2003. 'Teotihuacan and early classic interaction: a perspective from outside the Maya region.' In Braswell (ed.), pp. 315–35.

—. 2008. 'An update on Teotihuacan.' *Antiquity* 82: 962–75.

—. 2015. *Ancient Teotihuacan. Early Urbanism in Central Mexico*. Cambridge: Cambridge University Press.

Crawford, Harriet (ed.). 2013. *The Sumerian World*. Abingdon; New York: Routledge.

Crazzolara, Joseph Pasquale. 1951. *The Lwoo, Part II: Lwoo Traditions*. Verona: Missioni Africane.

Crema, Enrico R. 2013. 'Cycles of change in Jomon settlement: a case study from eastern Tokyo Bay.' *Antiquity* 87 (338): 1169–81.

Cro, Stelio. 1990. *The Noble Savage: Allegory of Freedom*. Waterloo, Ontario: Wilfred Laurier University Press.

Crosby, Alfred. W. 1972. *The Columbian Exchange: Biological and Cultural Consequences of 1492*. Westport, CT: Greenwood Press.

—.1986. *Ecological Imperialism: The Biological Expansion of Europe, 900–1900 BC*. Cambridge: Cambridge University Press.

Croucher, Karina. 2012. *Death and Dying in the Neolithic Near East*. Oxford: Oxford University Press.

—. 2017. 'Keeping the dead close: grief and bereavement in the treatment of skulls from the Neolithic Middle East.' *Mortality* 23 (2): 103–20.

Crown, Patricia L. et al. 2012. 'Ritual Black Drink consumption at Cahokia.' *Proceedings of the National Academy of Sciences of the United States* 109 (35): 13944–9.

Crumley, Carole. 1995. 'Heterarchy and the analysis of complex societies.' *Archaeological Papers of the American Anthropological Association* 6 (1): 1–5.

Crüsemann, Nicola et al. (eds). 2019. *Uruk: City of the Ancient World.* Los Angeles: The J. Paul Getty Museum.

Cunliffe, Barry (ed.). 1998. *Prehistoric Europe: An Illustrated History.* Oxford: Oxford University Press.

Cushing, Frank Hamilton. 1896. 'Exploration of the ancient key dwellers' remains on the Gulf Coast of Florida.' *Proceedings of the American Philosophical Society* 35: 329–448.

Cushner, Nicholas P. 2006. *Why Have You Come Here? The Jesuits and the First Evangelization of Native America.* Oxford: Oxford University Press.

Dalley, Stephanie. 2000. *Myths from Mesopotamia: Creation, The Flood, Gilgamesh, and Others.* Oxford: Oxford University Press.

D'Altroy, Terence N. 2015. *The Incas* (2nd edn). Chichester: Wiley-Blackwell.

Davis, E. N. 1995. 'Art and politics in the Aegean: the missing ruler.' In Paul Rehak (ed.), *The Role of the Ruler in the Prehistoric Aegean* (Aegaeum 11). Liège: University of Liège, pp. 11–20.

Day, John W. et al. 2007. 'Emergence of complex societies after sea level stabilized.' *EOS* 88 (15): 169–76.

De Waal, Frans. 2000. *Chimpanzee Politics: Power and Sex Among Apes.* Baltimore: Johns Hopkins University Press.

DeBoer, Warren R. 1993. 'Like a rolling stone: the Chunkey game and political organization in Eastern North America.' *Southeastern Archaeology* 12: 83–92.

—. 1997. 'Ceremonial centers from the Cayapas to Chillicothe.' *Cambridge Archaeological Journal* 7: 225–53.

—. 2001. 'Of dice and women: gambling and exchange in Native North America.' *Journal of Archaeological Method and Theory* 8 (3): 215–68.

Deino, Alan L. et al. 2018. 'Chronology of the Acheulean to Middle Stone transition in eastern Africa.' *Science* 360 (6384): 95–8.

Delâge, Denys (transl. Jane Brierley). 1993. *Bitter Feast: Amerindians and Europeans in Northeastern North America, 1600–64.* Vancouver: UBC Press.

Denevan, William M. 1992. *The Native Population of the Americas in 1492.* Wisconsin: University of Wisconsin Press.

Denham, Timothy et al. 2003. 'Origins of agriculture at Kuk Swamp in the highlands of New Guinea.' *Science* 301 (5630): 189–93.

Dennis, Matthew. 1993. *Cultivating a Landscape of Peace: Iroquois-European Encounters in Seventeenth-Century America*. Ithaca, NY: Cornell University Press.

Descola, Philippe. 1994. 'Pourquoi les Indiens d'Amazonie n'ont-ils pas domestiqué le pécari? Genéalogie des objets et anthropologie de l' objectivation.' In Bruno Latour and Pierre Lemonnier (eds), *De la préhistoire aux missiles balistiques: L'intelligence sociale des techniques*. Paris: La Découverte, pp. 329–44.

—. 2005. *Par-delà nature et culture*. Paris: Éditions Gallimard.

Detienne, Marcel. 1994. *The Gardens of Adonis. Spices in Greek Mythology*. Princeton, NJ: Princeton University Press.

Diamond, Jared. 1987. 'The worst mistake in the history of the human race.' *Discover Magazine* (May 1987).

—. 1997. *Guns, Germs and Steel: The Fates of Human Societies*. New York; London: W. W. Norton.

—. 2012. *The World Until Yesterday: What Can We Learn from Traditional Societies?* London: Allen Lane.

Dickason, Olive Patricia. 1984. *The Myth of the Savage and the Beginnings of French Colonialism in the Americas*. Alberta: University of Alberta Press.

Dickson, D. Bruce. 2006. 'Public transcripts expressed in theatres of cruelty: the Royal Graves at Ur in Mesopotamia.' *Cambridge Archaeological Journal* 16 (2): 123–44.

Dietrich, Laura et al. 2019. 'Cereal processing at Early Neolithic Göbekli Tepe, southeastern Turkey.' *PLoS ONE* 14 (5): e0215214.

Dietrich, O., L. Dietrich and J. Notroff. 2019. 'Anthropomorphic imagery at Göbekli Tepe.' In J. Becker, C. Beuger and B. Müller-Neuhof (eds), *Human Iconography and Symbolic Meaning in Near Eastern Prehistory*. Vienna: Austrian Academy of Sciences, pp. 151–66.

Dietrich, Olivier, Manfred Heun, Jens Notroff and Klaus Schmidt. 2012. 'The role of cult and feasting in the emergence of Neolithic communities. New evidence from Göbekli Tepe, south-eastern Turkey.' *Antiquity* 86 (333): 674–95.

Dodds Pennock, Caroline. 2008. *Bonds of Blood: Gender, Lifecyle, and Sacrifice in Aztec Culture*. New York: Palgrave Macmillan.

—. 2017. 'Gender and Aztec life cycles.' In D. L. Nichols and E. Rodríguez-Algería (eds), *The Oxford Handbook of the Aztecs*. Oxford: Oxford University Press, pp. 387–98.

Domenici, Davide. 2018. 'Beyond dichotomies: Teotihuacan and the Mesoamerican urban tradition.' In D. Domenici and N. Marchetti, *Urbanized*

Landscapes in Early Syro-Mesopotamia and Prehispanic Mesoamerica. Wiesbaden: Harrassowitz Verlag, pp. 35–70.

Donald, Leland. 1997. *Aboriginal Slavery on the Northwest Coast of North America.* Berkeley: University of California Press.

—. 2003. 'The Northwest Coast as a study area: natural, prehistoric, and ethnographic issues.' In G. G. Coupland, R. G. Matson, and Q. Mackie (eds), *Emerging from the Mist: Studies in Northwest Coast Culture History.* Vancouver: UBC Press, pp. 289–327.

Douglas, Mary. 1966. *Purity and Danger: An Analysis of Concepts of Pollution and Taboo.* London: Routledge.

Driver, Harold. E. 1938. 'Culture element distributions VIII: The reliability of culture element data.' *Anthropological Records* 1: 205–20.

—. 1962. 'The contribution of A. L. Kroeber to culture area theory and practice.' *Indiana University Publications in Anthropology and Linguistics* (Memoir 18).

Drucker, Philip. 1981. 'On the nature of Olmec polity.' In Elizabeth P. Benson (ed.), *The Olmec and Their Neighbors: Essays in Memory of Matthew W. Stirling.* Washington: Dumbarton Oaks, pp. 29–47.

Du Bois, Cora. 1935. 'Wintu Ethnography.' *University of California Publications in American Archaeology and Ethnology* 36 (1): 1–142. Berkeley: University of California Press.

Duchet, Michèle. 1995. *Anthropologie et histoire au siècle des Lumières.* Paris: A. Michel.

Dumont, Louis. 1972. *Homo Hierarchicus: The Caste System and its Implications.* London: Weidenfeld and Nicolson.

—. 1992. *Essays on Individualism: Modern Ideology in Anthropological Perspective.* Chicago: University of Chicago Press.

Dunbar, Robin I. M. 1988. *Primate Social Systems.* London and Sydney: Croom Helm.

—. 1992. 'Neocortex size as a constraint on group size in primates.' *Journal of Human Evolution* 20: 469–93.

—. 1996. *Grooming, Gossip, and the Evolution of Language.* London: Faber and Faber.

—. 2010. *How Many Friends Does One Person Need? Dunbar's Number and Other Evolutionary Quirks.* Cambridge, MA: Harvard University Press.

Dunbar, Robin I. M., Clive Gamble and John A. K. Gowlett (eds). 2014. *Lucy to Language: The Benchmark Papers.* Oxford: Oxford University Press.

Durkheim, Émile. 1915 [1912]. *The Elementary Forms of Religious Life.* London: Allen and Unwin.

Eastman, Charles A. 1937. *Indian Boyhood*. Boston: Little, Brown and Company.

Edinborough, Kevan et al. 2017. 'Radiocarbon test for demographic events in written and oral history.' *PNAS* 114 (47): 12436–41.

Edmonson, Munro S. 1982. *The Ancient Future of the Itza: The Book of Chilam Balam of Tizimin*. Austin: University of Texas Press.

Edwards, Tai S. 2010. 'Osage Gender: Continuity, Change, and Colonization, 1720s-1870s.' Doctoral dissertation, University of Kansas.

Eliade, Mircea. 1959. *The Sacred and the Profane: The Nature of Religion*. New York: Harcourt, Brace.

Elias, Norbert. 1969. *The Court Society*. Oxford: Blackwell.

Ellingson, Ter. 2001. *The Myth of the Noble Savage*. Berkeley: University of California Press.

Emerson, Thomas. 1997a. *Cahokia and the Archaeology of Power*. Tuscaloosa: University of Alabama Press.

—. 1997b. 'Reflections from the countryside on Cahokian hegemony.' In Pauketat and Emerson (eds), pp. 167–89.

—. 2007. 'Cahokia and the evidence for Late Pre-Columbian war in the North American midcontinent.' In R. J. Chacon and R. G. Mendoza (eds), *North American Indigenous Warfare and Ritual Violence*. Tucson: University of Arizona Press, pp. 129–48.

—. 2012. 'Cahokia interaction and ethnogenesis in the northern Midcontinent.' In Timothy R. Pauketat (ed.), *The Oxford Handbook of North American Archaeology*. Oxford: Oxford University Press, pp. 398–409.

Emerson, Thomas E. and R. Barry Lewis (eds). 1991. *Cahokia and the Hinterlands: Middle Mississippian Cultures of The Midwest*. Urbana: University of Illinois Press.

Emerson, Thomas. E. and Kristin. M. Hedman. 2014. 'The dangers of diversity: the consolidation and dissolution of Cahokia, Native America's first urban polity.' In Ronald K. Faulseit (ed.), *Beyond Collapse: Archaeological Perspectives on Resilience, Revitalization, and Transformation in Complex Societies*. Carbondale: Center for Archaeological Investigations, Southern Illinois University Press, Occasional Paper no. 42, pp. 147–75.

Emerson, Thomas et al. 2016. 'Paradigms lost: reconfiguring Cahokia's Mound 72 Beaded Burial.' *American Antiquity* 81 (3): 405–25.

Emerson, Thomas E., Brad H. Koldeho and Tamira K. Brennan (eds). 2018. *Revealing Greater Cahokia, North America's First Native City: Rediscovery and Large-Scale Excavations of the East St. Louis Precinct*. Studies in Archaeology 12. Urbana: Illinois State Archaeological Survey, University of Illinois.

Enajero, Samuel. 2015. *Collective Institutions in Industrialized Nations*. New York: Page.

Engelbrecht, William. 2003. *Iroquoia: The Development of a Native World*. Syracuse, NY: Syracuse University Press.

Englund, Robert K. 1988. 'Administrative timekeeping in ancient Mesopotamia.' *Journal of the Economic and Social History of the Orient* 31 (2): 121–85.

—. 1998. 'Texts from the Late Uruk period.' In J. Bauer, R. K. Englund and M. Krebernik (eds), *Mesopotamien. Späturuk-Zeit und Frühdynastische Zeit*. Göttingen: Vandenhoeck and Ruprecht.

—. 2009. 'The smell of the cage.' *Cuneiform Digital Library* 4.

Erdal, Yilmaz. 2015. 'Bone or flesh: defleshing and post-depositional treatments at Körtik Tepe (Southeastern Anatolia, PPNA Period).' *European Journal of Archaeology* 18 (1): 4–32.

Erickson, Clark L. 2008. 'Amazonia: the historical ecology of a domesticated landscape.' In H. Silverman and W. Isbell (eds), *Handbook of South American Archaeology*. New York: Springer, pp. 157–83.

Erlandson, Jon et al. 2007. 'The Kelp Highway hypothesis: marine ecology, the coastal migration theory, and the peopling of the Americas.' *The Journal of Island and Coastal Archaeology* 2 (2): 161–74.

Ethridge, Robbie. 2010. *From Chicaza to Chickasaw: The European Invasion and the Transformation of the Mississippian World, 1540–1715*. Chapel Hill: University of North Carolina Press.

Étienne, Louis. 1871. 'Un Roman socialiste d'autrefois.' *Revue des deux mondes*, 15 Juillet.

Eyre, Christopher J. 1999. 'The village economy in Pharaonic Egypt.' In Alan K. Bowman and Eugene Rogan (eds), *Proceedings of the British Academy, 96: Agriculture in Egypt from Pharaonic to Modern Times*. Oxford: Oxford University Press, pp. 33–60.

Fairbairn, Andrew et al. 2006. 'Seasonality (Çatalhöyük East).' In Ian Hodder (ed.), *Çatalhöyük Perspectives: Themes from the 1995–9 Seasons*. Cambridge: McDonald Institute Monographs, British Institute of Archaeology at Ankara, pp. 93–108.

Fairbanks, Charles. 1979. 'The function of Black Drink among the Creeks.' In C. Hudson (ed.), pp. 120–49.

Fargher, Lane, Richard E. Blanton and Verenice Y. Heredia Espinoza. 2010. 'Egalitarian ideology and political power in prehispanic Central Mexico: the case of Tlaxcalan.' *Latin American Antiquity* 21 (3): 227–51.

Fargher, Lane, Verenice Y. Heredia Espinoza and Richard E. Blanton. 2011. 'Alternative pathways to power in late Postclassic Highland Mesoamerica.' *Journal of Anthropological Archaeology* 30: 306–26.

Fargher, Lane et al. 2011. 'Tlaxcallan: the archaeology of an ancient republic in the New World.' *Antiquity* 85: 172–86.

Fausto, Carlo. 1999. 'Of enemies and pets: warfare and shamanism in Amazonia.' *American Ethnologist* 26 (4): 933–56.

—. 2008. 'Too many owners: mastery and ownership in Amazonia.' *Mana* 14 (2): 329–66.

Fausto, Carlo and Eduardo G. Neves. 2009. 'Was there ever a Neolithic in the Neo-tropics? Plant familiarization and biodiversity in the Amazon.' *Antiquity* 92 (366): 1604–18.

Federici, Silvia. 1998. *Caliban and the Witch*. New York: Autonomedia.

Fenton, William N. 1949. 'Seth Newhouse's traditional history and constitution of the Iroquois Confederacy.' *Proceedings of the American Philosophical Society* 93 (2): 141–58.

—. 1998. *The Great Law and the Longhouse: A Political History of the Iroquois Confederacy*. Norman: University of Oklahoma Press, 1998.

Ferguson, R. Brian. 1989. 'Do Yanomamo killers have more kids?' *American Ethnologist* 16 (3): 564–5.

Ferguson, R. Brian and Neil L. Whitehead. 1992. *War in the Tribal Zone: Expanding States and Indigenous Warfare*. Santa Fe: School for Advanced Research.

Field, Stephanie, Anne Goldberg and Tina Lee. 2005. 'Gender, status, and ethnicity in the Scioto, Miami, and Northeastern Ohio Hopewellian regions, as evidenced by mortuary practices.' In Carr and Case (eds), pp. 386–404.

Fischer, Claude S. 1977. 'Comment on Mayhew and Levinger's "Size and the density of interaction in human aggregates".' *American Journal of Sociology* 83 (2): 452–5.

Fitzhugh, William W. 1985. *Cultures in Contact: the Impact of European Contacts on Native American Cultural Institutions A.D. 1000–1800*. Washington: Smithsonian Institution Press.

Flannery, Kent and Joyce Marcus. 2012. *The Creation of Inequality: How our Prehistoric Ancestors Set the Stage for Monarchy, Slavery, and Empire*. Cambridge, MA: Harvard University Press.

Fleming Daniel. E. 2009. *Democracy's Ancient Ancestors: Mari and Early Collective Governance*. Cambridge: Cambridge University Press.

Fletcher, Alice C. and Francis La Flesche. 1911. 'The Omaha tribe.' *Twenty-seventh Annual Report of the Bureau of American Ethnology, 1905–6*. Washington: Bureau of American Ethnology, pp. 17–654.

Fletcher, Roland. 1995. *The Limits of Settlement Growth: A Theoretical Outline*. Cambridge: Cambridge University Press.

Flew, Antony. 1989. *An Introduction to Western Philosophy: Ideas and Argument from Plato to Popper.* London: Thames and Hudson.

Fogelson, Raymond D. 1984. 'Who were the Aní-Kutánî? An excursion into Cherokee historical thought.' *Ethnohistory* 31 (4): 255–63.

Formicola, Vincenzo. 2007. 'From the Sungir children to the Romito dwarf: aspects of the Upper Palaeolithic funerary landscape.' *Current Anthropology* 48: 446–53.

Foster, Benjamin. 1981. 'A new look at the Sumerian temple state.' *Journal of the Economic and Social History of the Orient* 24 (3): 225–41.

Fowler, Melvin. L. 1997. *The Cahokia Atlas: A Historical Atlas of Cahokia Archaeology*. Urbana: University of Illinois Press.

Fowler, Melvin L. et al. 1999. *The Mound 72 Area: Dedicated and Sacred Space in Early Cahokia.* Reports of Investigation no. 54. Springfield: Illinois State Museum.

Frangipane, Marcella. 2006. 'The Arslantepe "Royal Tomb": new funerary customs and political changes in the Upper Euphrates valley at the beginning of the third millennium BC.' In G. Bartoloni and M. G. Benedettini (eds), *Buried Among the Living*. Rome: Università degli studi di Roma 'La Sapienza', pp. 169–94.

—. 2012. 'Fourth millennium Arslantepe: the development of a centralized society without urbanization.' *Origini* 34: 19–40.

Frankfort, Henri. 1948. *Kingship and the Gods: A Study of Ancient Near Eastern Religion as the Integration of Society and Nature.* Chicago: Chicago University Press.

—. 1951. *The Birth of Civilization in the Near East.* Bloomington: Indiana University Press.

Franklin, Benjamin. 1961 [1753]. Letter to Peter Collinson, 9 May 1753. In Leonard W. Labaree (ed.), *The Papers of Benjamin Franklin.* New Haven, CT and London: Yale University Press, vol. 4, pp. 481–3.

Franks, C. E. S. 2002. 'In search of the savage *Sauvage*: an exploration into North America's political cultures.' *The American Review of Canadian Studies* 32 (4): 547–80.

Frazer, James G. 1911 [1890]. *The Dying God* (book 3 of *The Golden Bough: A Study in Magic and Religion*) (3rd edn). London: Macmillan.

Freidel, David A. and Linda Schele. 1988. 'Kingship in the Late Preclassic Maya Lowlands: the instruments and places of ritual power.' *American Anthropologist* 90 (3): 547–67.

Frenez, Dennys. 2018. 'Private person or public persona? Use and significance of standard Indus seals as markers of formal socio-economic identities.' In D. Frenez et al. (eds), *Walking with the Unicorn: Social*

Organization and Material Culture in Ancient South Asia. Oxford: Archaeopress, pp. 166–93.

Friedman, Renée F. 2008. 'Excavating Egypt's early kings: recent discoveries in the elite cemetery at Hierakonpolis.' In B. Midant-Reynes and Y. Tristant (eds), *Egypt at its Origins 2. Proceedings of the International Conference Origin of the State. Predynastic and Early Dynastic Egypt, Toulouse, 5th–8th September 2005*. Orientalia Lovaniensia Analecta 172. Leuven: Peeters, pp. 1157–94.

—. 2011. 'Hierakonpolis.' In Teeter (ed.), pp. 33–44.

Froese, Tom, Carlos Gershenson and Linda R. Manzanilla. 2014. 'Can government be self-organized? A mathematical model of the collective social organization of ancient Teotihuacan, Central Mexico.' *PLoS ONE* 9 (10): e109966.

Fukuyama, Francis. 2011. *The Origins of Political Order: From Prehuman Times to the French Revolution*. London: Profile.

Fuller, Dorian Q. 2007. 'Contrasting patterns in crop domestication and domestication rates: recent archaeobotanical insights from the Old World.' *Annals of Botany* 100: 903–9.

—. 2010. 'An emerging paradigm shift in the origins of agriculture.' *General Anthropology* 17 (2): 8–12.

Fuller, Dorian Q. and Robin G. Allaby. 2010. 'Seed dispersal and crop domestication: shattering, germination and seasonality in evolution under cultivation.' In L. Ostergaard (ed.), *Fruit Development and Seed Dispersal* (Annual Plant Reviews vol. 38). Oxford: Wiley-Blackwell, pp. 238–95.

Fuller, Dorian Q. et al. 2010. 'Domestication as innovation: the entanglement of techniques, technology and chance in the domestication of cereal crops.' *World Archaeology* 42 (1): 13–28.

Fuller, Dorian Q. and Leilani Lucas. 2017. 'Adapting crops, landscapes, and food choices: Patterns in the dispersal of domesticated plants across Eurasia.' In N. Boivin et al. (eds), *Complexity: Species Movements in the Holocene*. Cambridge: Cambridge University Press, pp. 304–31.

Furholt, Martin. 2018. 'Massive migrations? The impact of recent aDNA studies on our view of third millennium Europe.' *European Journal of Archaeology* 21 (2): 159–91.

Gage, Matilda Joslyn. 1893. *Woman, Church, and State. A Historical Account of the Status of Woman through the Christian Ages: with Reminiscences of Matriarchate*. Chicago: C. H. Kerr.

Gardner, Peter M. 1991. 'Foragers' pursuit of individual autonomy.' *Current Anthropology* 32: 543–72.

Garth, Thomas R. Jr. 1976. 'Emphasis on industriousness among the Atsugewi.' In Bean and Blackburn, pp. 337–54.

Geertz, Clifford 1980. *Negara: The Theatre State in Nineteenth-Century Bali.* Princeton, NJ: Princeton University Press.

—. 2001. 'Life among the anthros.' *New York Review of Books* 48 (2): 18–22.

Geertz, Hildred and Clifford Geertz. 1978. *Kinship in Bali.* Chicago: Chicago University Press.

Gelb, Ignace J. 1973. 'Prisoners of war in early Mesopotamia.' *Journal of Near Eastern Studies* 32 (1/2): 70–98.

Gell, Alfred. 1993. *Wrapping in Images: Tattooing in Polynesia.* Oxford: Clarendon Press.

Georgescu-Roegen, Nicholas. 1976. *Energy and Economic Myths: Institutional and Analytical Economic Essays.* New York: Pergamon.

Gerth, Hans H. and C. Wright Mills (eds). 1946. *From Max Weber: Essays in Sociology.* New York: Oxford University Press.

Gibson, Charles. 1952. *Tlaxcala in the Sixteenth Century.* New Haven, CT and London: Yale University Press.

Gibson, Jon L. 2000. *Ancient Mounds of Poverty Point: Place of Rings.* Gainesville: University Press of Florida.

Gibson, Jon. L. and Philip J. Carr. 2004. 'Big mounds, big rings, big power.' In J. L. Gibson and P. J. Carr (eds), *Signs of Power: The Rise of Complexity in the Southeast.* Tuscaloosa: University of Alabama Press, pp. 1–9.

Gimbutas, Marija. 1982. *The Goddesses and Gods of Old Europe.* London: Thames and Hudson.

Giosan, Liviu et al. 2012. 'Fluvial landscapes of the Harappan civilization.' *PNAS* : E1688–E1694.

Godoy, Ricardo A. 1986. 'The fiscal role of the Andean Ayllu.' *Man* (N.S.) 21 (4): 723–41.

Goldman, Irving. 1975. *The Mouth of Heaven: An Introduction to Kwakiutl Religious Thought.* New York: John Wiley and Sons.

Goldschmidt, Walter. 1951. 'Ethics and the structure of society: an ethnological contribution to the sociology of knowledge.' *American Anthropologist* 53 (4): 506–24.

Goldstein, Paul S. 2003. 'From stew-eaters to maize-drinkers: the *Chicha* economy and Tiwanaku.' In Tamara L. Bray (ed.), *The Archaeology and Politics of Food and Feasting in Early States and Empires.* New York: Plenum, pp. 143–71.

Golitko, Mark and Lawrence H. Keeley. 2007. 'Beating ploughshares back into swords: warfare in the Linearbandkeramik.' *Antiquity* 81: 332–42.

Golla, Susan. 1987. '"He has a name": History and Social Structure among the Indians of Western Vancouver Island.' PhD. dissertation, Columbia University.

Golson, Jack. et al. (eds). 2017. *Ten Thousand Years of Cultivation at Kuk Swamp in the Highlands of Papua New Guinea*. Acton, ACT: Australian National University Press.

Gombrich, Richard F. 1988. *Theravāda Buddhism. A Social History from Ancient Benares to Modern Colombo*. London and New York: Routledge.

Gómez-Cano, Grisel. 2010. *The Return to Caotlicue: Goddesses and Warladies in Mexican Folklore*. Bloomington, Indiana: XLibris.

González González, Enrique. 2014. 'A humanist in the New World: Francisco Cervantes de Salazar (*c*.1514–75).' In L. Dietz et al. (eds), *Neo-Latin and the Humanities: Essays in Honour of Charles E. Fantazzi*. Toronto: Center for Reformation and Renaissance Studies, pp. 235–57.

Goodman, Jordan, Paul E. Lovejoy and Andrew G. Sherratt (eds). 1995. *Consuming Habits: Drugs in Anthropology and History*. London and New York: Routledge.

Gose, Peter. 1996. 'Oracles, divine kingship, and political representation in the Inka State.' *Ethnohistory* 43 (1): 1–32.

—.2000. 'The state as a chosen woman: brideservice and the feeding of tributaries in the Inka Empire.' *American Anthropologist* 102 (1): 84–97.

—.2016. 'Mountains, kurakas and mummies: transformations in indigenous Andean sovereignty.' *Población & Sociedad* 23 (2): 9–34.

Gough, Kathleen. 1971. 'Nuer Kinship: a re-examination.' In T. O. Beidelman (ed.), *The Translation of Culture*. London: Tavistock, pp. 79–122.

Graeber, David. 1996. 'Love magic and political morality in Central Madagascar, 1875–1990.' *Gender & History* 8 (3): 416–39.

—. 2001. *Toward an Anthropological Theory of Value: The False Coin of Our Own Dreams*. New York: Palgrave.

—. 2006. 'Turning modes of production inside out: or, why capitalism is a transformation of slavery.' *Critique of Anthropology* 26 (1): 61–85.

—. 2007a. *Possibilities: Essays on Hierarchy, Rebellion, and Desire*. Oakland, CA: AK Press.

—. 2007b. 'There never was a West: or, democracy emerges from the spaces in between.' In Graeber 2007a, pp. 329–74.

—. 2011. *Debt: The First 5,000 Years*. New York: Melville House.

Graeber, David and Marshall Sahlins. 2017. *On Kings*. Chicago: HAU Books.

Graffigny, Françoise de. 1747. *Lettres d'une Péruvienne*. Paris: A. Peine. (English translation by Jonathan Mallason. 2009. *Letters of a Peruvian Woman*. Oxford: Oxford University Press.)

Green, Adam S. 2020. 'Killing the priest-king: addressing egalitarianism in the Indus civilization.' *Journal of Archaeological Research*. https://doi.org/10.1007/s10814-020-09147-9

Green, Richard. E., Johannes Krause et al. 2010. 'A draft sequence of the Neanderthal genome.' *Science* 328: 710–22.

Gresky, Julia, Juliane Haelm and Lee Clare. 2017. 'Modified human crania from Göbekli Tepe provide evidence for a new form of Neolithic skull cult.' *Science Advances* 3: e1700564.

Grier, Colin. 2017. 'Expanding notions of hunter-gatherer diversity: identifying core organizational principles and practices in Coast Salish Societies of the Northwest Coast of North America.' In Graeme Warren and Bill Finlayson (eds), *The Diversity of Hunter-Gatherer Pasts*. Oxford: Oxbow Press, pp. 16–33.

Grinde, Donald A. 1977. *The Iroquois and the Founding of the American Nation*. San Francisco: Indian Historian Press.

Groenewegen-Frankfort, Henriette. 1951. *Arrest and Movement: An Essay on Space and Time in the Representational Art of the Ancient Near East*. London: Faber and Faber.

Gron, Kurt J. et al. 2018. 'A meeting in the forest: hunters and farmers at the Coneybury "Anomaly", Wiltshire.' *Proceedings of the Prehistoric Society* 84: 111–44.

Grube, Nikolai, Eva Eggebrecht and Matthias Seidel (eds). 2001. *Maya: Divine Kings of the Rainforest*. Cologne: Könemann.

Haak, Wolfgang et al. 2005. 'Ancient DNA from the first European farmers in 7,500-year-old Neolithic sites.' *Science* 310: 1016–18.

—. 2010. 'Ancient DNA from European early Neolithic farmers reveals their Near Eastern affinities.' *PLoS Biology* 8: 1–16.

—. 2015. 'Massive migration from the steppe was a source for Indo-European languages in Europe.' *Nature* 522: 207–11.

Haas, Jonathan and Matthew Piscitelli. 2013. 'The prehistory of warfare: Misled by ethnography.' In Douglas P. Fry (ed.), *War, Peace, and Human Nature: The Convergence of Evolutionary and Cultural Views*. New York: Oxford University Press, pp. 168–90.

Habu, Junko. 2004. *Ancient Jomon of Japan*. Cambridge: Cambridge University Press.

Habu, Junko and Clare Fawcett. 2008. 'Science or narratives? Multiple interpretations of the Sannai Maruyama site, Japan.' In Junko Habu, Clare Fawcett and John M. Matsunaga (eds), *Evaluating Multiple Narratives: Beyond Nationalist, Colonialist, Imperialist Archaeologies*. New York: Springer, pp. 91–117.

Haddon, Alfred. C. and W. H. R. Rivers. 1902. 'A method of recording string figures and tricks.' *Man* 109: 146–53.

Hahn, Steven C. 2004. *The Invention of the Creek Nation, 1670–1763*. Lincoln: University of Nebraska Press.

Hajda, Yvonne P. 2005. 'Slavery in the Greater Lower Columbia region.' *Ethnohistory* 64 (1): 1–17.

Haklay, Gil and Avi Gopher. 2020. 'Geometry and architectural planning at Göbekli Tepe, Turkey.' *Cambridge Archaeological Journal* 30 (2): 343–57.

Hall, Anthony J. 2003. *The American Empire and the Fourth World*. Montreal and Kingston: McGill-Queen's University Press.

Hall, Robert L. 1997. *An Archaeology of the Soul: North American Indian Belief and Ritual*. Chicago: University of Illinois Press.

Halligan, Jessi J., Michael R. Waters et al. 2016. 'Pre-Clovis occupation 14,550 years ago at the Page-Ladson site, Florida, and the peopling of the Americas.' *Scientific Advances* 2 (5): e1600375.

Hamilton, Marcus et al. 2007. 'The complex structure of hunter-gatherer social networks.' *Proceedings of the Royal Society B* 274: 2195–2202.

Harari, Yuval N. 2014. *Sapiens: A Brief History of Humankind*. London: Harvill Secker.

Harris, David R. (ed.). 1996. *The Origins and Spread of Agriculture and Pastoralism in Eurasia*. London: UCL Press.

Harris, Mary. 1997. *Common Threads: Women, Mathematics, and Work*. Stoke on Trent: Trentham.

Harvati, Katerina et al. 2019. 'Apidima cave fossils provide earliest evidence of Homo sapiens in Eurasia.' *Nature* 571: 500–504.

Harvey, David. 2012. *Rebel Cities: From the Right to the City to the Urban Revolution*. London and New York: Verso.

Harvey, David Allen. 2012. *The French Enlightenment and its Others: The Mandarin, the Savage, and the Invention of the Human Sciences*. London: Palgrave.

Hassett, Brenna R. and Haluk Sağlamtimur. 2018. 'Radical "royals"? New evidence from Başur Höyük for radical burial practices in the transition to early states in Mesopotamia.' *Antiquity* 92: 640–54.

Hassig, Ross. 2001. 'Xicotencatl: rethinking an indigenous Mexican hero.' *Estudios de cultura náhuatl* 32: 29–49.

Havard, Gilles (transl. Phyllis Aronoff and Howard Scott). 2001. *The Great Peace of Montreal of 1701: French-Native Diplomacy in the Seventeenth Century*. Montreal: McGill-Queen's University Press.

Hayden, Brian. 1990. 'Nimrods, piscators, pluckers, and planters: the emergence of food production.' *Journal of Anthropological Archaeology* 9: 31–69.

—. 2014. *The Power of Feasts*. Cambridge: Cambridge University Press.

He, Nu. 2013. 'The Longshan period site of Taosi in Southern Shanxi Province.' In A. P. Underhill (ed.), *A Companion to Chinese Archaeology*. Chichester: Wiley, pp. 255–78.

—. 2018. 'Taosi: an archaeological example of urbanization as a political center in prehistoric China.' *Archaeological Research in Asia* 14: 20–32.

Headrick, Annabeth. 2007. *The Teotihuacan Trinity: The Sociopolitical Structure of an Ancient Mesoamerican City*. Austin: University of Texas Press.

Healy, George R. 1958. 'The French Jesuits and the idea of the noble savage.' *William and Mary Quarterly* 15: 143–67.

Heard, Joseph Norman. 1977. 'The Assimilation of Captives on the American Frontier in the Eighteenth and Nineteenth Centuries.' Doctoral thesis, Louisiana State University: LSU Digital Commons.

Heckenberger Michael J. et al. 2008. 'Pre-Columbian urbanism, anthropogenic landscapes, and the future of the Amazon.' *Science* 321: 1214–17.

Heckenberger Michael J. and Eduardo G. Neves. 2009. 'Amazonian archaeology.' *Annual Review of Anthropolology* 38: 251–66.

Helwig, Barbara. 2012. 'An age of heroes? Some thoughts on Early Bronze Age funerary customs in northern Mesopotamia.' In H. Niehr et al. (eds), *(Re-)constructing Funerary Rituals in the Ancient Near East*. Wiesbaden: Harrassowitz, pp. 47–58.

Henige, David. 1999. 'Can a myth be astronomically dated?' *American Indian Culture and Research Journal* 23 (4): 127–57.

de Heusch, Luc (transl. Roy Willis). 1982. *The Drunken King, or the Origin of the State*. Bloomington: Indiana University Press.

Hill, Kim et al. 2011. 'Co-residence patterns in hunter-gatherer societies show unique human social structure.' *Science* 331: 1286–9.

Hill, Warren and John E. Clark. 2001. 'Sports, gambling, and government: America's first social compact?' *American Anthropologist* 103 (2): 331–45.

Hillman, Gordon C. and Stuart Davies. 1990. 'Measured domestication rates in wild wheats and barley under primitive cultivation, and their archaeological implications.' *Journal of World Prehistory* 4 (2): 157–222.

Hobbes, Thomas. 1651. *Leviathan: Or the Matter, Forme and Power of a Commonwealth, Ecclesiasticall and Civil*. London: Andrew Crooke.

Hocart, Arthur. M. 1954. *Social Origins*. London: Watts.

—. 1969 [1927]. *Kingship*. London: Oxford University Press.

—. 1970 [1936]. *Kings and Councillors: An Essay in the Comparative Anatomy of Human Society*. Chicago: University of Chicago Press.

Hodder, Ian. 2003. 'The lady and the seed: some thoughts on the role of agriculture in the Neolithic Revolution.' In Mehmet Özdoğan, Harald Hauptmann and Nezih Başgelen (eds), *From Villages to Cities: Early Villages in the Near East – Studies Presented to Ufuk Esin*. Istanbul: Arkeoloji ve Sanat Yayinlari, pp. 155–61.

—. 2004. 'Women and men at Çatalhöyük.' *Scientific American* 290 (1): 76–83.

—. 2006. *Çatalhöyük. The Leopard's Tale. Revealing the Mysteries of Turkey's Ancient 'Town'*. London: Thames and Hudson.

Hodder, Ian and Craig Cessford. 2004. 'Daily practice and social memory at Çatalhöyük.' *American Antiquity* 69 (1): 17–40.

Hodgen, Margaret Trabue. 1964. *Early Anthropology in the Sixteenth and Seventeenth Centuries*. Philadelphia: University of Pennsylvania Press.

Hoffmann, Dirk et al. 2018. 'U-Th dating of carbonate crusts reveals Neanderthal origin of Iberian cave art.' *Science* 359 (6378): 912–15.

Holdaway, Simon J. and Willeke Wendrich (eds). 2017. *The Desert Fayum Reinvestigated: The Early to Mid-Holocene Landscape Archaeology of the Fayum North Shore, Egypt*. Los Angeles: UCLA Cotsen Institute of Archaeology.

Holm, Tom. 2002. 'American Indian warfare: the cycles of conflict and the militarization of Native North America.' In Philip J. Deloria and Neal Salisbury (eds), *A Companion to American Indian History*. Oxford: Blackwell, pp. 154–72.

Holt, Julie Zimmerman. 2009. 'Rethinking the Ramey state: was Cahokia the center of a theater state?' *American Antiquity* 74 (2): 231–54.

Hornborg, Alf. 2005. 'Ethnogenesis, regional interaction, and ecology in prehistoric Amazonia: toward a systems perspective.' *Current Anthropology* 46 (4): 589–620.

Hornborg, Alf and Jonathan D. Hill (eds). 2011. *Ethnicity in Amazonia: Reconstructing Past Identities from Archaeology, Linguistics, and Ethnohistory*. Boulder: University Press of Colorado.

Hrdy, Sarah Blaffer. 2009. *Mothers and Others: The Evolutionary Origins of Mutual Understanding*. Cambridge, MA: Harvard University Press.

Hudson, Charles. 1976. *The Southeastern Indians*. Knoxville: University of Tennessee Press.

—. (ed.) 1979. *Black Drink: A Native American Tea*. Athens: University of Georgia Press.

Hudson, Michael. 2018. *And Forgive Them Their Debts: Lending, Foreclosure, and Redemption from Bronze Age Finance to the Jubilee Year*. Dresden: ISLET-Verlag.

Hudson, Michael and Baruch A. Levine (eds). 1996. *Privatization in the Ancient Near East and the Classical World.* Cambridge, MA: Peabody Museum of Archaeology and Ethnology.

Humphrey, Louise and Chris Stringer. 2018. *Our Human Story.* London: Natural History Museum.

Hunter, Graeme. 1989. 'The fate of Thomas Hobbes.' *Studia Leibnitiana* 21 (1): 5–20.

Hutchinson, Sharon. 1996. *Nuer Dilemmas: Coping with Money, War, and the State*. Berkeley: University of California Press.

Hyland, Sabine. 2016. 'How khipus indicated labour contributions in an Andean village: An explanation of colour banding, seriation and ethnocategories.' *Journal of Material Culture* 21 (4): 490–509.

—. 2017. 'Writing with twisted cords: the inscriptive capacity of Andean *khipus*.' *Current Anthropology* 58 (3): 412–19.

Iakovleva, Lioudmila. 2015. 'The architecture of mammoth bone circular dwellings of the Upper Palaeolithic settlements in Central and Eastern Europe and their socio-symbolic meanings.' *Quaternary International* 359–60: 324–34.

Ingold, Tim, David Riches and James Woodburn (eds). 1998. *Hunters and Gatherers 1: History, Evolution and Social Change*. Oxford: Berg.

Insoll, Timothy (ed.). 2017. *The Oxford Handbook of Prehistoric Figurines*. Oxford: Oxford University Press.

Isaac, Barry. L. 1983. 'The Aztec "Flowery War": a geopolitical explanation.' *Journal of Anthropological Research* 39 (4): 415–32.

Isakhan, Benjamin. 2011. 'What is so "primitive" about "primitive democracy"? Comparing the ancient Middle East and classical Athens.' In B. Isakhan and S. Stockwell (eds), *The Secret History of Democracy*. London: Palgrave Macmillan, pp. 19–34.

Isbell, William H. 2008. 'Wari and Tiwanaku: international identities in the Central Andean Middle Horizon.' In H. Silverman and W. H. Isbell (eds), *The Handbook of South American Archaeology.* New York: Springer, pp. 731–59.

Jacobs, Jane. 1969. *The Economy of Cities*. New York: Knopf Doubleday.

Jacobs, Ken. 1995. 'Returning to Oleni' ostrov: social, economic, and skeletal dimensions of a Boreal Forest Mesolithic cemetery.' *Journal of Anthropological Archaeology* 14 (4): 359–403.

Jacobsen, Thorkild. 1943. 'Primitive democracy in ancient Mesopotamia.' *Journal of Near Eastern Studies* 2 (3): 159–72.

—. 1976. *The Treasures of Darkness: A History of Mesopotamian Religion.* New Haven, CT and London: Yale University Press.

James, Hannah V. A. and Michael D. Petraglia. 2005. 'Modern human origins and the evolution of behaviour in the Later Pleistocene record of South Asia.' *Current Anthropology* 46: 3–27.

Jamieson, Susan M. 1992. 'Regional interaction and Ontario Iroquois evolution.' *Canadian Journal of Archaeology/Journal Canadien d'Archéologie* 16: 70–88.

Jansen, Michael. 1993. 'Mohenjo-daro, type site of the earliest urbanization process in South Asia; ten years of research at Mohenjo-daro Pakistan and an attempt at a synopsis.' In A. Parpola and P. Koskikallio (eds), *South Asian Archaeology*. Helsinki: Suomalainen Tiedeakatemia, pp. 263–80.

Jaubert, Jacques et al. 2016. 'Early Neanderthal constructions deep in Bruniquel Cave in southwestern France.' *Nature* 534: 111–15.

Jeunesse, C. 1997. *Pratiques funéraires au néolithique ancien: Sépultures et nécropoles danubiennes 5500–4900 av. J.-C.* Paris: Errance.

Johansen, Bruce E. 1982. *Forgotten Founders: Benjamin Franklin, the Iroquois, and the Rationale for the American Revolution.* Ipswich, MA: Gambit, Inc.

—. 1998. *Debating Democracy: Native American Legacy of Freedom.* Santa Fe: Clear Light Publishers.

Johansen, Bruce Elliot and Barbara Alice Mann (eds). 2000. *Encyclopedia of the Haudenosaunee (Iroquois Confederacy).* Westport, CT: Greenwood Press.

Johnson, Douglas. 1997. *Nuer Prophets: A History of Prophecy from the Upper Nile in the Nineteenth and Twentieth Centuries.* Oxford: Clarendon Press.

Johnson, Gregory A. 1982. 'Organizational structure and scalar stress.' In Colin Renfrew, Michael Rowlands and Barbara A. Segraves-Whallon (eds), *Theory and Explanation in Archaeology.* New York: Academic Press, pp. 389–421.

Jones, Jana et al. 2014. 'Evidence for prehistoric origins of Egyptian mummification in Late Neolithic burials.' *PLoS ONE* 9 (8): e103608.

Jones, Martin et al. 2011. 'Food globalization in prehistory.' *World Archaeology* 43 (4): 665–75.

Jordan, Peter et al. 2016. 'Modelling the diffusion of pottery technologies across Afro-Eurasia: emerging insights and future research.' *Antiquity* 90 (351): 590–603.

Jorgensen, Joseph G. 1980. *Western Indians: Comparative Environments, Languages and Cultures of 172 Western American Indian Tribes.* San Francisco: W. H. Freeman and Co.

Jung, Carl G. 1958. *The Undiscovered Self.* Boston: Little, Brown and Co.

Kan, Sergei (ed.). 2001. *Strangers to Relatives: The Adoption and Naming of Anthropologists in North America*. Lincoln: University of Nebraska Press.

Kanjou, Youssef et al. 2013. 'Early human decapitation, 11,700–10,700 cal BP, within the Pre-Pottery Neolithic village of Tell Qaramel, North Syria.' *International Journal of Osteoarchaeology* 25 (5): 743–52.

Karahashi, Fumi. 2016. 'Women and land in the Presargonic Lagaš Corpus.' In B. Lyon and C. Michel (eds), *The Role of Women in Work and Society in the Ancient Near East*. Boston and Berlin: De Gruyter, pp. 57–70.

Karsten, Per and Bo Knarrström. 2001. 'Tågerup – fifteen hundred years of Mesolithic occupation in western Scania, Sweden: a preliminary view.' *European Journal of Archaeology* 4 (2): 165–74.

Kashina, E. and A. Zhulnikov. 2011. 'Rods with elk heads: symbols in ritual context.' *Estonian Journal of Archaeology* 15: 18–31.

Kavanagh, Thomas M. 1994. 'Reading the moment and the moment of Reading in Graffigny's *Lettres d'une Péruvienne*.' *Modern Language Quarterly* 55 (2): 125–47.

Kehoe, Alice Beck. 2007. 'Osage texts and Cahokia data.' In F. Kent Reilly III and James F. Garber (eds), *Ancient Objects and Sacred Realms: Interpretations of Mississippian Iconography*. Austin: University of Texas Press, pp. 246–62.

Keightley, David N. 1999. 'The Shang: China's first historical dynasty.' In M. Loewe and E. L. Shaughnessy (eds), pp. 232–91.

Kelly, John E. 2008. 'Contemplating Cahokia's collapse.' In Jim A. Railey and Richard Martin Reycraft (eds), *Global Perspectives on the Collapse of Complex Systems*. Anthropological Papers no. 8. Albuquerque: Maxwell Museum of Anthropology.

Kelly, Raymond C. 2000. *Warless Societies and the Origins of War*. Ann Arbor: University of Michigan Press.

Kelly, Robert L. 2013. *The Lifeways of Hunter-Gatherers: The Foraging Spectrum*. Cambridge: Cambridge University Press.

Kemp, Barry. 2006. *Ancient Egypt: Anatomy of a Civilization* (2nd edn). London: Routledge.

Kenoyer, J. M. 1992. 'Harappan craft specialization and the question of urban segregation and stratification.' *The Eastern Anthropologist* 45 (1–2): 39–54.

—. 1998. *Ancient Cities of the Indus Valley*. Karachi: Oxford University Press.

Kerig, T. 2003. 'Von Gräbern und Stämmen: Zur Interpretation bandkeramischer Erdwerke.' In U. Veit, T. L. Kienlin, C. Kümmel et al. (eds), *Spuren und Botschaften: Interpretationen materieller Kultur*. Münster: Waxmann, pp. 225–44.

Kidder, Tristram R. 2018. 'Poverty Point.' In Timothy R. Pauketat (ed.), *The Oxford Handbook of North American Archaeology*. Oxford: Oxford University Press, pp. 464–9.

Kilian, Klaus. 1988. 'The emergence of wanax ideology in the Mycenaean palaces.' *Oxford Journal of Archaeology* 7 (3): 291–302.

King, Adam. 2003. *Etowah: The Political History of a Chiefdom Capital*. Tuscaloosa: University of Alabama Press.

—. 2004. 'Power and the sacred: Mound C and the Etowah chiefdom.' In Richard F. Townsend and Robert V. Sharp (eds), *Hero, Hawk, and Open Hand: American Indian Art of the Ancient Midwest and South*. New Haven, CT: The Art Institute of Chicago, pp. 151–65.

—. 2007. 'Mound C and the Southeastern ceremonial complex in the history of the Etowah site.' In A. King (ed.), *Southeastern Ceremonial Complex Chronology, Content, Context*. Tuscaloosa, University of Alabama Press, pp. 107–33.

Kirch, Patrick V. 1990. 'Specialization and exchange in the Lapita Complex of Oceania (1600–500 B.C.).' *Asian Perspectives* 29 (2): 117–33.

Kirleis, Wiebke and Marta Dal Corso. 2016. 'Trypillian subsistence economy: animal and plant exploitation.' In Müller et al. (eds), pp. 195–206.

Knight, Chris. 1991. *Blood Relations. Menstruation and the Origins of Culture*. New Haven, CT and London: Yale University Press.

Knight, Vernon J. Jr. 1986. 'The institutional organization of Mississippian religion.' *American Antiquity* 51: 675–87.

—. 1989. 'Some speculations on Mississippian monsters.' In Patricia Galloway (ed.), *The Southeastern Ceremonial Complex: Artifacts and Analysis*. Lincoln: University of Nebraska Press, pp. 205–10.

—. 2001. 'Feasting and the emergence of platform mound ceremonialism in Eastern North America.' In Michael Dietler and Brian Hayden (eds), *Feasts: Archaeological and Ethnographic Perspectives on Food, Politics and Power*. Washington: Smithsonian Institution Press, pp. 311–33.

—. 2006. 'Symbolism of Mississippian mounds.' In Gregory A. Waselkov, Peter H. Wood and Tom Hatley (eds), *Powhatan's Mantle: Indians in the Colonial Southeast* (revised and expanded edn). Lincoln: Nebraska University Press, pp. 421–34.

Kobayashi, Tatsuo. 2004. *Jomon Reflections: Forager Life and Culture in the Prehistoric Japanese Archipelago*. Oxford: Oxbow.

Koch, Alexander et al. 2019. 'Earth system impacts of the European arrival and Great Dying in the Americas after 1492.' *Quaternary Science Reviews* 207 (1): 13–36.

Köksal-Schmidt, Çiğdem and Klaus Schmidt. 2010. 'The Göbekli Tepe "Totem Pole." A first discussion of an autumn 2010 discovery (PPN, Southeastern Turkey).' *Neo-Lithics* 1 (10): 74–6.

Kolata, Alan. 1992. 'In the realm of the Four Quarters.' In Alvin M. Josephy (ed.), *America in 1492*. New York: Knopf, pp. 215–47.

—. 1997. 'Of kings and capitals: principles of authority and the nature of cities in the Native Andean state.' In D. L. Nichols and T. H. Charlton (eds), *The Archaeology of City States: Cross-Cultural Approaches*. Washington: Smithsonian Institution Press, pp. 245–54.

Kononenko, Nina et al. 2016. 'Detecting early tattooing in the Pacific region through experimental usewear and residue analyses of obsidian tools.' *Journal of Archaeological Science*, Reports 8: 147–63.

Kopenawa, Davi and Bruce Albert. 2013. *The Falling Sky: Words of a Yanomami Shaman*. London and Cambridge, MA: The Belknap Press of Harvard University Press.

Kornienko, Tatiana V. 2015. 'On the problem of human sacrifice in Northern Mesopotamia in the Pre-Pottery Neolithic.' *Archaeology, Ethnology and Anthropology of Eurasia* 43 (3): 42–9.

Kowaleski, Jeff. 2003. 'Evidence for the functions and meanings of some northern Maya palaces.' In J. J. Christie (ed.), *Maya Palaces and Elite Residences*. Austin: University of Texas Press, pp. 204–52.

Kristiansen, Kristian. 1993. 'The strength of the past and its great might: an essay on the use of the past.' *Journal of European Archaeology* 1 (1): 3–32.

Kroeber, Alfred L. 1925. *Handbook of the Indians of California*. Bureau of American Ethnology Bulletin 78. Washington: Smithsonian Institution.

—. 1944. *Configurations of Cultural Growth*. Berkeley: University of California Press.

Kubler, George. 1962. *The Shape of Time. Remarks on the History of Things*. New Haven, CT and London: Yale University Press.

Kuijt, Ian. 1996. 'Negotiating equality through ritual: a consideration of Late Natufian and Prepottery Neolithic A period mortuary practices.' *Journal of Anthropological Archaeology* 15: 313–36.

Kuper, Rudoplh and Stefan Kroepelin. 2006. 'Climate-controlled Holocene occupation in the Sahara: motor of Africa's evolution'. *Science* 313: 803–7.

La Flesche, Francis. 1921. 'The Osage tribe: rite of the chiefs: sayings of the ancient men.' *Thirty-sixth Annual Report of the Bureau of American Ethnology* (1914–15), pp. 35–604. Washington.

—. 1930. 'The Osage tribe: rite of the Wa-xo'-be.' *Forty-fifth Annual Report of the Bureau of American Ethnology* (1927–28), pp. 529–833. Washington.

—. 1939. *War Ceremony and Peace Ceremony of the Osage Indians*. Bureau of American Ethnology Bulletin 101. Washington: US Government.

Lagrou, Els. 2009. *Arte Indígena no Brasil: Agência, Alteridade e Relação*. Belo Horizonte: C/Arte.

Lahontan, Louis Armand de Lom d'Arce (ed. Réal Ouellet and Alain Beaulieu). 1990 [1702a]. *Mémoires de l'Amérique septentrionale, ou la suite des voyages de Mr. le Baron de Lahontan*. Montreal: Presses de l'Université de Montréal.

—. (ed. Réal Ouellet and Alain Beaulieu). 1990 [1702b]. *Nouveaux Voyages de Mr. Le Baron de Lahontan, dans l'Amérique Septentrionale*. Montreal: Presses de l'Université de Montréal.

—. (ed. Réal Ouellet and Alain Beaulieu).1990 [1703]. *Supplément aux Voyages du Baron de Lahontan, ou l'on trouve des dialogues curieux entre l'auteur et un sauvage de bon sens qui a voyagé*. Montreal: Presses de l'Universite´ de Montréal.

—. 1735. *New Voyages to North America Giving a Full Account of the Customs, Commerce, Religion, and Strange Opinions of the Savages of That Country, With Political Remarks upon the Courts of Portugal and Denmark, and the Present State of the Commerce of Those Countries*. London: J. Walthoe.

Lang, Caroline et al. 2013. 'Gazelle behaviour and human presence at early Neolithic Göbekli Tepe, south-east Anatolia.' *World Archaeology* 45: 410–29.

Langley, Michelle C., Christopher Clarkson and Sean Ulm. 2008. 'Behavioural complexity in Eurasian Neanderthal populations: a chronological examination of the archaeological evidence.' *Cambridge Archaeological Journal* 18 (3): 289–307.

Lansing, J. Stephen. 1991. *Priests and Programmers: Technologies of Power in the Engineered Landscapes of Bali*. Princeton, NJ: Princeton University Press.

Larson, Greger. et al. 2007. 'Ancient DNA, pig domestication, and the spread of the Neolithic into Europe.' *Proceedings of the National Academy of Sciences* 104: 15276–81.

Larsson, Lars. 1990. 'The Mesolithic of southern Scandinavia.' *Journal of World Prehistory* 4 (3): 257–309.

Lattas, Andrew. 2006. 'The utopian promise of government.' *The Journal of the Royal Anthropological Institute* 12 (1): 129–50.

Lattimore, Owen. 1962. *Studies in Frontier History*. Collected Papers 1929–58. London: Oxford University Press.

Lazarovici, Cornelia-Magda. 2010. 'Cucuteni ceramics: technology, typology, evolution, and aesthetics.' In Anthony (ed.). pp. 128–61.

Le Guin, Ursula K. 1993 [1973]. *The Ones Who Walk Away from Omelas*. Mankato, MN: Creative Education.

Leach, Edmund. 1976. *Culture and Communication*. Cambridge: Cambridge University Press.

Leacock, Eleanor. 1978. 'Women's status in egalitarian society: implications for social evolution.' *Current Anthropology* 19: 247–76.

Lee, Richard B. and Irven DeVore (eds). 1968. *Man the Hunter*. Chicago: Aldine.

Lehner, Mark. 2015. 'Labor and the pyramids. The Heit el-Ghurab "workers town" at Giza.' In Steinkeller and Hudson (eds), pp. 397–522.

Lepper, Bradley T. 1995. 'Tracking Ohio's Great Hopewell Road.' *Archaeology* 48 (6): 52–6.

Lesure, Richard. 1998. 'The constitution of inequality in Yurok society.' *Journal of California and Great Basin Anthropology* 20 (2): 171–94.

Lévi-Strauss, Claude. 1963. 'Do dual organizations exist?' In C. Lévi-Strauss, *Structural Anthropology*. Harmondsworth: Penguin, pp. 132–63.

—. 1966. *The Savage Mind*. Chicago: University of Chicago Press.

—. 1967 [1944]. 'The social and psychological aspects of chieftainship in a primitive tribe: the Nambikwara of northwestern Mato Grosso.' In R. Cohen and J. Middleton (eds), *Comparative Political Systems*. Austin and London: University of Texas Press, pp. 45–62.

—. 1982 [1976]. *The Way of the Masks*. Seattle: University of Washington Press.

—. 1987. *Anthropology and Myth: Lectures 1951–1982*. Oxford: Blackwell.

Levy, Philip A. 1996. 'Exemplars of taking liberties: the Iroquois influence thesis and the problem of evidence.' *William and Mary Quarterly* 53 (3): 587–604.

Li, Jaang, Zhouyong Sun, Jing Shao and Min Li. 2018. 'When peripheries were centres: a preliminary study of the Shimao-centred polity in the loess highland, China.' *Antiquity* 92 (364): 1008–22.

Lightfoot, Kent G. and Otis Parrish. 2009. *Californian Indians and their Environment*. Berkeley: University of California Press.

Lincoln, Charles K. 1994. 'Structural and philological evidence for divine kingship at Chichén Itzá, Yucatan, México.' In Hanns J. Prem (ed.), *Hidden Among the Hills*. Acta Mesoamericana, vol. 7. Möckmühl: Verlag von Flemming, pp. 164–96.

Lipson, M. et al. 2017. 'Parallel palaeogenomic transects reveal complex genetic history of early European farmers.' *Nature* 551: 368–72.

Liu, Li and Xingcan Chen. 2012. *The Archaeology of China: From the Late Paleolithic to the Early Bronze Age*. Cambridge: Cambridge University Press.

Liverani, Mario (ed. and transl. Z. Bahrani and M. Van De Mieroop). 1998. *Uruk. The First City*. Sheffield: Equinox.

Lockhart, James. 1985. 'Some Nahua concepts in postconquest guise.' *History of European Ideas* 6 (4): 465–82.

Lockhart, James, Frances Berdan and Arthur J. O. Anderson. 1986. *The Tlaxcalan Actas: A Compendium of the Records of the Cabildo of Tlaxcala (1545–1627)*. Salt Lake City: University of Utah Press.

Loeb, Edwin M. 1931. 'The religious organizations of North-Central California and Tierra del Fuego.' *American Anthropologist* 33 (4): 517–56.

Loewe, Michael and Edward. L. Shaughnessy (eds). 1999. *The Cambridge History of Ancient China*. Cambridge: Cambridge University Press.

Lombardo, Umberto et al. 2020. 'Early Holocene crop cultivation and landscape modification in Amazonia.' *Nature* 581 (2020): 190–93.

Lorenz, Karl G. 1997. 'A re-examination of Natchez sociopolitical complexity: a view from the grand village and beyond.' *Southeastern Archaeology* 16 (2): 97–112.

Lovejoy, Arthur O. and George Boas. 1935. *Primitivism and Related Ideas in Antiquity*. Baltimore: Johns Hopkins University Press.

Lowie, Robert H. 1928. 'Incorporeal property in primitive society.' *Yale Law Journal* 37 (5): 551–63.

—. 1948. 'Some aspects of political organisation among the American Aborigines.' *Journal of the Royal Anthropological Institute of Great Britain and Ireland* 78: 11–24.

McBrearty, Sally and Alison S. Brooks. 2000. 'The revolution that wasn't: a new interpretation of the origin of modern human behaviour.' *Journal of Human Evolution* 39: 453–563.

MacCormack, Carol P. and Marilyn Strathern (eds). 1980. *Nature, Culture, and Gender.* Cambridge: Cambridge University Press.

McGregor, Gaile. 1988. *The Noble Savage in the New World Garden: Notes Toward a Syntactics of Place*. Toronto: University of Toronto Press.

McIntosh, Roderick. 2005. *Ancient Middle Niger: Urbanism and the Self-Organizing Past*. Cambridge: Cambridge University Press.

McIntosh, Susan Keech. 2009. *Beyond Chiefdoms: Pathways to Complexity in Africa*. Cambridge: Cambridge University Press.

MacLachlan, Colin M. 1991. *Spain's Empire in the New World: The Role of Ideas in Institutional and Social Change*. Berkeley: University of California Press.

MacLeod, William C. 1928. 'Economic aspects of indigenous American slavery.' *American Anthropologist* 30 (4): 632–50.

—. 1929. 'The origin of servile labor groups.' *American Anthropologist* 31 (1): 89–113.

MacPherson, C. B. 1962. *The Political Theory of Possessive Individualism*. Oxford: Oxford University Press.

Madgwick, Richard et al. 2019. 'Multi-isotope analysis reveals that feasts in the Stonehenge environs and across Wessex drew people and animals from throughout Britain.' *Science Advances* 5 (3): eaau6078.

Maeda, Osamu et al. 2016. 'Narrowing the harvest: increasing sickle investment and the rise of domesticated cereal agriculture in the Fertile Crescent.' *Quaternary Science Reviews* 145: 226–37.

Maine, Henry Sumner. 1893 [1875]. *Lectures on the Early History of Institutions*. London: John Murray.

Malinowski, Bronisław. 1922. *Argonauts of the Western Pacific: An Account of Native Enterprise and Adventure in the Archipelagoes of Melanesian New Guinea*. London: Routledge.

Mann, Barbara Alice. 1997. 'The lynx in time: Haudenosaunee women's traditions and history.' *American Indian Quarterly* 21 (3): 423–50.

—. 1998. 'Haudenosaunee (Iroquois) women, legal and political status.' In Bruce Elliott Johansen (ed.), *The Encyclopedia of Native American Legal Tradition*. Westport, CT: Greenwood Press, pp. 112–31.

—. 2000. *Iroquoian Women: The Gantowisas*. New York: Peter Lang.

—. 2001. 'Are you delusional? Kandiaronk on Christianity.' In B. A. Mann (ed.), *Native American Speakers of the Eastern Woodlands: Selected Speeches and Critical Analysis*. Westport, CT: Greenwood Press, pp. 35–82.

Mann, Barbara A. and Jerry L. Fields. 1997. 'A sign in the sky: dating the League of the Haudenosaunee.' *American Indian Culture and Research Journal* 21 (2): 105–63.

Mann, Charles C. 2005. *1491: The Americas before Columbus*. London: Granta.

Mann, Michael. 1986. *The Sources of Social Power*, vol. 1: *A History of Power from the Beginning to AD 1760*. Cambridge: Cambridge University Press.

Manning, Joseph G. 2003. *Land and Power in Ptolemaic Egypt: The Structure of Land Tenure*. Cambridge: Cambridge University Press.

Manzanilla, Linda R. 1993. 'Daily life in Teotihuacan apartment compounds.' In Berrin and Pasztory (eds), pp. 90–99.

—. 1996. 'Corporate groups and domestic activities at Teotihuacan.' *Latin American Antiquity* 7 (3): 228–46.

—. 2015. 'Cooperation and tensions in multi-ethnic corporate societies using Teotihuacan, Central Mexico, as a case study.' *Proceedings of the National Academy of Sciences* 112 (30): 9210–15.

—. 2017. 'Discussion: the subsistence of the Teotihuacan metropolis.' *Archaeological and Anthropological Sciences* 9: 133–40.

Manzura, Igor. 2005. 'Steps to the steppe: or, how the North Pontic region was colonised.' *Oxford Journal of Archaeology* 24(4): 313–38.

Marchesi, Gianni. 2004. 'Who was buried in the royal tombs of Ur? The epigraphic and textual data.' *Orientalia* 73 (2): 153–97.

Marean, Curtis. W. 2014. 'The origins and significance of coastal resource use in Africa and Western Eurasia.' *Journal of Human Evolution* 77: 17–40.

Martin, Simon. 2001. 'The power in the west: the Maya and Teotihuacan.' In Grube et al. (eds), pp. 98–113.

—. 2020. *Ancient Maya Politics: A Political Anthropology of the Classic Period 150–900 CE*. Cambridge: Cambridge University Press.

Martin, Simon and Nikolai Grube. 2000. *Chronicle of the Maya Kings and Queens: Deciphering the Dynasties of the Ancient Maya*. London: Thames and Hudson.

Matar, Nabil. 2009. *Europe Through Arab Eyes, 1578–1727*. New York: Columbia University Press.

Matsui, Akira and Masaaki Kanehara. 2006. 'The question of prehistoric plant husbandry during the Jomon period in Japan.' *World Archaeology* 38 (2): 259–73.

Matthews, Wendy. 2005. 'Micromorphological and microstratigraphic traces of uses and concepts of space.' In Ian Hodder (ed.), *Inhabiting Çatalhöyük: Reports from the 1995–1999 Seasons*. Cambridge: McDonald Institute for Archaeological Research, British Institute of Archaeology at Ankara, pp. 355–98.

Mauss, Marcel. 1968–9. *Oeuvres*, vols 1–3. Paris: Éditions de Minuit.

—. 2016 [1925]. *The Gift* (expanded edn, selected, annotated and transl. Jane I. Guyer). Chicago: HAU Books.

Mauss, Marcel and Henri Beuchat. 1979 [1904–5]. *Seasonal Variations of the Eskimo: A Study in Social Morphology*. London: Routledge.

Maybury-Lewis, David (ed.). 1979. *Dialectical Societies: The Gê and Bororo of Central Brazil*. Cambridge, MA: Harvard University Press.

Meek, Ronald. 1976. *Social Sciences and the Ignoble Savage*. Cambridge: Cambridge University Press.

Meillassoux, Claude (transl. Alide Dasnois). 1996. *The Anthropology of Slavery: The Womb of Iron and Gold*. Chicago: University of Chicago Press.

Mellaart, James. 1967. *Çatal Hüyük: A Neolithic Town in Anatolia*. London: Thames and Hudson.

Mellars, Paul et al. (eds). 2007. *Rethinking the Human Revolution: New Behavioural and Biological Perspectives on the Origin and Dispersal of Modern Humans*. Cambridge: McDonald Institute.

Meller, Harald and Michael Schefik. 2015. *Krieg: Eine Archäologische Spurensuche*. Halle (Saale): Landesmuseum für Vorgeschichte.

Menotti, Francesco and Aleksey G. Korvin-Piotrovskiy (eds). 2012. *The Tripolye Culture. Giant-Settlements in Ukraine. Formation, Development and Decline*. Oxford: Oxbow Books.

Merrick, Jeffrey. 1991. 'Patriarchalism and constitutionalism in eighteenth-century parliamentary discourse.' *Studies in Eighteenth-Century Culture* 20: 317–30.

Meskell, Lynn, Carolyn Nakamura, Rachel King and Shahina Farid. 2008. 'Figured lifeworlds and depositional practices at Çatalhöyük.' *Cambridge Archaeological Journal* 18 (2): 139–61.

Mieroop, Marc Van De. 1989. 'Women in the economy of Sumer.' In B. S. Lesko (ed.), *Women's Earliest Records: Western Asia and Egypt*. Atlanta, GA: Scholars Press, pp. 53–66.

—. 1997. *The Ancient Mesopotamian City*. Oxford: Oxford University Press.

—. 1999. 'The government of an ancient Mesopotamian city: what we know and why we know so little.' In K. Watanabe (ed.), *Priests and Officials in the Ancient Near East*. Heidelberg: Universitätsverlag C. Winter, pp. 139–61.

—. 2013. 'Democracy and the rule of law, the assembly, and the first law code.' In Crawford (ed.), pp. 277–89.

Migliano, Andrea et al. 2017. 'Characterization of hunter-gatherer networks and implications for cumulative culture.' *Nature Human Behaviour* 1 (2): 43.

Miller, Daniel. 1985. 'Ideology and the Harappan civilization.' *Journal of Anthropological Archaeology* 4: 34–71.

Miller, Heather. M. L. 2000. 'Reassessing the urban structure of Harappa: evidence from craft production distribution.' *South Asian Archaeology* (1997), pp. 207–47.

Miller, Joaquin. 1873. *Life Amongst the Modocs: Unwritten History*. London: Richard Bentley and Son.

Miller, Mary Ellen and Stephen D. Houston. 1987. 'The classic Maya ballgame and its architectural setting.' *RES* 14: 47–65.

Millon, René. 1964. 'The Teotihuacan mapping project.' *American Antiquity* 29 (3): 345–52.

—. 1970. 'Teotihuacan: completion of map of giant ancient city in the Valley of Mexico.' *Science* 170: 1077–82.

—. 1976. 'Social relations at ancient Teotihuacan.' In E. Wolf (ed.), *The Valley of Mexico: Studies in Pre-Hispanic Ecology and Society*. Albuquerque: University of New Mexico Press, pp. 205–48.

—. 1988. 'Where do they all come from? The provenance of the Wagner murals at Teotihuacan.' In Berrin (ed.), pp. 16–43.

—. 1993. 'The place where time began: an archaeologist's interpretation of what happened in Teotihuacan history.' In Berrin and Pasztory (eds), pp. 16–43.

Milner, George R. 1998. *The Cahokia Chiefdom: The Archaeology of a Mississippian Society*. Washington: Smithsonian Institution Press.

Milner, George R., George Chaplin and Emily Zavodny. 2013. 'Conflict and societal change in Late Prehistoric Eastern North America.' *Evolutionary Anthropology* 22(3): 96–102.

Mitchell, Donald. 1985. 'A demographic profile of Northwest Coast slavery.' In M. Thompson, M. T. Garcia and F. J. Kense (eds), *Status, Structure and Stratification: Current Archaeological Reconstructions*. Calgary, Alberta: Archaeological Association of the University of Calgary, pp. 227–36.

Mithen, Steven. J. 2003. *After the Ice: A Global Human History 20,000–5000 BC*. London: Weidenfeld and Nicolson.

Mittnik, Alissa et al. 2016. 'A molecular approach to the sexing of the triple burial at the Upper Paleolithic site of Dolní Věstonice.' *PLoS ONE* 11 (10): e0163019.

Moore, Andrew M. T. and Gordon C. Hillman. 1992. 'The Pleistocene to Holocene transition and human economy in Southwest Asia: the impact of the Younger Dryas.' *American Antiquity* 57 (3): 482–94.

Moorey, P. R. S. 1964. 'The "Plano-Convex Building" at Kish and Early Mesopotamian Palaces.' *Iraq* 26 (2): 83–98.

—. 1977. 'What do we know about the people buried in the royal cemetery?' *Expedition* 20 (1): 24–40.

Moreno García, Juan Carlos. 2014. 'Recent developments in the social and economic history of Ancient Egypt.' *Journal of Ancient Near Eastern History* 1 (2): 231–61.

Morgan, Lewis Henry. 1851. *League of the Ho-de-no-sau-nee, or Iroquois*. New York: Dodd, Mead and Co.

—. 1877. *Ancient Society, or Researches in the Lines of Human Progress, from Savagery through Barbarism to Civilization*. New York: Henry Holt and Co.

Morphy, Howard. 1991. *Ancestral Connections: Art and an Aboriginal System of Knowledge*. Chicago: University of Chicago Press.

Morris, Craig and Adriana von Hagen. 2011. *The Incas: Lords of the Four Quarters*. London: Thames and Hudson.

Morris, Ellen. 2007. 'Sacrifice for the state: First Dynasty royal funerals and the rites at Macramallah's triangle.' In Nicola Laneri (ed.), *Performing Death: Social Analysis of Funerary Traditions in the Ancient Near East and Mediterranean*. Chicago: Oriental Institute of Chicago, pp. 15–38.

—. 2014. '(Un)dying loyalty: meditations on retainer sacrifice in ancient Egypt and elsewhere.' In Campbell (ed.), pp. 61–93.

Morris, Ian. 2015. *Foragers, Farmers, and Fossil Fuels: How Human Values Evolve*. Princeton, NJ: Princeton University Press.

Moss, Madonna. 1993. 'Shellfish, gender, and status on the Northwest Coast: reconciling archaeological, ethnographic, and ethnohistorical records of the Tlingit.' *American Anthropologist* 95 (3): 631–52.

Motolinía, Fr. Toribio de Benavente. 1914 [1541]. *Historia de los Indios de la Neuva España*. Barcelona: Herederos de Juan Gili.

Muhlberger, Steven and Phil Pain. 1996. 'Democracy's place in world history.' *Journal of World History* 4 (1): 23–45.

Müller, Johannes. 2016. 'Human structure, social space: what we can learn from Trypyllia.' In Müller et al. (eds), pp. 301–4.

Müller, Johannes et al. 2016. 'Chronology and demography: how many people lived in a mega-site.' In Müller et al. (eds), pp. 133–70.

Müller, Johannes, Knut Rassmann and Mykhailo Videiko (eds). 2016. *Trypillia Mega-Sites and European Prehistory, 4100–3400 BCE*. London and New York: Routledge.

Murdock, George P. 1937. 'Comparative data on the division of labour by sex.' *Social Forces* 15: 551–3.

Murdock, George P. and Caterina Provost. 1973. 'Factors in the division of labour by sex: a cross-cultural analysis.' *Ethnology* 12: 203–26.

Murra, John Victor. 1956. *The Economic Organization of the Inca State*. Chicago: Department of Anthropology. PhD dissertation, republished as *The Economic Organization of the Inka State* by JAI Press, Stamford, CT 1980).

—. 1982. 'The Mit'a obligations of ethnic groups to the Inka state.' In G. Collier, R. Rosaldo and J. Wirth (eds), *The Inca and Aztec States, 1400–1800*. New York: Academic Press, pp. 237–62.

Muthu, Sankar. 2003. *Enlightenment Against Empire*. Princeton, NJ: Princeton University Press.

Nabokov, Peter. 1996. 'Native views of history.' In Bruce G. Trigger and Wilcomb E. Washburn (eds), *The Cambridge History of the Native Peoples of the Americas*. Cambridge: Cambridge University Press, pp. 1–60.

Nash, June. 1978. 'The Aztecs and the ideology of male dominance.' *Signs* 4 (2): 349–62.

Nebbia, Marco et al. 2018. 'The making of Chalcolithic assembly places: Trypillia megasites as materialized consensus among equal strangers?' *World Archaeology* 50: 41–61.

Neitzel, Robert S. 1965. *Archaeology of the Fatherland Site: The Grand Village of the Natchez*. New York: American Museum of Natural History.

—. 1972. *The Grand Village of the Natchez Revisited: Excavations at the Fatherland Site, Adams County, Mississippi*. New York: American Museum of Natural History.

Neumann, Hans et al. 2014. *Krieg und Frieden im Alten Vorderasien*. Münster: Ugarit-Verlag.

Nichols, Deborah L. 2016. 'Teotihuacan.' *Journal of Archaeological Research* 24: 1–74.

Nieboer, Herman J. 1900. *Slavery as an Industrial System: Ethnological Researches*. The Hague: Martinus Nijhoff.

Nilsson Stutz, L. 2010. 'A Baltic way of death? A tentative exploration of identity in Mesolithic cemetery practices.' In Å.M. Larsson and L. Papmehl-Dufay (eds), *Uniting Sea II; Stone Age Societies in the Baltic Sea Region*. Borgholm: Uppsala University.

Nisbet, Robert A. 1966. *The Sociological Tradition*. London: Heineman.

Nissen, Hans. 2002. 'Uruk: key site of the period and key site of the problem.' In N. Postgate (ed.), *Artefacts of Complexity: Tracking the Uruk in the Near East*. London: British School of Archaeology in Iraq, pp. 1–16.

Nissen, Hans, Peter Damerow and Robert Englund. 1993. *Archaic Bookkeeping: Early Writing and Techniques of Economic Administration in the Ancient Near East*. Chicago: University of Chicago Press.

Noble, William C. 1978. 'The Neutral Indians.' In William Englebrecht and Donald Grayson (eds), *Essays in Northeastern Anthropology in Memory of Marian E. White*. Occasional publication in Northeastern Anthropology 5. Department of Anthropology, Franklin Pierce College, Rindge, pp. 152–64.

—. 1985. 'Tsouharissen's chiefdom: an early historic 17th Century Neutral Iroquoian ranked society.' *Canadian Journal of Archaeology/Journal Canadien d'Archéologie* 9 (2): 131–46.

Noddings, Nel. 1984. *Caring: A Feminine Approach to Ethics and Moral Education*. Berkeley: University of California Press.

Notroff, Jens, Oliver Dietrich and Klaus Schmidt. 2016. 'Gathering of the dead? The early Neolithic sanctuaries at Göbekli Tepe, southeastern Turkey.' In Colin Renfrew, Michael J. Boyd and Iain Morley (eds), *Death Rituals, Social Order, and the Archaeology of Immortality in the Ancient World*. Cambridge: Cambridge University Press, pp. 65–81.

Nussbaum, Martha. 2011. *Creating Capabilities: The Human Development Approach*. Cambridge, MA: Harvard University Press.

Nuttall, Zelia. 1921. 'Francisco Cervantes de Salazar. Biographical notes.' *Journal de la Société des Américanistes* 13 (1): 59–90.

O'Curry, Eugene. 1873. *On the Manners and Customs of the Ancient Irish*. London: Williams and Norgate.

O'Meara, Walter. 1968. *Daughters of the Country*. New York: Harcourt, Brace.

O'Shea, John and Marek Zvelebil. 1984. 'Oleneostrovski Mogilnik: reconstructing the social and economic organization of prehistoric foragers in Northern Russia.' *Journal of Anthropological Archaeology* 3: 1–40.

Oates, Joan et al. 2007. 'Early Mesopotamian urbanism: a new view from the north.' *Antiquity* 81 (313): 585–600.

Ohlrau, René et al. 2016. 'Living on the edge? Carrying capacities of Trypillian settlements in the Buh-Dnipro Interfluve.' In Müller et al. (eds), 2016, pp. 207–20.

Oppenheim, A. Leo. 1977. *Ancient Mesopotamia: Portrait of a Dead Civilization*. Chicago and London: University of Chicago Press.

Ostrom, Elinor. 1990. *Governing the Commons: The Evolution of Institutions for Collective Action*. Cambridge: Cambridge University Press.

Ott, Sandra. 1981. *The Circle of Mountains: A Basque Shepherding Community*. Oxford: Clarendon Press.

Ouellet, Réal. 1990. 'Jésuites et philosophes lecteurs de Lahontan.' *Saggi e ricerche di letteratura francese* 29: 119–64.

—. 1995. 'A la découverte de Lahontan.' *Dix-Huitième Siècle* 27: 323–33.

Owen, Linda R. 1994. 'Gender, crafts, and the reconstruction of tool use.' *Helinium* 34: 186–200.

—. 1996. 'Der Gebrauch von Pflanzen in Jungpaläolithikum Mitteleuropas.' *Ethnographisch-Archäologische Zeitschrift* 37: 119–46.

Özbek Metin. 1988. 'Culte des cranes humains a Çayönü.' *Anatolica* 15: 127–37.

—. 1992. 'The human remains at Çayönü.' *American Journal of Archaeology* 96 (2): 374.

Özkaya, Vecihi and Aytaç Coşkun. 2009. 'Körtik Tepe, a new Pre-Pottery Neolithic A site in south-eastern Anatolia.' *Antiquity* 83 (320).

Pagden, Anthony. 1982. *The Fall of Natural Man: The American Indian and the Origins of Comparative Ethnology*. Cambridge: Cambridge University Press.

—. 1983. 'The savage critic: some European images of the primitive.' *The Yearbook of English Studies* 13 (Colonial and Imperial Themes special number): 32–45.

—. 1993. *European Encounters with the New World: From Renaissance to Romanticism*. New Haven, CT and London: Yale University Press.

Parker, Arthur C. 1912. *The Code of Handsome Lake, the Seneca Prophet*. New York State Museum Bulletin 163, Education Department Bulletin 530. Albany: University of the State of New York.

—. 1916. *The Constitution of the Five Nations, or the Iroquois Book of the Great Law*. New York State Museum Bulletin 184. Albany: University of the State of New York.

—. 1919. *The Life of General Ely S. Parker, Last Grand Sachem of the Iroquois and General Grant's Military Secretary*. Buffalo Historical Society Publications 23. Buffalo, NY: Buffalo Historical Society.

Parker Pearson, Mike (and the Stonehenge Riverside Project). 2012. *Stonehenge: Exploring the Greatest Stone Age Mystery*. London: Simon and Schuster.

Pärssinen, Martti et al. 2009. 'Pre-Columbian geometric earthworks in the upper Purús: a complex society in western Amazonia.' *Antiquity* 83: 1084–95.

Pascoe, Bruce. 2014. *Dark Emu: Aboriginal Australia and the Birth of Agriculture*. London: Scribe.

Pasztory, Esther. 1988. 'A reinterpretation of Teotihuacan and its mural painting tradition.' In Berrin (ed.), pp. 45–77.

—. 1992. 'Abstraction and the rise of a utopian state at Teotihuacan.' In J. C. Berlo (ed.), *Art, Ideology, and the City of Teotihuacan*. Washington: Dumbarton Oaks Research Library and Collection, pp. 281–320.

—. 1997. *Teotihuacan. An Experiment in Living*. Norman: University of Oklahoma Press.

Patterson, Orlando. 1982. *Slavery and Social Death: A Comparative Study*. Cambridge, MA: Harvard University Press.

Pauketat, Timothy R. 1994. *The Ascent of Chiefs: Cahokia and Mississippian Politics in Native North America*. Tuscaloosa: University of Alabama Press.

—. 2004. *Ancient Cahokia and the Mississippians*. Cambridge: Cambridge University Press.

—. 2007. *Chiefdoms and Other Archaeological Delusions*. Lanham, MD: AltaMira Press.

—. 2009. *Cahokia: Ancient America's Great City on the Mississippi*. Penguin Library of American Indian History. New York: Viking.

—. 2013. *An Archaeology of the Cosmos: Rethinking Agency and Religion in Ancient Times*. London: Routledge.

Pauketat, Timothy R. and Thomas E. Emerson (eds). 1997. *Cahokia: Domination and Ideology in the Mississippian World*. Lincoln: University of Nebraska Press.

Pauketat, Timothy R., Susan M. Alt and Jeffery D. Kruchten. 2015. 'City of earth and wood: New Cahokia and its material-historical implications.' In Norman Yoffee (ed.), *The Cambridge World History*. Cambridge: Cambridge University Press, pp. 437–54.

Paulinyi, Zoltán. 1981. 'Capitals in pre-Aztec Central Mexico.' *Acta Orientalia Academiae Scientiarum Hungaricae* 35 (2/3): 315–50.

Pearson, Jessica et al. 2013. 'Food and social complexity at Çayönü Tepesi, southeastern Anatolia: stable isotope evidence of differentiation in diet according to burial practice and sex in the early Neolithic.' *Journal of Anthropological Archaeology* 32 (2): 180–89.

Pennington, B. et al. 2016. 'Emergence of civilization, changes in fluvio-deltaic style and nutrient redistribution forced by Holocene seal-level rise.' *Geoarchaeology* 31: 194–210.

Peters, Joris et al. 2017. 'The emergence of livestock husbandry in Early Neolithic Anatolia.' In U. Albarella et al. (eds), *The Oxford Handbook of Zooarchaeology*. Oxford: Oxford University Press, pp. 247–65.

Peterson, Jane. 2016. 'Woman's share in Neolithic society: a view from the Southern Levant.' *Near Eastern Archaeology* 79 (3): 132–9.

Pettitt, Paul. 2011. *The Palaeolithic Origins of Human Burial*. London: Routledge.

Pfaffenberger, Bryan. 1988. 'Fetishised objects and humanised nature: towards an anthropology of technology.' *Man* (N.S.) 23 (2): 236–52.

Piccaluga, Giulia. 1977. 'Adonis, i cacciatori falliti e l'avvento dell'agricoltura.' In Bruno Gentili and Giuseppe Paioni (eds), *Il mito greco*. Rome: Edizioni dell'Ateneo, Bizzarri, pp. 33–48.

Pilling, Arnold R. 1989. 'Yurok Aristocracy and "Great Houses".' *American Indian Quarterly* 13 (4): 421–36.

Pinette, Susan. 2006. 'The importance of the literary: Lahontan's dialogues and primitivist thought.' *Prose Studies* 28 (1): 41–53.

Pinker, Steven. 2012. *The Better Angels of Our Nature: A History of Violence and Humanity*. London: Penguin.

—. 2018. *Enlightenment Now: The Case for Science, Reason, Humanism and Progress*. London: Allen Lane.

Piperno, Dolores R. 2011. 'The origins of plant cultivation and domestication in the new world tropics: patterns, process, and new developments.' *Current Anthropology* 52 (4): 453–70.

Plunket, Patricia and Gabriela Uruñuela. 2005. 'Recent research in Puebla prehistory.' *Journal of Archaeological Research* 13 (2): 89–127.

—. 2006. 'Social and cultural consequences of a late Holocene eruption of Popocatépetl in central Mexico.' *Quarternary International* 151 (1): 19–28.

Pomeau, René. 1967. *Voyages et lumières dans la littérature française du XVIIIe siècle*. SVEC LVII 22: 1269–89.

Pool, Christopher A. 2007. *Olmec Archaeology and Early Mesoamerica*. Cambridge: Cambridge University Press.

Possehl, G. L. 1996. *Indus Age: The Writing System*. Philadelphia: University of Pennsylvania Press.

—. 2002. *The Indus Civilization: A Contemporary Perspective*. Walnut Creek, CA: AltaMira Press.

Postgate, Nicholas. 1992. *Early Mesopotamia: Society and Economy at the Dawn of History*. London: Routledge.

Pournelle, Jennifer. 2003. *Marshland of Cities: Deltaic Landscapes and the Evolution of Mesopotamian Civilization*. San Diego: University of California.

Powell, Adam, Shennan Stephen and Mark G. Thomas. 2009. 'Late Pleistocene demography and the appearance of modern human behaviour.' *Science* 324: 1298–301.

Powell, Marvin A (ed.). 1987. *Labor in the Ancient Near East*. New Haven, CT: American Oriental Society.

Powers, Stephen. 1877. *Tribes of California*. Washington: Government Printing Office.

Prendergast, Amy L., Alexander J. E. Pryor, Hazel Reade and Rhiannon E. Stevens. 2018. 'Seasonal records of palaeoenvironmental change and resource use from archaeological assemblages.' *Journal of Archaeological Science*: Reports 21.

Preziosi, Donald and Louise A. Hitchcock. 1999. *Aegean Art and Architecture*. Oxford: Oxford University Press.

Provinse, John. 1937. 'Plains Indian culture.' In Fred Eggan (ed.), *Social Anthropology of North American Tribes*. Chicago: University of Chicago Press.

Pryor, Alexander J. E. et al. 2020. 'The chronology and function of a new circular mammoth-bone structure at Kostenki 11.' *Antiquity* 94 (374): 323–41.

Quilter, Jeffrey. 2002. 'Moche politics, religion, and warfare.' *Journal of World Prehistory* 16 (2): 145–95.

Ranere, Anthony J. et al. 2009. 'The cultural and chronological context of early Holocene maize and squash domestication in the Central Balsas

River Valley, Mexico.' *Proceedings of the National Academy of Sciences* 106 (13): 5014.

Ratnagar, Shereen. 2004. *Trading Encounters: From the Euphrates to the Indus in the Bronze Age*. New Delhi: Oxford University Press.

—. 2016. *Harappan Archaeology: Early State Perspectives*. Delhi: Primus Books.

Redding, Richard W. 1998. 'Ancestoral pigs: a New (Guinea) model for pig domestication in the Middle East.' *MASCA Research Papers in Science and Archaeology* 15: 65–76.

Rehak, Paul. 2002. 'Imag(in)ing a women's world in Bronze Age Greece.' In Nancy Sorkin Rabinowitz and Lisa Auanger (eds), *Among Women from the Homosocial to the Homoerotic in the Ancient World*. Austin: University of Texas Press, pp. 34–59.

Reich, David, Richard E. Green et al. 2010. 'Genetic history of an archaic hominin group from Denisova Cave in Siberia.' *Nature* 468 (7327): 1053.

Renfrew, Colin. 1972. *The Emergence of Civilization: The Cyclades and the Aegean in the Third Millennium BC*. London: Methuen.

—. 1987. *Archaeology and Language: The Puzzle of Indo-European Origins*. Cambridge: Cambridge University Press.

—. 2007. *Prehistory: The Making of the Human Mind*. London: Weidenfeld and Nicolson.

Renfrew, Colin and Bin Liu. 2018. 'The emergence of complex society in China: the case of Liangzhu.' *Antiquity* 92 (364): 975–90.

Renger, Johannes M. 1995. 'Institutional, communal, and individual ownership or possession of arable land in ancient Mesopotamia from the end of the fourth to the end of the first millennium BC.' *Chicago-Kent Law Review* 71 (1): 269–319.

Reynolds, Natasha and Felix Riede. 2019. 'House of cards: cultural taxonomy and the study of the European Upper Palaeolithic.' *Antiquity* 371: 1350–58.

Richardson, Seth. 2012. 'Early Mesopotamia: the presumptive state.' *Past and Present* 215: 3–49.

Richerson, Peter J. and Robert Boyd. 2001. 'Institutional evolution in the Holocene: the rise of complex societies.' *Proceedings of the British Academy* 110: 197–234.

Richerson, Peter J., Robert Boyd and Robert L. Bettinger. 2001. 'Was agriculture impossible during the Pleistocene but mandatory during the Holocene? A climate change hypothesis.' *American Antiquity* 66 (3): 387–411.

Richter, Daniel K. 1972. 'Lahontan dans l'Encyclopédie et ses suites.' In Jacques Proust (ed.), *Recherches nouvelles sur quelques écrivains des Lumières*. Geneva: Librairie Droz, pp. 163–200.

—. 1992. *The Ordeal of the Longhouse: The Peoples of the Iroquois League in the Era of European Colonization*. Chapel Hill: University of North Carolina Press.

Rick, John W. 2017. 'The nature of ritual space at Chavín de Huántar.' In Silvana Rosenfeld and Stefanie L. Bautista (eds), *Rituals of the Past: Prehispanic and Colonial Case Studies in Andean Archaeology*. Boulder: University Press of Colorado, pp. 21–50.

Ringle, William A. 2004. 'On the political organization of Chichen Itza.' *Ancient Mesoamerica* 15: 167–218.

Rissman, Paul. 1988. 'Public displays and private values: a guide to buried wealth in Harappan archaeology.' *World Archaeology* 20: 209–28.

Ritchie, Morgan, Dana Lepofsky et al. 2016. 'Beyond culture history: Coast Salish settlement patterning and demography in the Fraser Valley, BC.' *Journal of Anthropological Archaeology* 43: 140–54.

Rodrigues, Teresa. 2005. 'Gender and social differentiation within the Turner Population, Ohio, as evidenced by activity-induced musculoskeletal stress markers.' In Carr and Case (eds), pp. 405–27.

Roebroeks, Wil et al. 2000. *Hunters of the Golden Age: The Mid-Upper Palaeolithic of Eurasia 30,000–20,00 BP*. Leiden: University of Leiden Press.

Rollings, Willard H. 1992. *The Osage: An Ethnohistorical Study of Hegemony on the Prairie-Plains*. Columbia: University of Missouri Press.

Roosevelt, Anna. 2013. 'The Amazon and the Anthropocene: 13,000 years of human influence in a tropical rainforest.' *Anthropocene* 4: 69–87.

Rosenwig, Robert M. 2010. *The Beginnings of Mesoamerican Civilization: Inter-Regional Interaction and the Olmec*. Cambridge: Cambridge University Press.

—. 2017. 'Olmec globalization: a Mesoamerican archipelago of complexity.' In Tamar Hodos (ed.), *The Routledge Handbook of Archaeology and Globalization*. London: Routledge, pp. 177–93.

Roth, Anne Macy. 1991. *Egyptian Phyles in the Old Kingdom: The Evolution of a System of Social Organization*. Chicago: Oriental Institute.

—. 2002. 'The meaning of menial labour: "servant statues" in Old Kingdom serdabs.' *Journal of the American Research Center in Egypt* 39: 103–21.

Rousseau, Jean-Jacques (transl. Maurice Cranston). 1984 [1754]. *A Discourse on Inequality*. London: Penguin.

Rowe, John. 1982. 'Inca policies and institutions relating to cultural unification of the Empire.' In G. A. Collier, R. I. Rosaldo and J. D. Wirth (eds), *The Inca and Aztec States, 1400–1800*. New York: Academic, pp. 93–117.

Rowley-Conwy, Peter. 2001. 'Time, change, and the archaeology of hunter-gatherers: how original is the "Original Affluent Society"?' In P. R.

Rowley-Conwy, R. Layton and C. Panter-Brick (eds), *Hunter-Gatherers: An Interdisciplinary Perspective*. Cambridge: Cambridge University Press, pp. 39–72.

Sablin, Mikhail, Natasha Reynolds et al. (in preparation). 'The Epigravettian site of Yudinovo, Russia: mammoth bone structures as ritualized middens.'

Sagard, Gabriel. 1939 [1632]. *Le Grand Voyage du Pays des Hurons*. Paris: Denys Moreau.

Sahlins, Marshall. 1968. 'Notes on the Original Affluent Society.' In Lee and DeVore (eds), pp. 85–9.

—. 1985. *Islands of History*. Chicago: Chicago University Press.

—. 1996. 'The sadness of sweetness: the native anthropology of western cosmology.' *Current Anthropology* 37 (3): 395–428.

—. 1999. 'Two or three things I know about culture.' *Journal of the Royal Anthropological Institute* 5 (3): 399–421.

—. 2004. *Apologies to Thucydides*. Chicago: University of Chicago Press.

—. 2008. 'The stranger-king, or, elementary forms of political life.' *Indonesia and the Malay World* 36 (105): 177–99.

—. 2017 [1972]. *Stone Age Economics*. Abingdon and New York: Routledge.

—. 2017. 'The stranger-kingship of the Mexica.' In Graeber and Sahlins, pp. 223–48.

Sahlins, Marshall D. and Elman R. Service. 1960. *Evolution and Culture*. Ann Arbor: University of Michigan Press.

Saller, Richard P. 1984. '*Familia*, *domus*, and the Roman conception of the family.' *Phoenix* 38 (4): 336–55.

Salomon, Frank. 2004. *The Cord Keepers. Khipus and Cultural Life in a Peruvian Village*. Durham, NC and London: Duke University Press.

Sanday, Peggy R. 1981. *Female Power and Male Dominance: On the Origins of Sexual Inequality*. Cambridge: Cambridge University Press.

—. 2002. *Women at the Center: Life in a Modern Matriarchy*. Ithaca, NY: Cornell University Press.

Santana, Jonathan et al. 2012. 'Crania with mutilated facial skeletons: a new ritual treatment in an early Pre-Pottery Neolithic B cranial cache at Tell Qarassa North (South Syria).' *American Journal of Physical Anthropology* 149: 205–16.

Santos-Granero, Fernando. 2009. *Vital Enemies: Slavery, Predation, and the Amerindian Political Economy of Life*. Austin: University of Texas Press.

Sassaman, Kenneth E. 2005. 'Poverty Point as structure, event, process.' *Journal of Archaeological Method and Theory* 12 (4): 335–64.

—. (ed.). 2008. 'The New Archaic.' *The Society for American Archaeology: Archaeological Record* 8 (5), Special Issue.

—. 2010. *The Eastern Archaic, Historicized*. Lanham, MD: AltaMira Press.

Sassaman, Kenneth E. and Michael J. Heckenberger. 2004. 'Roots of the theocratic Formative in the Archaic Southeast.' In G. Crothers (ed.), *Hunter-Gatherers in Theory and Archaeology*. Carbondale: Center for Archaeological Investigations, Southern Illinois University Press, pp. 423–44.

Sayre, Gordon M. 1997. *Les Sauvages Américains: Representations of Native Americans in French and English Colonial Literature*. Chapel Hill: University of North Carolina Press.

Scarry, Elaine. 1985. *The Body in Pain: The Making and Unmaking of the World*. Oxford: Oxford University Press.

Scerri, Eleanor M. L. 2017. 'The North African Middle Stone Age and its place in recent human evolution.' *Evolutionary Anthropology* 26 (3): 119–35.

Scerri, Eleanor M. L., Mark G. Thomas et al. 2018. 'Did our species evolve in subdivided populations across Africa, and why does it matter?' *Trends in Ecology & Evolution* 33 (8): 582–94.

Scheffler, Thomas. 2003. '"Fertile Crescent", "Orient", "Middle East": the changing mental maps of Southwest Asia.' *European Review of History* 10 (2): 253–72.

Scheidel, Walter. 2017. *The Great Leveller: Violence and the History of Inequality from the Stone Age to the Twenty-First Century*. Princeton, NJ: Princeton University Press.

Schirmer, Wulf. 1990. 'Some aspects of building at the "aceramic-neolithic" settlement of Çayönü Tepesi.' *World Archaeology* 21 (3): 363–87.

Schlanger, Nathan. 2006. *Marcel Mauss: Techniques, Technology, and Civilisation*. New York and Oxford: Durkheim Press/Berghahn Books.

Schmandt-Besserat, Denise. 1992. *Before Writing*. Austin: University of Texas Press.

Schmidt, Isabell and Andreas Zimmermann. 2019. 'Population dynamics and socio-spatial organization of the Aurignacian: scalable quantitative demographic data for western and central Europe.' *PLoS ONE* 14 (2): e0211562.

Schmidt, Klaus. 1998. 'Frühneolithische Silexdolche.' In Güven Arsebük et al. (eds), *Light on Top of the Black Hill. Studies Presented to Halet Çambel*. Istanbul: Yeni, pp. 681–92.

—. 2006. *Sie bauten die ersten Tempel. Das rätselhafte Heiligtum der Steinzeitjäger*. Munich: C. H. Beck.

Schmidt, Morgan J. et al. 2014. 'Dark earths and the human built landscape in Amazonia: a widespread pattern of anthrosol formation.' *Journal of Archaeological Science* 42: 152–65.

Schnapp, Alain. 1996. *The Discovery of the Past: The Origins of Archaeology*. London: British Museum Press.

Schneider, Thomas. 2008. 'Periodizing Egyptian history: Manetho, convention, and beyond.' In Klaus-Peter Adam (ed.), *Historiographie in der Antike*. Berlin and New York: De Gruyter, pp. 183–97.

Schoep, Ilse. 2008. 'Building the labyrinth: Arthur Evans and the construction of Minoan civlisation.' *American Journal of Archaeology* 122 (1): 5–32.

Schor, Juliet B. 1991. *The Overworked American: The Unexpected Decline of Leisure*. New York: Basic Books.

Schrakamp, I. 2010. *Krieger und Waffen im frühen Mesopotamien. Organisation und Bewaffnung des Militärs in frühdynastischer und sargonischer Zeit*. Marburg: Philipps-Universität.

Schulting, Rick J. 2006. 'Antlers, bone pins and flint blades: the Mesolithic cemeteries of Téviec and Hoëdic, Brittany.' *Antiquity* 70 (268): 335–50.

Schulting, Rick J. and Michael P. Richards. 2001. 'Dating women and becoming farmers: new palaeodietary and AMS dating evidence from the Breton Mesolithic cemeteries of Téviec and Hoëdic.' *Journal of Anthropological Archaeology* 20: 314–44.

Schulting, Rick J. and Linda Fibiger (eds). 2012. *Sticks, Stones, and Broken Bones: Neolithic Violence in a European Perspective*. Oxford: Oxford University Press.

Schultz, James W. 1935. *My Life as an Indian*. Boston: Houghton Mifflin Company.

Scott, James C. 1990. *Domination and the Arts of Resistance*. New Haven, CT and London: Yale University Press.

—. 2009. *The Art of Not Being Governed: An Anarchist History of Upland Southeast Asia*. New Haven, CT and London: Yale University Press.

—. 2017. *Against the Grain: A Deep History of the Earliest States*. New Haven, CT and London: Yale University Press.

Seeman, Mark F. 1979. 'Feasting with the dead: Ohio Hopewell charnel house ritual as a context for redistribution.' In David S. Brose and N'omi B. Greber (eds), *Hopewell Archaeology: The Chillicothe Conference*. Kent, Ohio: Kent State University Press, pp. 39–46.

—. 1988. 'Ohio Hopewell trophy-skull artifacts as evidence for competition in Middle Woodland societies circa 50 B.C.–A.D. 350.' *American Antiquity* 53 (3): 565–77.

—. 2004. 'Hopewell art in Hopewell places.' In Richard F. Townsend and Robert V. Sharp (eds), *Hero, Hawk, and Open Hand: American Indian Art of the Ancient Midwest and South*. New Haven, CT and London: Yale University Press, pp. 57–71.

Seidlmayer, Stephan J. 2000. 'The First Intermediate Period (*c.*2160–2055 BC).' in I. Shaw (ed.), pp. 118–47.

Seligman, Adam B., Robert P. Weller, Michael J. Puett and Bennett Simon. 2008. *Ritual and its Consequences: An Essay on the Limits of Sincerity.* Oxford: Oxford University Press.

Sen, Amartya. 2001. *Development as Freedom.* Oxford: Oxford University Press.

Servet, Jean-Michel. 1981. 'Primitive order and archaic trade. Part I.' *Economy and Society* 10 (4): 423–50.

—. 1982. 'Primitive order and archaic trade. Part II.' *Economy and Society* 11 (1): 22–59.

Severi, Carlo (transl. Janet Lloyd, Foreword by David Graeber). 2015. *The Chimera Principle: An Anthropology of Memory and Imagination.* Chicago: HAU Books.

Shady Solis, Ruth, Jonathan Haas and Winifred Creamer. 2001. 'Dating Caral, a Preceramic site in the Supe Valley on the central coast of Peru.' *Science* 292 (5517): 723–6.

Sharan, M. K. 1983. 'Origin of republics in India with special reference to the Yaudheya tribe.' In B. N. Mukherjee et al. (eds), *Sri Dinesacandrika – Studies in Indology*. Delhi: Sundeep Prakashan, pp. 241–52.

Sharer, Robert J. 2003. 'Founding events and Teotihuacan connections at Copán, Honduras.' In Braswell (ed.), pp. 143–65.

Sharma, J. P. 1968. *Republics in Ancient India, 1500 B.C. to 500 B.C.* Leiden: Brill.

Shaughnessy, Edward. L. 1989. 'Historical geography and the extent of the earliest Chinese kingdom.' *Asia Minor* (2) 2: 1–22.

—. 1999. 'Western Zhou history.' In Loewe and Shaughnessy (eds), pp. 288–351.

Shaw, Ian (ed.). 2000. *The Oxford History of Ancient Egypt.* Oxford: Oxford University Press.

Shennan, Stephen. 2009. 'Evolutionary demography and the population history of the European Early Neolithic.' *Human Biology* 81: 339–55.

—. 2018. *The First Farmers of Europe: An Evolutionary Perspective.* Cambridge: Cambridge University Press.

Shennan, Stephen and Kevan Edinborough. 2006. 'Prehistoric population history: from the Late Glacial to the Late Neolithic in Central and Northern Europe.' *Journal of Archaeological Science* 34: 1339–45.

Shennan, Stephen et al. 2013. 'Regional population collapse followed initial agriculture booms in mid-Holocene Europe.' *Nature* 4: 1–8.

Sheppard, Peter J. 2011. 'Lapita colonization across the Near/Remote Oceania boundary.' *Current Anthropology* 52 (6): 799–840.

Sherratt, Andrew. 1980. 'Water, soil and seasonality in early cereal cultivation.' *World Archaeology* 11: 313–30.

—. 1995. 'Reviving the grand narrative: archaeology and long-term change.' *Journal of European Archaeology* 3 (1): 1–32.

—. 1997. *Economy and Society in Prehistoric Europe. Changing Perspectives*. Edinburgh: Edinburgh University Press.

—. 1999. 'Cash crops before cash: organic consumables and trade.' In Chris Gosden and John G. Hather (eds), *The Prehistory of Food: Appetites for Change*. London: Routledge, pp. 13–34.

—. 2004. 'Fractal farmers: patterns of Neolithic origins and dispersal.' In C. Scarre et al. (eds), *Explaining Social Change: Studies in Honour of Colin Renfrew*. Cambridge: McDonald Institute, pp. 53–63.

—. 2007. 'Diverse origins: regional contributions to the genesis of farming.' In Colledge and Conolly (eds), pp. 1–20.

Sherratt, Susan. 2000. *Arthur Evans, Knossos, and the Priest-King*. Oxford: Ashmolean Museum.

—. 2001. 'Potemkin palaces and route-based economies.' In Voutsaki and Killen (eds), pp. 214–38.

Sherwood, Sarah C. and Tristam R. Kidder. 2011. 'The DaVincis of dirt: geoarchaeological perspectives on Native American mound building in the Mississippi River basin.' *Journal of Anthropological Archaeology* 30: 69–87.

Shimony, Annemarie. 1961. *Conservatism Among the Six Nations Iroquois Reservation*. Yale University Publications in Anthropology 65. New Haven, CT and London: Yale University Press.

—. 1970. 'Iroquois witchcraft at Six Nations.' In Deward E. Walker, Jr (ed.), *Systems of Witchcraft and Sorcery*. Moscow, ID: Anthropological Monographs of the University of Idaho, pp. 239–65.

Shipton, Ceri et al. 2018. '78,000-year-old record of Middle and Later Stone Age innovation in an East African tropical forest.' *Nature Communications* 9: 1832.

Shklar, Judith. 1964. 'Rousseau's images of authority.' *The American Political Science Review* 58 (4): 919–32.

Shumilovskikh, Lyudmila S., Elena Novenko and Thomas Giesecke. 2017. 'Long-term dynamics of the East European forest-steppe ecotone.' *Journal of Vegetation Science* 29 (3): 416–26.

Sikora, Martin et al. 2017. 'Ancient genomes show social and reproductive behavior of early Upper Paleolithic foragers.' *Science* 358 (6363): 659–62.

Silver, Morris. 2015. 'Reinstating classical Athens: the production of public order in an ancient community.' *Journal on European History of Law* 1: 3–17.

Silverblatt, Irene. 1987. *Moon, Sun and Witches: Gender Ideologies and Class in Inca and Colonial Peru*. Princeton, NJ: Princeton University Press.

Silverman, Eric K. 2001. *Masculinity, Motherhood, and Mockery: Psychoanalyzing Culture and the Iatmul Naven Rite in New Guinea*. Ann Arbor: University of Michigan Press.

Simonse, Simon. 1992. *Kings of Disaster: Dualism, Centralism, and the Scapegoat-king in Southeastern Sudan*. Leiden: Brill.

—. 2005. 'Tragedy, ritual and power in Nilotic regicide: the regicidal dramas of the Eastern Nilotes of Sudan in contemporary perspective.' In D. Quigley (ed.), *The Character of Kingship*. Oxford: Berg, pp. 67–100.

Sioui, Georges. 1972. 'A la réflexion des Blancs d'Amérique du Nord et autres étrangers.' *Recherches amérindiennes au Quebec* 2 (4–5): 65–8.

—. 1992. *For an Amerindian Autohistory: An Essay on the Foundations of a Social Ethic*. Montreal: McGill-Queen's University Press.

—. 1999. *Huron-Wendat. The Heritage of the Circle*. Vancouver: British Columbia University Press.

Sipilä, Joonäs and Antti Lahelma. 2006. 'War as a paradigmatic phenomenon: endemic violence and the Finnish Subneolithic.' In T. Pollard and I. Banks (eds), *Studies in the Archaeology of Conflict*. Leiden: Brill, pp. 189–209.

Skoglund, Pontus et al. 2016. 'Genomic insights into the peopling of the Southwest Pacific.' *Nature* 538 (7626): 510–13.

Skousen, B. Jacob. 2016. 'Pilgrimage and the Construction of Cahokia: A View from the Emerald Site'. Doctoral dissertation, University of Illinois at Urbana-Champaign.

Slon, Viviane et al. 2014. 'The plastered skulls from the Pre-Pottery Neolithic B Site of Yiftahel (Israel) – a computed tomography-cased analysis.' *PLoS ONE* 9 (2): e89242.

Smith, Bruce D. 1992. 'Mississippian elites and solar alignments: a reflection of managerial necessity, or levers of social inequality?' In A. W. Barker and T. R. Pauketat (eds), *Lords of the Southeast: Social Inequality and the Native Elites of Southeastern North America*. Washington: Archaeological Papers of the American Anthropological Association, pp 11–30.

—. 2001. 'Low-level food production.' *Journal of Archaeological Research* 9 (1): 1–43.

Smith, De Cost. 1888. 'Witchcraft and demonism of the modern Iroquois.' *The Journal of American Folklore* 1 (3): 184–94.

Smith, Michael E. 1984. 'The Aztlan migrations of the Nahuatl chronicles: myth or history?' *Ethnohistory* 31 (3): 153–86.

—. 2012. *The Aztecs* (3rd edn). Oxford: Wiley-Blackwell.

—. 2015. 'Neighborhood formation in semi-urban settlements.' *Journal of Urbanism* 8 (2): 173–98.

—. 2019. 'Energized crowding and the generative role of settlement aggregation and urbanization.' In Attila Gyucha (ed.), *Coming Together: Comparative Approaches to Population Aggregation and Early Urbanization*. New York: SUNY Press, pp. 37–58.

Smith, Michael. E. et al. 2017. 'The Teotihuacan anomaly: the historical trajectory of urban design in ancient Central Mexico.' *Open Archaeology* 3: 175–93.

—. 2019. 'Apartment compounds, households, and population in the ancient city of Teotihuacan, Mexico.' *Ancient Mesoamerica* 1–20. doi:10.1017/S0956536118000573.

Snow, Dean. 1991. 'Dating the emergence of the League of the Iroquois: a reconsideration of the documentary evidence.' In Nancy-Anne McClure Zeller (ed.), *A Beautiful and Fruitful Place: Selected Rensselaerswijck Seminar Papers*. Albany, NY: New Netherland Publishing, pp. 139–43.

Soffer, Olga. 1985. *The Upper Palaeolithic of the Central Russian Plain*. London: Academic Press.

Soffer, Olga, James M. Adovasio and David C. Hyland. 2000. 'Textiles, basketry, gender and status in the Upper Paleolithic.' *Current Anthropology* 41 (4): 511–37.

Soustelle, Jacques. 1962. *The Daily Life of the Aztecs on the Eve of the Spanish Conquest*. New York: Macmillan.

Sperber, Dan. 2005. *Explaining Culture: A Naturalistic Approach*. Oxford: Blackwell.

Spier, Leslie. 1930. *Klamath Ethnography*. Berkeley: University of California Press.

Spott, Robert and Alfred L. Kroeber. 1942. 'Yurok narratives.' *University of California Publications in American Archaeology and Ethnology* 35 (9): 143–256.

Spretnak, Charlene. 2011. 'Anatomy of a backlash: concerning the work of Marija Gimbutas.' *Journal of Archaeomythology* 7: 25–51.

Spriggs, Matthew. 1995. 'The Lapita culture and Austronesian prehistory in Oceania.' In Peter Bellwood et al. (eds), *The Austronesians: Historical and Comparative Perspectives*. Canberra: ANU Press, pp. 112–33.

—. 1997. *The Island Melanesians*. Oxford: Blackwell.

Starna, William A. 2008. 'Retrospecting the origins of the League of the Iroquois.' *Proceedings of the American Philosophical Society* 152 (3): 279–321.

Steckley, John. 1981 'Kandiaronk: a man called Rat.' In J. Steckley, *Untold Tales: Four Seventeenth-Century Hurons*. Toronto: Associated Heritage Publishing, pp. 41–52.

—. 2014. *The Eighteenth-Century Wyandot: A Clan-Based Study*. Waterloo, Ontario: Wilfrid Laurier University Press.

Steinke, Kyle and Dora C. Y. Ching (eds). 2014. *Art and Archaeology of the Erligang Civilization*. Princeton, NJ: Princeton University Press.

Steinkeller, Piotr. 2015. 'The employment of labour on national building projects in the Ur III period.' In Steinkeller and Hudson (eds), pp. 137–236.

Steinkeller, Piotr and Michael Hudson (eds). 2015. *Labor in the Ancient World*. Dresden: ISLET-Verlag.

Stern, Jessica Yirush. 2017. *The Lives in Objects: Native Americans, British Colonists, and Cultures of Labor and Exchange in the Southeast*. Chapel Hill: University of North Carolina Press.

Stevens, Chris and Dorian Q. Fuller. 2012. 'Did Neolithic farming fail? The case for a Bronze Age agricultural revolution in the British Isles.' *Antiquity* 86 (333): 707–22.

Stone, Elizabeth C. and Paul Zimansky. 1995. 'The tapestry of power in a Mesopotamian city.' *Scientific American* 272 (4): 118–23.

Stordeur, Danielle. 2000. 'Jerf el Ahmar et l'émergence du Néolithique au Proche-Orient.' In Jean Guilaine (ed.), *Premiers paysans du monde: Naissances des agricultures, Séminaire du Collège de France*. Paris: Errance, pp. 33–60.

Straus, Lawrence G. 1977. 'The Upper Palaeolithic cave site of Altamira (Santander, Spain).' *Quaternaria* 19: 135–48.

Straus, Lawrence. G. et al. (eds). 1990. *Humans at the End of the Ice Age: The Archaeology of the Pleistocene-Holocene Transition*. New York and London: Plenum Press.

Strehlow, T. G. H. 1947. *Aranda Traditions*. Carlton: Melbourne University Press.

Strudwick, Nigel. 1985. *The Administration of Egypt in the Old Kingdom: The Highest Titles and their Holders*. London: KPI.

Stuart, David. 2000. 'The arrival of strangers: Teotihuacan and Tollan in Classic Maya history.' In D. Carrasco, L. Jones and S. Sessions (eds), *Mesoamerica's Classic Heritage: From Teotihuacan to Aztecs*. Boulder: University Press of Colorado, pp. 465–513.

Styring, A. et al. 2017. 'Isotope evidence for agricultural extensification reveals how the world's first cities were fed.' *Nature Plants* 3: 17076.

Subramanian T. S. 2010. 'The rise and fall of a Harappan city.' *Frontline* 27 (12).

Sugiyama, Nawa et al. 2019. 'Artistas mayas en Teotihuacan?' *Arqueología Mexicana* 24 (142): 8.

Sugiyama, Nawa, Saburo Sugiyama and Alejandro G. Sarabia. 2013. 'Inside the Sun Pyramid at Teotihuacan, Mexico: 2008–2011 excavations and preliminary results.' *Latin American Antiquity* 24 (4): 403–32.

Sugiyama, Saburo. 2005. *Human Sacrifice, Militarism, and Rulership: Materialization of State Ideology at the Feathered Serpent Pyramid, Teotihuacan.* Cambridge: Cambridge University Press.

Sugiyama, Saburo and Rubén Cabrera Castro. 2007. 'The Moon Pyramid project and the Teotihuacan state polity.' *Ancient Mesoamerica* 18: 109–25.

Suttles, Wayne. 1968. 'Coping with abundance.' In Lee and DeVore (eds), pp. 56–68.

Swanton, John Reed. 1931. 'Source material for the social and ceremonial life of the Choctaw Indians.' Bureau of American Ethnology Bulletin 103: 1–282.

Takahashi, Ryuzaburo and Leo Aoi Hosoya. 2003. 'Nut exploitation in Jomon society.' In S. L. R. Mason and J. G. Hather (eds), *Hunter-Gatherer Archaeobotany: Perspectives from the Northern Temperate Zone.* London: Institute of Archaeology, pp. 146–55.

Tambiah, Stanley J. 1973. 'The galactic polity in Southeast Asia.' In S. J. Tambiah, *Culture, Thought, and Social Action.* Cambridge, MA: Harvard University Press, pp. 3–31.

Tate, Carolyn E. 2012. *Reconsidering Olmec Visual Culture: The Unborn, Women, and Creation.* Austin: University of Texas Press.

Taube, Karl. A. 1986. 'The Teotihuacan cave of origin: the iconography and architecture of emergence mythology in Mesoamerica and the American Southwest.' *RES: Anthropology and Aesthetics* 12: 51–82.

—. 1992. 'The temple of Quetzalcoatl and the cult of sacred war at Teotihuacan.' *Anthropology and Aesthetics* 21: 53–87.

—. 2000. 'The writing system of ancient Teotihuacan.' *Ancient America*, vol. 1. Bardarsville, NC: Center for Ancient American Studies.

Teeter, Emily (ed.). 2011. *Before the Pyramids: The Origins of Egyptian Civilization.* Chicago: Oriental Institute.

Teschler-Nicola, M. et al. 1996. 'Anthropologische Spurensicherung – Die traumatischen und postmortalen Veränderungen an den linearbandkeramischen Skelettresten von Asparn/Schletz.' In H. Windl (ed.), *Rätsel um Gewalt und Tod vor 7.000 Jahren. Eine Spurensicherung.* Asparn: Katalog des Niederösterreichischen Landesmuseum, pp. 47–64.

Testart, Alain. 1982. 'The significance of food storage among hunter-gatherers.' *Current Anthropology* 23 (5): 523–37.

—. 2008. 'Des crânes et des vautours ou la guerre oubliée.' *Paléorient* 34 (1): 33–58.

Thapar, Romila. 1984. *From Lineage to State: Social Formations in the Mid-First Millennium BC in the Ganga Valley*. Oxford: Oxford University Press.

Thomas, Chad R., Christopher Carr and Cynthia Keller. 2005. 'Animal-totemic clans of Ohio Hopewellian peoples.' In Carr and Case (eds), pp. 339–85.

Thomas, Keith. 1978. *Religion and the Decline of Magic. Studies in Popular Beliefs in Sixteenth- and Seventeenth-Century England.* Harmondsworth: Penguin.

Thompson, Andrew et al. 2015. 'New dental and isotope evidence of biological distance and place of origin for mass burial groups at Cahokia's Mound 72.' *American Journal of Physical Anthropology* 158: 341–57.

Thwaites, Reuben Gold (ed.). 1896–1901. *The Jesuit Relations and Allied Documents: Travels and Explorations of the Jesuit Missionaries in New France, 1610–1791*. 73 vols. Cleveland, OH: Burrows Brothers.

Tilley, Lorna. 2015. 'Accommodating difference in the prehistoric past: revisiting the case of Romito 2 from a bioarchaeology of care perspective.' *International Journal of Paleopathology* 8: 64–74.

Tisserand, Roger. 1936. *Les Concurrents de J.-J. Rousseau à l'Académie de Dijon pour le prix de 1754*. Paris: Boivin.

Tooker, Elisabeth. 1964. *An Ethnography of the Huron Indians, 1615–1649*. Washington: Bureau of Ethnology, Bulletin number 190.

—. 1971. 'Clans and moieties in North America.' *Current Anthropology* 12 (3): 357–76.

—. 1978. 'The League of the Iroquois: its history, politics, and ritual.' In Bruce G. Trigger (ed.), *Handbook of North American Indians*, vol. 15: *Northeast*. Washington: Smithsonian Press, pp. 418–41.

—. 1984. 'Women in Iroquois society.' In M. K. Foster, J. Campisi and M. Mithun (eds), *Extending the Rafters: Interdisciplinary Approaches to Iroquoian Studies*. Albany: State University of New York Press, pp. 109–23.

—. 1988. 'The United States Constitution and the Iroquois League.' *Ethnohistory* 35: 305– 36.

—. 1990. 'Rejoinder to Johansen.' *Ethnohistory* 37: 291–7.

Torres, Constantino Manuel. 2008. 'Chavín's psychoactive pharmacopoeia: the iconographic evidence.' In Conklin and Quilter (eds), pp. 237–57.

Townsend, Camilla. 2006. 'What in the world have you done to me, my lover? Sex, servitude, and politics among the pre-Conquest Nahuas as seen in the Cantares Mexicanos.' *The Americas* 62 (3): 349–89.

Trautmann, Thomas R. 1992. 'The revolution in ethnological time.' *Man* 27 (2): 379–97.

Treherne, Paul. 1995. 'The warrior's beauty: the masculine body and self-identity in Bronze Age Europe.' *Journal of European Archaeology* 3 (1): 105–44.

Trigger, Bruce G. 1976. *The Children of Aataentsic: A History of the Huron People to 1660*. Montreal: McGill-Queen's University Press.

—. 1985. *Natives and Newcomers: Canada's 'Heroic Age' Reconsidered*. Montreal: McGill-Queen's University Press.

—. 1990. 'Maintaining economic equality in opposition to complexity: an Iroquoian case study.' In S. Upham (ed.), *The Evolution of Political Systems: Sociopolitics in Small-Scale Sedentary Societies*. Cambridge: Cambridge University Press, pp. 119–45.

—. 2006. *A History of Archaeological Thought* (2nd edn). Cambridge: Cambridge University Press.

Trinkaus, Erik. 2018. 'An abundance of developmental anomalies and abnormalities in Pleistocene people.' *PNAS* 115 (47): 11941–6.

Trinkaus, Erik and Alexandra P. Buzhilova. 2018. 'Diversity and differential disposal of the dead at Sunghir.' *Antiquity* 92 (361): 7–21.

Trouillot, Michel-Rolph. 2003. 'Anthropology and the savage slot: the poetics and politics of otherness.' In M. R. Trouillot, *Global Transformations: Anthropology and the Modern World*. New York: Palgrave Macmillan, pp. 7–28.

Tuck, James A. 1978. 'Northern Iroquoian prehistory.' In Bruce G. Trigger (ed.), *Handbook of North American Indians*, vol. 15: *Northeast*. Washington: Smithsonian Press, pp. 322–33.

Tuerenhout, Dirk Van. 2002. 'Maya warfare: sources and interpretations.' *Civilisations* 50: 129–52.

Tully, James. 1994. 'Aboriginal property and Western theory: recovering a middle ground.' *Social Philosophy and Policy* 11 (2): 153–80.

Turnbull, Colin M. 1982. 'The ritualization of potential conflict between the sexes in Mbuti.' In Eleanor B. Leacock and Richard B. Lee (eds), *Politics and History in Band Societies*. Cambridge: Cambridge University Press, pp. 133–55.

Turner, Nancy J. and Dawn C. Loewen. 1998. 'The original "free trade": exchange of botanical products and associated plant knowledge in Northwestern North America.' *Anthropologica* 40 (1): 49–70.

Turner, Victor. 1969. *The Ritual Process: Structure and Anti-Structure.* Chicago: Aldine.

Tushingham, Shannon and Robert L. Bettinger. 2013. 'Why foragers choose acorns before salmon: storage, mobility, and risk in Aboriginal California.' *Journal of Anthropological Archaeology* 32 (4): 527–37.

Tylor, Edward. B. 1879. 'Remarks on the geographical distribution of games.' *Journal of the Anthropological Institute* 9 (1): 26.

Underhill, Anne P. et al. 2008. 'Changes in regional settlement patterns and the development of complex societies in southeastern Shandong, China.' *Journal of Anthropological Archaeology* 27 (1): 1–29.

Ur, Jason. 2014. 'Households and the emergence of cities in Ancient Mesopotamia.' *Cambridge Archaeological Journal* 24: 249–68.

Uriarte, María Teresa. 2016. 'The Teotihuacan ballgame and the beginning of time.' *Ancient Mesoamerica* 17 (1): 17–38.

Urton, Gary and Carrie J. Brezine. 2005. 'Khipu accounting in ancient Peru.' *Science* 309 (5737): 1065–7.

Ussher, James. 1650. *The Annals of the Old and New Testament with the Synchronisms of Heathen Story to the Destruction of Hierusalem by the Romanes.* London: J. Crook and G. Bedell.

Valentine, B. et al. 2015. 'Evidence for patterns of selective urban migration in the Greater Indus Valley (2600–1900 BC): a lead and strontium isotope mortuary analysis.' *PLoS ONE* 10 (4): e0123103.

van der Velde, Pieter. 1990. 'Banderamik social inequality: A case study.' *Germania* 68: 19–38.

Vanhaeren, Marian and Francesco D'Errico. 2005. 'Grave goods from the Saint-Germain-de-la-Rivière burial: evidence for social inequality in the Upper Palaeolithic.' *Journal of Anthropological Archaeology* 24: 117–34.

Vayda, Andrew P. 1967. 'Pomo trade feasts.' In G. Dalton (ed.), *Tribal and Peasant Economies.* Garden City, NY: Natural History Press, pp. 495–500.

Veil, Stephan., K. et al. 2012. 'A 14,000-year-old amber elk and the origins of northern European art.' *Antiquity* 86: 660–63.

Vennum, Jr, Thomas. 1988. *Wild Rice and the Ojibway People.* St Paul: Minnesota History Society Press.

Vidale, Massimo. 2000. *The Archaeology of Indus Crafts: Indus Craftspeople and Why We Study Them.* Rome: IsIAO.

—. 2010. 'Aspects of palace life at Mohenjo-Daro.' *South Asian Studies* 26 (1): 59–76.

—. 2013. 'T-Shaped pillars and Mesolithic "chiefdoms" in the prehistory of Southern Eurasia: a preliminary note.' In Dennys Frenez and Maurizio Tosi (eds), *South Asian Archaeology 2007. Proceedings of the 19th International*

Conference of the European Association of South Asian Archaeology. Oxford: BAR, pp. 51–8.

Videiko, Mikhail. 1996. 'Die Grossiedlungen der Tripol'e-Kultur in der Ukraine.' *Eurasia Antiqua* 1: 45–80.

Viner, Sarah et al. 2010. 'Cattle mobility in prehistoric Britain: strontium isotope analysis of cattle teeth from Durrington Walls (Wiltshire, Britain).' *Journal of Archaeological Science* 37: 2812–20.

Von Dassow, Eva. 2011. 'Freedom in ancient Near Eastern societies.' In Karen Radner and Eleanor Robson (eds), *The Oxford Handbook of Cuneiform Culture.* Oxford: Oxford University Press, pp. 205–28.

Voutsaki, Sofia. 1997. 'The creation of value and prestige in the Aegean Late Bronze Age.' *Journal of European Archaeology* 5 (2): 34–52.

Voutsaki, Sofia and John Killen (eds). 2001. *Economy and Politics in the Mycenaean Palatial States.* Cambridge: Cambridge Philological Society.

Walens, Stanley. 1981. *Feasting with Cannibals: An Essay on Kwakiutl Cosmology.* Princeton, NJ: Princeton University Press.

Wallace, Anthony F. C. 1956. 'Revitalization movements.' *American Anthropologist* 58 (2): 264–81.

—. 1958. 'Dreams and the wishes of the soul: a type of psychoanalytic theory among the seventeenth century Iroquois.' *American Anthropologist* (N.S.) 60 (2): 234–48.

Warren, Carol. 1993. *Adat and Dinas: Balinese Communities in the Indonesian State.* Kuala Lumpur: Oxford University Press.

Weber, Max (transl. Talcott Parsons). 1930 [1905]. *The Protestant Ethic and the Spirit of Capitalism.* London: Unwin.

Weismantel, Mary. 2013. 'Inhuman eyes: looking at Chavín de Huantar.' In Christopher Watts (ed.), *Relational Archaeologies: Humans, Animals, Things.* London: Routledge, pp. 21–41.

Wengrow, David. 1998. 'The changing face of clay: continuity and change in the transition from village to urban life in the Near East.' *Antiquity* 72: 783–95.

—. 2001. 'The evolution of simplicity: aesthetic labour and social change in the Neolithic Near East.' *World Archaeology* 33 (2): 168–88.

—. 2003. 'Interpreting animal art in the prehistoric Near East.' In T. Potts, M. Roaf and D. Stein (eds), *Culture through Objects. Ancient Near Eastern Studies in Honour of P. R. S. Moorey.* Oxford: Griffith Institute, pp. 139–60.

—. 2006. *The Archaeology of Early Egypt: Social Transformations in North-East Africa, 10,000 to 2650 BC.* Cambridge: Cambridge University Press.

—. 2008. 'Prehistories of commodity branding.' *Current Anthropology* 49 (1): 7–34.

—. 2010a. *What Makes Civilization? The Ancient Near East and the Future of the West.* Oxford: Oxford University Press.

—. 2010b. 'The voyages of Europa: ritual and trade in the Eastern Mediterranean, *c.*2300–1850 BC.' In William A. Parkinson and Michael L. Galaty (eds), *Archaic State Interaction: The Eastern Mediterranean in the Bronze Age.* Santa Fe: School for Advanced Research Press, pp. 141–60.

—. 2011. 'Archival and sacrificial economies in Bronze Age Eurasia: an interactionist approach to the hoarding of metals.' In T. Wilkinson, D. J. Bennet and S. Sherratt (eds), *Interweaving Worlds*. Oxford: Oxbow Books, pp. 135–44.

—. 2015. 'Cities before the State in early Eurasia' (the Jack Goody Lecture). Halle: Max Planck Institute for Social Anthropology.

Wengrow, David et al. 2014. 'Cultural convergence in the Neolithic of the Nile Valley: a prehistoric perspective on Egypt's place in Africa.' *Antiquity* 88: 95–111.

Wengrow, David and David Graeber. 2015. 'Farewell to the childhood of man: ritual, seasonality, and the origins of inequality' (the 2014 Henry Myers Lecture). *Journal of the Royal Anthropological Institute* 21 (3): 597–619.

—. 2018. 'Many seasons ago: slavery and its rejection among foragers on the Pacific Coast of North America.' *American Anthropologist* 120 (2): 237–49.

Wernke, Stephen. 2006. 'The politics of community and Inka statecraft in the Colca Valley, Peru.' *Latin American Antiquity* 17 (2): 177–208.

White, Christine, T. Douglas Price and Fred J. Longstaffe. 2007. 'Residential histories of the human sacrifices at the Moon Pyramid, Teotihuacan: evidence from oxygen and strontium isotopes.' *Ancient Mesoamerica* 18 (1): 159–72.

White, Christine D. et al. 2002. 'Geographic identities of the sacrificial victims from the Feathered Serpent Pyramid, Teotihuacan: implications for the nature of state power.' *Latin American Antiquity* 13 (2): 217–36.

—. 2008. 'The Teotihuacan dream: an isotopic study of economic organization and immigration.' *Ontario Archaeology* 85–8: 279–97.

White, Randall. 1985. *Upper Palaeolithic Land Use in the Périgord: A Topographical Approach to Subsistence and Settlement.* Oxford: British Archaeological Reports.

—. 1999. 'Intégrer la complexité sociale et opérationnelle: la construction matérielle de l'identité sociale à Sungir.' In M. Julien et al. (eds),

Préhistoire d'os: recueil d'études sur l'industrie osseuse préhistorique offert à Henriette Camps-Faber. Aix-en-Provence: L'Université de Provence, pp. 319–31.

White, Richard. 1991. *The Middle Ground: Indians, Empires, and Republics in the Great Lakes Region, 1650–1815*. Cambridge: Cambridge University Press.

Whitelaw, Todd. 2001. 'Reading between the tablets: assessing Mycenaean palatial involvement in ceramic production and consumption.' In Voutsaki and Killen (eds), pp. 51–79.

—. 2004. 'Estimating the population of Neopalatial Knossos.' In Gerald Cadogan, Eleni Hatzaki and Antonis Vasilakis (eds), *Knossos: Palace, City, State*. London: The British School at Athens, pp. 147–58.

Widmer, Randolph J. 1988. *The Evolution of the Calusa: A Nonagricultural Chiefdom on the Southwest Florida Coast*. Tuscaloosa and London: University of Alabama Press.

Wild, Eva. M. et al. 2004. 'Neolithic massacres: local skirmishes or general warfare in Europe?' *Radiocarbon* 46: 377–85.

Wilk, Richard. 2004. 'Miss Universe, the Olmec and the Valley of Oaxaca.' *Journal of Social Archaeology* 4 (1): 81–98.

Wilkinson, Toby. 2014. 'The Early Transcaucasian phenomenon in structural-systemic perspective: cuisine, craft and economy.' *Paléorient* 40 (2): 203–29.

Wilkinson, Tony J. 2010. 'The Tell: social archaeology and territorial space.' In D. Bolger and L. Maguire (eds), *The Development of Pre-state Communities in the Ancient Near East: Studies in Honour of Edgar Peltenburg*. Oxford: Oxbow, pp. 55–62.

Will, Manuel, Nicholas J. Conard and Christian A. Tryon. 2019. 'Timing and trajectory of cultural evolution on the African continent 200,000–30,000 years ago.' In Y. Sahle et al. (eds), *Modern Human Origins and Dispersal*. Tübingen: Kerns Verlag, pp. 25–72.

Willcox, George. 2005. 'The distribution, natural habitats and availability of wild cereals in relation to their domestication in the Near East: multiple events, multiple centres.' *Vegetation History and Archaeobotany* 14: 534–41.

—. 2007. 'The adoption of farming and the beginnings of the Neolithic in the Euphrates valley: cereal exploitation between the 12th and 8th millennia cal. BC.' In Colledge and Conolly (eds), pp. 21–36.

—. 2012. 'Searching for the origins of arable weeds in the Near East.' *Vegetation History and Archaeobotany* 21 (2): 163–7.

Willcox, G., Sandra Fornite and Linda Herveux. 2008. 'Early Holocene cultivation before domestication in northern Syria.' *Vegetation History and Archaeobotany* 17: 313–25.

Williams, Stephen. 1990. 'The Vacant Quarter and other late events in the Lower Valley.' In D. H. Dye (ed.), *Towns and Temples Along the Mississippi*. Tuscaloosa: University of Alabama Press, pp. 170–80.

Wills, John H. 1970. 'Speaking arenas of ancient Mesopotamia.' *Quarterly Journal of Speech* 56 (4): 398–405.

Wissler, Clark. H. 1922. *The American Indian*. New York: Douglas C. McMurtrie.

—. 1927. 'The culture-area concept in social anthropology.' *American Journal of Sociology* 32 (6): 881–91.

Wolf, Eric. R. 1982. *Europe and the People Without History*. Berkeley: University of California Press.

—. 1999. *Envisioning Power: Ideologies of Dominance and Crisis*. Berkeley: University of California Press.

Wood, Andrée. R. 1992. 'The detection, removal, storage, and species identification of prehistoric blood residues from Çayönü.' *American Journal of Archaeology* 96 (2): 374.

Woodburn, James. 1982. 'Egalitarian societies.' *Man* (N.S.) 17: 431–51.

—. 1988. 'African hunter-gatherer social organization: is it best understood as a product of encapsulation?' In T. Ingold, D. Riches and J. Woodburn (eds), *Hunters and Gatherers*, vol. 1: *History Evolution and Social Change*. Oxford: Berg, pp. 43–64.

—. 2005. 'Egalitarian societies revisited.' In T. Widlok and W. G. Tadesse (eds), *Property and Equality*, vol. 1: *Ritualisation, Sharing, Egalitarianism*. New York: Berghahn Books, pp. 18–31.

Woods, William I. 2004. 'Population nucleation, intensive agriculture, and environmental degradation: the Cahokia example.' *Agriculture and Human Values* 21: 255–61.

Woods, William I. et al. (eds). 2009. *Amazonian Dark Earths: Wim Sombroak's Vision*. Dordrecht and London: Springer.

Wright, Emily et al. 2014. 'Age and season of pig slaughter at Late Neolithic Durrington Walls (Wiltshire, UK) as detected through a new system for recording tooth wear.' *Journal of Archaeological Science* 52: 497–514.

Wright, Katherine I. 2007. 'Women and the emergence of urban society in Mesopotamia.' In S. Hamilton and R. D. Whitehouse (eds), *Archaeology and Women: Ancient and Modern Issues*. Walnut Creek, CA: Left Coast Press, pp. 199–245.

Wright, Katherine I. et al. 2008. 'Stone bead technologies and early craft specialization: insights from two Neolithic Sites in eastern Jordan.' *Levant* 40 (2): 131–65.

Wright, Rita P. 2010. *The Ancient Indus: Urbanism, Economy, and Society.* New York: Cambridge University Press.

Yen, Douglas. E. 1995. 'The development of Sahul agriculture with Australia as bystander.' *Antiquity* 69 (265): 831–47.

Yerkes, Richard W. 2005. 'Bone chemistry, body parts, and growth marks: evaluating Ohio Hopewell and Cahokia Mississippian seasonality, subsistence, ritual, and feasting.' *American Antiquity* 70 (2): 241–65.

Yoffee, Norman. 1995. 'The political economy of early Mesopotamian states.' *Annual Review of Anthropology* 24: 281–311.

—. 2005. *Myths of the Archaic State: Evolution of the Earliest Cities, States, and Civilizations.* Cambridge: Cambridge University Press.

Younger, John. 2016. 'Minoan women.' In S. L. Budin and J. M. Turfa (eds), *Women in Antiquity: Real Women Across the Ancient World.* London and New York: Routledge, pp. 573–94.

Yuan, Jing and Rowan Flad. 2005. 'New zooarchaeological evidence for changes in Shang Dynasty animal sacrifice.' *Journal of Anthropological Archaeology* 24 (3): 252–70.

Zagarell, Allen. 1986. 'Trade, women, class, and society in Ancient Western Asia.' *Current Anthropology* 27 (5): 415–30.

Zedeño, María Nieves. 2008. 'Bundled worlds: the roles and interactions of complex objects from the North American Plains.' *Journal of Archaeological Method and Theory* 15: 362–78.

Zeder, Melinda A. and Brian Hesse. 2000. 'The initial domestication of goats (*Capra hircus*) in the Zagros Mountains 10,000 years ago.' *Science* 287: 2254–7.

Zheltova, Maria N. 2015. 'Kostenki 4: Gravettian of the east – not Eastern Gravettian.' *Quaternary International* 359–60: 362–71.

Zhilin, Mikhail et al. 2018. 'Early art in the Urals: new research on the wooden sculpture from Shigir.' *Antiquity* 92 (362): 334–50.

Zvelebil, Marek. 2006. 'Mobility, contact, and exchange in the Baltic Sea basin, 6000–2000 BC.' *Journal of Anthropological Archaeology* 25: 178–92.

Index